Communications in Computer and Information Science 1483

More information about this series at http://www.springer.com/series/7899

K. R. Venugopal · P. Deepa Shenoy ·
Rajkumar Buyya · L. M. Patnaik ·
Sitharama S. Iyengar (Eds.)

Data Science and Computational Intelligence

Sixteenth International Conference
on Information Processing, ICInPro 2021
Bengaluru, India, October 22–24, 2021
Proceedings

 Springer

Editors
K. R. Venugopal
Bangalore University
Bengaluru, India

Rajkumar Buyya
University of Melbourne
Melbourne, VIC, Australia

Sitharama S. Iyengar
Florida International University
Miami, FL, USA

P. Deepa Shenoy
Bangalore University
Bengaluru, India

L. M. Patnaik
NIAS
Bengaluru, India

ISSN 1865-0929   ISSN 1865-0937 (electronic)
Communications in Computer and Information Science
ISBN 978-3-030-91243-7   ISBN 978-3-030-91244-4 (eBook)
https://doi.org/10.1007/978-3-030-91244-4

This Springer imprint is published by the registered company Springer Nature Switzerland AG
The registered company address is: Gewerbestrasse 11, 6330 Cham, Switzerland

Preface

This volume contains the proceedings of the Sixteenth International Conference on Information Processing (ICInPro 2021) held in Bengaluru, India, during October 22–24, 2021. The International Conference on Information Processing aims to provide an opportunity for academicians, scientists, industry experts, and researchers engaged in teaching, research, and development to present and discuss ideas and share their views to solve real-world complex challenges.

Following the success of the past fifteen years of conference events, ICInPro 2021 was devoted to novel methods in the fields of the Internet of Things, data science, network and security, computing, and intelligence. Some of the best researchers delivered keynote addresses on the themes of the conference. This gave delegates the opportunity to interact with these experts and to address some of the challenging problems in advanced areas of research. These areas have been recognized to be the key technologies poised to shape the modern society in the next decade. ICInPro 2021 attracted over 175 submissions. Through rigorous peer reviews, 40 high-quality papers were recommended by the international Technical Program Committee for inclusion in the proceedings.

On behalf of the organizing committee, we would like to acknowledge the support from sponsors who helped us to achieve our goals for the conference. We wish to express our appreciation to Springer for publishing the proceedings of ICInPro 2021. We also wish to acknowledge the dedication and commitment of the Springer staff. We would like to thank the authors for submitting their work, as well as the Technical Program Committee members and reviewers for their enthusiasm, time, and valuable suggestions. The contributions from the organizing committee in setting up and maintaining the online submission systems, assigning papers to the reviewers, and preparing the camera-ready version of the proceedings is highly appreciated. We would like to profusely thank them for making ICInPro 2021 a success.

October 2021

K. R. Venugopal
P. Deepa Shenoy
Rajkumar Buyya
S. S. Iyengar
L. M. Patnaik

Organization

The Sixteenth International Conference on Information Processing (ICInPro 2021) held in Bengaluru, India during October 22–24, 2021, was organized by the Computing Professionals Charitable Trust (CPCT) and the Department of Computer Science and Engineering, University Visvesvaraya College of Engineering (UVCE), Bangalore University, India.

Advisory Committee

R. L. Kashyap	Purdue University, USA
Sajal K. Das	Missouri University of Science and Technology, USA
N. R. Shetty	Central University of Karnataka, India
Marimuthu Palaniswami	University of Melbourne, Australia

Conference Chair

L. M. Patnaik	NIAS, Indian Institute of Science, India

Technical Program Chairs

S. S. Iyengar	Florida International University, USA
Rajkumar Buyya	University of Melbourne, Australia
P. Deepa Shenoy	University Visvesvaraya College of Engineering, India

Publication Chair

Venugopal K. R.	Bangalore University, India

Finance Chairs

P. Deepa Shenoy	UVCE, India
Thriveni J.	UVCE, India

Technical Program Committee

A. Anny Leema	Vellore Institute of Technology, India
Abdallah Meraoumia	Larbi Tebessi University, Algeria
Abdelhakim Moutaouakil	Cadi Ayyad University, Morocco
Abhishek Appaji	B.M.S. College of Engineering, India
Adolfo Antonio-Bravo	Universidad Politécnica de Madrid, Spain

Ahmed Bouraiou	Renewable Energy Development Center (CDER), Algeria
Ajit Danti	CHRIST (Deemed to be University), India
Aksana Naranovich	Baranovichi State University, Belarus
Ali Yahyaouy	Université Sidi Mohamed Ben Abdellah de Fès, Morocco
Anand Nayyar	Duy Tan University, Vietnam
Aruna Govada	U.T. Administration of Dadra and Nagar Haveli and Daman and Diu, India
Ayush Goyal	Texas A&M University-Kingsville, USA
Badre Bossoufi	Université Sidi Mohamed Ben Abdellah de Fès and Université Mohammed Premier Oujda, Morocco
Balaji M.	Indian Space Research Organization, India
Benny Thomas	CHRIST (Deemed to be University), India
Biswanath Dutta	Indian Statistical Institute, India
C. Mala	NIT Tiruchirappalli, India
Celia Shahnaz	BUET, Bangladesh
Chandrashekaran K.	NITK Suratkal, India
Chouaib Ennawaoui	National School of Applied Sciences, Chouaib Doukkali University, Morocco
Daya Sagar B. S.	Indian Statistical Institute, India
Devika P. Madalli	Indian Statistical Institute, India
Devottam Gaurav	IIT Delhi, India
Dinesh K. Anvekar	Sapthagiri College of Engineering, India
El Mehdi Abdelmalek	Université Mohammed Premier Oujda, Morocco
Fatima Bennouna	Université Sidi Mohamed Ben Abdellah de Fès, Morocco
Fatima Zahra Fagroud	Université Hassan II de Casablanca, Morocco
Fernando Ortiz-Rodriguez	Universidad Autónoma de Tamaulipas, Mexico
G. K. Patra	CSIR-NAL, India
Hassan Belahrach	Royal School of Aeronautics, Marrakech, Morocco
Hiran Nandy	Qualcomm India, India
J. Thomas	CHRIST (Deemed to be University), India
Jadli Aissam	ENSAM Casablanca, Morocco
Jennifer D'Souza	TIB and L3S, Germany
Jitendra Kumar	C-DAC Bengaluru, India
Karthik Devaraj	Indian Space Research Organization, India
Kumudini Ravindra	Inytu Inc., India
L. Hamsaveni	University of Mysore, India
Levin Varghese	One Plus R&D, India
Lotfi Houam	Larbi Tebessi University, Algeria
M. A. Jabbar	Vardhaman College of Engineering, India
M. Pratheepa	ICAR-National Bureau of Agricultural Insect Resources
Ahmed Taieb	Institut Supérieur des Études Technologiques de Kélibia, Tunisia

M. Raja	BITS Pilani, Dubai, United Arab Emirates
Makhad Mohamed	Mohammed V University in Rabat, Morocco
Manimaran Govindarasu	Iowa State University, USA
Marlene Goncalves	Universidad Simon Bolívar, Venezuela
Mausumi Goswami	CHRIST (Deemed to be University), India
Mohammed Saber	National School of Applied Sciences, Mohammed First University, Morocco
Moncef Garouani	Université du Littoral Côte d'Opale, France, and ENSAM Casablanca, Morocco
Moses Ekpenyong	University of Uyo, Nigeria
Mounir Gouiouez	Université Sidi Mohammed Ben Abdullah de Fès, Morocco
Narendran Rajagopalan	NIT Puducherry, India
Nasim Matar	University of Petra, Jordan
Naveen J.	CHRIST (Deemed to be University), India
Neha Sharma	Tata Consultancy Services, UK
Omar Hussain Alhazmi	Taibah University, Saudi Arabia
P. Shivakumara	University of Malaya, Malaysia
P. Sreenivasa Kumar	IIT Madras, India
Patience Usoro Usip	University of Uyo, Nigeria
Preeth R.	IIITDM, Kurnool, India
Priya Philip	Don Bosco College, Sulthan Bathery, India
Raghu H. V.	C-DAC, India
Rajib Mall	IIT Kharagpur, India
Randima Dinalankara	University of Sri Jayewardenepura, Sri Lanka
Ravikiran A.	Numocity Technologies, India
Roshan G. Ragel	University of Peradeniya, Sri Lanka
Saad Motahhir	Ecole Nationale des Sciences Appliquées de Fès, Université Sidi Mohamed Ben Abdellah, Morocco
Saber Mohammed	National School of Applied Sciences, Mohammed First University, Morocco
Safae El Abkari	ENSAM, Mohammed V University in Rabat, Morocco
Saif Serag	Université Ibn Tofail, Morocco
Salhi Mourad	Mohammed First University, Morocco
Samiksha Sukhla	CHRIST (Deemed to be University), Lavasa, India
Samir Amraqui	Université Mohamed Premier Oujda, Morocco
Sanasam Ranbir Singh	IIT Guwahati, India
Sanju Tiwari	Universidad Autonoma de Tamaulipas, Mexico
Santhanavijayan A.	NIT Tiruchirapalli, India
Sarika Jain	NIT Kurukshetra, India
Saroj Kr. Biswas	NIT Silchar, India
Seddik Khemaissia	Larbi Tebessi University, Algeria
Shaji Thorn Blue	Radisys India, India
Shyamala Doraisamy	Universiti Putra Malaysia, Malaysia
Skander Bensegueni	National Polytechnic School of Constantine, Algeria
Soma Pandey	Reliance Jio Infocomm Ltd, India

Sonali Agarwal	IIIT-Allahabad, India
Soufian Lakrit	EMI, Mohammed V University of Rabat, Morocco
Srikanth M.	Intel, India
Sumith Shankar	Indian Space Research Organization, India
Toufouti Riad	Université Mohamed Chérif Messaadia de Souk-Ahras, Algeria
Vinai George B.	CHRIST (Deemed to be University), India
W. G. C. W. Kumara (Chinthaka)	South Eastern University of Sri Lanka, Sri Lanka
Wahabi Aicha	Université Hassan II de Casablanca, Morocco
Waseem Ahmed	King Abdul Aziz University, Saudi Arabia
Weam El Merrassi	Faculté des Sciences et Techniques de Béni Mellal, Morocco
Y. V. S. Laxmi	CDoT, India
Yahya Dbaghi	Ibn Zohr University, Morocco
Yasir Alani	Teesside University, UK
Younes Balboul	Sidi Mohamed Ben Abdellah University, Morocco
Youssef El Afou	Sidi Mohamed Ben Abdellah University, Morocco
Yusniel Hidalgo Delgado	Universidad de las Ciencias Informáticas, Cuba
Zameer Gulzar	B S Abdur Rahman Institute of Science and Technology, India

Organizing Committee

P. Deepa Shenoy	UVCE, India
K. B. Raja	UVCE, India
K. Suresh Babu	UVCE, India
J. Thriveni	UVCE, India
S. H. Manjula	UVCE, India
Arunalatha J. S.	UVCE, India
S. M. Dilipkumar	UVCE, India
Champa H. N.	UVCE, India
Tanuja R.	UVCE, India
Dharmendra Chouhan	UVCE, India
Lata B. T.	UVCE, India
Pushpa C. N.	UVCE, India
Kiran K.	UVCE, India
Venkatesh	UVCE, India
Sunil Kumar G.	UVCE, India
Kumaraswamy S.	UVCE, India
Prathibavani P. M.	UVCE, India
Roopa	UVCE, India
Vibha Lakshmikantha	Bangalore Institute of Technology, India
Anita Kanavalli	M. S. Ramaiah Institute of Technology, India
Shaila K.	Vivekananda Institute of Technology, India
Prashanth C. R.	Dr. Ambedkar Institute of Technology, India

Ramachandra A. C.	Nitte Meenakshi Institute of Technology, India
Srikantaiah K. C.	SJB Institute of Technology, India
T. Shiva Prakash	Vijaya Vittala Institute of Technology, India
Vidya A.	Vivekananda Institute of Technology, India

Tutorial Chairs

Arunalatha J. S.	UVCE, India
Shaila K.	Vivekananda Institute of Technology, India
Sujatha D. N.	B.M.S. College of Engineering, India

Local Arrangement Chairs

| K. B. Raja | UVCE, India |
| Kiran K. | UVCE, India |

Co-ordinators

Srikantaiah K. C.	SJB Institute of Technology, India
Pushpa C. N.	UVCE, India
Sunil Kumar G.	UVCE, India

Sponsoring Organizations

Computing Professionals Charitable Trust (CPCT),
Bengaluru, India

Department of Computer Science and Engineering,
University Visvesvaraya College of Engineering,
Bangalore University, India

Contents

Data Science

Intelligence and IOT

Computing and Network Security

Runtime Program Semantics Based Malware Detection in Virtual Machines of Cloud Computing

Harsha Dudeja[✉] and Chirag Modi

Department of Computer Science and Engineering,
National Institute of Technology Goa, Ponda, India
cnmodi@nitgoa.ac.in

Abstract. A virtual machine (VM) can be a major target to affect the security of cloud resources, and thus, it is required to re-investigate or extend malware detection system to satisfy the VM security requirements. In this paper, an efficient malware detection framework based on runtime program semantics of executables for securing the VMs is designed. It installs an agent on each VM to capture the running program semantics of VMs. An agent captures the behavior of portable executable (both *.exe* and *.dll* files) running in the VMs. In addition, DLL files are considered with executable to prevent manipulation of loading them into the system memory. The dumps of the running processes are checked for any attack behavior using the signature based detection. For anomaly detection, the running processes are disassembled and vectored into *1*-gram features. From these features, optimal features are selected and applied to the anomaly detection module at hypervisor level, where an ensemble of Random forest, Extra trees and Kernel SVM with majority voting technique is used to detect any malware. The alerts generated by signature based technique and anomaly detection from all the VMs are analyzed to check any distributed attack. The functional validation of the proposed framework is done using cloud testbed at NIT Goa and recent malware. In addition, the security needs satisfied by the proposed framework are compared with the existing approaches.

Keywords: Virtual machines · Cloud computing · Malware detection · Random forest · Extra trees · Support vector machine

1 Introduction

Virtualization play a vital role in cloud, which allows to run multiple OSs and applications in a distributed manner. Although cloud offers various benefits to end users and businesses, the security is a major concern due to the existence of various vulnerabilities [1,2]. As per our observation in [1,3], a VM is a major target to affect the security of cloud resources and services. Center for Internet Security (CIS) [4] has announced the top 10 malware in 2019, such as evasive variants of

© Springer Nature Switzerland AG 2021
K. R. Venugopal et al. (Eds.): ICInPro 2021, CCIS 1483, pp. 3–16, 2021.
https://doi.org/10.1007/978-3-030-91244-4_1

Botnets, Trojans and Ransomware. There have been many such vulnerabilities reported in NVD [5]. More than 350000 new malware excluding the existing signatures have been registered at the AVTEST institute [6]. The distributed and dynamic nature of the cloud cause the challenges to the traditional approaches. Signature based approaches need regular updates to maintain the attack signature database. In addition, attack signature databases are dependent on the type of OSs. Dynamic analysis based malware detection approaches rely on the run time statistics of the processes. Here, intrusive behavior of a process is determined based on deviation from initially built baseline profile of all the running processes, such as "Immediate System Call Sequence (ISCS)" [7], "key-value pairs" and "Bag of system calls" [8]. Immediate System Call Sequence (ISCS) may end up with infinite traces, while 'Bag of system calls' are not considering the ordering of system calls, and thus, it may generate more false alerts.

In this paper, we propose an efficient malware detection framework based on runtime program semantics of the executable. It installs an agent on each VM to capture the running program semantics of VMs. An agent captures the behavior of portable executables (both .exe and .dll files). In addition, DLL files are considered with executables to prevent manipulation of these files after loading them into the system memory. The dumps of the running processes are checked for any attack behaviour using the signature based technique. For anomaly detection, an agent disassembles and vectors the running processes into 1-gram features. From these features, optimal are features are selected and applied to the anomaly detection module at hypervisor level, where an ensemble of Random forest, Extra trees and Kernel SVM with majority voting technique is used to detect any malware. The alerts generated by signature based technique and anomaly detection from each VM are correlated to check any distributed attack. In addition, the alerts from anomaly detection are updated as signature in the signature database of each VM to avoid further analysis of the same malware in the future. The functional validation of the proposed framework is done using cloud testbed at NIT Goa and recent malware. The VM security requirements satisfied by the proposed framework are compared with the existing approaches.

In following, Sect. 2 background on executables and DLL files. It investigates the existing frameworks for malware detection in cloud. The proposed malware detection framework is presented in Sect. 3, while the experimental results and comparative analysis of the proposed framework are given in Sect. 4. In Sect. 5, our work is concluded followed by the references.

2   Background and Related Work

DLLs have the Portable Executable (PE) file format as Windows EXE files (please refer Fig. 1). As like EXEs, DLLs have a combination of code, data, and resources. In case of Windows API, the organization of DLL files is in the form of sections like PEs. Each section carries their own set of attributes such as editable, writable or read only; executable (code section) or non-executable (data section) etc. The PE format is a universal format for .exe files, DLLs and object code. Table 1 presents the headers and sections of PE file format.

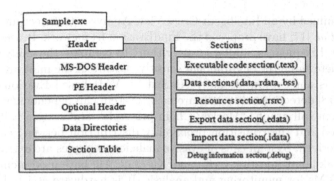

Fig. 1. File format of portable executable.

Table 1. Headers/sections and related information in PE file format.

Header/section	Information included
MS-DOS header	Compatibility information of the file with reference to the Windows and MS-DOS versions
PE header	Binary's sections, symbols, linking time, and whether the file is an executable, object file or a DLL
Optional header	Loading of the binary and in turn about the linker version, starting address, the data segment size, the amount of stack to reserve
Data directory	Virtual addresses to data directories residing within section bodies
Section table	Sectioning of the binary- section's name, offset and size
PE Sections	Details on executing a program loaded into the memory at runtime

The information available in headers and sections can help to detect the malware [9]. Angelos et al. [10] have used an "ensemble empirical mode decomposition (E-EMD)" technique to secure the VMs from intrusions. However, the feasibility for handling multiple VMs is not reported yet. Fattori et al. [11] have proposed an Access Miner that focuses on interactions taking place to and from the benign applications in the VM. Based on this, any deviation is considered as the anomalous behavior. Watson et al. [12] have applied support vector machine (SVM) on the system and network level features of the VM to detect the attacks. Mishra et al. [13] have analyzed the system call sequences of processes running in a VM. Here, decision tree classifier is applied to detect anomalies. It is tested using UNM dataset [14] which lacks in recent malware. Xie and Wang [15] have examined the functionality of the DLL files. Here, the authors have focused on relative virtual addresses (RVAs) of the functions. Ajay and Jaidhar [16] have utilized a lightweight in-VM agent for gathering state information. The in-VM component supplies the process related data to the hypervisor, where it is periodically examined by an Intelligent Cross View Analyzer to detect intrusive processes. Jia *et al.* [17] have proposed the FindEvasion framework. It uses a "Multiple Behavioral Sequences Similarity algorithm" to compare the analyzed behaviors and to determine the environment sensitive nature of the malware. However, this process is not computationally efficient. A VMI based Evasion Detection framework [18] detects evasive malware using system call dependency. To generate the system call graph, "Markov Chain principle" and system call orderings are monitored and the transition probability distribution between each system call pair is derived. System calls of evasive malicious samples are extracted using software break point injection technique. Xu *et al.* [19] have used memory access patterns of VMs for monitoring and analysis. It is hardware assisted approach. It emphasizes on the fingerprints left on the memory access by malware when trying to alter the control flow and data structures. To detect the malicious behavior on virtual memory access, machine learning is used. However, it has high number of false alerts. Joseph and Mukesh [20] have focused on analyzing VM memory snapshots for malware detection. To extract the feasible features, the API calls of VM instances are considered. The classification of snapshots is based on the reduced feature set. In [3], malware detection is performed using static features of the executables. The known malware are detected using the signature based detection through an agent at VM. From the unknown executables, the static features are extracted using an agent and the profile for each executable is derived. To detect anomaly, a random forest classifier is applied on such profile. This framework lacks in detecting any encrypted malware. Mathew and Kumara [21] have used a "Recurrent Neural Network-Long Short Term Memory (RNN-LSTM)" for distinguishing the semantic behaviour of malware samples from the benign ones. Here, the malware features are derived using the "n-gram" technique from API calls, while "Term Frequency-Inverse Document Frequency (TF-IDF)" is used for feature selection. Mishra *et al.* [22] have performed a KVM introspection for capturing dynamic behaviour of the running programs. It uses n-gram technique for vectoring the in-VM and out-VM system call traces and exploits Recursive Feature Elimination (RFE) to select the features. It applies an ensemble of different classifiers for malware detection. As per our observation, malware detection needs to handle large scale computing systems, avoiding re-analysis of the examined files, detecting variety of malware, resistance to compromise and high accuracy and low false alerts.

3 Malware Detection Framework for VMs in Cloud

3.1 Objective

The proposed framework aims at securing the VMs by detecting malicious executables, while fulfilling the VM security needs.

3.2 Design of the Proposed Malware Detection Framework

The proposed framework is deployed in the cloud, as shown in Fig. 2. It includes the virtualized physical servers through Type-1 hypervisor [23], VMs and control VM (CVM). Here, the agents are configured on each VM, while anomaly detection is performed on the control VM. The proposed framework is flexible to correlate the malware alerts at the correlation unit. An agent on each VM performs process dumping, signature based detection, mnemonics extraction, n-grams generation and feature extraction, as shown in Fig. 3.

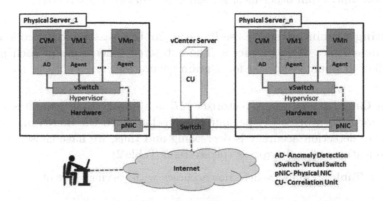

Fig. 2. Deployment of the proposed framework in cloud.

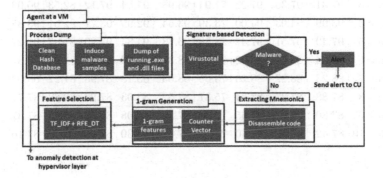

Fig. 3. Agent design in the proposed framework.

Process Dumping. Process dump [24] dumps all the packed and obfuscated malware files along with benign processes running in the system mode. It captures state of processes before their termination. During dumping, a clean hash database just after VM deployment ensures the skipping of clean files like kernel.dll. The process dump returns majority of the intrusive executable and *.dll* files. It handles the dumping of hidden processes and loose executable code chunks. Thus, the resulting dump can be of a hidden module, a code chunk or the complete process. After dumping both benign and intrusive processes (in both DLL and EXE format), the dumps are sent for signature based detection.

Signature Based Detection. For signature based detection, we use VirusTotal [25] which takes the dumped executables and DLL files as input and returns information about their behavior with malware type. In the proposed framework, the known malware are detected at VM level using VirusTotal and alerts are generated. For detecting the distributed attacks, the alerts are correlated at correlation unit. From other files, system call behavior is extracted.

Extracting Mnemonics. We use Pydasm [26] to disassemble the code and to extract mnemonics of each process. The extracted mnemonics of each process are applied to CountVectorizer for *1*-gram generation.

1-gram Generation. CountVectorizer [27] is a library that generates n-gram features for each process. As per our observation, with the increasing value of n, malware detection accuracy is decreasing and thus, we have chosen $n = 1$ to generate features of each process (please refer Table 2).

Table 2. Malware detection accuracy with varying value of n.

n	RF	ET	DT	NB	SVM	MLP	KNN	LR	GB
n = 1	**98.12**	**98.11**	97.99	78.12	**98.06**	95.98	97.84	89.8	98.01
n = 2	98.05	98.04	97.95	72.95	97.79	96.78	97.64	89.63	97.69
n = 3	97.31	97.30	97.22	71.94	96.98	95.84	97.12	82.53	96.93
n = 4	93.99	94.03	94.00	73.39	93.54	92.99	93.71	82.01	93.85
n = 5	97.49	97.51	97.47	73.39	96.73	94.84	97.29	81.67	97.27
n = 6	88.64	88.66	88.62	74.15	88.32	86.03	86.77	81.96	88.37
n = 7	88.23	88.22	88.20	73.65	88.03	85.18	87.96	81.22	88.04
n = 8	87.88	87.87	87.84	75.37	87.52	83.03	87.23	80.88	87.77
n = 9	87.80	87.7	87.65	74.93	87.30	83.98	87.48	80.63	87.37
n = 10	87.42	87.42	87.40	75.33	87.01	83.30	87.26	80.38	87.16

Feature Selection. For each process, we have considered 100 most frequent features in *1*-gram based on the "term frequency-inverse document frequency" [28]. Term frequency *tf(t,d)* is calculated using the raw count $(f_{t,d})$ of a term (t) in document (d) implying the number of times 't' occurred in 'd'.

$$tf(t, d) = f_{t,d} \tag{1}$$

The "Inverse Document Frequency" *idf(t,D)* measures the importance of the term in corpus. If total number of documents (D) in the corpus equals to N, i.e. $N = |D|$ and the number of documents with the term 't' appearing in them (i.e. $tf(t, d) \neq 0$) is denoted by $|d \in D : t \in d|$,

$$idf(t, D) = log \frac{N}{|d \in D : t \in d|} \tag{2}$$

Then "Term frequency-Inverse document frequency" is calculated as:

$$Tfidf(t, d, D) = tf(t, d).idf(t, D) \qquad (3)$$

In addition, we reduce the feature set by 40% using Recursive Feature Elimination (RFE) with Decision Trees Classifier [29]. Finally, the profile of executables and DLL files with the reduced 1-gram features are transferred to the CVM via SSH connection for anomaly detection.

Anomaly Detection. As shown in Fig. 4, the profiles of executables and DLL files are applied to an ensemble of different classifiers for malware detection at hypervisor layer. The malware alerts are processed and analyzed to detect any distributed attack. For anomaly detection, we have tested different well-known classifiers. From the results (please refer Table 2), we have observed that the Random forest (RF) [30], Extra trees (ET) [31] and Kernel SVM [32] classifiers perform well in malware detection. Random forest selects the locally optimal feature/split combination, whereas Extra trees classifier considers a random value for split, leading to more diversified trees and less splitters. SVM classifier is a discriminative classifier determining optimal classification in the form of a separating hyperplane. These classifiers are trained using the profiles of executables and DLL files, and later the trained classifiers are used independently to decide that the applied profiles are from malware or benign executable. The final decision about malware is done using a majority voting of each classifier. For the detected malware, an alert is generated and processed for further analysis.

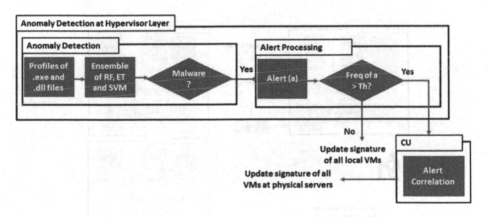

Fig. 4. Design of an anomaly detection in the proposed framework.

Alert Generation and Processing. An alert of malware is generated in the form of 1-gram features of the respective process. The alerts received for a VM are processed to generate the attack signature. The alert processing checks the alert frequency, and if the alert frequency crosses a threshold (Th) during time interval (e.g. 2 s), then the signature is updated to all the VMs. In addition, such alert is transferred to correlation unit for detecting distributed attack.

Correlation Unit (CU). It receives the frequent alerts generated from multiple VMs on different servers and check for any distributed attack using alert majority factor (AMF).

$$AMF = \frac{A_i}{A} \qquad (4)$$

Where, A_i is the number of VM instances from different servers generating same alert and A is the total number of VM instances deployed. If AMF crosses 0.5, then that alert is considered as an alert of distributed attack and attack signature is generated. Such attack signature is updated in signature database of all the agents installed in VMs at different servers. This helps to eliminate the re-analysis of the same attack and also to reduce the communication cost.

4   Results and Analysis

4.1   Experimental Testbed

The performance and functional validation of the proposed framework is done using a cloud testbed at NIT Goa. The testbed consists of three servers. On these servers, VMware Esxi is installed for running the VMs. Each server includes two VMs with different OSs viz; Windows 7 and Ubuntu 14.04, as shown in Fig. 5.

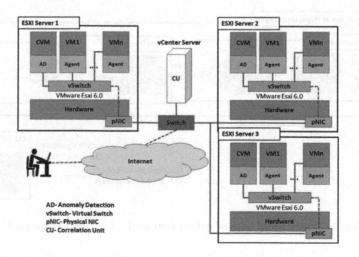

Fig. 5. Experimental testbed for the validation of the proposed framework.

We have carried out real time and offline experiments, as shown in Table 3 by considering different malware collected from [33, 34]. The performance results are derived in terms of False Positive Rate (FPR), True Positive Rate (TPR), False Negative Rate (FNR), True Negative Rate (TNR), Accuracy and Detection time.

Table 3. List of experiments for the evaluation of the proposed framework.

Test case	No of training profiles	No of testing profiles	Type of malware considered
Offline validation	22212 (8841 benign and 13371 Malware)	10- Cross fold validation	Trojans (Banking, Wacatac, Atraps, Unknown and potential categories), Viruses (Artemis, Nuke), Rootkits, Riskware, Adware and potentially malicious programs
Real time validation	17769 (7090 benign and 10679 Malware)	4443 (1751 benign and 2692 Unknown)	Trojans, viruses, worms, rootkits, Trojans and viruses

4.2 Experimental Results

The experimental results of the proposed framework in real time and offline validation are given in Table 4. The proposed framework achieves malware detection accuracy more than 98% in real time as well as offline. Table 4 and graphical representation (please refer Fig. 6, 7, 8, 9, 10 and 11) show that more than 98% malicious processes are correctly classified as malware. The false positive rate is less than 1.5%. The proposed framework achieves higher TNR (>98%), while false negatives is less than 2%. The malware detection time requirement is around 0.04 ms per profile, and thus it can examine around 2500 profiles in a second.

Table 4. Experimental results in offline and real time validation.

Model	Validation mode	TPR (%)	FPR (%)	TNR (%)	FNR (%)	Accuracy (%)	Detection time (ms/profile)
RF	Offline	99.69	0.25	99.75	0.31	99.71	0.002
	Real time	97.94	1.57	98.43	2.06	98.13	0.006
ET	Offline	99.99	0.00	100	0.01	99.95	0.0009
	Real time	97.94	1.57	98.43	2.06	98.13	0.0005
SVM	Offline	98.94	0.28	99.72	1.06	99.24	0.046
	Real time	97.95	1.50	98.50	2.05	98.06	0.082
Proposed work	Offline	99.99	0.00	100	0.01	99.96	0.043
	Real time	97.95	1.50	98.50	2.05	98.17	0.071

(RF- Random Forest, ET- Extra Trees, SVM- Support Vector Machines)

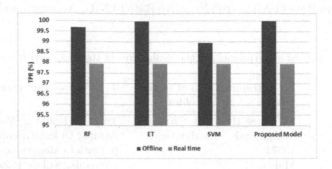

Fig. 6. True positive rate (TPR) of the proposed framework.

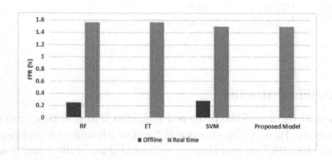

Fig. 7. False positive rate (FPR) of the proposed framework.

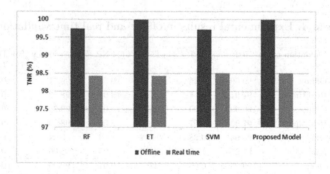

Fig. 8. True negative rate (TNR) of the proposed framework.

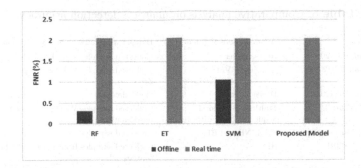

Fig. 9. False negative rate (FPR) of the proposed framework.

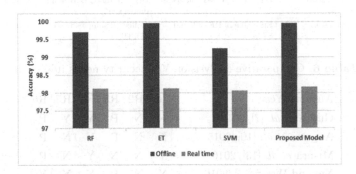

Fig. 10. Malware detection accuracy of the proposed framework.

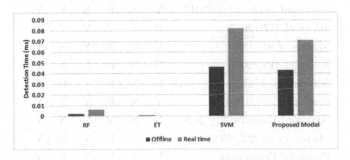

Fig. 11. Malware detection time requirement of the proposed framework.

4.3   Comparative Analysis

As shown in Table 5, the proposed framework achieves higher detection accuracy than the existing approaches. It fulfills the VM security needs as follows: To cope up with large scale computing systems (R1), Avoiding re-analysis of the examined files (R2), Detecting different malware (R3), Resistance to compromise (R4), High scalability (R5) and High accuracy and low false alerts (R6), as shown in Table 6.

Table 5. Comparative analysis on malware detection accuracy.

Authors/Year	Anomaly detection model	Dataset and malware types	Accuracy (%)
Fattori et al. [11]/2015	Access miner	Collected samples -Rootkits	90
Watson et al. [12]/2016	One class SVM	Collected samples-Kelihos, Zeus	90
Mishra et al. [13]/2016	Decision Trees	UNM dataset	91.16
Jia et al. [17]/2017	Multiple Behavioral Sequences Similarity	Collected samples-Win32 based malware	92.40
Joseph and Mukesh [20]/2018	SVM, NB and RF	Collected samples	93
Patil et al. [3]/2019	RF	Collected samples from many sources	97.61
Ajay and Kumara [21]/2020	RNN-LSTM	3000 malware and various benign traces	92
Mishra et al. [22]/2020	DT, RF, AB, NB and SVM	UNM dataset, Barecloud dataset [35]	81.25–99.92
Proposed Framework	Ensemble of RF, ET and SVM	Collected samples from [33, 34]	98.13–99.96

Table 6. Comparative analysis on VM security needs fulfillment.

Authors/Year	R1	R2	R3	R4	R5	R6
Gupta et al. [7]/2015	N	N	P	Y	Y	P
Angelos et al. [10]/2015	N	N	P	Y	N	P
Mishra et al. [13]/2016	Y	N	N	Y	N	P
Xie and Wang [15]/2016	Y	N	P	Y	Y	Y
Watson et al. [12]/2016	N	N	P	Y	N	P
Ajay and Jaidhar [16]/2017	Y	N	P	N	Y	Y
Jia et al. [17]/2017	N	N	Y	Y	Y	Y
Mishra et al. [18]/2017	N	N	N	Y	N	Y
Xu et al. [19]/2017	N	N	P	N	N	Y
Joseph and Mukesh [20]/2018	N	N	P	Y	P	P
Patil et al. [3]/2019	Y	N	P	Y	Y	Y
Ajay and Kumara [21]/2020	N	N	Y	Y	N	Y
Mishra et al. [22]/2020	N	N	P	Y	N	Y
Proposed Framework	Y	Y	Y	Y	Y	Y

(Y-Fulfilled, N-Not Fulfilled, P-Partially Fulfilled)

5   Conclusion

To fulfill the VM security needs, the proposed framework detects malware in VMs using runtime program semantics, and thus, it can detect static as well as dynamic malware. It applies signature based technique before anomaly detection, and thus reducing the computational overhead at anomaly detection as it has to process and analyze only unknown executables. The alerts generated from multiple VMs are correlated, and thus the distributed attacks can be detected using the proposed framework. From the functional validation, it is found that the ensemble of RF, SVM and ET on the reduced 1-gram features using TF-IDF and RFE (DT) improves detection accuracy. The deployment of anomaly detection at hypervisor offers the resistance to compromise capability. As per our analysis, the proposed framework satisfies the VM security needs.

Acknowledgment. The research work reported in this paper is a part of project "Designing out-of-VM Monitoring based Virtual Machine Introspection Framework for Securing Virtual Environment of Cloud Computing", funded by SERB, DST, GOI. Grant No: ECR/2017/001221.

References

1. Modi, C., Acha, K.: Virtualization layer security challenges and intrusion detection/prevention systems in cloud computing: a comprehensive review. J. Supercomput. **73**(3), 1192–1234 (2017)
2. Patil, R., Modi, C.: An exhaustive survey on security concerns and solutions at different components of virtualization. **52**(1) (2019)
3. Patil, R., Dudeja, H., Modi, C.: Designing in-VM-assisted lightweight agent-based malware detection framework for securing virtual machines in cloud computing. J. Supercomput. **19**, 1–16 (2019)
4. Top 10 Malware January (2019). https://www.cisecurity.org/blog/top-10-malware-january-2019
5. National vulnerability database—search and statistics (2017). https://nvd.nist.gov/vuln/data-feeds/
6. Malware statistics (2019). https://www.av-test.org/en/statistics/malware/
7. Gupta, S., Kumar, P.: An immediate system call sequence based approach for detecting malicious program executions in cloud environment. **81**(1), 405–425 (2015)
8. Alarifi, S.S., Wolthusen, S.D.: Detecting anomalies in IaaS environments through virtual machine host system call analysis. In: 2012 International Conference for Internet Technology and Secured Transactions, pp. 211–218 (2012)
9. Shafiq, M.Z., Tabish, S.M., Mirza, F., Farooq, M.: PE-miner: mining structural information to detect malicious executables in realtime. In: Kirda, E., Jha, S., Balzarotti, D. (eds.) RAID 2009. LNCS, vol. 5758, pp. 121–141. Springer, Heidelberg (2009). https://doi.org/10.1007/978-3-642-04342-0_7
10. Marnerides, A.K., Spachos, P., Chatzimisios, P., Mauthe, A.U.: Malware detection in the cloud under ensemble empirical mode decomposition. In: 2015 International Conference on Computing, Networking and Communications, pp. 82–88 (2015)
11. Fattori, A., Lanzi, A., Balzarotti, D., Kirda, E.: Hypervisor-based malware protection with accessminer. Comput. Secur. **52**(C), 33–50 (2015)
12. Watson, M.R., Shirazi, N., Marnerides, A.K., Mauthe, A., Hutchison, D.: Malware detection in cloud computing infrastructures. IEEE Trans. Dependable Secure Comput. **13**(2), 192–205 (2016)
13. Mishra, P., Pilli, E.S., Varadharajan, V., Tupakula, U.: Securing virtual machines from anomalies using program-behavior analysis in cloud environment. In: 2016 IEEE 18th International Conference on High Performance Computing and Communications, pp. 991–998 (2016)
14. UNM dataset (1998). https://www.cs.unm.edu/immsec/systemcalls.htm

15. Xie, X., Wang, W.: Lightweight examination of DLL environments in virtual machines to detect malware. In: Proceedings of the 4th ACM International Workshop on Security in Cloud Computing, pp. 10–16 (2016)
16. Ajay Kumara, M., Jaidhar, C.: Leveraging virtual machine introspection with memory forensics to detect and characterize unknown malware using machine learning techniques at hypervisor. Digit. Investig. **23**, 99–123 (2017)
17. Jia, X., Zhou, G., Huang, Q., Zhang, W., Tian, D.: FindEvasion: an effective environment-sensitive malware detection system for the cloud. In: Matoušek, P., Schmiedecker, M. (eds.) ICDF2C 2017. LNICST, vol. 216, pp. 3–17. Springer, Cham (2018). https://doi.org/10.1007/978-3-319-73697-6_1
18. Mishra, P., Pilli, E.S., Varadharajan, V., Tupakula, U.: VAED: VMI-assisted evasion detection approach for infrastructure as a service cloud. Concurr. Comput.: Pract. Exp. **29**(12), e4133 (2017)
19. Xu, Z., Ray, S., Subramanyan, P., Malik, S.: Malware detection using machine learning based analysis of virtual memory access patterns. In: Design, Automation Test in Europe Conference Exhibition (DATE), pp. 169–174 (2017)
20. Joseph, L., Mukesh, R.: Detection of malware attacks on virtual machines for a self-heal approach in cloud computing using VM snapshots. J. Commun. Softw. Syst. **14**(3), 249–257 (2018)
21. Mathew, J., Ajay Kumara, M.A.: API call based malware detection approach using recurrent neural network—LSTM. In: Abraham, A., Cherukuri, A.K., Melin, P., Gandhi, N. (eds.) ISDA 2018 2018. AISC, vol. 940, pp. 87–99. Springer, Cham (2020). https://doi.org/10.1007/978-3-030-16657-1_9
22. Mishra, P., Verma, I., Gupta, S.: KVMInspector: KVM based introspection approach to detect malware in cloud environment. J. Inf. Secur. Appl. **51**, 102,460 (2020)
23. Barham, P., et al.: Xen and the art of virtualization. In: Proceedings of the Nineteenth ACM Symposium on Operating Systems Principles, pp. 164–177 (2003)
24. Github repository (2019). https://github.com/glmcdona/Process-Dump
25. Virustotal (2019). https://developers.virustotal.com/reference
26. Code disassembly with Pydasm (2017). https://axcheron.github.io/codedisassembly-with-pydasm/
27. Pedregosa, F., et al.: Scikit-learn: machine learning in python. J. Mach. Learn. Res. **12**, 2825–2830 (2011)
28. Jones, K.S.: A statistical interpretation of term specificity and its application in retrieval. J. Doc. **60**(5), 493–502 (2004)
29. Lian, W., Nie, G., Jia, B., Shi, D., Fan, Q., Liang, Y.: An intrusion detection method based on decision tree-recursive feature elimination in ensemble learning. Math. Probl. Eng. (2020)
30. Breiman, L.: Random forests. Mach. Learn. **45**(1), 5–32 (2001)
31. Geurts, D.E., Wehenkel, L.: Extremely randomized trees. Mach. Learn. **63**(1), 3–42 (2006)
32. Cortes, C., Vapnik, D.E.: Support-vector networks. Mach. Learn. **20**(3), 273–297 (1995)
33. Github repository (2020). https://github.com/ytisf/theZoo
34. Microsoft malware classification challenge dataset (2015). http://arxiv.org/abs/1802.10135
35. Kirat, D., Vigna, G., Kruegel, C.: BareCloud: bare-metal analysis-based evasive malware detection. In: 23rd USENIX Security Symposium (USENIX Security 14), pp. 287–301 (2014)

Topic Models for Re-Id from Curbed Video Traces

N. A. Deepak$^{(\boxtimes)}$, D. J. Deepak, and G. Savitha

RV Institute of Technology and Management, Bengaluru, Karnataka, India
deepakna.rvitm@rvei.edu.in

Abstract. The problem of pedestrian re-identification is important. It is an essential pre-requisite for most of the video based surveillance studies. The re-identification will be the concrete platform for the several post-high level activities of surveillance (like recognition, analysis, tracking, tracing, monitoring, etc.). The re-identification includes, recognizing the pedestrian footprints over the frames of occluded video traces. The video traces, maybe captured from a single or network of surveillance cameras, housed at different places in non-overlapping fashion. Driven by the recent developments in topic models and its application in video/image domain, the proposed problem is tackled by finding the basic building blocks for topic model, called as words. In experimental analyses the newly introduced person re-identification method is compared with several popular re-identification methods based on re-identification perspective. Several experiments were carried-out on well-known public re-identification datasets, and the superiority of the proposed novel re-identification technique is proved with the rigorous analysis on these datasets.

Keywords: Occlusion · Re-identification · Surveillance · Topic-model · Words

1 Introduction

Pedestrian re-identification, mainly aims to identify an individual and re-identify the same person over the successive frames. That is, re-identifying the person after occlusion or out-of- focus. The idea behind this work is to solve the most challenging re-identification problem, by re-identifying the person over the successive frames. When a person determined within a surveillance camera view/coverage area, an image glimpse or path highlighting an object/person is selected, by eliminating the image background. Then the visual features of the selected target are extracted either manually using several techniques like histograms of color or texture, CNN, Linear Discriminate Analysis, Radial Basis functions, SVM, Fisher Discriminate Analysis etc. or automatically using deep learning concepts, like machine learning, deep latent features, deep metric learning etc. After the occlusion or lost-in-focus the person being tracked is re-identified, by extracting the similar set features from the successive frames.

© Springer Nature Switzerland AG 2021
K. R. Venugopal et al. (Eds.): ICInPro 2021, CCIS 1483, pp. 17–28, 2021.
https://doi.org/10.1007/978-3-030-91244-4_2

The extracted features, before and after occlusion is evaluated, and the target is identified as the one with highest feature similarity, as the re-identification out-come. The most of the contemporary studies extract the visual features manually, whereas some others extracts it automatically. The multivariate statistics, encompassing the simultaneous observations and analysis on more than one hash features [1], includes histograms, means, covariance and co-occurrences. They are integrated using Gabour filter to create combined feature that results in better re-identification rate. The kanade-Lucas-Tomasi (KLT) features tracker [2] uses spatial intensity information to search for the position which yields the closest match, to solve problem of re-identification during occlusion. The optical flow, color and texture histograms of well-known KLT feature tracker are used to differentiate the identities. The scale-invariant feature (SIFT) algorithm is used to detect and describe the local features, in the segmented image frames [3]. Further, this is, incorporated to solve the complex re-identification problem, its performance is boosted by adding extra features like color and visual co-occurrences with SIFT features. The most contemporary deep learning models or deep latent features are not so easy to be trained, because of vanishing gradients problem. Furthermore the deep learning method requires huge set of training data to re-identify the person in more efficient way. As most of the private or public dataset available may have this major pitfall.

The proposed work tries in solving the problem with the minimal training set and manual extraction of required features. This helps in matching the extracted features before and after the occlusion or lost in focus from the camera view. This will be bit difficult to achieve, in contemporary deep learning methods.

2   Related Works

Since the newly introduced topic model based person re-identification from the occluded video traces, is closely related to those methods that extracts features manually and also automatically. Hence most of the methods discussed in this section talks about the person re-identification, using similar platform as discussed above. Person re-identification by comparing the images captured in daytime visible modality and night-time infrared modality is discussed in [4]. A trimodal based learning method is used, which preserves structure information of visible images and approximates image style of night-time infrared modality. The tensor based multiple-task model for person re-identification is discussed in [5]. The captured image is linearly classified as one-v/s-all, associating a classifier to a specific person. Re-identification is aimed to solve the problem of matching the extracted features before and after the occlusion across the camera views [6], their study involves in integrating the attention mechanism, hard sample acceleration and similarity optimization techniques. A multilevel feature fusion model [7] combines the global features with local features extracted from images using deep learning networks, to re-identify the person from multi-camera views. A long short-term memory [8] uses an end-to-end way to model person from head to foot. The contextual information and discriminative ability of local feature are integrated for identification and ranking the re-identification task.

3   Implementation

The feature extraction and its analysis to re-identify person is a more complex and challenging task in the field of image processing. This is due to variations in human body shapes, articulations of human body, cluttered image background, camera views and occlusions. The feature extraction and its analysis are the key concept. The Latent Dirichlet Allocation (LDA), [9,10] designed for text classification, text analysis or text mining, has some latent topics, and is one of the potential semantic approaches in this study. The proposed topic model based algorithm uses LDA, a 'Topic Model' for re-identification over the different camera views.

3.1   Pre-processing

Most of the image processing algorithms used in re-identification, will undergo pre-processing, this includes 1) Image background subtraction 2) Binary image construction and 3) Tracking.

Background Subtraction and Binary Image. A simple background subtraction method is used to separate the foreground from the background scene. This method uses the well-known mean, variance and threshold value method to subtract the foreground images from background scene.

Feature Extraction and Tracking. The primary goal of feature extraction process is to project the data onto a lower dimension space. It's an important step and mainly relies on linear transformation. The dimensionality of the features can be reduced by selecting some of the important features based on the application. The feature extraction is the 'key concept' in the proposed person re-identification algorithm. The suitable features are selected by tracking the person under the camera view. This tracking is restored from other camera views once the person under surveillance is detected, after the lost in focus from the earlier camera eye or being occluded. During tracking the features like pace, (a single step taken when walking or running), step-width (vertical motion), cardinality (pixels within an ROI), Below-Area-Left (BAL) and Below-Area-Right (BAR) (pixels corresponding to lower parts of human body) are selected, which will provide the necessary feature values used in re-identification between the different camera views.

3.2   Generation of Words - The Basic Building Blocks of LDA

The words are the basic building blocks for LDA. The performance of LDA depends completely on the words generated. The words are generated by analysing the sequences extracted during tracking. Increase in words generated will be proportional to the classification rate of proposed algorithm. Hence the generation of words gains importance in person re-identification. The proposed algorithm, generate words by processing the elements (δ_i; δ_{i+1}; δ_{i+2} and δ_{i+3}), as shown below. To generate words, we consider only four elements at each time, from the selected sequence. Initially, the variables (δ_i; δ_{i+1}; δ_{i+2} and δ_{i+3}), are assigned with the initial values found in the sequence and during the successive iterations these elements (δ_i; δ_{i+1}; δ_{i+2} and δ_{i+3}) are assigned with the new values found by moving the elements towards right by one position within the sequence. This process is repeated until all the words associated within the sequence are generated.

3.3 Proportionality of Words

After the words are generated, they are fed as an input to LDA algorithm, to produce Word-probability matrix (β) and Topic-probability matrix (α). The matrix (β) in Eq. 1 is the result of summand value of word posteriors (θ) and the diagonal matrix (λ). The matrix (β) specifies the distribution of words in the topic [9,10]. The (t) in Eq. 1 refers to a document.

$$\beta(i,j) = \beta(i,j) + \lambda(t) * \theta \tag{1}$$

The topic probability matrix or topic distribution matrix (α) in Eq. 2 is represented as a matrix of Dirichlet posteriors. The matrix (α) specifies the distribution of topics in each document. The (h) in Eq. 2 refers to the vector of ploy-gamma function [9,10].

$$\alpha(i) = \alpha(i) - h(i) \tag{2}$$

The proportionality words are found using the matrix (β). Let the words and topics are the rows and columns of matrix (β). For example, consider the matrix (β) of size 5×4, let the rows (words w) be $w_1 = 0.3, 0, 0.4, 0$; $w_2 = 0.15, 0.35, 0.2, 0.05$; $w_3 = 0, 0, 0.5, 0.5$; $w_4 = 0, 0.18, 0.6, 0$; $w_5 = 0, 0, 0.1, 0$; and the columns (topics t) will be found as $t_1 = 0.3, 0.15, 0, 0, 0$; $t_2 = 0, 0.35, 0, 0.180$; $t_3 = 0.4, 0.2, 0.5, 0.6, 0.1$; $t_4 = 0, 0.05, 0.5, 0, 0$; The proportionality of each word in an document is found as shown below.

★ First-of-all we calculate the row-sum (i.e., summation of elements row-wise), by considering the elements in matrix (β), for the given example above, the row-sum will be (0.7; 0.75; 0.71; 0.78; 0.1). Then, each element within the row is divided by the summation value, found from the elements of that row Eq. 3

$$w(i,j) = \frac{\beta(i,j)}{\sum_{i=1}^{m} \lambda(i)} \tag{3}$$

For example, let the proportionality of word-occurrences be ($w_1 = 1$, $w_2 = 1$, $w_3 = 1$, $w_4 = 1$ and $w_5 = 1$) found within a document.

★ Then we find the summation of word occurrences as Eq. 4

$$\sigma = \sum_{k=1}^{N} \nabla(k) \tag{4}$$

In the above example $N = 5$, represents words considered in the document.

★ The proportionality of each word in a document is found by dividing the word-occurrence by the words considered within the document. Equation 5

$$\cup (i, j) = \frac{\omega(i, j)}{\sigma} \tag{5}$$

★ Finally, the per-topic, word distribution, is found as the product of two normalized factors (ω) and $(\cup)$ Eq. 6. The variables (ω) and $(\cup)$ are the normalized values. The (γ) refers to per-topic, word distribution.

$$\gamma(i, j) = \omega(i, j) * \cup(i, j) \tag{6}$$

3.4   Person Re-identification

The re-identification is the final step in the proposed algorithm. This involves in analysing the matrices generated in the previous steps.

■ Initially, the videos of the dataset considered are broadly divided into two groups, namely test and training.
■ The words - the building blocks of LDA and its occurrences are found from the test set.
■ The LDA algorithm is executed on the training set, to produces the word-topic distribution matrix (β) and topic - document distribution matrix (α).
■ The problem of re-identification involves in selecting some of the words and its occurrences. To do this, we first calculate the standard deviation value (δ_{test}), by working on the test set Eq. 7

$$\delta_{test} = \delta[_{j=1}^{N}(w_j, w_{j+1}, w_{j+2}...w_n] \tag{7}$$

The value (N), in Eq. 7, represents the number of words found in the test-set and (w_j) represents the count of (j^{th}) word-occurrence. Once we find the (δ_{test}), we filter-out some words based on the condition as shown in the Eq. 8. The words are selected, only if its occurrence value is more than the standard deviation calculated.

$$\gamma_{test(j)} = |_{j=1}^{n} (w_j - \delta_{test}) \geq 0 | \tag{8}$$

For each word in $\gamma_{test(j)}$ found in Eq. 8, we find the corresponding topic, using the word-topic distribution matrix (β). The selected topic would be the maximum value in the row. Similarly, for each selected topic, the corresponding document is found, from topic-document distribution matrix (α). Here the document selected, would be the maximum value. Finally, we re-identify the person by selecting a single document from the training set, which is the maximum value found by adding the elements column wise.

For an example, consider the word-occurrences (w_j), where $(j = 1, 2...9)$,

✓ Initially, we find the standard deviation ($\delta = 0.2603$), found from the values of Table 1.

Table 1. Words and its Occurrences

Words	w_1	w_2	w_3	w_4	w_5	w_6	w_7	w_8	w_9
Occurrences	0.2	0.5	0.1	0.7	0.5	0.4	0.2	0.9	0.6

✓ The Table 2 shows the words selected based on the standard deviation - using Eq. 8.

Table 2. Words Selected - based on condition Eq. 8

Words	w_2	w_4	w_5	w_6	w_8	w_9
Occurrences	0.5	0.7	0.5	0.4	0.9	0.6

✓ Let the Table 3 and 4 represents the word-topic distribution matrix (β) and topic-document distribution matrix (α), generated from the training-set:

Table 3. Word-Topic Distribution Matrix (β)

Word/Topic	T_1	T_2	T_3	T_4	T_5
w_1	0.1	0.5	0.6	0.4	0.3
w_2	0.1	0.2	0.5	0.8	0.2
w_3	0.2	0.3	0.4	0.1	0.3
w_4	0.9	0.8	0.2	0.7	0.5
w_5	0.7	0.1	0.5	0.2	0.9
w_6	0.4	0.1	0.3	0.8	0.7
w_7	0.2	0.5	0.7	0.4	0.5
w_8	0.5	0.3	0.3	0.3	0.4
w_9	0.1	0.2	0.1	0.5	0.1

✓ In Table 3, the row represents the words and column represents the topics, similarly, in Table 4, the topics and documents are represented as rows and columns respectively. In the example the matrix beta (β) is of size 9×5 and matrix alpha (α) of size 5×5. Based on the selection condition Eq. 8, some of the words are filtered-out, as shown in Table 5. We find the topics, from Table 5, by selecting the value which is maximum in the row (indicated in boldface). Similarly, for the selected topics in Table 5, we find the corresponding documents using matrix alpha (α). The condition, to select the documents is same as the condition used to select the topics as shown in Table 6.

Table 4. Topic-Document Distribution Matrix (α)

Topic/Document	D_1	D_2	D_3	D_4	D_5
T_1	0.5	0.8	0.4	0.1	0.6
T_2	0.2	0.6	0.1	0.1	0.8
T_3	0.2	0.3	0.1	0.2	0.2
T_4	0.1	0.1	0.4	0.9	0.1
T_5	0.3	0.1	0.2	0.1	0.7

Table 5. Selected words and Topics

Word/Topic	**T1**	T_2	T_3	**T4**	**T5**
w_2	0.1	0.2	0.5	**0.8**	0.2
w_4	**0.9**	0.8	0.2	0.7	0.5
w_5	0.7	0.1	0.5	0.2	**0.9**
w_6	0.4	0.1	0.3	**0.8**	0.7
w_8	**0.5**	0.3	0.3	0.3	0.4
w_9	0.1	0.2	0.1	**0.5**	0.1

Table 6. Selected Documents - Re-identified

Topic/Document	D_1	**D2**	D_3	**D4**	**D5**
T_1	0.5	**0.8**	0.4	0.1	0.6
T_4	0.1	0.1	0.4	**0.9**	0.1
T_5	0.3	0.1	0.2	0.1	**0.7**
Sum(Column-Sum)	0.9	1.0	1.0	1.1	**1.4**

✓ Finally, we re-identify the person by selecting a single document (**D5**) as the result, shown in Table 6, which is the maximum value found by adding the elements column-wise

4   Experiments and Results

Several literature on person re-identification deals with methods that use deep learning, shallow learning and non-machine machine learning concepts. In shallow machine leaning algorithm, learning happens from the data described by the pre-defined features, the deep learning algorithm learns features automatically, whereas in non-machine learning algorithms, features are extracted manually from the exhibited behaviour of the individuals. The motivation to compare the proposed unsupervised machine learning algorithm with the other machine learning and non-machine learning algorithms is that, all the methods used in comparison including proposed algorithm divides the input data into two sets namely, test and training. In proposed algorithm, the learning happens with sets of observations to be explained by unobserved groups that explain why some parts of the data are similar. For example, if observations are words collected into documents, it posits that each document is a mixture of a small number of topics and that each word's presence is attributable to one of the document's topics. LDA is an example of a topic model and belongs to the unsupervised machine learning technique. The performance of the proposed re-identification algorithm is evaluated on public datasets namely CUHK01 dataset [11], PRID2011 dataset [12], iLIDS-VID dataset [13], and MARS re-identification dataset [14]. For PRID2011 and iLIDS-VID datasets the Cumulative Match Characteristics (CMC) curve is used for evaluation.

The same partition rule given in the dataset is considered to create the test and training samples. As mentioned earlier the dataset PRID2011 uses 2 non-overlapping cameras, out of which, 385 identities are captured by camera A and 749 identities are captured by camera B. The initial set of 200 identities found in both the views. The remaining identities in each camera view correspond to gallery set of the corresponding view. Here the evaluation consists of searching the union set of identities found in one camera view with all other identities of the other view. Hence this results in two possible evaluation, firstly, the probe set is taken from camera A (200 identities, present in both the views) and 749 identities of camera B will be considered as training, and in successive evaluation 385 identities captured by camera A will be the gallery set and union identities of camera B will be considered as probe set. Similarly, iLIDS-VID dataset comprises 600 image sequences of 300 distinct individuals, with one pair of image sequences from two camera views for each person. The image sequences length varies between the bands 23 to 192. This is a more challenging dataset, due to environment variations found in the images. The proposed technique is evaluated by making equal partition of probe and gallery set, consisting of 150 identities on both the sides. The MARS dataset is a huge dataset with 1261 identities here we use fixed partitioning of test and training samples, with 625 ID's corresponds to 8298 tracklets, considered as gallery set, whereas 636 ID's corresponds to 12,180 tracklets, are taken as probe set. The dataset of CUHK01 is an example for balanced learning problem, which contains exact number of images for each individual identity. As mentioned earlier the CUHK01 dataset contains images captured from 2 non-overlapping cameras.

To evaluate proposed algorithm, the images captured from the single camera view is taken as probe set and images from another camera as gallery set. It can be summarized that, the proposed algorithm uses 3 deep learning-based methods, 3 shallow learning-based methods and 3 non-machine learning algorithms for comparison. The motivation to compare proposed method with DRML+CKT is that, DRML+CKT uses deep learning concepts and extracts features automatically, the deep learning algorithms will be more effective, when we have access to large amount of labeled training data. It is also an extremely complex and time consuming, to obtain high quality labels for re-identification. Along with this the DRML+CKT exhibits a narrow structure during the recognition. Therefore increase in extracted features, will significantly make the proposed algorithm become wider in recognizing the results. This is concept is demonstrated using precession-recall curve, where we compare the proposed algorithm and DRML+CKT on dataset. The Normalized+X-Correlation and Fused method are the two well-known deep learning based methods to solve re-identification problem. These methods are capable of handling the difficult situations like illumination change, partial occlusions, and wide view point changes in the re-identification in more appropriate way. The major differences between these two algorithms are: The Normalized+X-Correlation use inexact matching techniques, whereas fused algorithm uses both inexact and exact matching techniques to solve the re-identification problem. The shallow learning algorithms like Grey Level Co-occurrence Matrix (GLCM) with Support Vector Machine (SVM), GLCM with Discriminative Component Analysis (GLCM+DCA), and GLCM with Relevant Component Analysis (GLCM + RCA) are commonly used algorithms, to solve person re-identification problems more effectively. It has been shown that, the proposed algorithm out performs the re-identification rate considerably through the precession-Recall curve in Fig. 1. This paper also compares the proposed method with non-machine learning algorithms. There are several non-machine learning algorithms used to solve re-identification problem. A semi- supervised clustering algorithms [15], Semi-supervised learning using Bayesian attributes [16] and Null space for Re-identification [17] are used here to highlight the superiority of machine learning algorithms over the non-machine learning algorithms in solving the re-identification problem.

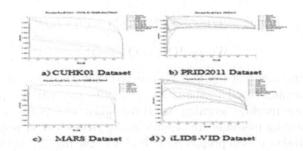

Fig. 1. Precision Recall Curve - Person Re-identification on different datasets.

In Fig. 1 we show the precision-recall curves for re-identification using datasets CUHK01, PRID2011, iLIDS-VID and MARS respectively. The area under each curve gives the mean average precision obtained from each method. It is observed from the Fig. 1a, Fig. 1b and Fig. 1c the performance of proposed algorithm is better than other methods used in comparison. This indicates the newly proposed LDA algorithm for re-identification, performs better in both balanced and unbalanced learning situations (that is, image samples of different identities found in the dataset are distributed between test and training samples equally). The CHUK01 dataset is an example of balanced dataset whereas

MARS and PRID2011 dataset are unbalance dataset. In Fig. 1d, the precision-and-recall curve of proposed algorithm is not up-to the mark. It lags in identifying the results more appropriately. This is because the other algorithms used in comparison achieves higher re-identification rate on the identities of iLIDS-VID unbalanced dataset. With the analysis carried-out above qualitatively, the proposed algorithm shows better performance, with most of the dataset used. But we still proceed further to evaluate re-identification accuracies quantitatively using F-measure Eq. 9. This is for imbalanced data classification and provides a way to combine the precision and recall into single measure with both the properties.

$$F - measure = 2 * \frac{precision * recall}{precision + recall} \tag{9}$$

In Fig. 2 and Fig. 3, the box-and-whisker plots using F-measure, is used to evaluate the re-identification accuracies, which provides a way to combine the precision and recall properties. The outcomes of all methods used for comparison with different datasets CUHK01, PRID2011, MARS and iLIDS-VID are demonstrated respectively. The vertical line drawn at the center of each rectangle represents the median of F-space which is also known as middle quartile (Q2). The upper and lower quartiles of each rectangle is corresponds to vertical boundaries of each rectangle (Quartile Q1 and Q3). The horizontal dashed lines are drawn from quartile Q1 and Q3 to their extreme data points, are within inter-quartile range (IQR). The inter-quartile range for each rectangle in box and whisker plot is found using Eq. 10. The outliers beyond IQR are marked using red dot.

$$Range = [Q1 - 1.5 * IQR, Q3 + 1.5 * IQR] \tag{10}$$

Where, IQR = Q3 − Q1

In box and whisker plot Fig. 3, it is observed that, the boxes of deep learning and shallow learning-based methods are generally towards the positive value than those of non-machine learning based methods. This is the indication that the machine learning methods are capable and provides better re-identification. The box and whisker plot of proposed method is at the higher level in Fig. 3. This indicates that even the deep learning based methods have poor performance, if the training samples considered are less. This results in inappropriate training of algorithm, and results in poor performance on unbalanced dataset.

In Fig. 3, we may clearly notice that the performance of the proposed algorithm is not at the highest level. The methods Norm+X+Corr, DRML+CKT and fused algorithms have better performance than the proposed algorithm. This is because the fused algorithm uses inexact and exact matching techniques to reduce false matches that are the major concern in reduction in performance among the other algorithms used in comparison. Figure 3 shows the performance of different methods using MARS dataset. Here we compare the proposed algorithm with other methods. It is noted that the performance of proposed

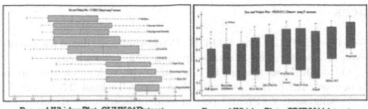

Box and Whisker Plot–CUHK01Dataset   Box and Whisker Plot – PRID2011dataset

Fig. 2. Box and Whisker Plot - CUHK01 dataset & PRID2011 dataset.

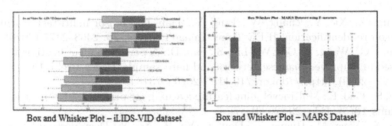

Box and Whisker Plot – iLIDS-VID dataset   Box and Whisker Plot – MARS Dataset

Fig. 3. Box and Whisker Plot - iLIDS-VID dataset & MARS Dataset.

method is similar to fused method. The other machine learning algorithm performs equally well, whereas the non-machine learning algorithm lags in their performance.

5   Conclusion

The described algorithm and the projected result in this paper are used to assess the quality of re-identification. As the results show the new technique achieves its objectives in terms of superior performance but implementation section in this paper is huge and its essential contributions tend to get submerged in the details. It is therefore, imperative to summarize the distinguishing features of this paper over the contemporary work and then provide the concluding statements. Put succinctly algorithm for re-identification in contemporary literature extract the gait sequences. This is fed to the standard package of LDA and then the output is computed. Thus, the first basic contribution is extraction of these sequences from the segmented video frames. The second basic contribution is in the generation of words from these sequences and the third basic contribution is the re-identification from occluded scenes using LDA. With these three new elements, the results become qualitatively superior which is not surprising. However, limitations of the present work should be outlined and they are the following: On occlusion, the re-identification performance degrades if the captured image before and after occlusion is from multiple camera views with variations in angle and noise, whereas this problem is not found in the single camera view even after heavy occlusion or noise. It may be pointed out that these limitations are going to stay for quite some time to come.

References

1. Fang, W., Hu, H., Hu, Z., Liao, S., Li, B.: Perceptual hash-based feature description for person re-identification. Neuro-computing **272**, 520–531 (2018)
2. Camps, O., et al.: From the lab to the real world: re-identification in an airport camera network. IEEE Trans. Circuits Syst. Video Technol. **27**(3), 540–553 (2017)
3. Zhang, Z., Saligrama, V.: PRISM-person re-identification via structured matching. IEEE Trans. Circuits Syst. Video Technol. **27**(3), 499–512 (2017)
4. Ye, M., Shen, J., Shao, L.: Visible-infrared person re-identification via homogeneous augmented tri-modal learning. IEEE Trans. Inf. Forensics Secur. **16**, 728–739 (2021)
5. Zhang, Z., Xie, Y., Zhang, W., Tang, Y., Tian, Q.: Tensor multi-task learning for person re-identification. IEEE Trans. Image Process. **29**, 2463–2477 (2020)
6. Wang, G., Wang, S., Chi, W., Liu, S., Fan, D.: A person re-identification algorithm based on improved Siamese network and hard sample. Math. Probl. Eng. 3731848, 11–22 (2020). https://doi.org/10.1155/2020/3731848
7. Wu, S., Gao, L.: Multi-level joint feature learning for person re-identification. Algorithms **13**(5), 111–129 (2020)
8. Bai, X., Yang, M., Huang, T., Dou, Z., Yu, R., Xu, Y.: Deep-person: learning discriminative deep features for person re-identification. Pattern Recogn. **98**(107036), 88–95 (2020)
9. Blei, D.M., Ng, A.Y., Jordan, M.I.: Latent Dirichlet allocation. J. Mach. Learn. Res. **3**, 993–1022 (2003)
10. Blei, D.M.: Probabilistic topic models. Commun. Associ. Comput. Mach. **55**(4), 77–84 (2012)
11. Li, W., Zhao, R., Wang, X.: Human reidentification with transferred metric learning. In: Lee, K.M., Matsushita, Y., Rehg, J.M., Hu, Z. (eds.) ACCV 2012. LNCS, vol. 7724, pp. 31–44. Springer, Heidelberg (2013). https://doi.org/10.1007/978-3-642-37331-2_3
12. Martin, H., Csaba, B., Peter, M., R., Horst, B.: Proceeding Scandinavian Conference on Image Analysis (SCIA) (2011)
13. Xiatian, Z.: Target re-identification and multi-target multi-camera tracking. In: Proceedings IEEE Conference on Computer Vision and Pattern Recognition (CVPR) (2019)
14. Zheng, L., et al.: MARS: a video benchmark for large-scale person re-identification. In: Leibe, B., Matas, J., Sebe, N., Welling, M. (eds.) ECCV 2016. LNCS, vol. 9910, pp. 868–884. Springer, Cham (2016). https://doi.org/10.1007/978-3-319-46466-4_52
15. Xin, X., Wang, J., Xie, R., Zhou, S., Huang, W., Zheng, N.: Semi-supervised person re-identification using multi-view clustering. Pattern Recogn. **88**, 285–297 (2019). ISSN 0031-3203
16. Liu, W., Chang, X., Chen, L., Yang, Y.: Semi-supervised Bayesian attribute learning for person re-identification. In: Proceedings 32nd Association for the Advancement of Artificial Intelligence Conference on Artificial Intelligence (AAAI 2018), pp. 7162–7169 (2018)
17. Zhang, L., Xiang, T., Gong, S.: Learning a discriminative null space for person re-identification. In: Proceedings IEEE Conference on Computer Vision and Pattern Recognition (CVPR), pp. 1239–1248 (2016)

Effectual Single Image Dehazing with Color Correction Transform and Dark Channel Prior

Jeena Thomas[ID] and Ebin Deni Raj[✉][ID]

Indian Institute of Information Technology, Kottayam, Kerala, India
{jeenaphd2019,ebindeniraj}@iiitkottayam.ac.in

Abstract. Image acquisition during bad weather conditions like haze may adversely affect the quality and information processing. The image degradation affects object detection, which plays a prominent role in computer vision applications. Restoration of hazy images is necessary for exact identification and location of objects. This paper proposes a new prior based dehazing algorithm named Color Correction Transform Dark Channel Prior (CCTDCP). The proposed algorithm utilizes white balance color correction transform, dark channel prior and gamma correction for retaining the originality of the image. The experimental results on benchmark images exhibit superior performance of CCTDCP algorithm over the state-of-the-art methods.

Keywords: Dehazing · White balance · Dark channel prior · Gamma correction

1 Introduction

Information processing from images with computer vision techniques helps to analyse, understand and extract relevant high dimensional data. Object detection is observed as the significant subfield of computer vision. The detection of objects from outdoor scenes is challenging under various weather conditions [1,2]. The presence of various particles in the atmosphere may results in diverse weather conditions and affects the image capturing. The various particles such as water droplets, aerosol and molecules lead to fog, haze, cloudy, rain and wind climatic conditions are considered as the adverse factors which affects the object detection. The existence of haze in the atmosphere may degrades the quality of the image [1–4]. Haze occurs due to the scattering and absorption of atmosphere, caused by particles or water droplets. The open-air and indoor images are generally degraded by impure medium due to the presence of particles and water droplets in the atmosphere. Haze is a resultant of scattering and atmospheric absorption and creates whitening effect which distort as well occlude the objects. The sunlight and environmental illumination captured by the camera from the outdoor scene is diminished along the sightline. The scattering of light

© Springer Nature Switzerland AG 2021
K. R. Venugopal et al. (Eds.): ICInPro 2021, CCIS 1483, pp. 29–41, 2021.
https://doi.org/10.1007/978-3-030-91244-4_3

is caused by floating particles in the atmosphere which causes image degradation [1,2]. This defect cannot be avoided because it occurs in the atmosphere beforehand influenced by the capturing device. Haze is an adverse factor which decreases the visibility and contrast of outdoor scenes. Therefore, removing hazing i.e.; dehazing of images is critical and extensively demanded subject matter in computer vision domain [1–3]. Dehazing involves the process of separating haze components from scenes and regain the original color with better clarity.

Haze elimination plays a significant role in restoration of images and dehazing is an inverse problem which estimates the unidentified depth of the haze. Lot of research has been carried out to estimate the transmission map which indicates unknown depth of the haze. Various methods were proposed which includes multiple image and single image dehazing techniques. The methods formulated on varying atmospheric conditions, polarization and depth falls under first category. Dark object subtraction, visibility maximization and independent component analysis are the examples of single image dehazing techniques [1]. In most of the dehazing approaches, Atmospheric Scattering Model (ASM) is considered as the backbone of representing hazy images, whereas the hazy images can be expressed as two components such as direct attenuation and airlight [1,2,4–6]. The major steps involved in dehazing approach is estimating transmission map and atmospheric light.

The dehazing operations which involves statistical property or assumptions are prior based approaches. Among them, Dark Channel Prior (DCP) is considered as the most prominent one [1]. But sometimes the DCP fails to dehaze the images which have bright sky regions. Xiao et al. [6] proposed an enhanced DCP model which can effectively dehaze the images based on sky region segmentation with Otsu algorithm. For estimating the atmospheric light, this strategy has involved, if there is larger sky regions in the image.

Haze-line prior is built on color consistency approach and this prior obtain haze free image by estimating the transmission map. In this method, the image is represented as different clusters and assumes the pixels located in the cluster are global. These discrete set of clusters in RGB space indicates the color distribution in non-haze image. But for hazy images, haze-lines are generated which indicates the different lines in RGB space. The transmission level can be identified with the help of location of pixels in haze-line [7].

Ju et al. [8] proposed a novel prior based dehazing approach is Gamma Correction Prior (GCP) which aids lesser processing time. The novelty of this method involves the transmission map is estimated with the help of obtained depth ratio of input image and virtual image. For estimating the atmospheric light, quad tree subdivision approach is employed. Most of the research work for dehazing is carried out in homogenous haze images. FRIDA, D-Hazy [9], O-Haze, I-Haze, RGB-NIR, DENSE-HAZE and RESIDE are the prominent examples for homogenous haze image benchmarking datasets during daytime. NightHaze-1 [10], NightHaze-2 [10], YellowHaze-1 and Yellow-Haze-2 are nighttime haze image datasets. NH-Haze represents the non homogenous haze image dataset [7].

In addition to these prior based approaches, various dehazing mechanisms which involves machine learning and deep learning techniques are proposed. But all the existing methods are ineffective to work with dense haze images and especially the deep learning methods are computationally complex. The major motivation of our approach is to better dehaze the image scenes with effective prior based approach with minimal involvement of highly computational equipment. This paper implements a new and improved prior based approach named Color Correction Transform Dark Channel Prior (CCTDCP). This method provides effectual haze removal by employing the color correction approaches in pre-processing and enhancement.

The primary contributions of this work is summarized as follows:

1. We propose an improved prior based dehazing mechanism, Color Correction Transform Dark Channel Prior (CCTDCP).
2. The CCTDCP method is designed to consider the whitening effect of haze. We introduce white balance color correction transform to reduce this effect.
3. The CCTDCP is an effective method to dehaze the image scenes with least involvement of extremely computational equipment. Our method notably outperforms the prevailing methods in terms of PSNR.

The remaining part of the paper is structured as: Sect. 2 specifies the various existing techniques used for dehazing. Section 3 illustrates about the proposed technique and corresponding algorithm for haze removal. Section 4 discusses about the experimental results obtained. Section 5 concludes our study.

2 Related Works

The effects of the haze should be removed and the original details of the image should be preserved for object detection. The haze removing operations can be applied to various contexts such as nighttime images, daytime images, indoor scenes and aquatic images. The dehazing is essential for numerous applications such as remote sensing, aerial images, medical diagnosis, under water imaging, intelligent surveillance, self-driving cars, face recognition and anomaly detection. Single image dehazing techniques can be mainly categorised as: a) Prior based methods b) Learning based methods.

2.1 Prior Based Methods

Prior based approaches work on the underlying assumptions and dehazing with single images is very important in computer vision research because the priors indicate the accomplishment. He et al. [1] proposed a novel prior known as Dark Channel Prior (DCP), which assumes that in a cloudless day photograph, the intensity of each pixel will be near to zero, at a minimum in one color channel, excluding the sky parts. This property is known as dark channel prior and corresponding channel is dark channel. The presence of shadows, colorful and dark objects contribute the small intensities in dark channel. The dark pixels present in the dark channel affords exact estimation of haze spread. The depth of the haze is described by the transmission map and gaining transmission is substantial for dehazing images. The thickness of haze can be roughly represented by the intensity of the dark channel, which is used to estimate atmospheric light and the transmission map [1].

There are different refinement mechanisms used for transmission map with DCP, which is a patch based approach for dehazing of single images. He et al. [1] used soft matting approach with DCP for filtering the transmission map, but it is computationally complex and not work well with dense haze images. They have randomly chosen some images from flickr.com especially the outdoor landscape and cityscape images during daytime. Song et al. [11] adopted guided filtering mechanism for refinement using flickr images as input. In guided filtering, edge conserving smoothing operation is based on the content of the guidance image. The guidance image can be the input image itself, a diverse variant or entirely varied image. They analysed dehazing results on different patch sizes and utilized gamma correction for enhancing the final results.

Joint trilateral filter was effectively used for refining the transmission map, where this technique has the capability to control the gradient reversal artifacts of haze free photographs. Singh et al. [2] used this approach for effectively dehazing the remote sensing images especially from QUICKBIRD, IKONOS and MODIS sensors. Xiao et al. [6] employed median filter for better refining the transmission map along with guided filtering mechanism. A smaller value of the gamma coefficients was used with gamma correction for improving the final results. This approach is helpful for avoiding the distortions present in the sky regions and removes the irregularities related with the edges.

Nair et al. [3] deployed Gaussian filter for obtaining the transmission map, which could automatically deliver a smoothing effect. The experiment has been conducted along with DCP on IVCDehazed dataset. Borkar et al. [4] proposed a novel technique for single image dehazing using nearest neighbour regularization to obtain haze free smooth transmission map. This technique will substitute haze pixels with moderate intensities in RGB space, so that image uniqueness can be maintained. They have used D-HAZY dataset to compute the result, which provides better outcome especially the images with sky regions.

Haze removal from nighttime images is more difficult when compared to daytime image dehazing. Nighttime image haze occurs along with various artificial light sources which causes uneven illumination. Tang et al. [10] utilized bilateral filtering algorithm to get the final transmission map. They have used two hazy image datasets NightHaze-1 and NightHaze-2, in which the images mainly contain external light sources such as street lamp and city light. Gradient Channel Prior (GCP) is another prior based dehazing approach where the transmission map and atmospheric light estimation can be obtained. In this approach, gradient of the images is the key principle and to preserve the texture details, the transmission map is polished using guided L0 filter [5].

2.2 Learning Based Methods

The learning wise strategy, makes use of deep neural network architectures for dehazing mechanism. Convolutional Neural Networks (CNN) and Generative Adversarial Networks (GAN) are mainly adopted for dehazing. AOD-Net [12] and DehazeNet [13] comes under the first category whereas DehazeGAN, ReviewNet and RI-GAN falls under another category. All in One Dehazing Network (AOD-Net) redeveloped atmospheric scattering model with the help of mapping function.

Mehra et al. [14] proposed ReviewNet which provides an end-to-end solution of dehazing image and this architecture works on four color spaces namely RGB, Lab, HSV and YCbCr. Differential programming and CycleGAN are employed for dehazing in DehazeGAN and RI-GAN respectively [4]. An end to end dehazing method is DehazeNet in which the transmission map is estimated with the help of CNN. Feature extraction, multiscale mapping, local extrema and non linear regression are major steps involved in this learning approach [13,15,16]. Another CNN based approach is Perceptual Pyramid Deep Network where learning is based on perceptual losses and mean squared error [7,17,18]. Park et al. [19] instigated the advantage of involving mixed deep neural network architectures for dehazing with the help of CycleGAN, CGAN and CNN.

All the aforementioned methods are experimented on the images with homogenous haze. The major drawback related with the prior based approach is generated transmission map may contain halos and artifacts. Increased training cost is the foremost problem with learning based approach while estimating the transmission map. But both of these methods failed for dehazing the images with dense haze.

3   Color Correction Transform Dark Channel Prior (CCTDCP) Dehazing Technique

Haze is considered as an additive component of the image and our aim is to effectively recover the originality of the image. The definition of a haze image during daytime can be written as

$$H(y) = S(y)t(y) + A(1 - t(y)) \tag{1}$$

where $H(y)$ is the observed haze image. S indicates the scene radiance or original color of the image. A denotes the atmospheric light, y indicates pixel coordinates position and t represents the amount of light transmitted and reaches the camera. The first term is called direct attenuation and second term as airlight. The goal of the dehazing algorithm is to regain the values of S, t and A from H.

For homogeneous atmosphere, the transmission t can be stated as

$$t(y) = e^{-\beta d(y)} \tag{2}$$

where β represents the scattering coefficient of the atmosphere and d indicates the scene depth. The components of the haze imaging equation such as atmospheric light, haze image and scene radiance are coplanar in RGB color space. But the endpoints of these vectors are treated as collinear. We have employed dark channel prior with effective white balancing and gamma correction methods for better removal of haze and the proposed CCTDCP method is shown in Fig. 1.

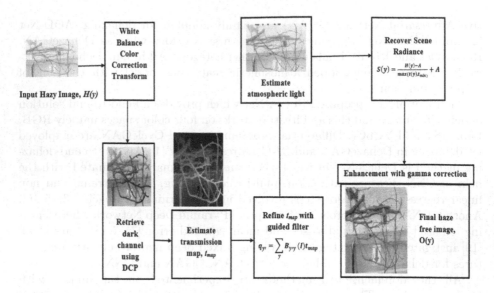

Fig. 1. Flowchart of proposed CCTDCP method

3.1 White Balance Color Correction Transform

Haze may affect the color consistency of images which leads to poor analysis due to the increased whitening effect. This drawback can be resolved by using color correction transform by incorporating *k nearest neighbour strategy* [20]. Let $H(y)$ be the hazy image, our aim is to generate color correction matrix which involves a polynomial kernel function and Euclidean norm. The color correction matrix $M(y)$ can be computed by minimizing the mapping between hazy image, $H(y)$ and ground truth image, $H_{gt}(y)$ with polynomial kernel function, Φ and represented in Eq. (3).

$$\arg \min_{M(y)} \parallel M(y)\Phi(H(y)) - H_{gt}(y) \parallel_2 \tag{3}$$

The final corrected white balanced image can be obtained by performing the combined operations with final correction matrix, kernel function and input hazy image. The final color correction matrix, $M(y)$ is generated by extracting the similar pixels by nearest neighbour approach. The ultimate white balance image, $W(y)$ can be computed as in Eq. (4).

$$W(y) = M(y)\Phi(H(y)) \tag{4}$$

3.2 Dark Channel Prior

Dark Channel Prior (DCP) is a patch based statistical property related with the images where it contains dark pixels and dark channel [1]. Dark pixel can be defined as, among the three RGB channels, the lowest intensity of the pixel is less than a small threshold. The dark pixel is mathematically represented in Eq. (5).

$$S(y) \equiv \min_{c \in \{r,g,b\}} (S_c(y)) \leq \delta \tag{5}$$

where c indicates the color channel and δ denotes the threshold value. Since $S(y)$ is a dark pixel, every pixels, $S(y')$ which is substantial to the below condition are also dark pixels.

$$\min_{c \in \{r,g,b\}} (S_c(y')) \leq \min_{c \in \{r,g,b\}} (S_c(y)) \tag{6}$$

where y' denotes the other coordinate pixels. This inequality proposes that in a patch which is centered at pixel y of the image, there occurs at least one dark pixel. DCP is a statistical property concerned with a patch of pixels where pixels' intensity is near to null, at least in one color channel. Given an image S, the dark channel of S can be defined in Eq. (7).

$$S^{dark}(y) = \min_{y' \in \Omega(y)} (\min_{c \in \{r,g,b\}} S^c(y')) \tag{7}$$

where $\Omega(y)$ denotes the patch centered at the pixel y.

3.3   Obtain Transmission Map

The transmission map indicates the depth of the haze penetrated in an image and assumes that transmission through a patch remains constant [1]. The map is obtained by performing minimum operation over the three color channels separately in the haze imaging equation. The transmission map, t_{map} can be calculated as

$$t_{map} = 1 - \alpha \min_{y' \in \Omega(y)} (\min_c \frac{H^c(y')}{A^c}) \tag{8}$$

α is the constant parameter and $0 < \alpha \leq 1$.

3.4   Refine Transmission Map

The transmission map estimated using DCP may contain halos and blocking artifacts, which hinder the exact detection of objects [1]. It is essential to use effective filtering mechanism that can reconstruct the noise free image. Guided filtering [21] mechanism is employed to remove the degradation and preserving of edges. This technique generates the output image by considering patch-based statistics especially the orientation of the pixels in an image. The filtering operation involves on the input image which is the transmission map, t_{map} and the guidance image, I can be expressed as

$$q_{y'} = \sum_y B_{y'y}(I)t_{map} \tag{9}$$

where $q_{y'}$ is the filtering output centered at the pixel y', $B_{y'y}$ is the filter kernel which is linear with respect to the transmission map.

Since the guidance image is same as the input image, guided filter is locally linear which maps guidance image and output image. The guided filtering approach on the guidance image, I with a window, W_s centred at pixel, x is represented in Eq. (10).

$$q_{y'} = a_x I_{y'} + b_x, \forall x \in W_s \tag{10}$$

where a_x and b_x are linear coefficients which is assumed to be constant in the window.

3.5   Estimation of Atmospheric Light

We adopt the DCP approach by choosing the pinnacle 0.1% brightest pixels in the dark channel. Among them, the pixels with highest intensity value in the input image are decided as the atmospheric light [1]. These pixels might not be brightest ones within the entire input photograph.

3.6   Recover Scene Radiance

The scene radiance or original color of the image, $S(y)$ can regained with Eq. (11).

$$S(y) = \frac{H(y) - A}{max(t(y), t_{min})} + A \tag{11}$$

The typical value of t_{min}, which is the lower bound of $t(y)$ is 0.1.

3.7   Enhancement with Gamma Correction

The contrast of image obtained through the pipeline of DCP, can be improved by applying the technique of gamma correction. The intensities of image are restricted over a dynamic range by applying a power function. Gamma correction eliminates the extremely darker and bright regions based on the selected gamma values [22]. The gamma transformation, G over the intensities of the image is represented as:

$$O(y) = G(S(y)) = S_{max}(S(y)/S_{max})^\gamma \tag{12}$$

where $O(y)$ is the enhanced final image, S_{max} is the maximum intensity of the image and γ is the varying gamma coefficients. By combining the aforementioned steps in Sect. 3, Algorithm 1 summarizes the steps for image dehazing.

Algorithm 1: Color Correction Transform Dark Channel Prior(CCTDCP) Algorithm

---

Input: Haze image, $H(y)$

Output: Enhanced image, $O(y)$

1: Apply white balance color transform by using the kernel function, Φ along with k nearest neighbour strategy.

2: Retrieve the dark channel, $S^{dark}(y)$

$$S^{dark}(y) = \min_{y' \in \Omega(y)}(\min_{c \in \{r,g,b\}} S^c(y'))$$

   2.1: Set the patch(Ω) size, as 15x15

3: Estimate the transmission map, t_{map} using equation (8).

   3.1: Set $\alpha = 0.95$

4: The refinement of t_{map} is performed using guided filter with equations (9) and (10).

5: Atmospheric light, A is estimated by considering the pixel with highest intensity value among the brightest pixels in dark channel.

6: Reacquire the scene radiance, $S(y)$ using equation (11).

   6.1: Set the value of t_{min} as 0.1.

7: Apply gamma correction over obtained haze free image.

$$O(y) = G(S(y)) = S_{max}(S(y)/S_{max})^{\gamma}.$$

8: Return $O(y)$.

To retrieve dark channel and local constraints, we set patch size of 15×15. We assign the fraction of the haze in images, $\alpha = 0.95$ to hold aerial perspective, after the removal of haze. To reduce the noise of the recovered image scene, we curb the value of $t(y)$ as 0.1.

4   Experimental Results and Discussion

For experimenting we used Middlebury set of D-Hazy dataset [9]. Middlebury dataset includes 23 pairs of indoor image scenes, which consists of ground truth and hazy images. All the computations are done using python and MATLAB codes by utilizing the OpenCV library. Every test was carried out with an Intel(R) Core (TM) i7-7700 CPU @ 3.60 GHz and 8 GB RAM processor. All the images of Middlebury set are experimented using the Algorithm 1.

The obtained results are also tested and verified using statistical quantitative metric. We used the Peak signal-to- noise ratio (PSNR) as the evaluation metric for quantitative assessment. The PSNR of an images with size $X \times Y$ can be calculated as

$$PSNR(a,b) = 10\log_{10}(255^2/MSE(a,b)) \tag{13}$$

where a and b are reference image and test image respectively.

$$MSE(a,b) = \frac{1}{XY} \sum_{i=1}^{X} \sum_{j=1}^{Y} (a_{ij} - b_{ij})^2 \qquad (14)$$

The images are resized and CCTDCP method achieves higher PSNR value of 16.2614 dB.

Table 1. Mean PSNR(dB) values on Middlebury dataset

Ref	Dehazing methods	PSNR(dB)
[23]	Wavelet Fusion	10.3249
[24]	CycleGAN	11.3037
[1]	DCP	12.0234
[12]	AOD-Net	13.3500
[13]	DehazeNet	13.5959
[25]	MSCNN	13.8200
[26]	BCCR	14.1302
[27]	CAP	14.1601
[28]	NLD	14.1827
[29]	Opening-Closing Net	14.7000
[30]	Disentangled dehazing network	14.9539
[31]	MOF	15.3070
[32]	Dehaze-GLCGAN	15.6802
	CCTDCP	16.2614

The higher PSNR value indicates the better quality of images. The Table 1, shows the mean PSNR value obtained on Middlebury subset of D-Hazy images. Our study focused on single image dehazing based on two major techniques:prior based and learning based. An analysis of prior based and learning based approaches are represented in Fig. 2. CCTDCP method has compared with existing prior and deep learning based methods. Our approach outperform these methods in terms of PSNR. Figure 2 (a) indicates the comparison of our approach with the prevailing prior based methods. Since our method is an improved version of DCP, CCTDCP returns much higher value than DCP. Figure 2 (b) illustrates the comparison of existing deep learning based dehazing methods. It also reveals that CCTDCP outstand the learning based methods. Our proposed CCTDCP algorithm, which is a prior based approach successfully dehaze the image scenes and outperforms the other dehazing methods.

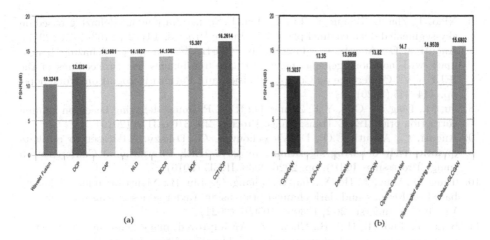

Fig. 2. PSNR values of various dehazing methods. (a) Comparison of CCTDCP method with prior based approaches (b) Learning based methods

5   Conclusion

A powerful and effective dehazing algorithm, CCTDCP, was proposed and validated using benchmarking datasets. The proposed prior based algorithm aids higher capability of restoration of hazy images. White balanced color correction transform with k nearest neighbour strategy is employed for processing hazy images. The obtained results using dark channel prior is enhanced using gamma correction technique. Experimentation analysis indicates that the proposed algorithm yielded higher outcome which apparently improves the quality of the image. The CCTDCP algorithm has a mean PSNR value of 16.2614 dB, which is better than other dehazing techniques. Extra evaluation metrics other than PSNR will be adopted in the future work.

Acknowledgement. This work is supported by IIT Palakkad Technology IHub Foundation Doctoral Fellowship IPTIF/HRD/DF/021.

References

1. He, K., Sun, J., Tang, X.: Single image haze removal using dark channel prior. IEEE Trans. Pattern Anal. Mach. Intell. **33**(12), 2341–2353 (2010)
2. Singh, D., Kumar, V.: Dehazing of remote sensing images using improved restoration model based dark channel prior. Imaging Sci. J. **65**(5), 282–292 (2017)
3. Nair, D., Sankaran, P.: Color image dehazing using surround filter and dark channel prior. J. Vis. Commun. Image Representation **50**, 9–15 (2018)
4. Borkar, K., Mukherjee, S.: Single image dehazing by approximating and eliminating the additional airlight component. Neurocomputing **400**, 294–308 (2020)
5. Kaur, M., Singh, D., Kumar, V., Sun, K.: Color image dehazing using gradient channel prior and guided L0 filter. Inf. Sci. **521**, 326–342 (2020)

6. Xiao, J., Zhu, L., Zhang, Y., Liu, E., Lei, J.: Scene-aware image dehazing based on sky-segmented dark channel prior. IET Image Process. 11(12), 1163–1171 (2017)
7. Ancuti, C.O., Ancuti, C., Timofte, R.: NH-HAZE: an image dehazing benchmark with non-homogeneous hazy and haze-free images. In: Proceedings of the IEEE/CVF Conference on Computer Vision and Pattern Recognition Workshops, pp. 444–445 (2020)
8. Ju, M., Ding, C., Guo, Y.J., Zhang, D.: IDGCP: image dehazing based on gamma correction prior. IEEE Trans. Image Process. 29, 3104–3118 (2019)
9. Ancuti, C., Ancuti, C.O., De Vleeschouwer, C.: D-hazy: a dataset to evaluate quantitatively dehazing algorithms. In: 2016 IEEE International Conference on Image Processing (ICIP), pp. 2226–2230. IEEE (2016)
10. Tang, Q., Yang, J., He, X., Jia, W., Zhang, Q., Liu, H.: Nighttime image dehazing based on Retinex and dark channel prior using Taylor series expansion. Comput. Vis. Image Underst. 202, 103086–103097 (2021)
11. Song, Y., Luo, H., Hui, B., Chang, Z.: An improved image dehazing and enhancing method using dark channel prior. In: The 27th Chinese Control and Decision Conference (2015 CCDC), pp. 5840–5845. IEEE (2015)
12. Li, B., Peng, X., Wang, Z., Xu, J., Feng, D.: AOD-Net: all-in-one dehazing network. In: Proceedings of the IEEE International Conference on Computer Vision, pp. 4770–4778 (2017)
13. Cai, B., Xu, X., Jia, K., Qing, C., Tao, D.: DehazeNet: an end-to-end system for single image haze removal. IEEE Trans. Image Process. 25(11), 5187–5198 (2016)
14. Mehra, A., Mandal, M., Narang, P., Chamola, V.: ReViewNet: a fast and resource optimized network for enabling safe autonomous driving in hazy weather conditions. IEEE Trans. Intell. Transp. Syst. 1–11 (2020, early access)
15. George, N., Thomas, J.: Authenticating communication of autonomous vehicles with artificial intelligence. In: IOP Conference Series: Materials Science and Engineering, vol. 396, no. 1, pp. 012017–12025. IOP Publishing (2018)
16. James, P., Thomas, J., Alex, N.: A survey on soft Biometrics and their application in person recognition at a distance. In: 2015 International Conference on Soft-Computing and Networks Security (ICSNS), pp. 1–5. IEEE (2015)
17. Samariya, D., Matariya, A., Raval, D., Dhinesh Babu, L.D., Raj, E.D., Vekariya, B.: A hybrid approach for big data analysis of cricket fan sentiments in Twitter. In: Satapathy, S.C., Joshi, A., Modi, N., Pathak, N. (eds.) Proceedings of International Conference on ICT for Sustainable Development. AISC, vol. 408, pp. 503–512. Springer, Singapore (2016). https://doi.org/10.1007/978-981-10-0129-1_53
18. Nivash, J.P., Raj, E.D., Babu, L.D., Nirmala, M., Manoj, K.V.: Analysis on enhancing storm to efficiently process big data in real time. In: Fifth International Conference on Computing, Communications and Networking Technologies (ICCCNT), pp. 1–5. IEEE (2014)
19. Park, J., Han, D.K., Ko, H.: Fusion of heterogeneous adversarial networks for single image dehazing. IEEE Trans. Image Process. 29, 4721–4732 (2020)
20. Afifi, M., Price, B., Cohen, S., Brown, M.S.: When color constancy goes wrong: correcting improperly white-balanced images. In: Proceedings of the IEEE/CVF Conference on Computer Vision and Pattern Recognition, pp. 1535–1544 (2019)
21. He, K., Sun, J., Tang, X.: Guided image filtering. IEEE Trans. Pattern Anal. Mach. Intell. 35(6), 1397–409 (2012)
22. Huang, S.C., Cheng, F.C., Chiu, Y.S.: Efficient contrast enhancement using adaptive gamma correction with weighting distribution. IEEE Trans. Image Process. 22(3), 1032–1041 (2012)

23. Nnolim, U.A.: Single image de-hazing using adaptive dynamic stochastic resonance and wavelet-based fusion. Optik **195**, 163111–163135 (2019)
24. Zhu, J.Y., Park, T., Isola, P., Efros, A.A.: Unpaired image-to-image translation using cycle-consistent adversarial networks. In: Proceedings of the IEEE International Conference on Computer Vision, pp. 2223–2232 (2017)
25. Ren, W., Liu, S., Zhang, H., Pan, J., Cao, X., Yang, M.-H.: Single image dehazing via multi-scale convolutional neural networks. In: Leibe, B., Matas, J., Sebe, N., Welling, M. (eds.) ECCV 2016. LNCS, vol. 9906, pp. 154–169. Springer, Cham (2016). https://doi.org/10.1007/978-3-319-46475-6_10
26. Meng, G., Wang, Y., Duan, J., Xiang, S., Pan, C.: Efficient image dehazing with boundary constraint and contextual regularization. In: IEEE International Conference on Computer Vision, pp. 617–624 (2013)
27. Zhu, Q., Mai, J., Shao, L.: A fast single image haze removal algorithm using color attenuation prior. IEEE Trans. Image Process. **24**(11), 3522–3533 (2015)
28. Berman, D., Avidan, S.: Non-local image dehazing. In: Proceedings of the IEEE Conference on Computer Vision and Pattern Recognition, pp. 1674–1682 (2016)
29. Mondal, R., Dey, M.S., Chanda, B.: Image restoration by learning morphological opening-closing network. Math. Morphol.-Theory Appl. **4**(1), 87–107 (2020)
30. Yang, X., Xu, Z., Luo, J.: Towards perceptual image dehazing by physics-based disentanglement and adversarial training. In: Proceedings of the AAAI Conference on Artificial Intelligence, vol. 32, no. 1, pp. 7485–7492 (2018)
31. Zhao, D., Xu, L., Yan, Y., Chen, J., Duan, L.Y.: Multi-scale optimal fusion model for single image dehazing. Signal Process. Image Commun. **74**, 253–265 (2019)
32. Anvari, Z., Athitsos, V.: Dehaze-GLCGAN: unpaired single image de-hazing via adversarial training. arXiv preprint arXiv:2008.06632 (2020)

Intelligent Computer Vision System for Detection of Tomatoes in Real Time

Navid A. Mulla, Shama Ravichandran$^{(\boxtimes)}$, and B. U. Balappa

M. S. Ramaiah University of Applied Sciences, Bangalore, India
shama.ee.et@msruas.ac.in

Abstract. In this paper, an intelligent computer vision system is developed for detection and classification of tomatoes based on maturity level. The proposed system is fabricated in real time using a slider mechanism. Fabrication was carried out by arriving at two design concepts for the slider mechanism. Evaluation of the design concepts resulted in selection of optimal design for real time implementation. Real time results show the efficacy of the proposed system. Moreover, the designed system resulted in an accuracy of 90%. The execution time taken for detection of a single tomato is 0.2 s.

Keywords: Deep learning · Tomato · CAD model

1 Introduction

Edible berry of the plant Solanum Lycopersicum is called a tomato [1]. Tomato is a vegetable that is found in majority of the recipes. Hence, there is an increase in demand for good quality tomatoes that require harvesting to be done at the right stage. Tomatoes don't ripen or mature at the same time, which is one of the biggest challenges faced during manual harvesting. This challenge can be overcome by means of an intelligent system that can detect maturity level of a tomato in agricultural fields.

Intelligent systems are usually designed to detect tomatoes based on its color, shape and size. In [2], tomatoes were classified into six stages based on its maturity using color image analysis. This method was useful in grading fresh tomatoes. A ripe tomato recognition technique is presented in [3] by merging HIS (Hue, saturation and intensity), RGB (Red, Green and Blue) and YIQ models. It is difficult to recognize and classify tomatoes based only on its color and shape. Hence automatic detection of ripe tomatoes by combining multiple features with a bilayer classification strategy is developed in [4]. A scheme for detecting individual tomatoes on plants using RGB digital camera and machine learning technique is also available in literature [5]. Arefi et al. proposed a method for classifying ripe and unripe tomatoes by integrating image processing with artificial neural networks [6]. It resulted in detection accuracies of 95.45% in case of ripe tomato and 90% in case of unripe tomato. A Computer Vision (CV) method for detecting three levels of maturity in a tomato is presented by Wan et al. [7]. In this method, a tomato maturity detection device based on CV technology was devised inside a laboratory to obtain tomato images. Hu et al. proposed a technique to detect a single

K. R. Venugopal et al. (Eds.): ICInPro 2021, CCIS 1483, pp. 42–51, 2021.
https://doi.org/10.1007/978-3-030-91244-4_4

ripe tomato using deep learning with edge contour detection [8]. Studies present various algorithms developed by several authors for detecting tomatoes [9–15].

From the literature it's clear that, not many works are fabricated and tested in real time. This is because, automatic detection of tomatoes in an agricultural environment is still a big challenge due to numerous perturbations present in the background of the image. The efficacy of the detection algorithms can be evaluated only by testing in real time. This requires the design of a mechanical system. The system presented in this work is a slider mechanism. Design of the mechanical system aids in smooth functioning of the developed algorithm.

The objective of this work is to fabricate an intelligent CV system to detect and classify tomatoes based on their maturity level in real time. The mechanical design aspects and modelling is presented in detail. This paper is organized as follows: Literature survey of the existing works available for detecting tomatoes is given in Sect. 1. Intelligent CV system configuration is explained in Sect. 2. This is followed by the slider mechanism design. CAD model for the slider mechanism is presented in Sect. 3. Experiment results are given in Sect. 4 and conclusion in Sect. 5.

2 Intelligent CV System

In this section, various steps involved in the development of an intelligent CV system is explained. Intelligent CV system comprises of a personal computer (PC), tomato maturity level detection algorithm, slider mechanism, vision sensor and an actuator. The maturity detection algorithm is implemented on a PC. The developed algorithm is tested in real time using a mechanical system. For any developed algorithm, fabrication of mechanical system plays a major role. The mechanical system considered here is a slider mechanism. Design of the slider mechanism requires specifications of tomato plants. The specifications of the slider mechanism is arrived based on the survey conducted on tomato plants. Once the specifications of the slider mechanism are determined, two slider mechanism designs are presented. The designs are evaluated and an optimal design is chosen. Based on the chosen design a functional model is developed in real time.

2.1 Maturity Level Detection Algorithm

A deep learning algorithm is developed to detect the maturity level of tomatoes using transfer learning. This is done by constructing a tomato database. The database helps the system to recognize tomatoes and classify it based on maturity level. Four stages of maturity are considered here. Database consists of four datasets containing images of red ripe tomato, cherry red tomato, yellow tomato and unripe tomato. The proposed algorithm makes use of two networks, one for feature extraction (ResNet-50) and another one for localization of the tomato (YOLO-V2).

2.2 Plant Specification

In India, the average height of a tomato plant is approximately 450–600 mm. Tomato plants vary in sizes, which depends on the type of family they belong to. To determine the specifications of a tomato plant, five tomato seedlings were planted in ceramic pots. After two months tomatoes began to appear on the plant. Once the tomatoes mature, the height and width of the plant was measured manually. The average height and width of the plants were determined as 600 mm and 400 mm respectively. The height and width of the plant is required to determine the dimensions the slider mechanism (Figs. 1 and 2).

400mm

Fig. 1. Average width of tomato plant

Fig. 2. Average height of tomato plant

2.3 Slider Mechanism

To detect tomatoes in real time there is a need for a stationary sliding structure. The basic idea behind the mechanical system design is to place the vision sensor on the stationary platform of the slider mechanism. Slider mechanism exhibits linear motion and hence has the following advantages (i) Reduced friction (ii) High rigidity and (iii) Accuracy. In this section, two slider mechanism designs are presented for a single tomato plant. The designs are evaluated and an optimal design is chosen. A functional model of the chosen design is then fabricated.

Design 1. Design 1 consists of two end supports, one linear actuator with belt and a vision sensor. Advantages are (i) smooth operation and (ii) low power consumption. Disadvantages are (i) more torque needed to drive the belt (ii) life time of belt is less and (iii) poor maintenance of belt results in belt failure (Fig. 3).

Design 2. Design 2 consists of two end supports, one linear actuator with screw nut mechanism and a vision sensor. Advantages of design 2 are given as follows (i) Low power consumption compared to design 1 and (ii) High precision. Disadvantages are (i) Damage to threads due to high voltage and high speed and (ii) Moving speed is limited (Fig. 4).

Fabrication of Slider Mechanism. Design 2 results in better precision and extended lifetime compared to design 1. Hence design 2 is chosen for fabrication.

Based on the specifications of the tomato plant, dimensions for the slider mechanism are arrived. Height of the slider mechanism is 300 mm, which is half of the plant height

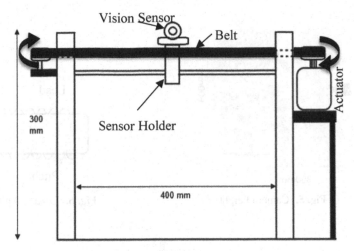

Fig. 3. Design 1

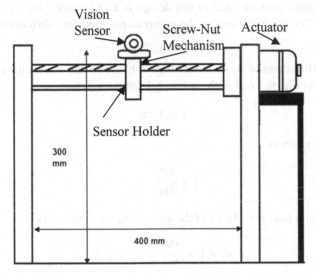

Fig. 4. Design 2

and length of the slider is 400 mm. A vision sensor holder is placed on the slider mechanism in such a way it covers the entire surface area of the plant as shown in Fig. 5. Vision sensor used here is a camera.

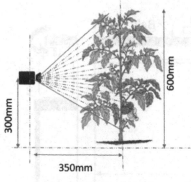

Fig. 5. Camera height

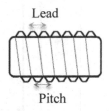

Fig. 6. Lead and pitch

Camera is placed on the holder. Camera moves in a straight line at a speed of $360°/500$ ms^{-1} parallel to the tomato plant. It is capable of moving through a distance of 400 mm horizontally. Linear actuator used in this design is a DC motor which requires a 12 V power supply. The specifications of the components used for fabrication are listed in Table 1.

Calculation. Movement of slider mechanism is based on threads on the rod. Lead and pitch is shown in Fig. 6. Lead of the screw is 1 cm per rotation

$$l = 1\,cm \tag{1}$$

Distance is given as

$$l \times \frac{\alpha^0}{360} = d \tag{2}$$

In one rotation (i.e., $\alpha = 360°$) of the shaft distance travelled is

$$d = 1 \times \frac{360}{360} = 0.01\,m/s \tag{3}$$

For a distance of 400 mm (0.4 m) time taken by the slider is

$$\frac{0.4\,m * 1\,s}{0.01\,m} = 40\,s \tag{4}$$

The slider moves 0.01 m in one second. DC motor enables the camera that is placed on the slider to travel a distance of 400 mm in 40 s and thus a distance of 1 mm is covered in approximately 0.1 s.

Table 1. Specification table.

Required models	Specifications
USB camera	Maximum resolution: 640 × 480 15 frames/s Power: 5V USB plug In
Raspberry Pi	Model: 3B+ RAM: 1 GB Memory: 16 GB
DC motor	Working voltage: DC 3–12 V Operating Current 40–180 mA
Batteries	Lithium ion battery Voltage: 12 V Current: 3 A
Bearing	8 mm inner diameter 20 mm outer diameter
Steel rod	8 mm diameter 400 mm length
Screw rod	1.25 mm pitch 8 mm diameter 450 mm length

3 CAD Model

3D model of the chosen design is developed using CATIA software. Different views of the model are shown in Fig. 7. 3D assembly model consists of support frame, screw rod, steel rod, camera and motor.

The support frame is constructed using wood. Linear bearing were used for smooth motion of the slider. Based on the bearing specifications, an 8 mm plain rod and screw rod were selected for fabrication. Finally, the camera is placed on the surface of the slider. Fabricated model of the intelligent CV system is shown in Fig. 8.

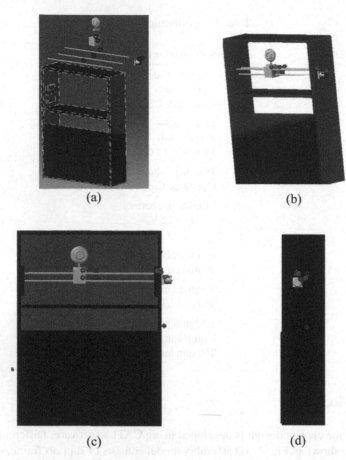

(a)

(b)

(c)

(d)

Fig. 7. 3D Assembly model: (a) & (b) Isometric view (c) front view and (d) side view

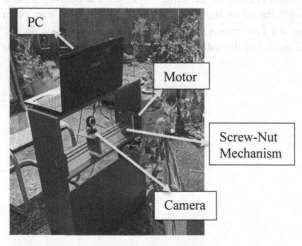

Fig. 8. Developed model

4   Results

Intelligent CV system is tested in real time and its performance is examined. Results are included to demonstrate the efficacy of the developed system. Figure 9 (a) shows the intelligent CV system setup view and Fig. 9 (b) shows the tomato plant in real time as viewed through the camera. The images are segregated based on maturity using the maturity detection algorithm. Few of the camera view snapshots are shown in Figs. 10 and 11. It can be seen that the test results are accurate.

(a)   (b)

Fig. 9. (a) Setup view and (b) camera view (real-time detection)

The execution time taken for the system to detect and classify the maturity of a tomato is approximately 0.2 s. Here, the execution time is measured considering recognition of tomato, image capture, classification of maturity followed by display of results on the PC. While calculating execution time, placement of the portable developed system in front of the plant is not considered. If the system is fixed to the ground and made permanent then, placement time will have no role in calculation of execution time as it is a one time task. From the above figures it can be clearly observed that the orientation of tomato on the plant varies widely. The slider mechanism design provided a constant smooth motion which improved the recognition of tomatoes. The rate of accurate detection of tomatoes is found to be 90%.

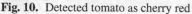

Fig. 10. Detected tomato as cherry red   **Fig. 11.** Detected Tomatoes Un-Ripe_Tomato and Red_Ripe_Tomato

5   Conclusion

Intelligent CV system is fabricated using a slider mechanism. Slider mechanism results in smooth operation which enables the vision system to detect and classify images accurately. Results show that the developed system is capable of working well in complex agricultural environments.

References

1. Boswell, V.R.: Improvement and genetics of tomatoes, peppers and eggplant (1937)
2. Choi, K., Lee, G., Han, Y.J., Bunn, J.M.: Tomato maturity evaluation using color image analysis. Trans. Am. Soc. Agric. Eng. **38**, 171–176 (1995)
3. Arefi, A., Motlagh, A.M., Mollazade, K., Teimourlou, R.F.: Recognition and localization of ripen tomato based on machine vision. Aust. J. Crop Sci. **5**, 1144–1149 (2011)
4. Wu, J., Zhang, B., Zhou, J., Xiong, Y., Gu, B., Yang, X.: Automatic recognition of ripening tomatoes by combining multi-feature fusion with a bi-layer classification strategy for harvesting robots. Sensors (Switzerland) **19** (2019)
5. Yamamoto, K., Guo, W., Yoshioka, Y., Ninomiya, S.: On plant detection of intact tomato fruits using image analysis and machine learning methods. Sensors (Switzerland) **14**, 12191–12206 (2014)
6. Arefi, A., Motlagh, A.M.: Development of an expert system based on wavelet transform and artificial neural networks for the ripe tomato harvesting robot. Aust. J. Crop Sci. **7**, 699–705 (2013)
7. Wan, P., Toudeshki, A., Tan, H., Ehsani, R.: A methodology for fresh tomato maturity detection using computer vision. Comput. Electron. Agric. **146**, 43–50 (2018)
8. Hu, C., Liu, X., Pan, Z., Li, P.: Automatic detection of single ripe tomato on plant combining faster R-CNN and intuitionistic fuzzy set. IEEE Access. **7**, 154683–154696 (2019)
9. Zhao, Y., Gong, L., Zhou, B., Huang, Y., Liu, C.: Detecting tomatoes in greenhouse scenes by combining adaboost classifier and colour analysis. Biosyst. Eng. **148**, 127–137 (2016)
10. Zhao, Y., Gong, L., Huang, Y., Liu, C.: Robust tomato recognition for robotic harvesting using feature images fusion. Sensors (Switzerland) **16** (2016)

11. Ji, C., Zhang, J., Yuan, T., Li, W.: Research on key technology of truss tomato harvesting robot in greenhouse. Appl. Mech. Mater. **442**, 480–486 (2014)
12. Lv, X.-L., Lv, X.-R., Lu, B.-F.: Identification and location of picking tomatoes based on machine vision. In: Proceedings of 4th International Conference on Intelligent Computation Technology and Automation, ICICTA 2011, vol. 2, pp. 101–107 (2011)
13. Malik, M.H., Zhang, T., Li, H., Zhang, M., Shabbir, S., Saeed, A.: Mature tomato fruit detection algorithm based on improved HSV and watershed algorithm. IFAC-Papers online, vol. 51, pp. 431–436 (2018)
14. Khoshroo, A., Arefi, A., Khodaei, J.: Detection of red tomato on plants using image processing techniques. Agric. Commun. **2**, 9–15 (2014)
15. Hosna, M.M., Alimardami, R., Omid, M.: Detection of red ripe tomatoes on stem using image processing techniques. J. Am. Sci. **7**, 376–379 (2011)

Intelligent Code Completion

Danish Waseem[✉], Pintu, and B. R. Chandavarkar

Department of Computer Science and Engineering, National Institute of Technology
Karnataka, Surathkal, Mangalore, India
{dan.181co116,pintu.181co139}@nitk.edu.in

Abstract. Auto complete suggestions for IDEs are widely used and often extremely helpful for inexperienced and expert developers alike. This paper proposes and illustrates an intelligent code completion system using an LSTM based Seq2Seq model that can be used in concert with traditional methods (Such as static analysis, prefix filtering, and tries) to increase the effectiveness of auto complete suggestions and help accelerate coding.

Keywords: LSTM · RNN · Seq2Seq · Deep learning · PBN

1 Introduction

Intelligent code completion is a context-cognizant code completion feature found in certain programming environments that expedites the coding process by minimizing typos and other prevalent errors. This is conventionally accomplished by auto completion pop-ups while typing, querying function parameters, query tips related to syntax errors, and so on. Intelligent code completion and related tools use reflection to provide documentation and disambiguation for variable denominations, functions, and methods [1].

Intelligent code completion, like other auto-completion systems, provides a convenient way to access function descriptions, especially parameter lists. It minimizes the amplitude of name memorization and keyboard input needed, which expedites software development. It additionally allows for less reliance on external documentation because interactive documentation on several symbols (i.e., variables and functions) in the active scope appears dynamically in the form of a tool-tip when programming.

Intelligent code completion utilizes an in-memory database of classes, variable names, and other constructs identified or referenced by the application being edited. Typically, the context is resolute by the static form of the variable on which the completion is triggered by the developer. Exhibiting the programmer hundreds of suggestions could be as inefficient as exhibiting none.

Modern Integrated Development Environments (IDEs) have code completion systems as a standard feature. Developers withal utilize them as a simple guide for the Application Programming Interface (API) since they denote which fields and methods can be utilized in a particular context.

K. R. Venugopal et al. (Eds.): ICInPro 2021, CCIS 1483, pp. 52–65, 2021.
https://doi.org/10.1007/978-3-030-91244-4_5

Many IDEs support prefix filtering to abbreviate the amplitude of dispensable suggestions. If completion is activated on a prefix, only proposals that commence with that prefix will be shown. This efficaciously fortifies developers who know what they are probing for but is of little benefit to developers who are unfamiliar with the API.

This paper is meant to propose and illustrate an intelligent code completion system using an LSTM based Seq2Seq model [2] that can be used in concert with traditional methods to minimize the quantity of typing needed and expedite the coding process.

Section 2 of this paper evaluates and synthesises the implementation of intelligent code completion using some of the pre-existing techniques. Section 3 describes overarching strategy, by studying the methods used for the execution of intelligent code completion. Section 4 outlines the contextual analysis and report of the model. Finally the conclusion along with the future work and references are provided at the end of the paper.

2   Literature Survey

Intelligent code completion can be found in a variety of programming environments, including Atom's auto-complete and Visual Studio's Intellisense. IntelliSense's "classic" implementation detects marker characters such as periods or other separator characters as shown in Fig. 1. IntelliSense suggests matches in a pop-up window when the user types one of these characters immediately after the name of an entity with one or more accessible members (such as contained variables or functions) [3].

Based on prior research that examined the effect of user interface delays on users, the timing constraint is that the response must be returned in less than a second because the developer's current thinking is disrupted otherwise, so it is important that predictions of the model be computed quickly so that the developer's workflow is not interrupted. Much before the developer hits the enter button for next line of code - predictions must be computed - so that it can save his time.

There has been a substantial amount of research done in recent years to enhance the pertinence of API recommendations through the utilization of conceptual understanding, machine learning, and statistical approaches.

Some of the pre-existing techniques include a frequency-based code completion system as shown in paper [4]. A feasible approach for intelligent CCS (Code Completion Systems) is to decide the consequentiality of each method predicated on its frequency of use in the example code. This technique is justified by the fact that the more often a method is utilized, the more likely it is that other programmers will utilize the same method again.

The next approach is the best-matching neighbor algorithm (BMN). BMN adapts the K-nearest- neighbors (KNN) algorithm to the problem of finding method calls to recommend for particular objects. The KNN algorithm comes from the pattern recognition research. The algorithm's intuition is based on a

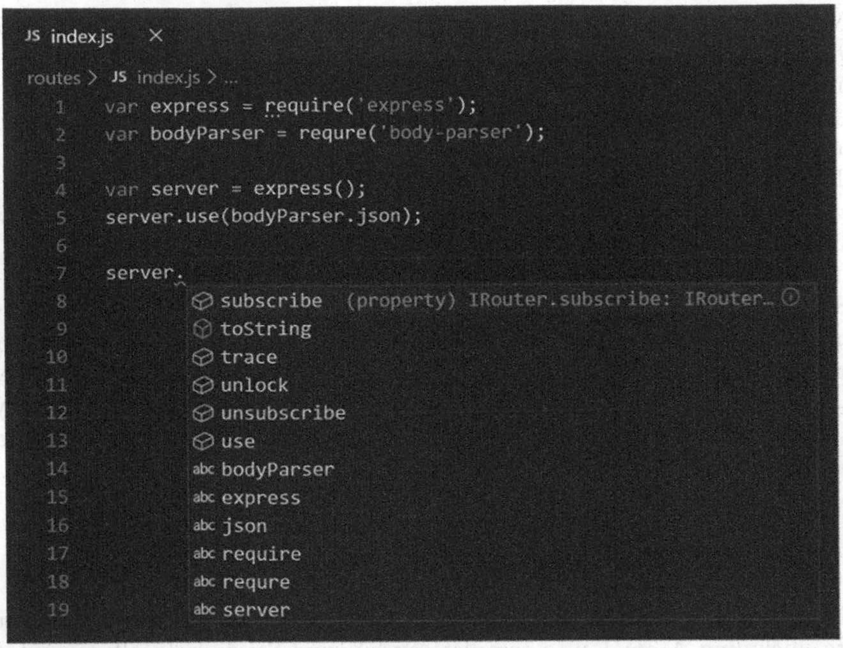

Fig. 1. IntelliSense in visual studio code [3]

common-sense rule that can be enounced as follows: to predict something according to some observation, let's find it in one's experience in a similar situation, and predict what actually happened at the end.

In the paper [5], a new approach for intelligent code completion called Pattern-Based Bayesian network (PBN) is described.

It is a technique to infer intelligent code completions that enables to reduce model sizes via clustering. They introduced a clustering approach for PBN that enables trade-off in model size for prediction quality.

They extended the state-of-the-art methodology for evaluating code completion systems and performed comprehensive experiments to investigate the correlation between prediction quality and different model sizes. They showed that clustering could decrease the model size by as much as 90%, with only a minor decrease in prediction quality. They also conducted a thorough examination of the impact of input data size on prediction quality, speed, and model size. The experiments show that prediction quality increases with increased input data and that both the model size and prediction speed scales better with the input data size for PBN compared to BMN.

A significant technical difference is that PBN internally stores floating-point values in multiple conditioned nodes, whereas BMN stores binary values in a table. A key consequence is that PBN allows to merge of different patterns and to denote probabilities - instead of boolean existence - for all context information.

They introduced a clustering approach for PBN that leverages this property and enables to trade-off model size for prediction quality.

The biggest drawback observed in the above paper is they use the F1 score as the metric for scoring the final model. We believe this is inherently flawed because this means that they are giving incorrect predictions made by the model at the beginning of the program the same importance as inaccurate predictions made towards the end of the program. At the commencement of a program, it's virtually infeasible to conjecture what the developer will type next. Still, it is more accessible towards the end of a program, so it's relatively more acceptable if the model guesses wrong in the beginning compared to the end. If the model continues to make many incorrect predictions towards the end of a program, the model should be penalized more.

We address this by not scoring the model based on the correctness of its predictions but rather by how helpful it is. We do so by scoring the final model using the total number of keystrokes it saves the developer.

3   Methodology

Our primary goal is to implement a highly advanced algorithm for this model that mimics the workings of the human encephalon in processing data to achieve high precision and immediate results for intelligent code completion. The outcome of the prediction synchronizes with the thought of human brain and its fast computing makes it look like they are coming out of the human brain. Model helps the user saving time and memorization overhead. This model is build using deep learning neural networks.

3.1   Working with a Dataset

To easily expand the dataset (Or shorten the dataset if training and evaluating takes too long), we will write a python script that looks in a particular folder and automatically builds the dataset and splits it into training and validation datasets. It searches this folder for all python scripts and uses the found code to build the dataset (Fig. 2). To make our work easier, the script recursively searches any folders in this location so we can easily clone Github repositories and place them without worrying about the file structure in the repositories.

Basic cleaning of the dataset happens at this stage (The script automatically does this while building the dataset). This is required in order for the model to work correctly.

3.2   Building and Training the Model

We have opted to utilize deep learning algorithms to implement this model. Deep learning is a branch of machine learning in artificial intelligence that involves networks that can learn unsupervised from unstructured or unlabeled data. Deep

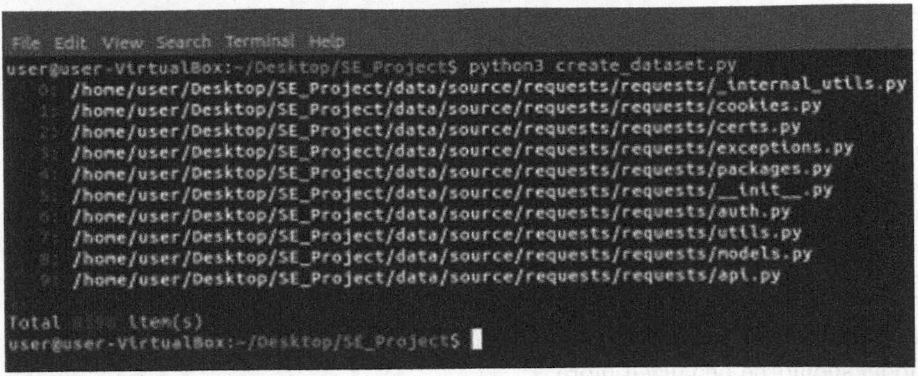

Fig. 2. Creating the dataset (output using labml logger)

neural networks or deep neural learning are other terms for equipollent. A recurrent neural network (RNN) is a form of artificial neural network that works with sequential or time series data. Text translation, natural language processing (nlp), verbalization apperception, and image captioning are all examples of quandaries wherever deep learning algorithms are utilized.

The initial thought process was to use a Sequence to Sequence model since the input is a sequence of tokens (Obtained from the Python code) and the output is another sequence of tokens (These are the suggestions). Sequence to Sequence (seq2seq) models are a subset of Recurrent Neural Network architectures that we customarily (but not exclusively) use to solve intricate language problems such as machine translation, question answering, building chatbots, text summarization, and so on [2]. It is immediately evident that the length of the input sequence will not be the same as the length of the output sequence. To overcome this, an encoder and decoder system is used. A RNN (Recurrent Neural Network) layer (Or multiple layers) acts as the encoder, it processes the input sequence and returns its own internal state (Fig. 4). Another RNN layer (Or multiple layers) acts as the decoder, it is trained to predict the next characters of the target sequence, given previous characters of the target sequence. Together they make up the entire Encoder-Decoder LSTM model that we used. A basic visual representation of the model can be found below in Fig. 3 [6].

Long Short-Term Memory (LSTM) networks are a form of recurrent neural network that can learn order dependence in sequence prediction quandaries. This is a demeanor expected in intricate quandary domains such as machine translation, verbalization apperception, and others. LSTMs are a difficult subset of deep learning. The definition of LSTMs can be difficult to understand and how words like bidirectional and sequence-to-sequence apply to the field.

Theoretically we could've used a normal RNN instead of an LSTM since an RNN too holds a memory of previous data inputs but practically RNNs are not great at learning long term dependencies whereas LSTMs were built to mitigate this issue [6].

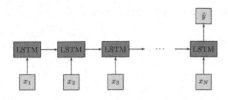

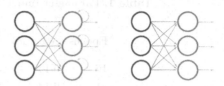

Fig. 3. LSTM model structure

Fig. 4. Comparison of recurrent neural networks (on the left) and feedforward neural networks (on the right)

To convert each word into a feature vector we use pre trained word embeddings from GloVe (Global Vectors for Word Representation) which labml does for us using PyTorch.

SoftMax function was used as the activation function (The activation function of a node defines the output of that cell/node given an input or set of inputs.) and the Log Loss (Also known as Cross Entropy Loss) function (Fig. 5) was used as the loss function.

$$y \log(p) + (1 - y) \log(1 - p)$$

Fig. 5. Log loss formula, y is the output and p is the probability

The "ADAM" optimiser was used. It is an extension of Gradient Descent, it is invariant to scaling and is therefore more efficient and leads to more accurate models.

3.3 Parameter Tuning

In deep learning neural network architecture we often require weights and biases for every layer. There are no preset values of them and need to be generated dynamically. We use many hyper-parameters to get this done. The model learns them using back-propagation.

Parameters Used:

- Epochs: 100 (As indicated in Table 1)
- Embedding Size (Number of Features): 512 (As indicated in Table 1)
- Number of Hidden Layers: 2 (As indicated in Table 2)
- Size of Hidden Layer: 512 (As indicated in Table 2)
- Batch Size: 16 (As indicated in Table 3)
- Dropout Rate: 0.35 (As indicated in Table 3)

Table 1. Parameter tuning for epochs and embedding size

Epochs	Accuracy	Embedding Size	Accuracy
16	0.13	64	0.06
21	0.14	128	0.12
50	0.14	256	0.14
64	0.14	384	0.14
100	0.15	512	0.15
128	0.14	640	0.15
150	0.15	768	0.13
200	0.15		
256	DNF		

Table 2. Parameter tuning for number of hidden layers and number of hidden layer size

Hidden Layers	Accuracy	Hidden Layer Size	Accuracy
1	0.11	16	0.13
2	0.15	32	0.13
3	0.15	64	0.13
4	0.15	128	0.15
		256	0.15
		384	0.15
		512	0.16
		640	0.16

During parameter tuning, the following patterns were observed (within margin of error) while training (Fig. 13) and evaluation (Fig. 14):

Epochs (Refer to Fig. 6):

- When the entire dataset is transferred forward and backward through the neural network once, this is referred to as an Epoch.
- More epochs would mean the model would be more accurate as the learning algorithm will not converge to the correct weights in just a few epochs.
- Once it has converged, increasing the number of epochs will not make a difference.
- More epochs also means the time taken for training will increase greatly.

Embedding Size (Refer to Fig. 7):

- A Word Embedding is a mapping from words to vectors (Feature vectors). The length of these feature vectors is referred to as embedding size. In other terms, it is the number of features that describes each word.

Table 3. Parameter tuning for batch size

Batch Size	Accuracy	Dropout Rate	Accuracy
4	0.13	0.5	0.14
8	0.15	0.4	0.14
16	0.16	0.35	0.18
32	0.15	0.3	0.17
64	0.16	0.25	0.17
128	0.16	0.2	0.16
256	0.15	0.15	0.12
512	DNF	0.1	0.13

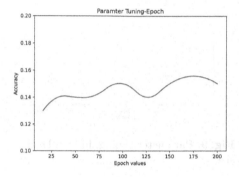

Fig. 6. Parameter tuning: epochs

Fig. 7. Parameter tuning: embedding size

- Using more dimensions to represent each word means the representation of words would be more accurate but would require more computational resources.
- Referencing [7], there exists a theoretical lower bound for the embedding size, above which model accuracy more or less stabilizes.
- The accuracy of the model increases exponentially as the embedding size increases from 0 to 180 and stabilises at 200. The accuracy of the model becomes constant for embedding size 200 to 400. But, we tried further and got slight higher accuracy for embedding size 600. Later, the model seems to become over-fitted and the accuracy falls.

No. of Hidden Layers (Refer to Fig. 8):

- The algorithm's input and output are dissevered by a hidden layer, through which the function applies weights to the inputs and guides them through an activation function as the output. In a nutshell, the hidden layers perform nonlinear transformations on the network's inputs.

- Generally, a model with zero hidden layers would only be capable of representing linearly separable functions. Because hidden layer extracts the complex features which are surely not linearly separable.
- A single hidden layer would allow the model to represent any function that contains a continuous mapping from one finite set to another.
- Two hidden layers would allow the model to represent most arbitrarily complex functions.
- Increasing the number of hidden layers would allow the model to represent increasingly complex functions at the expense of computational resources.

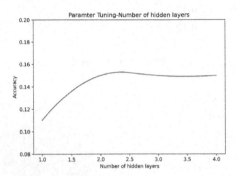

Fig. 8. Parameter tuning: number of hidden layers

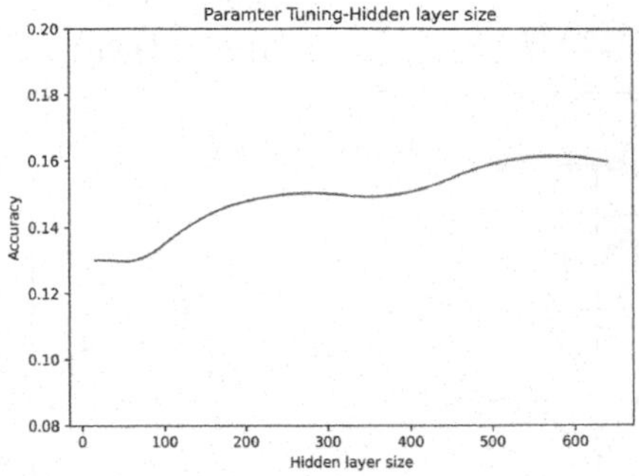

Fig. 9. Parameter tuning: hidden layer size

Hidden Layer Size (Refer to Fig. 9):

- Hidden layer size refers to the number of nodes in each hidden layer.
- Too few nodes in each hidden layer will cause the generalization and training error to be very large due to high bias and under-fitting. As the number of nodes are increased, the generalization will begin to fall and then increase once again while the training error keeps falling. When the number of nodes is huge (This scenario was never encountered due to the restriction on computational resources), the generalization error will be very high while the training error will be very low due to over-fitting as shown in Fig. 10.

Batch Size (Refer to Fig. 11):

- Since one epoch is too big to feed into the network at once (Due to memory restrictions), each epoch is divided into smaller batches. The batch size is the total number of training samples in one batch [8].
- Ideally each batch should be big enough so as to provide a stable estimate of what the gradient of the full dataset would be.
- Very large batch sizes tend to take up too much memory and are very prone to getting stuck in local minimas but take up much less time.

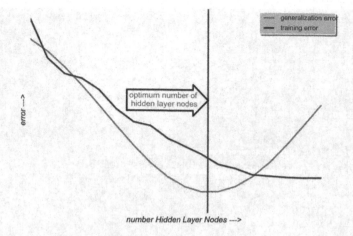

Fig. 10. Hidden layer size against error (expected behaviour)

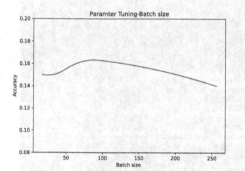

Fig. 11. Parameter tuning: batch size

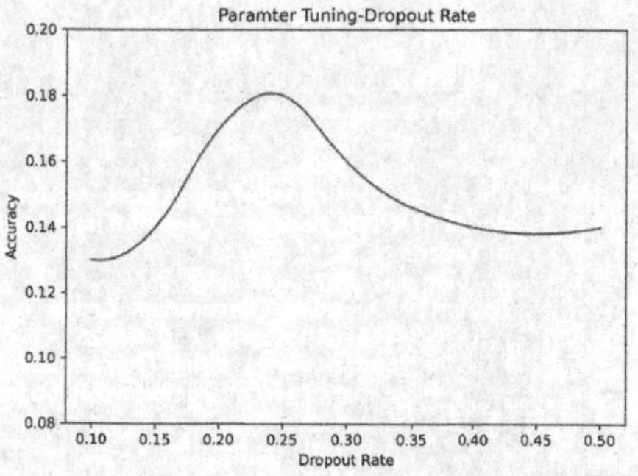

Fig. 12. Parameter tuning: dropout rate

- Very small batch sizes can cause the weights of the nodes in the model to jump around a lot if the data is very noisy causing it to converge very slowly (Requiring a large number of epochs) but is less likely to get stuck in local minimas.

Dropout Rate (Refer to Fig. 12):

- Dropout is a regularization technique (To prevent over-fitting) where randomly selected nodes are dropped out. If the dropout rate is 0.1 (10%), that means it is expected that 1 in every 10 nodes will be randomly excluded from each update cycle.
- A dropout rate of 0.5 provides the most amount of regularization but will cause training error to increase but will reduce regularization error.
- A dropout rate of 0 provides the least amount of regularization and will reduce the training error greatly but will increase regularization error (Overfitting).
- The accuracy of the model maximises at the dropout rate approximately equals to 23%.

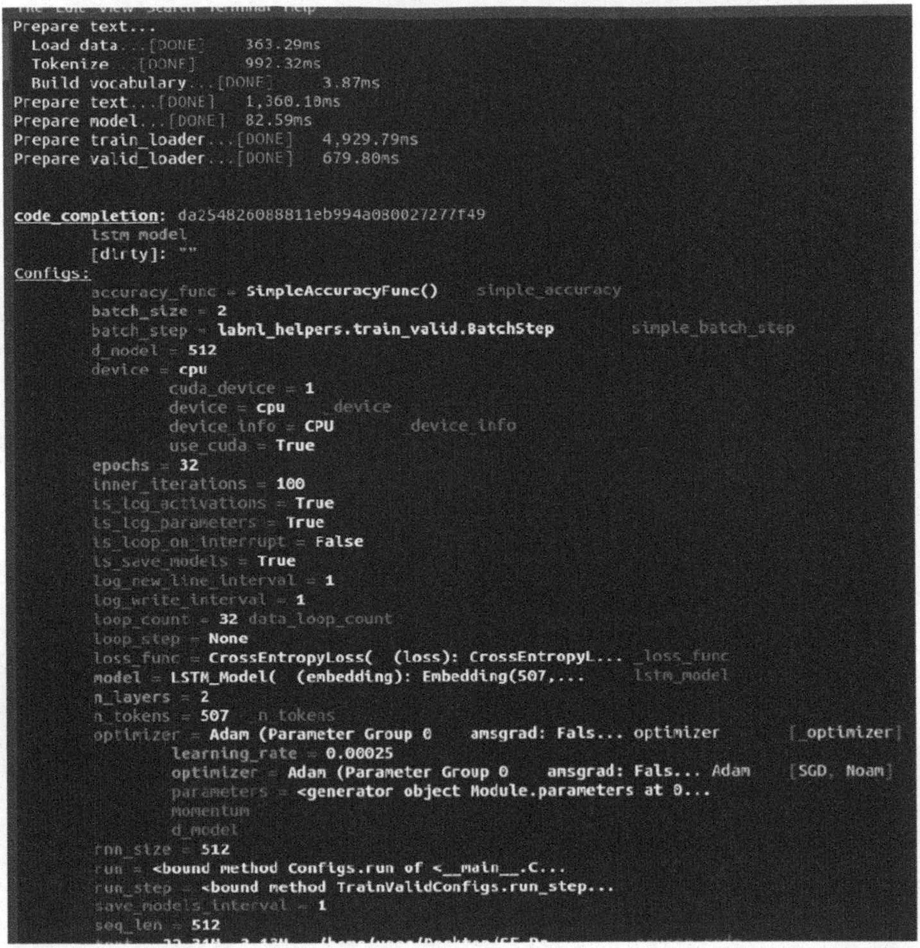

```
Prepare text...
  Load data...[DONE]      363.29ms
  Tokenize   [DONE]       992.32ms
  Build vocabulary...[DONE]     3.87ms
Prepare text...[DONE]   1,360.10ms
Prepare model...[DONE]  82.59ms
Prepare train_loader...[DONE]   4,929.79ms
Prepare valid_loader...[DONE]   679.80ms


code_completion: da254826088811eb994a080027277f49
        lstm model
        [dirty]: ""
Configs:
        accuracy_func = SimpleAccuracyFunc()      simple_accuracy
        batch_size = 2
        batch_step = labml_helpers.train_valid.BatchStep      simple_batch_step
        d_model = 512
        device = cpu
                cuda_device = 1
                device = cpu       device
                device_info = CPU         device_info
                use_cuda = True
        epochs = 32
        inner_iterations = 100
        is_log_activations = True
        is_log_parameters = True
        is_loop_on_interrupt = False
        is_save_models = True
        log_new_line_interval = 1
        log_write_interval = 1
        loop_count = 32 data_loop_count
        loop_step = None
        loss_func = CrossEntropyLoss( (loss): CrossEntropyL... _loss_func
        model = LSTM_Model( (embedding): Embedding(507,... lstm_model
        n_layers = 2
        n_tokens = 507    n_tokens
        optimizer = Adam (Parameter Group 0    amsgrad: Fals... optimizer    [_optimizer]
                learning_rate = 0.00025
                optimizer = Adam (Parameter Group 0    amsgrad: Fals... Adam    [SGD, Noam]
                parameters = <generator object Module.parameters at 0...
                momentum
                d_model
        rnn_size = 512
        run = <bound method Configs.run of <__main__.C...
        run_step = <bound method TrainValidConfigs.run_step...
        save_models_interval = 1
        seq_len = 512
```

Fig. 13. Different model parameters while training (output using labml logger)

3.4   Evaluating the Model

We can use a simple beam search to retain multiple highly probable choices instead of just picking the single most probable output (Token) but while testing it was more useful to only check the best (Most probable) output to validate the model otherwise predictions which have the correct token as the most probable will be "more correct" than predictions which have the correct token somewhere down the list. This will lead to confusion while calculating a "score" for the model.

Fig. 14. Evaluation of the model and its parameters (output using labml logger)

4 Results

The model was run on Ubuntu in Linux and training (With the current dataset) took about 7 h and the final evaluation took about 5 h.

System Configuration: i7 7700HQ with 16 GB RAM

Dataset used: All ".py" files from the top Github repositories with the keyword "python":

1. https://github.com/django/django
2. https://github.com/keras-team/keras
3. https://github.com/scikit-learn/scikit-learn
4. https://github.com/pallets/flask
5. https://github.com/ageitgey/face_recognition
6. https://github.com/TheAlgorithms/Python

7. https://github.com/httpie/httpie
8. https://github.com/ansible/ansible
9. https://github.com/psf/requests
10. https://github.com/scrapy/scrapy
11. https://github.com/pandas-dev/pandas and more.

80% of the dataset was used for the training set and 20% for the testing set. The training set is further divided into two parts, 80% for training and 20% for validation. The data was split this way was due to the Pareto principle. Apart from that, it is mostly an arbitrary choice since the more training data is present, the better the model will be and the more testing data is present, the lesser the variance in the model results and the 80/20 split is a well-known rule of thumb.

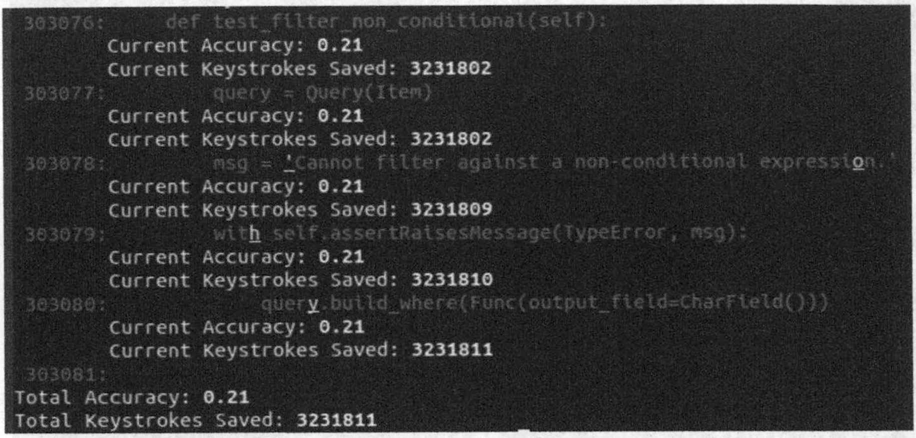

Fig. 15. Results obtained on the final run (Output using labml logger)

The accuracy obtained after final model evaluation: **0.21 (21%)** as shown in Fig. 15.

The number of keystrokes saved during the final model evaluation: **3231811 (Around 18% of the total keystrokes necessary to type the entire test set)**.

While on the surface this seems very poor, it is to be expected due to the relatively small dataset. An acceptable accuracy for code completion would be 25–30% since it would prevent roughly a quarter of the total keystrokes a programmer would have to make.

While accuracy is sufficient during the validation stage while training the model, for a final metric it is very weak since it is almost impossible to predict what the programmer is going to type next in the beginning of a program but it is easier towards the end of a program but accuracy does not take this into account. As mentioned earlier, we addressed this issue by using the total number of keystrokes saved as the metric for the final model.

5   Conclusion

Based on the accuracy obtained and the keystrokes saved, this approach definitely holds the promise that is important for the improvement of intelligent code completion. In comparison with PBN and frequency analysis models, our model has a larger probability of outperforming them if enhanced. Moreover PBN model uses F1 score to measure performance, which means in the beginning of the program, model will not be able to guess properly. If our approach was coupled with the conventional method of static analysis and a trie of keywords and tokens, not only would the correctness of the predictions increase, the number of false predictions would greatly fall. We presume that based on the shortcomings faced, the current model can be improved upon by keeping the dataset further extended, further tuning the hyper-parameters (This will probably yield a negligible increase in prediction quality), and processing and cleansing the data to help the model better understand the tokens (For example, "x = x + 1" and "x += 1" mean the same thing but the tokens they consist of are different and the model will therefore view them as different).

References

1. Intelligent code completion (2021). https://en.wikipedia.org/wiki/Intelligent_code_completion
2. Chollet, F.: A ten-minute introduction to sequence-to-sequence learning in keras (2017). https://blog.keras.io/a-ten-minute-introduction-to-sequence-to-sequence-learning-in-keras.html
3. Intellisense in visual studio code (2016). https://code.visualstudio.com/docs/editor/intellisense
4. Bruch, M., Monperrus, M., Mezini, M.: Learning from examples to improve code completion systems. In: Proceedings of the 7th Joint Meeting of the European Software Engineering Conference and the ACM Symposium on the Foundations of Software Engineering, Amsterdam, Netherlands (2009). https://doi.org/10.1145/1595696.1595728. https://hal.archives-ouvertes.fr/hal-01575348
5. Proksch, S., Lerch, J., Mezini, M.: Intelligent code completion with Bayesian networks. ACM Trans. Softw. Eng. Methodol. **25**, 1–31 (2015). https://doi.org/10.1145/2744200
6. Romaniuk, M.: Machine learning for code completion (2019). https://medium.com/@myroslavarm/machine-learning-for-code-completion-2583792997e3
7. Patel, K., Bhattacharyya, P.: Towards lower bounds on number of dimensions for word embeddings. In: IJCNLP (2017)
8. Sharma, S.: Epoch vs batch size vs iterations (2019). https://towardsdatascience.com/epoch-vs-iterations-vs-batch-size-4dfb9c7ce9c9

(t, k, n)-Deterministic Extended Visual Secret Sharing Scheme Using Combined Boolean Operations

L. Chandana Priya and K. Praveen[✉]

TIFAC-CORE in Cyber Security, Amrita School of Engineering,
Amrita Vishwa Vidyapeetham, Coimbatore, India
k_praveen@cb.amrita.edu

Abstract. In Visual Cryptographic Scheme (VCS), the shares of the secret image are viewed as random whereas in Extended Visual Cryptographic Scheme (EVCS), the shares are viewed as meaningful images. When the number of participants increases the graying effect problem in deterministic VCS is caused due to high pixel expansion. This graying effect was resolved using ideal contrast constructions in VCS, but not in EVCS. In ideal contrast VCS; combinations of Boolean operations are used during reconstruction. Here in this paper, we designed a deterministic construction for (t, k, n) access structure using the existing VCS constructions with contrast one as building block. In existing EVCS constructions the relative contrast will vary depending on the access structure, but in the proposed EVCS construction it is 0.25, which is better than existing schemes.

Keywords: Secret sharing · Visual cryptography · Contrast · Extended visual cryptography · Essential access structure

1 Introduction

In Visual cryptographic scheme, dealer constructs random shares from a secret image (SI) and then distributes shares to participants. During sufficient participants join their shares, the reconstructed image (RI) will be produced. Based on the pixel expansion and the contrast value [1], the quality of a VCS is measured. XOR, OR, AND and NOT are the Boolean operations used for reconstruction in VCS. In deterministic VCS all the black and white pixels of SI will be reconstructed in RI, but in probabilistic VCS [11] the correct reconstruction cannot be assured. VCS was pioneered by Naor *et al.* [1] in 1994. The deterministic VCS for general access structures are introduced in papers [2, 3, 12]. The constructions given in papers [3, 4] reconstruct the black pixel without distortion. For VCS constructions with contrast value one, the secret SI is generated without loss of resolution using combined Boolean operations [5–7, 27]. EVCS is a type of VCS developed by Ateniese *et al.* [13] where the encoded shares of SI are viewed as meaningful image. Schemes given in papers [13–20] are EVCS with pixel expansion and constructions given in papers [21–26] are EVCS without pixel expansion.

© Springer Nature Switzerland AG 2021
K. R. Venugopal et al. (Eds.): ICInPro 2021, CCIS 1483, pp. 66–77, 2021.
https://doi.org/10.1007/978-3-030-91244-4_6

Table 1. Notations and its description

Symbol	Description
$\otimes$	OR operation
$\oplus$	XOR operation
$\ominus$	AND operation
α_S	Relative contrast of meaningful share
α_{RI}	Relative contrast of reconstructed image
n	Total number of participants
m	Pixel expansion
k	Minimum participants needed to generate the secret in a (k, n) scheme
t	Number of mandatory participants needed to generate the secret in a (t, k, n) scheme
APE	Average pixel expansion [12]
$f(x) = \bar{x}$	NOT operation
Mp	Mp set contains distinct participants. For every $A \in \Gamma_{QM}$, at least one of the participant of $A \in Mp$
$COV_{(p_u, j)}(z, b)$	$n \times m$ Cover Images of size $p \times q$ where $1 \le u \le n, 1 \le j \le m, 1 \le z \le p$, $1 \le b \le q, p_u \in P$
$Sh_{(p_u, j)}(z, b, l)$	$n \times m$ Cover Shares of size $p \times 4q$ (generated using $n \times m$ Cover images) where $1 \le z \le p, 1 \le b \le q, 1 \le l \le 4, p_u \in P$
$M(z, b)$	Mask Cover Image of size $p \times q$ where $1 \le z \le p, 1 \le b \le q$
$MS(z, b, l)$	Mask Cover share of size $p \times 4q$ (generated using Mask Cover Image) where $1 \le z \le p, 1 \le b \le q, 1 \le l \le 4$
S^0 and S^1	Basis matrices of a perfect black VCS
$D_{(p_u, j)}$	$D_{(p_u, j)}$ is a block of size 1×4 in $Sh_{(u, j)}$ generated for a pixel s in $SI, p_u \in P$
MD	MD is a block of size 1×4 in MS generated for a pixel s in SI

Let $P = \{p_1, p_2, p_3, \ldots, p_t, p_{t+1}, p_{t+2}, \ldots, p_n\}$ be the set of participants, and 2^P denotes the power set of P. Let us denote Γ_{Qual} as qualified set and Γ_{Forb} as forbidden set where $\Gamma_{Qual} \cap \Gamma_{Forb} = \emptyset$. Any set $A \in \Gamma_{Qual}$ can recover SI whereas any set $A \in \Gamma_{Forb}$ cannot recover SI. Let $\Gamma_{QM} = \{A \in \Gamma_{Qual} : A' \notin \Gamma_{Qual}$ for all $A' \subseteq A, A' \ne A\}$ be the set of minimal qualified subset of P. Let Γ_{FM} be denoted as the maximum forbidden set of P. Then the pair $\Gamma = (\Gamma_{Qual}, \Gamma_{Forb})$ is denoted as the access structure of the scheme. Let S be a $n \times m$ Boolean matrix and $A \subseteq P$. The vector obtained by applying the Boolean operation (e.g.: OR) using rows of S corresponding to the elements of A is denoted by S_A. Then $w(S_A)$ is denoted as the Hamming weight of vector S_A. The definition for contrast and security of the VCS are given in [3].

Define two sets $L = \{p_1, p_2, p_3, \ldots, p_t\}$ and $R = \{p_{t+1}, p_{t+2}, \ldots, p_n\}$ which contains t respectively $(n - t)$ participants. For a (t, k, n)-EVCS the minimal qualified set is represented as $\Gamma_{QM} = \{A : A \subseteq P, L \in A \text{ and } |A| = k\}$. It is mandatory that all the participants in the set L and any k participants from set R need to involve in the reconstruction phase of a (t, k, n)-EVCS [8–10].

In this paper, we propose a deterministic EVCS for (t, k, n) access structure with a related contrast of 0.25. Our scheme is applicable only for sharing binary images. In the deterministic EVCS given in papers [13, 14, 16–20], each of the participants carry single meaningful image as share. Our proposed constructions are similar to the deterministic EVCS given in paper [15], where each of the participants carries multiple meaningful images as shares. Our EVCS has better reconstruction image quality than the deterministic EVCS constructions given in papers [13–20]. Table 1 describes the notations used in our algorithm.

2   A (t, k, n)-EVCS with Reduced Pixel Expansion

Let us define a set $L = \{p_1, p_2, p_3, \ldots, p_t\}$ which contains t essential participants and $R = \{p_{t+1}, p_{t+2}, \ldots, p_n\}$ which contains $(n - t)$ remaining participants. For a (t, k, n)-EVCS the minimal qualified set is represented as $\Gamma_{QM} = \{A : A \subseteq P, L \in A \text{ and } |A| = k\}$. It is mandatory that all the participants in the set L and any k participants from set R need to involve in the reconstruction phase of (t, k, n)-EVCS.

In today's world, there is a need for storing user's valuable private data like texts, images, passwords, or keys, on his/her computer or on any other electronic device. But in the current computing world all devices can fall prey to viruses, spyware, Trojan horses and other types of malware, exposing user data and breaching his/her privacy. Under such circumstances, it becomes paramount for a user or company to ensure the confidentiality of his/her data. One of the applications of essential access structure (1, k, n)-VCS in this scenario is that, user can store copies of the essential share in each of his/her own devices (home computer, mobile phones etc.) and outsource the remaining $n - 1$ shares to $n - 1$ trusted servers. So, an event of corruption or loss of his/her own single device will not result in the loss of data. Even on compromise of $k - 1$ servers, it will not expose the private information of user to public. A $(1, k, n)$ scheme is a (t, k, n) scheme where the value of t is 1. Figure 1 shows the experimental results and implies that our scheme has better contrast than Ateniese et al. [13] scheme. The following section shows our proposed algorithm for share generation and secret reconstruction for (t, k, n)-EVCS.

2.1   Share Generation and Distribution Phase

Input:

1. SI and a random image K of size $p \times q$.
2. The t cover images $COV_{(p_r, 1)}$ of size $p \times q$, where $p_r \in L$, $1 \leq r \leq t$.

    ($COV_{(p_r, 1)}$, used for generating meaningful shares for t mandatory participants in L).
3. The $(n - t) \times m$ cover images $COV_{(p_u, j)}$ of size $p \times q$, where $p_u \in R$, $(t + 1) \leq u \leq n$.

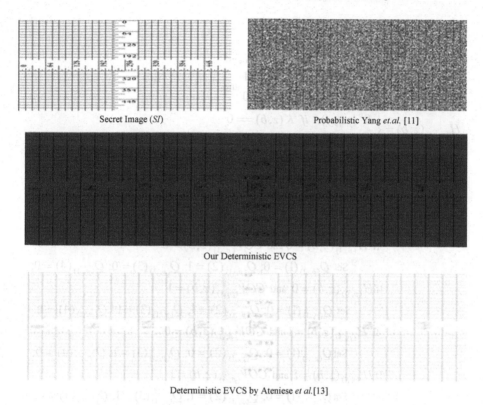

Fig. 1. Reconstructed image for (2, 3) scheme

($COV_{(p_u,j)}$, used for generating meaningful shares for remaining $(n - t)$ participants where $1 \leq j \leq m$, pixel expansion of a perfect black $(k - t, n - t)$ scheme is
m).

4. S^0 and S^1 are basis matrices of a perfect black $(k - t, n - t)$-VCS of size $(n - t) \times m$.
5. Let *Odd* (resp. *Even*) be column vectors of size $(t \times 1)$ which contain odd (resp.
   even) number of ones.
6. Set *Mp* (*Mp* contains any one essential participant, $|Mp| = 1$).

Algorithm:

Step 1. Let $E(z,b) = SI(z,b) \oplus K(z,b)$;

Step 2. Based on share generation given in [5, 6 and 7]

$$H_{(p_r,1)}(z,b) = \begin{cases} Odd(r) & if\ E(z,b) == 1 \\ Even(r) & if\ E(z,b) == 0 \end{cases} \text{and}$$

$$H_{(p_u,j)}(z,b) = \begin{cases} S^0(p_u,j) & if\ K(z,b) == 0 \\ S^1(p_u,j) & if\ K(z,b) == 1 \end{cases}$$

where, $1 \le r \le t, (t+1) \le u \le n, 1 \le j \le m$. While encoding each pixel, S^0 or S^1 is updated with same column permutation.

Step 3.

for $z = 1$ to p
  for $b = 1$ to q
    for $r = 1$ to t

If $H_{(p_r,1)}(z,b) = 0$ and $COV_{(p_r,1)}(z,b) = 0$

Set $Q_{(p_r,1)}(1) = 0$; $Q_{(p_r,1)}(2) = 1$; $Q_{(p_r,1)}(3) = 0$; $Q_{(p_r,1)}(4) = 0$;

If $H_{(p_r,1)}(z,b) = 0$ and $COV_{(p_r,1)}(z,b) = 1$

Set $Q_{(p_r,1)}(1) = 0$; $Q_{(p_r,1)}(2) = 1$; $Q_{(p_r,1)}(3) = 0$; $Q_{(p_r,1)}(4) = 1$;

If $H_{(p_r,1)}(z,b) = 1$ and $COV_{(p_r,1)}(z,b) = 0$

Set $Q_{(p_r,1)}(1) = 0$; $Q_{(p_r,1)}(2) = 0$; $Q_{(p_r,1)}(3) = 1$; $Q_{(p_r,1)}(4) = 0$;

If $H_{(p_r,1)}(z,b) = 1$ and $COV_{(p_r,1)}(z,b) = 1$

Set $Q_{(p_r,1)}(1) = 0$; $Q_{(p_r,1)}(2) = 1$; $Q_{(p_r,1)}(3) = 1$; $Q_{(p_r,1)}(4) = 0$;

end
  for $j = 1$ to m
    for $u = t+1$ to n

If $H_{(p_u,j)}(z,b) = 0$ and $COV_{(p_u,j)}(z,b) = 0$

Set $Q_{(p_u,j)}(1) = 0$; $Q_{(p_u,j)}(2) = 1$; $Q_{(p_u,j)}(3) = 0$; $Q_{(p_u,j)}(4) = 0$;

If $H_{(p_u,j)}(z,b) = 0$ and $COV_{(p_u,j)}(z,b) = 1$

Set $Q_{(p_u,j)}(1) = 0$; $Q_{(p_u,j)}(2) = 1$; $Q_{(p_u,j)}(3) = 0$; $Q_{(p_u,j)}(4) = 1$;

If $H_{(p_u,j)}(z,b) = 1$ and $COV_{(p_u,j)}(z,b) = 0$

Set $Q_{(p_u,j)}(1) = 0$; $Q_{(p_u,j)}(2) = 0$; $Q_{(p_u,j)}(3) = 1$; $Q_{(p_u,j)}(4) = 0$;

If $H_{(p_u,j)}(z,b) = 1$ and $COV_{(p_u,j)}(z,b) = 1$

Set $Q_{(P_u,j)}(1) = 0; Q_{(P_u,j)}(2) = 1; Q_{(P_u,j)}(3) = 1; Q_{(P_u,j)}(4) = 0;$

        end

    end

Apply same column permutation for all the $Q_{(P_u,j)}$ matrices and MD matrix of size 1×4.

    for $j = 1$ to m

        for $u = t + 1$ to n

            for $l = 1$ to 4

$$Sh_{(P_u,j)}(z,b,l) = Q_{(P_u,j)}(l);$$

            end

        end

        end

        for $l = 1$ to 4

$$MS(z,b,l) = MD(l);$$

            for $r = 1$ to t

$$Sh_{(p_r,1)}(z,b,l) = Q_{(p_r,1)}(l);$$

            end

        end

    end

end

Output:

1. Shares $\{Sh_{(p_r,1)}: 1 \leq r \leq t\}$. The t meaningful shares of size $p \times 4q$ are distributed to t mandatory participants in set L.
2. Shares $\{Sh_{(p_u,j)}: (t + 1) \leq u \leq n, 1 \leq j \leq m\}$. The $(n - t) \times m$ meaningful shares of size $p \times 4q$ are distributed to remaining participants in set R.
3. Share MS of size $p \times 4q$ is distributed to participants in the set Mp.

2.2 Secret Reconstruction Phase

Step 1. λ_j's are generated by OR-ing shares $Sh_{(p_u,j)}$ of any $(k - t)$ out of $(n - t)$ participants in $R = \{p_{t+1}, p_{t+2}, ..., p_n\}$, where $1 \leq j \leq m$.

Step 2.

The construction of K can be done using any one of the following schemes

a) Cimato *et al.* [5] (OR and NOT operation) **b)** Wang *et al.* [6] (OR and XOR operation) and **c)** Praveen *et al.* [7] (OR and AND operation).

a) Cimato *et al.* [5]

for $z = 1$ to p
  for $b = 1$ to q
    for $l = 1$ to 4

      $1. \sigma(z,b,l) = \bigotimes_{j=1}^{m} f(\lambda_j(z,b,l))$

      $2. K(z,b,l) = f(\sigma(z,b.,l));$

    end

  end

end

b) Wang *et al.* [6]

for $z = 1$ to p
  for $b = 1$ to q
    for $l = 1$ to 4

$$K(z,b,l) = \bigoplus_{j=1}^{m} \lambda_j(z,b,l);$$

    end

  end

end

c) Praveen *et al.* [7]

for $z = 1$ to p
  for $b = 1$ to q
    for $l = 1$ to 4

$$K(z,b,l) = \bigodot_{j=1}^{m} \lambda_j(z,b,l);$$

    end

  end

end

Step 3. XOR-ing the shares of all the participants in the set L along with K will generate RI. Then AND-ing RI with MS will reconstruct the secret OI.

$$
\begin{aligned}
&\text{for } z = 1 \text{ to } p \\
&\quad \text{for } b = 1 \text{ to } 4q \\
&\qquad \text{for } l = 1 \text{ to } 4 \\
&\qquad RI(z,b) = K(z,b) \oplus Sh_{(P_1,1)}(z,b) \oplus \text{........} \oplus Sh_{(p_r,1)}(z,b) \\
&\qquad OI(z,b,l) = RI(z,b,l) \;\; \ominus \;\; MS(z,b,l) \\
&\qquad\quad \text{end} \\
&\quad \text{end} \\
&\text{end}
\end{aligned}
$$

Tables 2, 3 and 4 shows the comparison results.

Table 2. Reconstruction operations count for (t, k, n)

Operation	Cimato *et al.* [5]	Wang *et al.* [6]	Praveen *et al.* [7]
OR	$4 \times (m(k - t) + (m - 1))$	$4 \times (m(k - t)))$	$4 \times (m(k - t))$
NOT	$4 \times (m + 1)$	0	0
AND	4	4	$4m$
XOR	$4t$	$4t + 4(m - 1)$	$4t$

Table 3. Comparison of deterministic $(1, 3, 4)$ scheme

Scheme	Operations	APE	α_{RI}
Ateniese *et al.* [13]	OR	10	0.100
Liu *et al.* [14]	OR	16	0.062
Wang *et al.* [16]	OR	16	0.062
Yan *et al.* [19]	OR	16	0.062
Our scheme (Sect. 2.1)	OR, XOR, AND	11	0.250

2.3   Analysis on the Pixel Expansion, Contrast and Security

The matrices S^0 (resp. S^1) are constructed based on the Definition 1. The proposed (t, k, n)-perfect black EVCS is valid only when the following three conditions meet.

Condition 1: It should not be possible for any $(k - t)$ participant in the set R out of $(n - t)$ participants to identify SI in the absence of participants in the set L.

Table 4. Comparison of deterministic (2, 4, 5) scheme

Scheme	Operations	APE	α_{RI}
Ateniese et al. [13]	OR	18	0.055
Liu et al. [14]	OR	25	0.040
Wang et al. [16]	OR	32	0.031
Yan et al. [19]	OR	32	0.031
Our scheme (Sect. 2.1)	OR, XOR, AND	9.6	0.250

Condition 2: It should not be possible for any participant less than $(k - t)$ in the set R out of $(n - t)$ participants to identify SI with participants in the set L.

Condition 3: It should be possible for any $(k - t)$ in the set R out of $(n - t)$ participants to identify SI in the presence of all participants in the set L. The first two conditions are for the security of the scheme and the third is for the correctness of reconstruction. Assume that variables q, b, b_1, b_2, z take the values either 0 or 1. Let $Pbr(q = b)$ denote the probability of occurrence of q equal to b. Let b_1 and b_2 be the two independent bits and $q = b_1 \oplus b_2$. Let $Pbr((q = b_2)/(b_1 = z))$ be the probability of $q = b_2$ given bit b_1 equal to z.

Lemma 1: Let b_1 (resp. b_2) be known (resp. unknown) bit and $q = b_1 \oplus b_2$ then

$$Pbr((q = b_2)/(b_1 = 0)) = Pbr((q = b_2)/(b_1 = 1)) = \frac{1}{2}.$$

Proof: Here the given information is b_1 and $q = b_1 \oplus b_2$. But b_2 is unknown then,
$Pbr((q = 0)/(b_1 = 0)) =$
$Pbr((q = 1)/(b_1 = 0)) = \frac{1}{2}$ and $Pbr((q = 0)/(b_1 = 1)) = Pbr((q = 1)/(b_1 = 1)) =$
$\frac{1}{2}$ accordingly,$Pbr((q = b_2)/(b_1 = 0)) = Pbr((q = b_2)/(b_1 = 1)) = \frac{1}{2}$.

Lemma 2: Given two bits b_1, b_2 and $q = b_1 \oplus b_2$ then $Pbr((q = b_2)/(b_1 = 0)) = 1$ and $Pbr((q = b_2)/(b_1 = 1)) = 0$.

Proof: Here the given information is b_1, b_2 and $q = b_1 \oplus b_2$ then,
$Pbr((q = 0)/(b_1 = 0)) = Pbr((q = 1)/(b_1 = 0)) = 1$ and
$Pbr((q = 0)/(b_1 = 1)) = Pbr((q = 1)/(b_1 = 1)) = 0$ accordingly,
$Pbr((q = b_2)/(b_1 = 0)) = 1$ and $Pbr((q = f(b_2))/(b_1 = 1)) = 1$.
which implies that $Pbr((q = b_2)/(b_1 = 1)) = 0$.

Let x be a bit obtained by combining the shares (MS is not taken) of any $(k - t)$ out of $(n - t)$ participants from the set R. Let y be a bit obtained by combining all the shares of participants in the set L and s is the secret bit in RI. Then $s = x \oplus y$. The security for Condition 1 and 2 can be proved using Lemma 1.

Condition 1: Here $b_1 = y$ (is given), $b_2 = x$ (is unknown bit either 0 or 1) and $q = s$ (either 0 or 1). $Pbr((s = x)/(y = 0)) = Pbr((s = x)/(y = 1)) = \frac{1}{2}$ (Lemma 1).

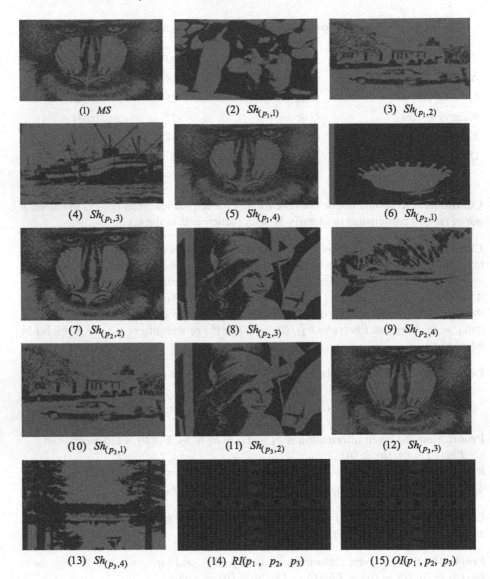

(1) *MS*   (2) $Sh_{(p_1,1)}$ (3) $Sh_{(p_1,2)}$

(4) $Sh_{(p_1,3)}$ (5) $Sh_{(p_1,4)}$ (6) $Sh_{(p_2,1)}$

(7) $Sh_{(p_2,2)}$ (8) $Sh_{(p_2,3)}$ (9) $Sh_{(p_2,4)}$

(10) $Sh_{(p_3,1)}$ (11) $Sh_{(p_3,2)}$ (12) $Sh_{(p_3,3)}$

(13) $Sh_{(p_3,4)}$ (14) $RI(p_1,\ p_2,\ p_3)$ (15) $OI(p_1,p_2,\ p_3)$

Fig. 2. Experimental results for our (2, 3) EVCS

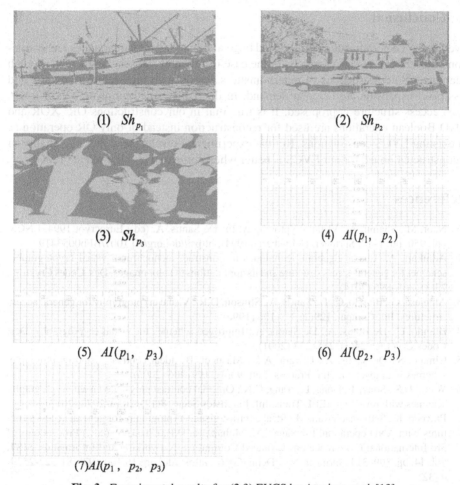

(1) Sh_{p_1}

(2) Sh_{p_2}

(3) Sh_{p_3}

(4) $AI(p_1, p_2)$

(5) $AI(p_1, p_3)$

(6) $AI(p_2, p_3)$

(7) $AI(p_1, p_2, p_3)$

Fig. 3. Experimental results for (2,3) EVCS by Ateniese *et al.* [13]

Condition 2: Here $b_1 = x$ (is given), $b_2 = y$ (is unknown bit either 0 or 1) and $q = s$ (either 0 or 1). $Pbr((s = y)/(x = 0)) = Pbr((s = y)/(x = 1)) = \frac{1}{2}$ (Lemma 1).

Condition 3: Here $b_1 = y$, $b_2 = x$ (is given) and $q = s$. $Pbr((s = x)/(y = 0)) = 1$, $Pbr((s = x)/(y = 1)) = 0$ (Lemma 2).

Figure 2 shows the experimental results of our scheme with meaningful shares (Sh) and reconstructed images (RI and OI). Figure 3 shows the experimental results of Ateniese *et al.* [13] scheme with meaningful shares (Sh) and reconstructed images (AI). It is clear that the contrast of AI is less than OI.

3 Conclusion

Even though deterministic schemes need huge amount of data, it guarantees reconstruction of all the secret data correctly. In the case of probabilistic scheme there is no such guarantee. So constructions of deterministic schemes with high relative contrast and less pixel expansion are of great demand. In this paper a deterministic EVCS for (t, k, n) access structure is proposed. It is true that in our constructions OR, XOR and AND Boolean operations are used for reconstruction instead of only OR operation as in existing EVCS constructions. But the experimental results show that, the quality of reconstructed image for our EVCS is better when compared to other related EVCS.

References

1. Naor, M., Shamir, A.: Visual cryptography. In: De Santis, A. (ed.) Eurocrypt 1994. LNCS, vol. 950, pp. 1–12. Springer, Heidelberg (1994). https://doi.org/10.1007/BFb0053419
2. Adhikari, A.: Linear algebraic techniques to construct monochrome visual cryptographic schemes for general access structure and its applications to color images. Des. Codes Cryptogr. **73**(3), 865–895 (2014)
3. Ateniese, G., Blundo, C., De Santis, A., Stinson, D.R.: Visual cryptography for general access structures. Inf. Comput. **129**(2), 86–106 (1996)
4. Blundo, C., De Bonis, A., De Santis, A.: Improved schemes for visual cryptography. Des. Codes Cryptogr. **24**(3), 255–278 (2001)
5. Cimato, S., De Santis, A., Ferrara, A.L., Masucci, B.: Ideal contrast visual cryptography schemes with reversing. Inf. Process. Lett. **93**(4), 199–206 (2005)
6. Wang, D.S., Song, T., Dong, L., Yang, C.N.: Optimal contrast grayscale visual cryptography schemes with reversing. IEEE Trans. Inf. Forensics Secur. **8**(12), 2059–2072 (2013)
7. Praveen, K., Sethumadhavan, M.: Ideal contrast visual cryptography for general access structures with AND operation. In: Nagar, A., Mohapatra, D.P., Chaki, N. (eds.) Proceedings of 3rd International Conference on Advanced Computing, Networking and Informatics. SIST, vol. 44, pp. 309–314. Springer, New Delhi (2016). https://doi.org/10.1007/978-81-322-2529-4_32
8. Arumugam, S., Lakshmanan, R., Nagar, A.K.: On (k, n) *-visual cryptographic scheme. Des. Codes Cryptogr. **71**(1), 153–162 (2014)
9. Guo, T., Liu, F., Wu, C.K., Ren, Y.W., Wang, W.: On (k, n) visual cryptography scheme with t essential parties. In: Padró, C. (ed.) ICITS 2013. LNCS, vol. 8317, pp. 56–68. Springer, Cham (2014). https://doi.org/10.1007/978-3-319-04268-8_4
10. Dutta, S., Rohit, R.S., Adhikari, A.: Constructions and analysis of some efficient t-(k, n)* - visual cryptographic schemes using linear algebraic techniques. Des. Codes Cryptogr. 1–32 (2016)
11. Yang, C.N.: New visual secret sharing schemes using probabilistic method. Pattern Recogn. Lett. **25**, 481–494 (2004)
12. Liu, F., Wu, C., Lin, X.: Step construction of visual cryptographic schemes. IEEE Trans. Inf. Forensics Secur. **5**(1), 25–34 (2010)
13. Ateniese, G., Blundo, C., De Santis, A., Stinson, D.R.: Extended capabilities for visual cryptography. Theor. Comput. Sci. **250**(1), 143–161 (2001)
14. Liu, F., Wu, C.: Embedded extended visual cryptography schemes. IEEE Trans. Inf. Forensics Secur. **6**(2), 307–322 (2011)
15. Zhou, Z., Arce, G.R., Di Crescenzo, G.: Halftone visual cryptography. IEEE Trans. Image Process. **15**(8), 2441–2453 (2006)

16. Wang, Z., Arce, G.R., Di Crescenzo, G.: Halftone visual cryptography with error diffusion. IEEE Trans. Inf. Forensics Secur. **4**(3), 383–396 (2009)
17. Wang, D., Yi, F., Li, X.: On general construction for extended visual cryptography schemes. Pattern Recogn. **42**(11), 3071–3082 (2009)
18. Yang, C.N., Yang, Y.Y.: New extended visual cryptography schemes with clearer shadow images. Inf. Sci. **271**, 246–263 (2014)
19. Yan, X., Wang, S., Niu, X., Yang, C.N.: Halftone visual cryptography with minimum auxiliary black pixels and uniform image quality. Digit. Sig. Process. **38**, 53–65 (2015)
20. Lu, S., Manchala, D., Ostrovsky, R.: Visual cryptography on graphs. J. Comb. Optim. **21**(1), 47–66 (2011)
21. Lee, K.H., Chiu, P.L.: An extended visual cryptography algorithm for general access structures. IEEE Trans. Inf. Forensics Secur. **7**(1), 219–229 (2012)
22. Guo, T., Liu, F., Wu, C.: k out of k extended visual cryptography scheme by random grids. Sig. Process. **94**, 90–101 (2014)
23. Chiu, P.L., Lee, K.H.: User-friendly threshold visual cryptography with complementary cover images. Sig. Process. **108**, 476–488 (2015)
24. Ou, D., Sun, W., Wu, X.: Non-expansible XOR-based visual cryptography scheme with meaningful shares. Sig. Process. **108**, 604–621 (2015)
25. Yan, X., Wang, S., Niu, X., Yang, C.N.: Generalized random grids-based threshold visual cryptography with meaningful shares. Sig. Process. **109**, 317–333 (2015)
26. Wang, S., Yan, X., Sang, J., Niu, X.: Meaningful visual secret sharing based on error diffusion and random grids. Multimedia Tools Appl. **75**(6), 3353–3373 (2016)
27. Praveen, K., Sethumadhavan, M., Krishnan, R.: Visual cryptographic schemes using combined Boolean operations. J. Discrete Math. Sci. Cryptogr. **20**(2), 413–437 (2017)

Improving Storage Efficiency with Multi-cluster Deduplication and Achieving High-Data Availability in Cloud

M. Bhavya[1]([⊠]), M. Prakash[1], J. Thriveni[1], and K. R. Venugopal[2]

[1] Department of Computer Science and Engineering, University of Visvesvaraya College of Engineering, Bangalore University, Bangalore 560001, India
[2] Bangalore University, Bangalore 560001, India

Abstract. The Cloud service providers (CSP) such as iCloud, Google Drive, Mega and others adopt secure data deduplication to remove redundancies of encrypted data. Earlier research in deduplication of encrypted files resulted in an idea of converging encryption, which generates identical cipher texts from identical plain text. While the use of converging encryption does not address security challenges since users can still determine the occurrence of deduplication. Few users access the data from the Cloud for a particular time period and later they do not need the data. In that case the user is charged for the unused data, which leads to wastage of storage space and cost for keeping unused data. To overcome the above limitations, in this paper, a multi cluster deduplication techniques is proposed. The proposed work use the erasure coding technique to achieve high data availability in the cloud, by keeping the encrypted files in multiple servers. The cloud server minimizes the storage space by deleting the unused files from the server. The implementation of RC4 (Rivest Cipher 4) symmetric encryption technique reduces the cipher text size over AES (Advanced Encryption Standard) and RC4 consumes less time to encrypt the data compare to AES. The performance analysis shows that the proposed scheme improves system efficiency by reducing storage space, minimizing storage cost, reduce the file uploading and downloading time and maintain data integrity and data security as compared to the existing scheme.

Keywords: Cloud storage server · Data deduplication · Erasure coding · Data security · Storage management

1 Introduction

In present era people are migrating to digitalization, it produces huge amount of data in daily basis, using relational database is too expensive to analyze, store and manage big data. Due to their inflexibility, accessing real-time data, storing unstructured data and to maintain scalability to support very large data volumes are cost-inefficient. Cloud storage provide feasible and efficient platform to store big data. The leading technologies in cloud storage results helps user to store and process huge volume of data with less cost [1].

© Springer Nature Switzerland AG 2021
K. R. Venugopal et al. (Eds.): ICInPro 2021, CCIS 1483, pp. 78–91, 2021.
https://doi.org/10.1007/978-3-030-91244-4_7

Sometimes there will be multiple copies of same data, which leads to data redundancy. The deduplication technique helps to remove duplicity and stores unique copy of the data. Data deduplication helps to minimize bandwidth requirements and storage space for uploaded data [2, 3]. To maintain privacy and security to the data, users encrypt their data using encryption key before outsourcing to cloud. Deduplication on encrypted data in single domain is achieved using different techniques. Some techniques are inefficient for the cross-user deduplication as a similar data will be encrypted by different cipher texts by different user keys. This is a challenging task to find duplicate data over different cipher texts. Hence, it is a crucial task to perform deduplication over encrypted data in multi-domain.

Many researchers proposed efficient methods to achieve data deduplication in the fast research work. These methods are implemented to perform deduplication over encrypted data. Convergent encryption [4] is one technique which allows deduplication on encrypted data efficiently. The cryptographic hash value of the data generates the convergent key is used to encrypts the data.

The convergent encryption is a type of symmetric encryption technique, which generate same cipher text and convergent key for the identical data and achieves deduplication on encrypted data successfully. Due to its deterministic property the convergent encryption technique is easily exposed to brute-force attacks. To prevent from such attacks, the server-aided encryption techniques [5] were proposed. In this technique dedicated Key Distribution Server (KDS) is introduced by a cluster or tenant (company or university) to help authorized users to have the corresponding tag. An inter-deduplication method has been constructed [6] recently to support cross-domain deduplication. The multi-domain deduplication scheme supports to perform both intra-deduplication (in the same cluster) and inter-deduplication (across multiple clusters) efficiently. As per the study, it is understood that the existing schemes cannot achieve the overall cloud storage efficiency in terms of storage space reduction, file uploading and downloading time and data security.

1.1 Motivation

The existing system implemented a data deduplication technique that keeps only one copy of the data, which leads to data loss or corruption [7, 8]. The existing system will not delete the files from the server when it not used by the users, it leads to more cloud storage space consumption and increase in the storage cost to the users. The usage of AES symmetric encryption technique to convert the file before uploading to the cloud server consumes more storage space.

1.2 Research Contribution

In this paper, we implemented erasure coding technique, in which the cloud server splits the file into multiple sub files and stores it in multiple servers. The proposed system provides delete operation to the users when the uploaded files are not used for the longer time, which minimizes the storage cost to the users. The RC4 symmetric encryption algorithm consumes less storage space in the cloud server than the existing technique, which improves the cloud server storage space efficiently.

1.3 Objectives

The main objectives of the proposed system are:

- Reduce the storage cost and time to the users and allow to use storage space efficiently.
- Provide privacy to the stored encrypted data.

1.4 Paper Organization

The paper is organized as follows. The related work is highlighted in Sect. 2. Section 3 presents the background work. Section 4 shows the system model which explains the main modules implemented in the proposed system. Section 5 describes the system design and explains the system model with description and workflow. Section 6 shows the performance evaluation of existing and proposed system with results. The conclusions are described in Sect. 7.

2 Related Work

Deduplication is a method used by cloud storage providers to save storage space by storing only one copy of each uploaded file on the server. Bellare *et al.* [9] presented the Message-Locked Encryption (MLE) method, which obtains the secret key for both encryption and decryption from the original message itself. The MLE provides a semantic security proof for dissimilar uploaded information, the MLE is leads to brute-force attack when the uploaded information is similar.

To overcome the brute-force attacks in Message-Locked Encryption (MLE) technique, Keelveedhi *et al.* [10] have implemented DupLESS technique. In this system, a key server (KS) generates symmetric keys based on uploaded information from the clients by using the Oblivious PRF (Pseudo-Random Function) protocol, and then the clients encrypt the data using keys received from a key-server. The KS) should be available always to provide keys to each user for data upload. If the KS is compromised, then it leads to the single point failure of the system.

If the main server is compromised, the DupLESS system will not work. Based on threshold blind signature, Miao *et al.* [3] have introduced a multi-server data deduplication system that can handle collusion attacks across multiple key servers and cloud servers. The secret key is obtained from the additional private key of key cloud service providers (K-CSP). To implement the blind signature, the user needs to communicate with multiple K-CSP. Only if the K-CSP is compromised, the private key can be exposed. Off line brute force attacks are eliminated and the server provides robust security.

For a multi-domain scheme, Yang *et al.* [11] have implemented an efficient data deduplication technique. By generating fixed set of ciphertexts and random tags, the system provides semantic security and guarantees that data can be decrypted by data owners. Only the CSP can determine whether duplicate data exists in other domains by checking the inter-tag created by each domain's agent, which eliminates the exposure of tag generation information, ensures data confidentiality, and it will not allow unauthorized user to modify the outsourced data.

Jiang *et al.* [12] have proposed a secure data deduplication technique that maintains adaptive ownership using a PoW system. It supports data deduplication at the file level between users as well as data deduplication at the block level inside users' files. It guarantees the confidentiality and accuracy of outsourced data, and it will handle data ownership management efficiently.

Li *et al.* [13] have introduced SecCloud and SecCloud+ techniques to achieve both integrity of the data and deduplication in cloud. SecCloud uses an auditing entity with MapReduce cloud maintenance to assist clients in generating information tags before uploading data and auditing the integrity of data uploaded and stored in the cloud. It uses a Proof of Ownership protocol to enable secure deduplication and prevent the leakage of side channel information during data deduplication. The speed at which users can upload data and perform auditing computations in SecCloud has been reduced. Users still want to encrypt their data before uploading, and users allow for secure deduplication and data integrity auditing directly on encrypted data with the SecCloud+ framework.

Encrypted data on a cloud server poses new challenges for cloud data deduplication, which is becoming increasingly necessary for big data storage and processing in the cloud. Yan *et al.* [14] have used ownership challenge and proxy re-encryption to deduplicate encrypted data stored in the cloud. It incorporates data deduplication and access control in the cloud. The framework suggested an ownership verification protocol based on a cryptoGPS identification to search for duplication. The symmetric keys for data decryption can only be obtained by registered data owners. Symmetric keys can be used to access encrypted data in a secure manner. To provide identity based cryptosystems and efficient reliable public key, D. Boneh *et al.* [15] have implemented a Hierarchical Identity Based Encryption (HIBE) technique in which the ciphertexts is made of sets of three elements and decryption consists of two bilinear map computations.

3   Background Work

3.1   Bilinear Groups of Composite Order

A composite bilinear parameter generator [16] $Gen(\kappa)$ outputs a tuple (p, q, G, GT, e) for the given input parameter κ, where

- $p \rightarrow$ k - bit prime number
- $q \rightarrow$ k - bit prime number
- G *and* GT are finite cyclic multiplicative groups of Composite order $N = pq$.
- e: $G \times G \rightarrow GT$ is a bilinear map with the following properties.

  - *Bilinear*: $e(x^a, y^b) = e(x, y)^{ab}$ for all $x, y \in G$, and $a, b \in Z_N$.
  - *Non-degeneracy*: In the event that g is a generator of G, then $e(g, g)$ is a generator of GT with the order N.
  - *Computability*: For all $x, y \in G$, there exists an efficient algorithm to compute $e(x, y) \in GT$.

3.2 Boneh-Goh-Nissim Cryptosystem

The key generation, data encryption, and data decryption are the three algorithms presents in Boneh-Goh-Nissim cryptosystem (BGN) [17] which are explained in detailed below.

- *Key generation*: For the given input κ, run the algorithm *Gen (κ)* to get a output tuple (p, q, G, GT, e) and set $N = pq$. Randomly select two generators $g, x \in G$ and set $y = xq$. The private key is p and set public key is $pk = (N, G, GT, e, g, y)$.
- *Encryption*: For the given message $m \in \{0, 1, ..., W\}$, pick an arbitrary number $r \in Z_N$, and generate the cipher text as $C = g^m y^r \in G$.
- *Decryption*: For the generated ciphertext C, select the private key p, and generate $c^p = (g^m y^r)^p = (g^p)^m$.

3.3 Symmetric Encryption Algorithm

Symmetric encryption is a kind of encryption where only one key is used to both encode and decode data. RC4 is a stream cipher that doesn't have a discrete block size. The RC4 operation is faster as compared to other encryption techniques. The size of the cipher text of RC4 technique is less compare to other encryption techniques.

4 System Model

4.1 System Model

To overcome the issues present in the existing work, new data storage techniques are implemented to protect against data loss or corruption and eliminate the storage redundancy by removing unused data. This scheme helps in easy storing of encrypted data between multiple cloud servers from different clusters.

Figure 1 shows the proposed architecture that contains four types of entities:

- Key Distribution Server (KDS) - It is authoritative for generating a secret key to perform inters deduplication by CSP and different keys for users from different cluster. The believed KDS is entrusted with the conveyance and the board of private keys for the framework.
- Cloud Service Provider (CSP) - The CSP with multiple servers provide information storage services for the customers CSP is designed to fulfill the data storage needs of customers and helps to decrease the data maintenance and management overheads. The CSPs have to execute inter-cluster deduplication at cluster level, and find out the duplicate data from various clusters if exist else CSP uses the erasure coding technique to split the encrypted file into sub files to store it in multiple servers.
- Cluster Manager (CM_i) - A cluster manager CM_i situated with clients and the Cluster Manager is recruited by clients to perform intra-cluster deduplication and convert an intra-cluster tag into an inter-cluster tag. The CM_i to perform deduplication within cluster to provide tag to search deduplication with multiple clusters.

- Clients (c) - Each client is affiliated with an area or cluster (e.g., organization or college). Customers transfer and store their information with the CSP so as to ensure their information protection and help the CSP to finish data deduplication over encoded data. For encrypting the data and deduplication, the clients generate an arbitrary convergent key and an intra-cluster tag respectively.

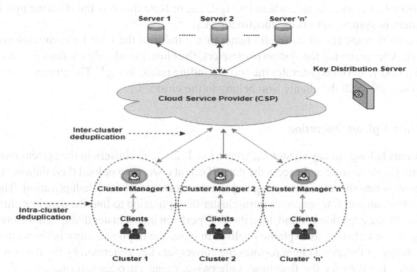

Fig. 1. System model

The general workflow is implemented as follows. At the point when a client c from cluster i need to transfer information m, Client creates an intra-cluster -tag and sends it to a cluster manager CM_i. At that point, an agent CM_i plays out the intra cluster -deduplication depending on the intra cluster-tag. If the duplicate discovered in the same cluster, a CM_i returns the corresponding information to the user otherwise, a CM_i will convert the intra cluster tag into an inter-cluster tag and transfer it to the CSP to perform inter cluster deduplication check. If the CSP doesn't find a copy, client have to encode message 'm' and transfer the encrypted data to the CSP. To resolve vulnerability to data loss or corruption present in the existing scheme, we keep multiple copies of the data on different servers using the erasure coding technique with data deduplication.

5 System Design

This section explains the proposed system, an efficient and secure data deduplication technique for a multi-cluster system with multiple servers. It provides an efficient data storage and privacy to the data owner. The proposed system protects data against data loss or corruption by using erasure coding technique. The proposed system includes four stages: Key generation operation, data upload operation, data delete operation and data download operation.

5.1 Key Generation Operation

In this section, the private keys for the clients and a secrete key for the CSP will be generated using the trusted KDS. The KDS outputs a tuple $(N = p, q, G, G_T, e)$ by executing the composite bilinear boundary generator $Gen(\kappa 0)$. Then the KDS randomly selects two generators g, $x \in G$ and two numbers α, $\gamma \in Z_N$, and for i = 2,..., n, n + 2,...,2n, it creates $y = x^q$, $v = g^\gamma$ and $g^i = g^{\alpha i}$. For all clients in the cluster C_i, compute the private key c_i as $c_i = g_i^\gamma$, where i = 1, 2, ..., n. Note that n is the absolute number of clusters in system and i is the identifier of C_i.

The KDS transfers the c_i to all clients in C_i and p to the CSP by communication channel. After receiving the system parameters, the cluster randomly chooses $a_i \in Z_N$ as the private key, and then generates the corresponding public key g^{ai}. The cluster manager CM_i sends g^{ai} to all the clients who belong to the cluster C_i.

5.2 Data Upload Operation

The clients belongs to any cluster C_i, where i = 1, 2, ..., n, clusters in the system model are permissible to store and access the data. If client c wants to upload their data m, they need to generate the intra-cluster tag to perform cluster level data deduplication. Then, the cluster manager CM_i executes intra-cluster deduplication to find the data in a cluster C_i. If duplicate copy doesn't find, then the CSP perform inter-cluster deduplication check among various clusters as explained below. In case, the duplicate copy is found in the same cluster, at that point client restores the convergent key generated by the data owner who uploaded data for the first time. Otherwise, client encodes information '*m*' and uploads the corresponding cipher texts to the CSP. The CSP will convert the uploaded cipher texts into equal size of data blocks by using the erasure coding technique and it will be stored in different servers.

- *Intra-cluster tag generation:* When a client c from Cluster C_i needs to transfer information m, the client picks a random number $r_m \in Z_N$, and produces random intra-cluster tag τ_m with the private key c_i and CM_i's public key g^{ai}. After generating the intra-cluster tag, client sends a message "TagUpload $\| (L_m, \tau_m)$" to cluster manager CM_i. The L_m is the size of the message m, and is also used a search keyword.
- *Intra-cluster duplicate check:* After receiving the message "Tag Upload $\| (L_m, \tau_m)$" from the client, the C_{Mi} executes the intra-cluster deduplication check based on the private key $_{ai}$ of the cluster manager and computes $d^{ih2(m)}$. After that, the C_{Mi} computes hash code of the $_d{}^{ih2(m)}$ as

$$T_m = h3\left(d^{ih2(m)}\right).$$

To check weather a message m exists or not in the CSP, T_m is compared with existing hash values from cluster C_i. If the tag T_m exists in the same cluster, e.g., $(T_m*, Bm*)$ then CM_i returns "duplication$\|B_m*$" to the client. The $Bm*$ is the cipher text helps to retrieve data of the convergent key.

If duplicate data is not found in the same cluster, the CM_i uses the tag T_m value to create a inter-cluster tag $^\wedge\tau_m$ randomly using the BGN encryption and randomly chosen $r \in Z_N$. After generating inter-cluster tag, the CM_i stores T_m in the hash file and transfer the information "Inter Tag Upload$\|$(i, L_m, $\hat{\tau}_m$)" to the CSP for the inter-cluster deduplication check, where "I" is the identifier of the Cluster C_i.

- *Inter-cluster*
  duplicate Check: After getting the message "Inter Tag Upload$\|$(i, L_m, $\hat{\tau}_m$)" from cluster C_i, the CSP executes the inter-cluster deduplication to remove the duplicate copy of data. The CSP is responsible to search a duplicate of L_m in inter cluster level. The CSP searches the copy using dissimilar data length L_m in the root node. If the similar value L_m* exists, then a check on whether duplicate exists or not is made for the adjacent stored data (j, $\hat{\tau}_m*$, B_m*, C_m*). If the duplicate data is found, then the CSP sends "duplication$\|$$link_m* \|B_m*$" to CM_i, where $link_m*$ is the link to retrieve ciphertext C_m*, else, the CSP sends "data upload" message to CM_i.

- *Data encryption/key restore operation:* After receiving the message "duplication$\|$ $link_m* \|B_m*$", the CM_i sends "duplication$\|B_m*$" to the client and stores (B_m*, $link_m*$) together with T_m, i.e., (T_m, B_{m*}, $link_{m*}$), where $m = m^*$. In other case, the CM_i sends the "data upload" message to the client. If "data upload" message received from CM_i, then client performs data encryption operations. If client received "duplication$\|B_m*$", message then they execute key restore operation. The client encrypts the data while uploading and randomly select $\beta_m \in Z_N$, set $K_m = e(g_{n+1}, g)^{\beta m}$, and generate the convergent key ck_m as $h1(m\|K_m)$. The value $e(g_{n+1}, g)$ can be calculated as $e(g_i, g_{n+1-i})$. The cipher text of m is generated as:

$$
\begin{aligned}
C_m &= RC4_{ckm}(m) \\
B_m &= (g^{\beta m}, v * (g_1 + g_2 + \dots g_n)^{\beta m})
\end{aligned}
\tag{1}
$$

$RC4_{ckm}(m)$ symmetric encryption technique is used to generate ciphertext from the information. The B_m is used to obtain the secrete key K_m for both encryption and decryption for all the clients from the different clusters, in which the group of ciphertexts is similar for any number of receivers from different clusters, the B_m is generated using the broadcast encryption technique [13], where g_1, $g_2 \dots g_n$ is used to represent 'n' different clusters. For each data 'm', the ciphertext tuple (B_m, C_m) is generated. The broadcast encryption technique enables all the clients to decrypt uploaded data correctly.

After data encryption, client sends (B_m, C_m) to the CM_i. After receiving (B_m, C_m) the CM_i sends it to the CSP and stores (T_m, B_m). Then CSP will convert the uploaded chipertexts into equal size of data blocks using the erasure coding technique and it will be stored in different servers. The details of the Reed-Solomon erasure coding is described in Algorithm 1.

Algorithm 1 : Algorithm for Erasure Coding

Input: chipertex C_m

Output: Data blocks with equal size

Step 1: Start

Step 2: num←total number of servers in which chipertext will be stored

Step 3: chipetext _size←Read length of the chipetext C_m

Step 4: Calculate the block_size = chipertext_size / num

Step 5: for i←0 to num-1

        $block_i$ ← move $chipetext_i$(block_size)

        calculate the parity bits ($block_i$)

        encode the parity bits ($block_i$)

Step 6: for end

Step 7: Write the data block into different servers

Step 8: end

The CSP stores $(i, L_m, \hat{\tau}_m, B_m, C_m)$ in the selected position of the DDT using Algorithm 1, and sends the $link_m$ to CM_i for storing. Finally, the CM_i stores $(T_m, B_m, link_m)$.

- *Key recovery:* Once *"duplication∥B_m*"* message is received, using private key c_i and the data m the client recovers the convergent key $ck_m* = h1(e(g_{n+1},g)^{\beta m} * \|m*)$. After that, client can recover ck_m* with the data m as: $ck_m* = h1(e(g_{n+1},g)^{\beta m} * \|m*)$. In this case, $m = m*$ i.e., the data m is the duplicate of the data $m*$.

5.3 Data Delete Operation

In the proposed system, to minimize the storage cost the clients can delete the uploaded data from the cloud by sending the delete request. The CSP will remove the data blocks that are not used and this operation helps to utilizes storage efficiently. This feature eliminates wastage of storage space in the cloud servers. When client wishes to delete the data, the client will find the tag T_m generated by the uploaded information. At this time, the client only deletes the link to the ciphertext and the tag T_m. The CSP erases the data blocks based on the *refCount* of the uploaded data blocks. The details of the delete operation are described in Algorithm 2.

Algorithm 2 : Algorithm for data delete operation

Step 1: **For** the stored data $(i,_{Lm}, \hat{\tau}_m)$, the CSP use *refCount* to delete L_m and data which has already been stored.

Step 2: if *refCount* == 0 **then**

        "delete data $(i, L_m, \hat{\tau}_m)$".

Step 3: **else**

Step 4:   Find the clusters in which the uploaded information *refCount* exists

Step 5:   update the *refCount* ← *refCount* - 1;

5.4 Data Download Operation

When the client c from the cluster C_i wish to download data m from CSP, client uses the $label_m$ to get stored tag T_m, and then forward a request "download$\|T_m$" to the CM_i. Once the request is received, CM_i finds the hash value and sends $link_m$ to client. After receiving the $link_m$, the client can download the ciphertext C_m. The original code word $c(x)$ plus the errors $e(x)$ is the received code word $r(x)$, i.e.,

$$r(x) = c(x) + e(x) \tag{2}$$

A Reed-Solomon decoder finds magnitude and position of 't' errors or '$2t$' erasures and tries to correct them. Finally the CSP will returns the ciphertext C_m to the client. After retrieving the ciphertext, client c decrypts C_m with the exisitng convergent key ck_m and check for data integrity.

6 Results and Performance Evaluations

In the proposed system, we have used a Reed-Solomon erasure coding technique that divides the uploaded ciphertexts into equal size of 'n' data blocks and it will be stored in different servers. In existing system, the data deduplication technique stores only single copy of the data in CSP which leads to data loss or corruption. We found that the proposed system overcomes the data loss or corruption by storing the data blocks in different servers. In the proposed system, we have used RC4 symmetric encryption technique over the AES encryption technique. A RC4 symmetric encryption technique will take less time to perform encrypt operation and the chiper text generated by the RC4 consumes less storage space compared to existing system. We proposed a delete operation, in which the CSP can delete the unused data. This feature reduce the storage cost and reduces the file upload and download time. We found that the efficiency and performance of the proposed system is improved compared to the existing system.

The performance of the proposed system was demonstrated through a test set executed on a laptop which was equipped with an Intel(R) Core™ i5 CPU [2-core 1.6–1.8 GHZ], 8 GB RAM using 64 bit Win 10 OS and NetBeans IDE 8.0.2.

6.1 Erasure Coding

Table 1 shows the erasure coding technique, in which the encrypted data is divided into 3 data blocks with equal size. The CSP will store these data blocks in different servers to protect data loss or corruption.

The encrypted file is split into 3 data blocks with equal size as shown in the Fig. 2.

6.2 RC4 Symmetric Encryption

Table 2 shows the time consumed to encrypt data using symmetric key for different numbers of documents with different size by the proposed and the existing system. Here we consider document size in kb and MB, and encryption time in seconds, encryption time consumed by RC4 technique is less compared to the AES technique.

Table 1. Erasure coding technique to generate 3 data blocks

Original file size	Data block 1 (KB)	Data block 2 (KB)	Data block 3 (KB)
10 KB	7	7	7
25 KB	17	17	17
50 KB	33	33	33
100 KB	66	66	66
200 KB	133	133	133
500 KB	333	333	333
750 KB	499	499	499
1 MB	665	665	665
2 MB	1,330	1,330	1,330
5 MB	3,325	3,325	3,325
10 MB	6,650	6,650	6,650
20 MB	13,300	13,300	13,300
50 MB	33,249	33,249	33,249
75 MB	49,873	49,873	49,873
100 MB	66,498	66,498	66,498

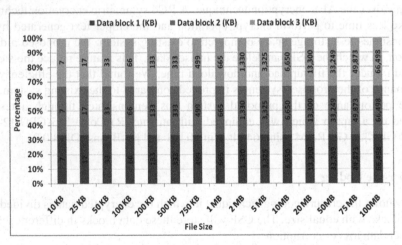

Fig. 2. Generated the data blocks using the erasure coding

In the existing system to provide data security, data is encrypted before uploading using AES algorithm. To reduce the cipher text size and speed up the encryption process we replaced AES with RC4 and achieved better efficiency. Figure 3 shows the time taken to encrypt the various file with different file size and proved RC4 performs better than AES in terms of encryption time.

Table 2. Data encryption time of RC4 and AES

File Size	AES	RC4
10 KB	0.352	0.296
25 KB	0.388	0.298
50 KB	0.393	0.313
100 KB	0.406	0.313
200 KB	0.412	0.328
500 KB	0.440	0.328
750 KB	0.510	0.344
1 MB	0.541	0.344
2 MB	0.570	0.344
5 MB	0.605	0.359
10 MB	0.638	0.422
20 MB	0.665	0.515
50 MB	0.916	0.812
75 MB	1.247	1.140
100 MB	1.284	1.312

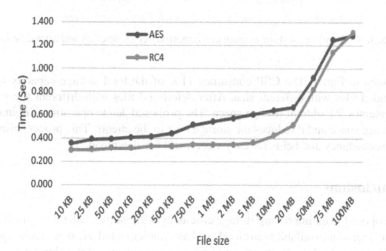

Fig. 3. Comparison of time taken for data encryption in existing and proposed system.

6.3 Delete Operation

Table 3 shows the space consumed for the uploaded chipertext in the server before performing delete operation and after performing delete operation, if the CSP stores 15 files with various file size, it consumes 527934 KB of the total 5 GB storage capacity.

Table 3. Storage space consumption before deletion and after deletion

Cloud storage space	Before deletion (KB)	After deletion (KB)
Total space	5000000	5000000
Used space	527934	333426
Available space	4472066	4666574

After performing deletion of 3 different files with different size, the server consumes 333426 KB of total 5 GB storage capacity.

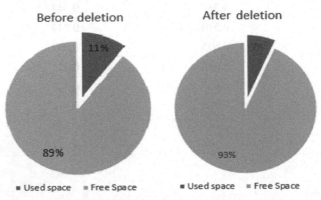

Fig. 4. Comparison of storage space consumption before deletion and after deletion

As seen in Fig. 4. The CSP consumes 11% of the total storage capacity to store 15 uploaded files with different size. After deleting 3 files with different file size, the CSP consumes 7% of total storage space. The proposed delete operation eliminates the junk storage space and it reduces the storage cost to the clients. This process eliminates storage redundancy and helps for easy search and file download.

7 Conclusions

In the proposed work improving storage efficiency with multi-cluster deduplication and achieving high-data availability in cloud, we have implemented advanced techniques that are applied to multiple applications and various requirements of the cloud data users. It supports efficient data storing and managing among the different clusters. It achieves the semantic security of the uploaded data by implementing RC4 and ensuring the stored encrypted information can be decrypted correctly by generating a constant number of cipher texts from clients. After deduplication, the uploaded data is prone to data loss or getting compromised. To resolve this we have implemented Reed-Solomon erasure coding which is not in directly readable format and achieved high data availability. The deletion operation removes the unused files and saves the storage space and cost. The performance analysis shows that the proposed system reduces file encryption time and data storage cost, improves data storage efficiency, data confidentiality and availability. In the future work we can provide a high level and better of security to the data. The proposed system performs deduplication on file level, it can be improved to block level, to improve storage efficiency.

References

1. Quick, D., Choo, K.-K.R.: Impacts of increasing volume of digital forensic data: a survey and future research challenges. Digit. Investig. **11**(4), 273–294 (2014)
2. Dutch, M.: Understanding data deduplication ratios (2009). http://www.chinabyte.com/ima gelist/2009/222/l3pm284d8r1s.pdf
3. Miao, M., Wang, J., Li, H., Chen, X.: Secure multi-server-aided data deduplication in cloud computing. Pervasive Mob. Comput. **24**, 129–137 (2015)
4. Duan, Y.: Distributed key generation for encrypted deduplication: achieving the strongest privacy. In: Proceedings of the 6th Edition of the ACM Workshop on Cloud Computing Security, CCSW 2014, Scottsdale, Arizona, USA, pp. 57–68, 7 November 2014
5. Shin, Y., Koo, D., Yun, J., Hur, J.: Decentralized server-aided encryption for secure deduplication in cloud storage. IEEE Trans. Serv. Comput. **2**(99), 1 (2017)
6. Stanek, J., Kencl, L.: Enhanced secure thresholded data deduplication scheme for cloud storage. IEEE Trans. Dependable Secure Comput. **4**(99), 1 (2016)
7. Yang, X., Lu, R., Choo, K.-K.R., Yin, F., Tang, X.: Achieving efficient and privacy-preserving cross-domain big data deduplication in cloud. IEEE Trans. Big Data **3**(99), 1 (2017)
8. Yang, X., Lu, R., Shao, J., Tang, X., Ghorbani, A.A.: Achieving efficient and privacy-preserving multi-domain big data deduplication in cloud. IEEE Trans. Serv. Comput. 1–14 (2018). https://doi.org/10.1109/TSC.2018.2881147
9. Bellare, M., Keelveedhi, S., Ristenpart, T.: Message-locked encryption and secure deduplication. In: Johansson, T., Nguyen, P.Q. (eds.) EUROCRYPT 2013. LNCS, vol. 7881, pp. 296–312. Springer, Heidelberg (2013). https://doi.org/10.1007/978-3-642-38348-9_18
10. Keelveedhi, S., Bellare, M., Ristenpart, T.: Dupless: server-aided encryption for deduplicated storage. In: Proceedings of the 22th USENIX Security Symposium, Washington, DC, USA, pp. 179–194, 14–16 August 2013
11. Yang, X., Lu, R., Shao, J., Tang, X., Ghorbani, A.A.: Achieving efficient and privacy-preserving cross-domain big data deduplication in cloud. IEEE Trans. Serv. Comput. 1–14 (2017). https://doi.org/10.1109/TSC.2018.2881147
12. Jiang, S., Jiang, T., Wang, L.: Secure and efficient cloud data deduplication with ownership management. IEEE Trans. Serv. Comput. (99), 1 (2017)
13. Li, J., Li, J., Xie, D., Cai, Z.: Secure auditing and deduplicating data in cloud. IEEE Trans. Comput. **65**(8), 2386–2396 (2016)
14. Yan, Z., Ding, W., Yu, X., Zhu, H., Deng, R.H.: Deduplication on encrypted big data in cloud. IEEE Trans. Big Data **2**(2), 138–150 (2016)
15. Boneh, D., Boyen, X., Goh, E.-J.: Hierarchical identity based encryption with constant size ciphertext. In: Cramer, R. (ed.) EUROCRYPT 2005. LNCS, vol. 3494, pp. 440–456. Springer, Heidelberg (2005). https://doi.org/10.1007/11426639_26
16. Lewko, A.: Tools for simulating features of composite order bilinear groups in the prime order setting. In: Pointcheval, D., Johansson, T. (eds.) EUROCRYPT 2012. LNCS, vol. 7237, pp. 318–335. Springer, Heidelberg (2012). https://doi.org/10.1007/978-3-642-29011-4_20
17. Benamara, O., Merazka, F.: A new distribution version of Boneh-Goh-Nissim cryptosystem: security and performance analysis. J. Discrete Math. Sci. Cryptogr. 1–15 (2020)

Handwritten Kannada Digit Recognition System Using CNN with Random Forest

G. Ramesh$^{(\boxtimes)}$, M. Tejas, Rakesh Thakur, and H. N. Champa

Department of Computer Science and Engineering, University Visvesvaraya College of Engineering, Bangalore University, K. R Circle, Bengaluru, India

Abstract. Handwritten digit recognition is a major role in the applications of pattern recognition field. Various applications of handwritten digit recognition are sorting postal mails, processing bank cheques, data entry form, etc. We proposed a widely usage of modern handwritten digit recognition, i.e. Kannada-MNIST (Modified National institute of Standards and Technology) database holds various handwritten digits from 0 to 9, also the MNIST dataset is divided into training and testing as suitable parameter. Initially two basic functions are used, namely feature extraction and feature classification for HDR (Handwritten Digit Recognition) is achieved by Convolutional Neural Network (CNN). This works with some specific algorithms based on the requirements. Deep Learning Python framework is used to implement CNN on the plain MNIST dataset gives 99.63%. Random Forest (RF) classifier is used to make the feature extracted output of CNN model to predict better accuracy, by combining with CNN. Automatic Feature extractor is one the main function of the CNN with the Random Forest as prior classifier. The main aim of this research is to make the implementation by integrating both CNN and RF for higher accuracy. This combined algorithms is in charge to gain the score upto 99.66%.

Keywords: Convolutional neural network · Digit recognition · Deep learning · Kannada digit · Random forest

1 Introduction

Handwritten Digit Recognition is active working research field in Handwriting recognition and pattern recognition Domain. In modern days, different handwritten digit recognition system has implemented for many real-time applications based on its functionality and high accuracy in recognition. However, basic enhancement of digit recognition accuracy have been achieved by different classifiers, Generally HDR is often challenging problem for prospect for new ideas that improves recognition accuracy of working scenario, process time and its complexity. This challenging problem has resolved with many research and various method, such as NN (Neural Networks), CNN (Convolutional Neural Networks), RF (Random Forest), SVM (Support Vector Machine) and KNN (K-Nearest Neighbors) have been implemented [1]. Nowadays compared to DNN

© Springer Nature Switzerland AG 2021
K. R. Venugopal et al. (Eds.): ICInPro 2021, CCIS 1483, pp. 92–104, 2021.
https://doi.org/10.1007/978-3-030-91244-4_8

(Deep Neural Network) [2] CNN works better by its ability of using very few number of hidden layers and parameters required is less as it is ultimate notable tool. Valuable results are obtained by different input dataset by various implementation Frameworks. HDR is particular for preprocessing the raw image of handwritten input datasets. Kannada MNIST is proposed as dataset as it is pre-qualified with the preprocessing and segmentation of images. Mainly two functionality are feature extraction and feature classifications are imposed with digit recognition. Kannada is state language of Karnataka, even the digits are represented in this language from 0–9. Finally the work is proposed with the Python Framework using the functionalities of CNN followed by RF for HDR. The work is done by extracting features from the Kannada MNIST using CNN then RF is imposed to the extracted feature for the better classification process. Subsequently, this work, a framework of CNN and Random Forest is proposed for manually written digit recognition. The center of the paper is to extracting the features from the given input transcribed numerals pictures of MNIST dataset utilizing CNN classifier. These highlights are at that point passed to the Random Forest classifier for the proposed written by hand digits recognition test. The handcrafted feature extraction strategy includes a trade-off between effectiveness and recognition exactness since the processing of unessential highlights may increment computation overhead driving to the poor execution of the recognition system. On the other hand, non-handcrafted features extraction prepare comprises of retrieving features specifically from the raw images. This strategy disposes of the need of collecting earlier information and design details of features.

The input image within the CNN moves directly over the nonlinear work, convolutional arrangements, pooling and totally associated layers and eventually yields us an output. In any case, the methodologies, as specified prior, express hindered computational time with a need of exactness for HDR action and does not give standard result. Our work will clear the way with respect to digitalization. Test result appear that the proposed strategy altogether moves forward the exactness rate, with accuracy upto 99.63% with CNN and 99.66% with CNN and Random Forest, comparison of our comes about with those of other considers on the Kannada MNIST transcribed digit dataset appears that our approach accomplishes the most elevated accuracy rate.

The contribution of the paper includes:

1. To incorporate a combination of techiques to create an kannada handwritten digit classifier that can be trained to retain high accuraciers on classes.
2. The main purpose of the enhancement is to get feature extraction from the handwritten Kannada MNIST input dataset using CNN. These extracted features are the inputs for the Random Forest classifier for greater accuracy.
3. The effect of the proposed work is calculated by comparing against Convolutional methods in Python Framework with ReLU activation.

The paper is organised in such a way that Sect. 2 gives the records of the of related work followed by Sect. 3 which describes the Problem Statement and Objective of our work. Section 4 describes the Proposed Methodology. Section 5 describes the architecture of CNN. The Sect. 6 explain the experimental results and discussion followed by Conclusion in the Sect. 7.

2   Related Work

Handwritten digits recognition has been persuaded by various authors, with significant success [3]. This paper presents the utilize of the classifier Optimum-Path Forest (OPF) connected in recognition of digits. Concurring to the comes about displayed, it shows up that the discovery and recognition of characters are being carried out palatably within the Manhattan separate stood out with an normal precision of 99.53%, and get preparing times and test lower than the other strategies such because It is the characteristic of OPF strategy.

The authors [4] in this paper moved forward the conventional BP neural arrange and tests with the MNIST information set on the MATLAB simulation stage. The test comes about appear that the progressed organize meet quicker and the classification is more exact. Not as it were the arrange preparing quick, and the organize acknowledgment rate than the standard BP neural arrange is additionally tall. As the organize planned in this paper is quick, it can be connected to the real-time preparing.

The technique used by [5] in this paper is propelled by the victory of the exceptionally profound state of the craftsmanship VGGNet, they proposed VGG-No for HDR. VGG-No is quick and solid, which progressed the classification execution successfully. VGG-No built by thirteen convolutional layers, two max-pooling layers, and three completely associated layers. A Cross-Validation investigation has been performed utilizing the 10-Fold Cross-Validation methodology, and 10-Fold classification correctnesses of 99.57% and 99.69% have been gotten for ADBase database and MNIST database individually. While [6] Phase Change Memory (PCM), one of the foremost develop rising non-volatile recollections, has picked up significant consideration over the a long time for utilize as electronic neural connections in organically propelled neuromorphic frameworks. The resistance float characteristics measured from tests and incorporate them into the spiking neural organize (SNN) for MNIST handwritten digits classification. In character segmentation, [10] algorithm is proposed for kannada handwriting recognition. In this author, a primary segmentation paths are obtained using structural property of characters, whereas overlapped and joined characters are separated using graph distance theory. Finally, segmentation results are validated using Support Vector Machine (SVM) classifier. Comprehensive simulation is carried out on different databases containing printed as well as handwritten texts.

Programmed recognition [7] of manually written digit string with obscure length has numerous potential genuine applications. The proposed strategy employments a unused cascade of crossover foremost component investigation organize (PCANet) and back vector machine (SVM) classifier called PCA-SVMNet. Each PCA-SVMNet classifier is prepared independently utilizing combinations of genuine and manufactured touching digits. The primary 1D-PCA-SVMNet arrange is prepared to recognize confined manually written digits

(0 . . . 9) whereas sending non-isolated digits to the following stages. Tests on the TP engineered database highlight the focal points of the proposed strategy by accomplishing more than 95% of acknowledgment rate.

Further down the lane, [8] present Deep Convolutional Neural Networks (DCNN) are as of now the transcendent method commonly utilized to memorize visual highlights from pictures. The cascade of convolutional SOM layers prepared successively to speak to different levels of features. The 2D SOM lattice is commonly utilized for either information visualization or highlight extraction. The work utilizes high dimensional outline estimate to make a modern profound arrange. The yield layer of the DCSOM network computes nearby histograms of each FII bank within the last convolutional SOM layer. A set of experiments utilizing MNIST written by hand digit database and all its variations are conducted to assess the strong representation of the proposed DCSOM organize. The Convolutional neural networks [14] have been known to have performed extremely well, on the vintage classification problem in the field of computer vision. Using the advantages of the architecture and leveraging on the preprocessing free deep learning techniques, dynamic and swift method to solve the problem of handwritten character recognition, for Kannada language. We discuss the performance of the network on two different approaches with the dataset. The obtained accuracy measured upto 93.20% and 78.73% for the two different types of datasets used in the work.

A new dataset of Arabic letters [9] composed solely by children matured 7–12 which is known as Hijja. Our dataset contains 47,434 characters composed by 591 members. In expansion, they propose an programmed handwriting recognition model based on convolutional neural networks (CNN). They prepared show on Hijja, as well as the Arabic Transcribed Character Dataset (AHCD) dataset. Comes about appear that our model's execution is promising, accomplishing correctnesses of 97% and 88% on the AHCD dataset and the Hijja dataset, individually, beating other models within the writing.

The detection of Farsi Handwritten Digits [15]. They used CNN, SVM, KNN and DNN as a classifiers which is Geometric and correlation based features. They observed that the Artificial neural network performs faster execution than the SVM but its accuracy is less and obtained the accuracy upto 99.45% by SVM and CNN.

3   Problem Statement and Objectives

The idea of Kannada handwritten digit recognition were arised by the reference of many papers of handwritten recognition of different languages. As the major porblem with the kannada digits is that the shapes are more curve compared to the english numerals for recognition. As the digits are not in the printed format, its in the handwritten varies with different handwritting styles. Many work have been implemented for recognition of each character. As many algorithm differs from various implementation but major ones follows the path from basic documentation to format of UNICODE.

The Objectives this work are:

1. To achieve better accuracy in training and validation dataset over various classes
2. To compare and choose the better classifier for the feature extractor of CNN that provides higher accuracies
3. To reduce the number of epochs required for the model to converge.
4. By average of the different outputs of each decision trees, RF decreases the possibility that can create errors in the Decision tree.

4   Proposed Methodology

The proposed work is the combination by the best qualities of CNN and RF for the classification of the Kannada Handwritten MNIST datasets. Supervised learning is the main advantage of the CNN that comprises a completely various associated layers. The work of the CNN is likely to the design its done by the humans and exceptionally well can learn different local features. Segregation is done foremost from the raw digits of images by CNN extraction. The hybrid combination is proposed because of the binary classification is done by the RF in which the softmax layer is replaced by the CNN.

The basic function of the CNN and RF is Extracting and classifying features respectively, so this combination has been proposed. Before CNN were introduced there were many specific algorithms that were utilized for feature extraction and classification separately. So after CNN came into existence it was like onset prefound learning, which doesn't require separate algorithms for classification and extraction of features separately. In Random Forest (RF) the hyperplane is used as class separator and directs to view multi-dimensional dataset with this hyperplane.

The reduction of generalization error and incomplete data is achieved by the RF. Even though RF is a great classifier, it doesn't works well for the noisy data. Also there are many challenges for the RF as it has shallow design for learning deep features. CNN has the ability to re-create the low level layers to High lever feature generation. By the convolutional layer and pooling layer of CNN, the dimension of the features are reduced at the time of propagation of features. However, implementation can be done by choosing the feature map for the input images for classification to improve accuracy.

The input of the fully connected layer is the output of the last layer of CNN. Softmax operation is proposed to achieve better classification results. In the softmax operation, x is represented as input sample of ith class, was weight vector and k as functions of distinct linear.

$$P(y = i/x) = \frac{exp(x^T w_i)}{\sum_{k=1}^{K} exp(x^T w_k)} \tag{1}$$

5   Convolutional Neural Networks Architecture

Image processing is achieved by the one of the function called image classifier that is built by best suited algorithm known as CNN. Architecture of CNN makes use of multi-layer neural network design, which needs a minimum degree of processing. As the input dataset is pre qualified with pre-processing and data cleaning there is no need of performing the same operation on CNN. CNN has ability to extract all the available features that one common from the group of input images used for learning. The classification of input images are done based on the features extracted. CNN Algorithm performs well when the size of the dataset is more. Without the input data CNN is null, computes better with the dependent of the input images. Various stage of resizing is a method to create better CNN algorithm with high computational mean. Initially very few features are extracted as the training is done only for the minimum number of data since its process faster. Later, in stages can increase the size of data accordingly. A new dataset is much required to increase the feature extraction by replacing initial convolutional layer with new layer. As this technique helps in preventing some of the problems like over fitting and so on. Maxpooling is one of the widely used algorithms for resizing the image from higher to model required size. This process helps to improve the computational speed.

5.1 Convolutional Neural Networks Approach

Most of the positional base 10 numeral frameworks within the world have begun from India, where the concept of positional numerology was to begin with created. The Indian numeral system is commonly alluded to within the West as the Hindu-Arabic numeral framework or indeed Arabic numerals, since it come to Europe through the Arabs. Hence, dataset of printed and written by hand Kannada numerals to 9 is made. Dataset of completely unconstrained manually written Kannada numerals 0 to 9 is made by collecting the manually written records from journalists having a place to distinctive callings. Scholars were inquired to compose the numerals on plain A4 estimate papers. No limitations were imposed on the utilize of ink or write. Journalists were chosen from schools, colleges and experts and the reason of collection was not unveiled to them. The collected archives were checked utilizing HP flatbed scanner at 300 dpi which abdicate low noise great quality gray scale pictures. It is guaranteed that the skew presented amid the archive scanning is insignificant and consequently disregarded. At last Kannada MNIST was created.

5.2 Convolutional Neural Networks Model

The proposed work is represented as shown in Fig. 1 which consist of 2 convolutional layer followed with maxpooling layer after each convolutional layer. The overflow of parameter can be reduced with the functionality of convolution and maxpooling layer. The kernel convolutional is achieved by using 5×5 input kernel in the convolutional layer and these combinations are repeated in the output of maxpooling layer, directed to flatten layer over dense layer. Flattening layer in used to minimize the channel count for dataflow. After Flattening layer is computed Feature Pick-up in obtained by the dense layer contribution. The output is directed to specific nodes in dense layer in which output is dropped

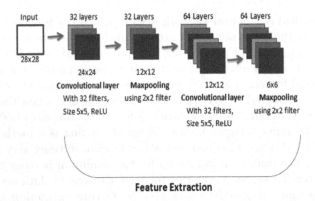

Feature Extraction

Fig. 1. Architecture diagram for the proposed model

from drop-out layer. The maximum probability of the output is achieved in the next dense layer that produces the output.

5.3   Random Forest

Extracting and classifying the features have been employed by many algorithms was a past implementation. As the feature classification in one of the application of CNN in DL (Deep Learning), hence not required to specify separately. RF is directed with the representation of multi-dimensional image dataset where each image belongs to different class in which hyperplane acts as a separator. The generalization error and invisible data is reduced by the RF classifier. Optimal hyperplane is one of the techniques in hyperplane separator. RF describes that it has shallow type of architecture which has difficulties in deep feature learning.

Gini index is often used while preforming RF for the classification on the data or formula is used on branch of decision tree as how the nodes are decided.

$$\text{Gini} = 1 - \sum_{i=1}^{c}(p_i)^2 \tag{2}$$

Gini of every branch of a node is determined by this formula which uses probability & class and determining the branch which occur most likely. Therefore pi is represented as relative frequency of classes in the dataset and number of class is represented by C.

$$\text{Entropy} = \sum_{i=1}^{c} -p_i * log_2(p_i) \tag{3}$$

Certain outcome of probability is achieved by entropy as decision is made for the method of branching of nodes. Entropy has better mathematical intensive compared to gini as the calculation of logarithmic function used in it.

6   Experimental Results and Discussion

6.1   Dataset Collection

The CNN model and Random Forest classifier is implemented in this work. Image size of 28 × 28 matrix is the input for the CNN architecture. Where the Kannada MNIST of Handwritten digit is taken and normalization is done. Filtering size of 5 × 5 and size 2 of strides is parameterized to the convolutional layer. The extracted value from the feature map layer of the input image is considered as distinguishing features. 10 epochs is computed for the CNN training to obtain the converges of training process. It is very difficult job to aim the NN for very small number of inputs. Not every functioning parameter of CC is required for the training but very few are needed. To overcome this problem mapping can be done of the suggested parameters by comparing the input and output of previous parametered values. Also aiming for the better accuracy, dataset mapping can be proposed by some required parameters. However, in this paper, better suitable python framework is implemented for Kannada Handwritten Digit recognition with the help of CNN. Sample for Kannada Handwritten digit MNIST dataset as shown in Fig. 2 is engaged to practice as it is standardized.

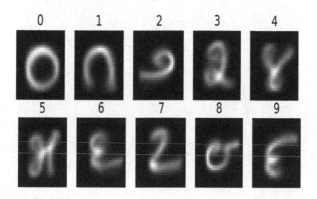

Fig. 2. Sample for the datasets

6.2   Results

The Kannada MNIST is directed to the CNN architecture for training and the accuracy rate is improved by 99.63% as 2 layer of convolutional and maxpooling layer followed by flattening and dense layer is applied. The extracted features from the CNN is given as input for the RF classifier which further improves the accuracy by 99.66% in Fig. 3.

Classification is the only task where metrics of accuracy can be applied. It gives the detail percentage of how many testing data that are predicted correctly

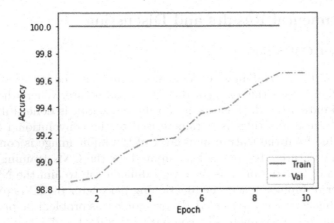

Fig. 3. CNN - random forest accuracy

based on the training. Validation and training score is calculated in the percentage defined by the every step of epochs. Even if the number of epochs increased the accuracy will be same as shown in Fig. 4.

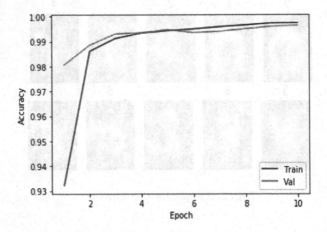

Fig. 4. CNN validation accuracy

Loss is referred as the error rate in which that cannot able to predict the actual value. Even the score of the loss is defined in the each step of epochs. In the initial epoch the loss rate will be more compared to the last epoch. Loss is also defined as sum of wrong predicted output of real values in the validation dataset as shown in Fig. 5.

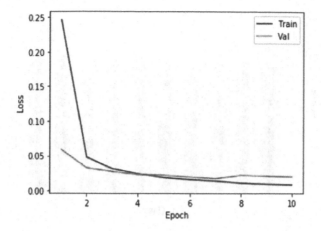

Fig. 5. CNN validation loss

6.3 Classification Report

The following are four parameters are used to check whether the prediction are correct or incorrect:

1. True Negative (TN): The case was negative and predicted negative
2. True Positive (TP): The case was positive and predicted positive
3. False Negative (FN): The case was positive but predicted negative
4. False Positive (FP): The case was negative but predicted positive

Precision is the function of classifier where it does not label the wrong value to the classes. It is calculated by the ratio of TP (True Positive) with sum of TP (True Positive) and FP(False Positive).

$$\text{Precision} = \frac{TP}{TP + FP}$$

Recall is one of the functions of classifier to find every positive instance. It is calculated by the ratio of TP (True Positive) with sum of TP (True Positive) and FN (False Negative).

$$\text{Recall} = \frac{TP}{TP + FN}$$

F1-Score is Combination of both precision & Recall i.e. percentage of catching the correct positive prediction is known as F1 –Score. The Score range from 0.0 to 1.0, where 1.0 is best score & 0.0 the least score. Mainly F1-Score is used for comparing the classification models than global accuracy as shown in Fig. 6.

$$\text{F1Score} = \frac{2 * Recall * Precision}{Recall + Precision}$$

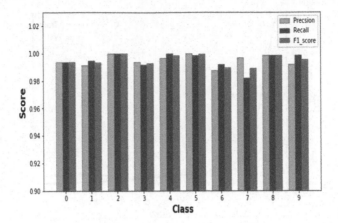

Fig. 6. Classification report

In Table 1 by comparing with the existing system the proposed methodology has many advantages. In one of the existing system it uses cloud based or online based system in which the end users may not have the access to online everywhere. So to overcome this we proposed offline system that can be access anywhere. In existing system java framework has been used. Implementing java in Deep Learning leads to difficulty in coding. So the proposed system is used with python framework with easy understanding of code. Many of the existing system use only CNN for both feature extraction and classification. Hence the proposed method uses extra classifier with CNN for further classification with better result.

Table 1. Represents the comparison between the proposed work and existing work.

Author	Method	Accuracy
Zeeshan et al. 2020 [16]	HDR (cloud-based) + DL4J	99.41%
Savita A et al. 2020 [17]	CNN + SVM	99.28%
Proposed work	**CNN + RF**	**99.66%**

7   Conclusion

In this work the combination of Convolution Neural Network and Random forest proposed for Kannada Handwritten Digit Recognition which includes include era utilizing CNN and Random forest utilized for output expectation. The combined show of CNN and Random forest employments classifiers for the recognition of manually written digit by integrating the Python framework with ReLU activation. For the expanding accuracy of Kannada MNIST digit classification, the recommended CNN design is effectively encouraged with an suitable parameters. We were able to achieve 99.63% in Fig. 4 accuracy for as it were CNN when we trained and tested dataset with 54000 28×28 gray-scale training images and 6000 test images as validation as it were with the CNN demonstrate and 99.66% in Fig. 3 of CNN and Random forest.

References

1. Tuba, E., Capor Hrosik, R., Alihodzic, A., Jovanovic, R., Tuba, M.: Support vector machine optimized by fireworks algorithm for handwritten digit recognition. In: Simian, D., Stoica, L.F. (eds.) MDIS 2019. CCIS, vol. 1126, pp. 187–199. Springer, Cham (2020). https://doi.org/10.1007/978-3-030-39237-6_13
2. Nouri, H.E.: Handwritten digit recognition by deep learning for automatic entering of academic transcripts. In: Silhavy, R., Silhavy, P., Prokopova, Z. (eds.) CoMeSySo 2020. AISC, vol. 1295, pp. 575–584. Springer, Cham (2020). https://doi.org/10.1007/978-3-030-63319-6_53
3. Lopes, G.S., da Silva, D.C., Rodrigues, A.W.O., Reboucas Filho, P.P.: Recognition of handwritten digits using the signature features and optimum-path forest classifier. In: IEEE Latin America Transactions, vol. 14, pp. 2455–2460. IEEE (2016)
4. Hou, Y., Zhao, H.: Handwritten digit recognition based on improved BP neural network. In: Jia, Y., Du, J., Zhang, W. (eds.) CISC 2017. LNEE, vol. 460, pp. 63–70. Springer, Singapore (2018). https://doi.org/10.1007/978-981-10-6499-9_7
5. Almodfer, R., Xiong, S., Mudhsh, M., Duan, P.: Very deep neural networks for Hindi/Arabic offline handwritten digit recognition. In: Liu, D., Xie, S., Li, Y., Zhao, D., El-Alfy, E.S. (eds.) Neural Information Processing. LNCS, vol. 10635, pp. 450–459. Springer, Heidelberg (2017). https://doi.org/10.1007/978-3-319-70096-0_47
6. Oh, S., Shi, Y., Liu, X., Song, J., Kuzum, D.: Drift-enhanced unsupervised learning of handwritten digits in spiking neural network with PCM synapses. IEEE Electron Device Lett. **13**, 1768–1771 (2018)
7. Aly, S., Mohamed, A.: Unknown-length handwritten numeral string recognition using cascade of PCA-SVMNet classifiers. IEEE Access **7**, 52024–52034 (2019). https://doi.org/10.1109/ACCESS.2019.2911851
8. Aly, S., Almotairi, S.: Deep convolutional self-organizing map network for robust handwritten digit recognition. Neural Comput. Appl. **8**, 107035–107045 (2020)
9. Altwaijry, N., Al-Turaiki, I.: Arabic handwriting recognition system using convolutional neural network. Neural Comput. Appl. **33**(7), 2249–2261 (2020). https://doi.org/10.1007/s00521-020-05070-8
10. Ramesh, G., Kumar, S., Champa, H.N.: Recognition of Kannada handwritten words using SVM classifier with convolutional neural network. In: 2020 IEEE Region 10 Symposium (TENSYMP), pp. 1114–1117 (2020)
11. Rabby, A.K.M.S.A., Abujar, S., Haque, S., Hossain, S.A.: Bangla handwritten digit recognition using convolutional neural network. In: Abraham, A., Dutta, P., Mandal, J.K., Bhattacharya, A., Dutta, S. (eds.) Emerging Technologies in Data Mining and Information Security. AISC, vol. 755, pp. 111–122. Springer, Singapore (2019). https://doi.org/10.1007/978-981-13-1951-8_11
12. Rao, A.S., Sandhya, S., Anusha, K., Arpitha, C.N., Meghana, S.N.: Exploring deep learning techniques for Kannada handwritten character recognition: a boon for digitization (2020)

13. Basly, H., Ouarda, W., Sayadi, F.E., Ouni, B., Alimi, A.M.: CNN-SVM learning approach based human activity recognition. In: El Moataz, A., Mammass, D., Mansouri, A., Nouboud, F. (eds.) ICISP 2020. LNCS, vol. 12119, pp. 271–281. Springer, Cham (2020). https://doi.org/10.1007/978-3-030-51935-3_29

14. Ramesh, G., Sharma, G.N., Balaji, J.M., Champa, H.N.: Offline Kannada handwritten character recognition using convolutional neural networks. In: IEEE International WIE Conference on Electrical and Computer Engineering (WIECON-ECE), pp. 1–5 (2019)

15. Nanehkaran, Y., Zhang, D., Salimi, S., Chen, J., Tian, Y., AlNabhan, N.: Analysis and comparison of machine learning classifiers and deep neural networks techniques for recognition of Farsi handwritten digits. J. Super Comput. 77(4), 3193–3222 (2021). https://doi.org/10.1007/s11227-020-03388-7

16. Shaukat, Z., Ali, S., Farooq, Q.A., Xiao, C., Sahiba, S., Ditta, A.: Cloud-based efficient scheme for handwritten digit recognition. Multimed. Tools Appl. 79(39), 29537–29549 (2020). https://doi.org/10.1007/s11042-020-09494-1

17. Ahlawat, S., Choudhary, A.: Hybrid CNN-SVM classifier for handwritten digit recognition. Procedia Comput. Sci. 167, 2554–2560 (2020)

Enterprise Systems and Emerging Technologies - A Futuristic Perspective and Recommendations for a Paradigm Shift and Sustainability

Arunkumar Narayanan[1]([✉]) and Meenakumari[2]

[1] International School of Management Excellence, Research Center,
University of Mysore, Bangalore, India
[2] International School of Management Excellence, Bangalore, India

Abstract. This study explores the enterprise systems technology solutions and various phases of its evolution that triggered through the wide adoption of product development practices and process involvement, the significance of enterprise systems and various technology evolution that enhanced the dynamics of the enterprise computing systems, with an improved performance, scalability are discussed. The emergence of hypervisor and virtualization platform solutions benefited the enterprise organizations to scale their infrastructure and application portfolio to meet the dynamic business needs without major challenges. Over the last few years web based cloud platform adoption is booming, that offers enterprise computing solutions "as a service" with subscription based pay as you utilize pricing model that brings nearly no CAPEX to avail the solution and enabled with the high-availability, scalability, elasticity, security and governance. More recently, with the proliferation in the industry ecosystem, customer demands across the board, and wide range of technology solutions adoption and frequent innovations, led us to, yet another industry revolution 4.0 - which references to various digital transformation paradigms and other pioneering solutions i.e., automation, artificial intelligence, machine learning etc., through this descriptive study paper assessed the key factors to realize AI, probable way to adopt AI, use cases, inferences, competitive advantages, current and futuristic perspectives and opportunities to address the contiguous inevitable challenges through adoption of artificial intelligence; further discussed on the emergence of AI, benefits, gaps and recommendations to achieve it, are explained in detail.

Keywords: Artificial intelligence · Information systems · Automation · Digital transformation · Technology adoption · Industry 4.0 · Product development

1 Introduction

The Technology is much broader than Information Technology - It refers to enterprise systems, servers, networks, security solutions, process, and application of tools, services

Research Guide Dr. Meenakumari

K. R. Venugopal et al. (Eds.): ICInPro 2021, CCIS 1483, pp. 105–117, 2021.
https://doi.org/10.1007/978-3-030-91244-4_9

and techniques. Through the technological innovation and advancements, many industry has evolved and outperformed in all scales, the technology evolution has not only benefitted other industries but also the information technology as a result more competitive solutions has emerged, such as cloud computing, DevOps, artificial intelligence, machine learning, blockchain, internet of things, data science, these solutions are being widely adopted by the organizations in recent times. IT or technology strategy is an approach to achieve a business and/or a common goal.

2 Technology Evolutionary Pattern

Often, witness the technology based products evolved through set of process, result of a scientific research discovery or merely an industry requirement that instrumented in inventing or enhancing to a new product with a commercial angle to it, at times it is adopted by the wide range of customers across the industry [1]. In some cases, the product deliverables are not realized on time due to its usability, easiness, various support and integration related reasons. as a result researchers and product engineers, enhances the product through various tools, development methods, and phases, that ensures the performance improvement, product usability for wide range of solutions, applications integration etc.

The improved product versions are nothing but a resultant of successful test cases, as the product gets better through different versions and generation of releases - that meets the wider range of potential customers and adopted by more users example: Office 365 products.

2.1 Product Development Process

Product life-cycle management process refers to the total duration of a product is made available to the wide range of consumers and to the global market, until it is removed from the marketplace shelves [1]. This approach is devised by the product owners, enterprise management stakeholders and the marketing professionals - as they're instrumental in defining when it is absolute to increase the advertising about the product, its features, to slash the prices, expand the horizons to newer markets and enhance or redesign according to the contemporary requirements at certain fields in the market [1].

1. A product life cycle is all about the overall life-time of a product.
2. There are four key phases, in product development life cycle – they are introduction, growth, maturity, and decline.
3. The product lifecycle approach distinctively helps to cater the business decision-making, defining pricing, promotion, expansion and cost-cutting.
4. Latest, and better products led the older product out of the market as a result of obsolete or lack of extensive features (Fig. 1).

Example. Scenario involving product lifecycle Over the last decade or more, industry and consumer expectations had raised extensively on the software products in terms of fixed features, capability to integrate and automate services, access virtualized environment seamlessly, elasticity etc., to develop Microsoft windows server and client software

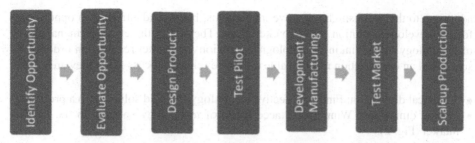

Fig. 1. The product development process. **Source:** Technology strategy for managers and entrepreneurs [1]

products with above mentioned, has become excessively complex for designers and engineering to implement major changes to the product and still ensured that the software works seamlessly.

As part of the product, Microsoft developed different versions for each windows server and client operating system environment and various other applications by integrating individual and tiny software drivers, codes written by different engineers [1, 2] - this led the software to act in unpredictable ways during and after the engineering, it becomes harder to embed new feature & functionality, in addition to that conducting daily tests of the code is not effective or nearly impossible, as a result manual tests and searches were performed for the millions of code lines to identify bugs.

To mitigate this Microsoft transformed into a modular approach software development practices for the Windows Vista release; with this different development & engineering teams were actively involved in parallel to craft various software programs according to the desired architecture design specifications, later the developed modules were integrated together according to the initial design framework [1]. Also the modular approach enabled extensively to customize the line products, components, drivers in accordance with the customer requirements, this breadth of alternatives, encourages to target new opportunity at various market segments to address rapidly changing customer needs.

Whilst, the modular approach enables the product development to be easier to do, as a result of smaller subsystems that formulates as a module; and with this approach it is simplified to develop than a complete non-modular; modularity reduces the errors, it simplifies the product development and easy to identify the required expertise which is necessary for successful product development.

Through these events, modular approach enables to lower the cost to develop a product, in addition, previously developed components that could be well utilized, and modularity allows to purchase off-the shelf components, which are often cheaper than developing new ones and the competitive battles is reduced toward nil.

Modularity doesn't work well unless the technical standards are defined as part of the technical architecture framework developed and also, must ensure the different components interfaces integrates correctly.

Technology Role in Achieving Transformation and Market Factors. Successful technological transformation involves in discovering new technology solutions and adopting, to the promisingly effective innovations, but also identifying an opportunity for the developed solution in the market-place. The key to the effective management of technology lies in linking technological solutions to market realities, in order to be successful an organization must be able to handle related process [2]... they are

- Technical dimension: Finding effective technology oriented solutions to a problem.
- Market dimension: Winning the acceptance of technology solution in the desired market (Fig. 2).

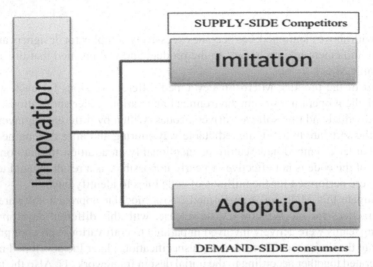

Fig. 2. Simplified process and market factors to achieve technology change. Source: Adapted from Robert Grant, Contemporary strategy Analysis, Basil Blackwell, 1995 [2].

3 Enterprise Computing System Evolution and Current Trends

Enterprise computing is the use of information systems, allied technology solutions and computers within an organization. The management of technology solutions through the datacenter is centralized to the IT functional unit of the organization.

> *"Thomas Watson, chairman of IBM, declared in 1943 "I think there is a world market for maybe five computers." He meant five IBM mainframes, the dominant form of computing in the 1960s and 1970s. His statement indicates that he thought five computers would be sufficient to serve all of the world's computer market needs. In 1980s, the personal computer revolutionized not only how businesses were able to utilize computers by making computing far more affordable but also they brought computers into individual households. Billions of PCs have been manufactured and made Watson's declaration a truly bad prediction."*

Many global organizations have shifted their focus towards an Enterprise IT portfolio as part of their infrastructure strategy during the end of 20th century, to handle plethora of technology and services requirement [3], the Datacenter infrastructure management and operations are reliable enough to manage the committed business need, however unable to scale the infrastructure to address the dynamic business needs due to various internal and external factors..., and also reliability, availability, security, compliance, governance, disaster recovery, high-availability were some of the key challenging areas for the IT management to have an alignment across the boundaries. The key tools and solution that are commonly used in enterprise computing include datacenter infrastructure management (DCIM), hypervisor and virtualization platforms, enterprise resource planning (ERP), customer relationship management (CRM), operational support systems (OSSs) etc.

In the recent times, cloud technology solutions have emerged as primary player in the IT industry, that doesn't replaced the datacenter completely however has a strong global market presence, some of the flagship services are Infrastructure-as-a-Service (IaaS), Platform-as-a-Service (PaaS), Software-as-a-Service (SaaS) etc... [3, 4].

3.1 Datacenter

A datacenter is the central and physical IT infrastructure landscape facility of the organization, managed by the Enterprise IT operations unit and the datacenter facility distinctively enables the enterprise computing possible, to meet the business needs through the repository of hardware's, network equipment and appliances, business-critical and functional servers, virtualization platform and server technologies operating system environment software's, various range of business-critical and functional applications, database systems, security servers, firewalls, - where most of the business data is stored, processed, and disseminated to users, most importantly managing the security aspects and reliability of the datacenter is highly essential in order to assure an enterprise systems with the BCP-business (operational) continuity plan and the ability to carry out business functions without interruption.

An enterprise range datacenter facility has wide range of equipment's installed and that supports the deployed infrastructure solutions to be operational 24×7 throughout the calendar years, following are the essential components that are part of a facility...

Cabinets, Racks. The data center equipment's such as networking hardware solutions - routers, switches, enterprise computing solutions - rack mounted and blade servers, KVM switches and monitor, is installed in the specially designed racks or cabinets shelves as per the design.

Environmental Controls. In order to keep the wide range of datacenter hardware's solutions up and running - the air conditioning, server cooling devices, and sensors plays a key role in monitoring the humidity, airflow, datacenter facility temperature at all times, that assures the temperature and humidity remain maintained within hardware (OEM) manufacturers specified ranges.

Power Supplies. The enterprise range datacenter facility is equipped with the primary power-supply source and subsystems that provisioned and supported by the power-grid facility (government, private) in the region, and power-supply failure or maintenance outages, larger generators are deployed to compensate.

Cabling and Cable Management Systems. The datacenter networking equipment's, servers, databases, console servers and client computer systems are interconnected, connectivity between the different datacenter facility, other business networks is enabled through fiber optic cable systems managed by the Internet service provider.

3.2 Emergence of Virtualization Platforms

The purpose of the virtualization technology is to enable a computing environment to run multiple independent systems at the same time.

"The concept of virtualization is generally believed to have its origins since the mainframe days in the late 1960s and early 1970s, where IBM invested great deal of time in developing robust time-sharing solutions – the term time-sharing refers to the shared usage of computer resources among a large group of users, aiming to increase the efficiency of both the users and the expensive computer resources they share."

The datacenter virtualization software products are offered by the software giant's Microsoft, VmWare, Oracle and Citrix - products from the OEM's provides a robust and resilient platform for virtualization server and application workloads, they played an important role in transforming the Enterprise computing from the decade old Mainframe technologies and stand-alone servers.

At the beginning of 21st century enterprise range datacenters were built and opted by the organizations widely, today the virtualization solution provides a robust and resilient platform for servers and application workloads and it delivers the abstraction out of the physical hardware and at the core of any virtualization platform is the hypervisor; The features available in the platform simplifies administering and managing VM resources, increase the availability of applications, and the performance of workloads deployed in the virtualized datacenter are assured.

Through the virtualization server (I.e., VmWare ESXi, Microsoft Hyper-V) platform, provides access to centralized management interface to manage and configure groups to deploy multiple virtual machines with various operating systems to run parallel [3], by sharing the resources of the underlying physical hardware, ability to create large aggregated pools of logical resources comprising CPU's, disks, memory, file-storage, ability to install and manage various applications, deploy and manage the database solutions, extensive networking capability etc., also distinctively offers agility, scalability, reliability, distributed resource scheduler (DRS), high availability (HA) and disaster recovery (DR) capabilities.

3.3 Emergence of Cloud Solutions, Dimensions and Benefits

Cloud is a natural extension of enterprise computing and it is one of the buzzwords keeps popping up in the information & communication industry.

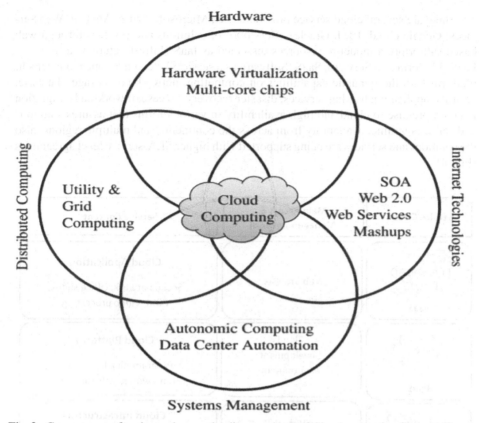

Fig. 3. Convergence of various advances leading to advent of Cloud computing **Source:** Cloud computing principles and paradigm [3]

The constant evolution of technology solutions in information technology sector, increases complexity to the organizations to effectively operate their datacenter and infrastructure systems as a result there are several impacts realized by the IT, business functions and global customers, with the recent technology innovations, and advancements, many organization's rapidly moving towards the cloud computing solution - Once again, need of transformation at the technology and strategy level become essential, whereas in mid 1980s through 2010 [3], most companies purchased own computing hardware equipment and software, today, with the extensive presence of cloud computing technology, capital investment to deploy the infrastructure is minimized to the greater extent.

The coexistence of various enterprise infrastructure platforms and computing systems, that serves for different purposes (i.e., multi-site infrastructure environment, disaster recovery datacenters, colocation datacenters etc.,) often stacked in isolation that creates mammoth of difficulties to integrate these systems, to support end-to-end business functions, consumers, and to implement, manage and effectively operate the infrastructure, governance, security, compliance and policies [4] (Fig. 3).

There are several cloud service providers (i.e., Microsoft Azure, Amazon Web Services, Google cloud, IBM, Oracle) offers pool of solutions and services through web based subscription model to resources dissected as IaaS (Infrastructure as a Service), PaaS (Platform as a Service), SaaS (Software as a services) etc., under one management platform with the dynamic capabilities on computing, networking, storage, database, security, application hosting services, disaster recovery, an ease of workloads migration from on-premise to cloud and high-availability in various forms that assures solutions and services uptime, availability from across the continents, and multiple regions; also the solutions and services are being supported with higher SLA-service level agreements (Fig. 4).

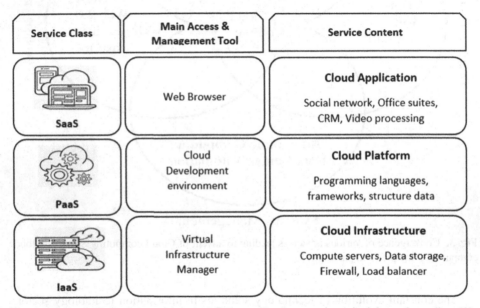

Fig. 4. The typical cloud computing stack **Source:** Cloud computing principles and paradigm [3]

The cloud technology has emerged to a state that assures increased interoperability, elasticity, reliability, usability, reduced capital expenditure and dynamic OPEX based on the utilization.

Benefits. Cloud solutions providers empowers economies of scale and the services are made available to the tenants, by equipped with latest and upgraded hardware in the cloud datacenter at all regions, competitive edge network locations, cutting edge security solutions, various compliance standards, better availability and resiliency. Some of the key benefits of cloud data centers include the following:

Efficient Use of Hardware Resources. In the public cloud architectures, multiple tenants/customer share the cloud service provider datacenter's physical infrastructure. This excludes a primary need to purchase dedicated hardware, datacenter facility, build and maintain the platform and workloads.

Rapid Deployment and Scalability. With the cloud solutions, new workloads could be easily deployed, with few configuration steps involved and has an ability to integrate various interfaces, real faster than on-premises.

Reduced Capital Expenditure - CAPEX. The opted cloud service economics works based on subscription model, for what you build and use for day to day operations, there is no upfront commitment needed to avail the services.

IT Operations Workforce. The cloud service provider owns the responsibility to manage, maintain and securing the hardware platform and also the cloud datacenter virtualization and operating system environment layer [3].

Interconnected - Global Network of Data Centers. The key big players in the cloud technology, are Microsoft, Amazon web services, google cloud, oracle and IBM, service providers had established their cutting-edge datacenter across the continent and multiple regions. This enables to optimize performance, high processing power, ease in addressing the security, compliance and regulatory requirements, closer the data resides and network edge facility, better the services will perform.

4 Bridging Gaps Through Industry 4.0

The first three industrial revolutions had deeper impacts to our lives, which is well beyond the work environment. Today, in a midst of fourth industrial revolution - an era of automation and digitization, which were started sometime during the year 2011 and 2015 [5]. Industries are fast pacing towards the digitization that includes new technologies like artificial intelligence, deep learning, neural network, data science, cloud computing, robotics, automation, edge computing, blockchain, Internet of Things (IoT), 3D printing, 5G wireless technologies and renewable power energy.

Digitization led to a change in the pace of production efficiency [6], rapidly industries utilizing the automation opportunities, introducing smart machines and automated proactive monitoring I.e., Chatbot's etc. (Fig. 5).

5 Emergence of Artificial Intelligence in the Enterprise Computing

Artificial intelligence is a tool that deals with enormous data, without data there is no AI and Machine Learning is the subset of artificial intelligence and it brings greatest potential that will create an impact, on the way the industries operate, and it could be leveraged in every product development lifecycle, business sector to realize the top-line growth and bottom line efficiencies [7]. This includes business operations, finance, marketing, employee engagement and management.

"As Google CEO Sundar Pichai vowed, "we will move from a mobile-first to an AI first world". In the year2016, Google implemented an AI and ML system to manage the cooling of its data centers and saw energy savings of up to 40%. The AI system was able to adapt and to manage the cooling of its data centers, power, and pump speeds and then predict future temperatures and pressures over time. This helped prevent over utilizing unnecessary energy by learning from past patterns"

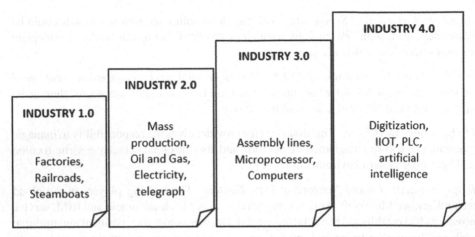

Fig. 5. Above figure illustrates the four industry revolutions. **Source:** Designing Internet of Things with Microsoft Azure [6]

As the organizations embrace AI, it is going to be critical to fragment as part of the existing IT and business strategy, to define and realize the long-term opportunities is immense. The transformation towards a production ready AI, will be basis of three developments [7].

Global Scale Infrastructure. Organizations that adopt cloud and edge computing solutions with a focus on enterprise range solutions (compute, storage, network, high-availability) with a global presence, will allow the developers, data scientists, to enable AI solutions in real-time.

Data. The rapid growth of structured and unstructured raw data, fuels in creating AI models. I.e., Infrastructure and application logs, events, IoT sensors, web search, social media, etc.

Reusable Algorithms. The development of automated functions, reusable models and algorithms for the cognitive services functions distinctively enables the access to AI solution. I.e., speech, understanding, vision, natural language processing, etc.

Interference in Adoption of AI. The key challenges in artificial intelligence is, enormously complex & continually evolving, as a result many organizations facing real-time issues in incorporating AI with their existing operating models and globally there is an emerging regulations being structured to protect data privacy, legal and responsible use of algorithms [7, 8].

Perceptions of AI – Through Research Analysis. The major asset of AI model is it brings competitive advantage over other emerging solutions, it helps to develop robust data based prediction system, Chabot's, solutions and has ability to develop new product lines, and services through AI integrated models, solutions and automation. Thus increases the efficiency through accuracy, speed, proactive monitoring, and availability [7].

Gaps and Influencing Factors. AI solution is highly complex to leverage and non-availability of AI specialized technical resources and integrating with the existing infrastructure, advocacy, change management remains challenged. There are many moving levers, due to its evolution and allied solutions, notably consequences will be likely high, when there is an event of data breaches, compromise on the integrity. With the algorithm based model, AI offers to address pressing issues in the mainstream industry, enterprise organizations functions and at social to create new paradigms, however the end results may be subjective or restricted to only what it was originally purposed or as defined. It is imperative to implement and revisit the regulatory compliance, standards and robust security mechanism at the layer of AI and its allied solutions.

6   Contribution and Recommendation

- Artificial intelligence could be effectively leveraged in wide range of industry and business units with an organization, that includes IT Enterprise, products development, operations, marketing, finance, and for society needs [7].
- The distinctive opportunity is to develop new technical capability streams, service lines, new products, enhancements, system-operated operations and services through effective automation and AI algorithms.
- Accelerating AI solutions readiness through the cloud infrastructure and edge computing ecosystem resides at the last mile of the internet, which could result on a transmission delay, thus, CDN services and gateway must be deployed for assured performance and reliability.
- Demonstrated practices in the cloud and software engineering, DevOps method continuous integration (CI), continuous delivery (CD) and staged deployments, would bring benefits such as focused deployment environments [8], increased ability to identify code errors that reduces reduced risk of deployment failure, vulnerability, ability to reproduce the errors.
- Artificial intelligence solution comprises of several modules and components of NLP, machine learning etc. which demands high compute clusters of GPU-backed VM's and long running jobs - DevOps CI/CD method, enables to deploy the code with various environment such as Dev, Test, Staging, Production etc. and this could be achieved in parallel without impacting other deployment stages [8].
- In the geo-model framework, infrastructure solution (i.e., compute, database, edge services) performance, security, privacy, monetary cost needs to be categorically defined to avoid overload [12].
- AI solution will greatly help the banking sector through automated algorithms and bot agents which could effectively identify fraudulent transactions, various type of transactions, peak trends, credit scores and automate intensive data operations and management tasks.
- For the Retail and e-commerce industry through AI, virtual and personalized shopping capabilities could be built and has an ability to offer stock statistics and purchase options [11].

- With the growing concerns in healthcare industry demands most robust and automated ways to address the patients queries, AI solutions could be a life coach and act as a personal health care assistant [9] and also has capability to provide personalized report, medicine recommendations etc., based on the health assessment of the individual and others historical data.
- The manufacturing and logistics industry deals with terabytes of structure and unstructured data, connected equipment's, demand and forecast – with the AI's deep learning capabilities sequential data could be assessed, programmed for results and automation of bot and engaging on the manufacturing unit helps to get the work done faster and with accuracy [10].
- The emergence of AI based computer vision and statistical and numerical methods helps to achieve time series forecasting and predictions analysis, also helps to improve the performance of existing and predict future values based on existing data points at the repository [12].
- Cognitive engagement through intelligent agent offers round the clock customer service, addressing a comprehensive and rising issues from password requests to answering employee questions, employee benefits, various internal policy, determining and solving problems, product and service recommendations.
- Cognitive insights provided by Artificial intelligence and machine learning solutions vary from traditional analytics solution, AI subset of tools attempts to imitate the human brain activity to recognize the patterns example recognizing speech, voice, text and images [13].
- Neural networks plays vital role to effectively handle digital clinical treatment and response, pattern recognition and image segmentation, patients classification, clinical data sets, [13] through Machine learning and compute vision – this will bridge the gaps in analysis and benefit the biotechnology organization and healthcare research labs.

7 Conclusion

As part of this descriptive study, enterprise systems evolutions and technology transformations and impacts were assessed in detail and further, more prominent research on artificial intelligence as a solution and adoption were being experimented across the industry. The future of information technology solutions (infrastructure, applications, database, security system), will be closely coupled with the Artificial intelligence capabilities and its subset of tools and utilities, as it works through algorithms, it functions differently than the way humans thinks - it augments with an ability to understand relationships between the data, process and various technology patterns. This will makes us better at what aim to do. Thus it is evident, that sooner or later, every organization will drive its own AI transformation journey, on their existing enterprise infrastructure systems and solutions to enhance the businesses and to help customers and also, the AI capabilities will broadly emerge to operationalize for the commercial application solutions, to help and solve the social, economic, and environmental problems for the benefit of society.

References

1. Shane, S.: Technology Strategy for Managers and Enterprise, pp. 9–23, 144–169. Pearson Education Inc., London (2009)
2. Narayanan, V.K.: Managing Technology and Innovation for Competitive Advantage, pp. 24–67. Pearson Education Inc., London (2006)
3. Buyya, R., Broberg, J., Goscinski, A.: Cloud computing principles and paradigms. Wiley, Hoboken (2011)
4. Ali, O., Shrestha, A., Osmanaj, V., Muhammed, S.: Cloud computing technology adoption: an evaluation of key factors in local governments, pp. 666–703. Emerald Publishing Ltd., Bingley (2021)
5. Lasi, H., Fettke, P., Kemper, H.G.: Industry 4.0. WIRTSCHAFTSINFORMATIK **56**(4), 261–264 (2014)
6. Bansal, N.: Designing Internet of Things Solutions with Microsoft Azure: A Survey of Secure and Smart Industrial Applications. Apress, Berkeley (2020)
7. Haq, R.: Enterprise Artificial Intelligence Transformation, pp. 17–25, 236–257. Wiley, Hoboken (2020)
8. Hummer, W., Muthusamy, V., Rausch, T., Dube, P., El Maghraoui, K.: ModelOps cloud-based lifecycle management for reliable and trusted AI. In: 2019 IEEE International Conference. IEEE, Prague (2019)
9. Gao, J.: Machine Learning: Applications for Data Center Optimization. Google, New York (2014)
10. Zhou, Z., Chen, X., Li, E., Zeng, L., Luo, K., Zhang, J.: Edge intelligence: paving the last mile of artificial intelligence with edge computing. Proc. IEEE (2019)
11. Yao, X., Zhou, J., Zhang, J., Boër, C.R.: From intelligent manufacturing to smart manufacturing for industry 4.0 driven by next generation artificial intelligence and further on. In: 2017 5th International Conference on Enterprise Systems (ES). IEEE, Beijing (2017)
12. Chen, H.: Business intelligence and analytics: from big data to big impact. JSTOR, Minnesota (2012)
13. Shah, P., Kendall, F., Khozin, S.: Artificial intelligence and machine learning in clinical development: a translational perspective. NPJ Digital Medicine, USA (2019)

Highly Accurate Optical Encryption for Image Security Applications Using FFT and Block Swapping

L. Anusree[1](✉) and M. Abdul Rahiman[2]

[1] LBSITW, Thiruvananthapuram, Kerala, India
[2] LBSCST, Thiruvananthapuram, Kerala, India

Abstract. Nowadays, safe data transmission is one of the challenging tasks due to intelligent hackers in the transmission medium. Especially, digital image transmission plays a major role in various applications such as biomedical engineering, Remote sensing, Robotics, and industrial automation, etc. Previously, various techniques have been proposed to perform the optical image encryption process with a high degree of security. The main issue of previous works are flexibility and computational cost. The Least Significant Bit (LSB) replacing technique is used to perform secret data hiding process in the cover image. Fast Fourier transform (FFT) and blocks the swapping technique is proposed in this work to increase the security level further with improved flexibility and reduced computational cost. The reverse process is performed on the receiver side for image decoding process. To evaluate the performance, standard performance measures such as Mean Square Error (MSE), Peak Signal to Noise Ratio (PSNR), and correlation coefficient (CC) are considered in various noise conditions. From the performance measures, it is clear that this work achieved 1.8s for compression and improved quality metrics when compared to previous works.

Keywords: Block swapping · Ciphertext · Correlation coefficient · Fast Fourier transform · Image encryption · LSB · Optical encryption

1 Introduction

In the fastest progress of information technology, particularly widespread internet use, image transmission, and storage have become more convenient. However, the network's transparency, image transmission, and access have revealed relevant security concerns. Encrypting these images is the most immediate and productive method to achieve protection. Strong encryption is a standard technique and can function as protected by encrypting all image files. With the exponential advancement of optical encryption processing, new methods and the system for information security have emerged. Optical encryption strategies have started the public's consideration since they take into account fast equal preparation of picture information just as safeguarding information in various measurements like phase, wavelength, spatial frequency, or polarization other than that, optical methods are the ideal mode for managing images or visualizations, and they

© Springer Nature Switzerland AG 2021
K. R. Venugopal et al. (Eds.): ICInPro 2021, CCIS 1483, pp. 118–128, 2021.
https://doi.org/10.1007/978-3-030-91244-4_10

profit by progressing improvements in electro-optic framework innovation. They are fundamental in computerized communication procedures like information on the board and protected innovation rights [1].

The most generally perceived system of optical encryption relies upon optical diffraction and fell phase veils. Until is the ciphertext, the input plain picture is optical contrarily Fourier changed and Fourier changed by the double-focal point 4f optical framework. At nearly the same time, the light field is adjusted by two random phase covers placed in the input plane and the Fourier plane. The double random phase encoding (DRPE) calculation can be expanded from the Fourier domain into different domains, like the fractional Fourier, Fresnel, and gyrator domains, to build the security of the standard DRPE conspire by utilizing structure boundaries as extra private keys to expand the key spaces [2].

The plaintext is bonded with random phase-only masks (RPMs) and joint power spectrum intensity distribution in the joint transform correlator JTC-based cryptosystem (JPS) input plane as the encrypted data can be registered using a standard power-law sensor, such as a charge-coupled device (CCD) [3]. The FFT widens signal portrayal prospects with the incorporation of the fractional-request, a, showing the domain into which, the sign is changed, while likewise including the frequency spatial domains as restricting models. Usually, optical encoding has a few methods for encoding two-dimensional data. The additional degrees of incentive it provides, primarily fractional instructions and scaling factors in both the x and y axes, improve framework execution, for example, assurance and efficiency to dazzle unscrambling attacks [4].

With a particular goal in a way, ghost imaging (GI), additionally recognized as connected imaging, is an entrancing optical strategy for imaging and finding objects in optically unpleasant or boisterous settings. A GI framework for the most part has two optical beams. One beam, known as the signed beam. It tends to cross the item and it was seen by the container finder with no spatial goal. The other beam, known as the reference beam, is seen by a spatially settling locator [5].

This work proposes a new technique to enhance the image quality, and robustness of optical encryption. FFT and 8×8 block swapping method is used to perform optical encryption. Finally, the performance accuracy, the correlation coefficient are performed on the natural dataset.

The rest of the paper is arranged as follows. Section 2, presents literature survey of previous studies in detail. In Sect. 3, the proposed method is clarified in detail. Results and discussions are in Sect. 4. Finally, Sect. 5 presents conclusion of this work.

2 Literature Survey

Various works have been proposed before accurate optical encryption for image security. All works are aimed to improve security and robustness. This section describes some of the earlier work proposed before optical encryption.

Sixing Xi et al. proposed θ modulation and a computer-generated hologram (CGH) are used to encrypt the multiple images. The input images are overlaid and converted into the binary value CGH after being moderated by DRPE in the Fresnel transform and modulation scheme. Decryption is realized by the use of spatial filtering of optical and a Fresnel transform unit in combination with a spatial light modulator. Multiple images may be encrypted at the same time using the suggested method, which is highly efficient [6].

Xiaogang Wang et al. developed a strong optical encryption with made on divergent illumination and two RPMs asymmetric encoding. It can be used to encrypt the input image in a diverging spherical one wave field at the input and the other at the conjugate plane. In contrast to equivalents that use planar lighting and symmetric keys, continuous movement of optical elements used for encryption is permitted, resulting in decryption keys that vary from the encryption keys (or their conjugates) and variable size display of encrypted/decrypted images [7].

Yi Qin and Yingying Zhang proposed using a customizable the exclusive-OR (XOR) operation and data container to solve security issues for the constituents, there are two forms of defense. The first stage entails converting the primary data into a binary signal (i.e., bitstream) through the ASCII norm and XOR. It is linked to the main bitstream. In the second stage, the first-level outcome is enforced into an intricate data container, which is then sent to the ghost imaging encryption scheme [8].

Lina Zhou et al. proposed that learning-based attacks are possible against optical encryption based on diffractive imaging. An attacker extract unidentified plain image from given cipher image using a machine-learning attack. An unauthorized user may use trained learning models to extract unknown plaintext from a given cipher text without the need for direct extraction or optical encryption key estimation. It can be derived from unknown plaintexts from given ciphertexts using trained learning models without the use of several optical encryption keys, thus providing a new optical encryption mechanism [9].

Shengmei Zhao et al. developed a new methodolgy of ghost imaging-based optical encryption strategy utilizing fractional Fourier transform (FrFT). In this technique, exceptional FrFT spots with arranged requests of fractional are delivered and used to enlighten an obscure objective, and a container identifier with no spatial goal is then used to recognize the sign. The objective picture just is reproduced by the approved client who has the key utilizing the three-step or four-venture phase move measure. Critically, the two FFT boundaries are utilized as the key, that altogether less the number of key pieces traded between enlisted clients when contrasted with current optical encryption frameworks depending on correlated ghost imaging (CGI). This plan can be reached out for more approved clients in a transmission style, where each approved client can get to the scrambled image [10].

From the above discussion, it is very clear that previously several works have been proposed to perform to improve the robustness. The main disadvantage of the previous works is the poor quality of the image and reduced accuracy. The following is the primary goal of this work:

1. To enhance the robustness of optical encryption.
2. To perform the encryption with less computational cost without compromising the quality of output.
3. To enhance the accuracy of the optical encryption system.

The details of the implementation of the proposed work are given below sections.

3 Highly Accurate Optical Encryption for Image Security Applications Using FFT and Block Swapping

In the highly accurate optical encryption for image security applications using FFT and block swapping, the plain image converts into a bit plane slice. At that time secret text converts to image conversion using the LSB replace technique. Apply FFT and convert it into 8x8 blocks. Then swapping the blocks using the key. Finally, perform inverse FFT and get the encrypted image.

3.1 Block Diagram

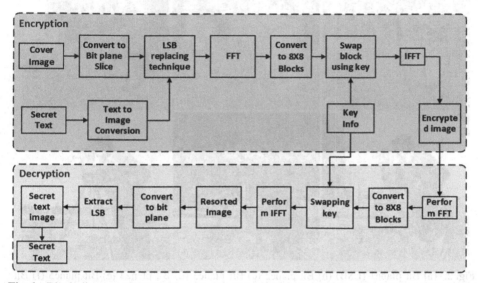

Fig. 1. Block diagram of the highly accurate optical encryption for image security applications using FFT and block swapping

For the decrypting method first, perform the FFT in the encrypted image and it can be converted into 8 × 8 blocks swapping using the secret key information. Then get the resorted image using inverse FFT. The resorted image can be converting into a bit plane. Extract the LSB to get the secret text image with secret text. Figure 1 shows the block diagram of the highly accurate optical encryption for image security applications using FFT and block swapping.

Bitplane Slicing. The image is first transformed to bit plane slice images depending on the amount of bits used to display it. The digital image is divided into sub-images depending on the number of bits used in this phase. Figure 2 shows the transformation of the input image from LSB first bit plane to Most Significant Bit (MSB).

LSB Replacing for Data Hiding. The single bit of the text is replaced with the LSB bit of the input image. To get the original image again the binary image is converted into

Fig. 2. (a) Bit plane1 (LSB) (b) Bit plane2 (c) Bit plane3 (d) Bit plane4 (e) Bit plane5 (f) Bit plane6 (g) Bit plane6 (h) Bit plane7 (i) Bit plane8 (MSB)

a decimal image. This is a simple and efficient technique for a fast data hiding process. Fourier transform is further used to perform the image encryption process. Figure 3 shows the LSB replacement scheme.

3.2 Fast Fourier Domain

To calculate an N-point FFT, log2N levels are required. The FFT technique is used to merge two-phase masks into a wide-scale. Phase mask to improve security.In this proposed method, the N-point FFT can be computed more efficient way. Initially, the input Random-access memory (RAM) contains the input, which can be extracted using the address generator. An address generator is connected to the multiplexer where select line 0 is sent directly to the input RAM. This pipelined system has a butterfly block that acts as a butterfly at each stage. The order, subtraction, and processing components for

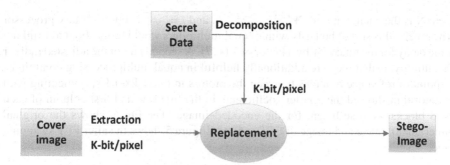

Fig. 3. LSB replacement scheme

multiplication twiddle factors are key components of the butterfly structure [11]. The technological element is an important component of this butterfly machine as it occupies a major proportion of the whole area. The fixed-point 16-bit binary demonstration is used to signify the real and imaginary portions of the model. It is considered as the inputs for the first stage of the FFT algorithm are valid, so the imaginary region is zero. The TFT equation is isolated into two shortened forms, one of which covers the sum over the initial N/2 information points, and the subsequent total contains the keep going N/2 information points.

$$Y[k, l] = \sum_{m=0}^{M-1} \sum_{n=0}^{N-1} i[m, n] W_N^{mnx} \tag{1}$$

Where $W\_N = e\hat{\ }(-j(2\pi/N))$. A 32-point discrete Fourier transform (DFT) should require more than 4N billions of complex qualities to be put away. For signals, $x[n]$ is a parallel grouping that decreases the signs dimensionality, all things being equal, it would not be plausible to store all 2N DFT yield prospects.

$$Y(k, l) = \sum_{m=0}^{M-1} \sum_{n=0}^{N-1} i[m, n] W_N^{mnx}, x = 0, \ldots, N-1 \tag{2}$$

From $y[k, l]$ whose input sequences, $y_m[k, l]$ is extracted by

$$Y(k, l) = \sum_{m=0}^{\frac{M}{2}-1} \sum_{n=0}^{\frac{N}{2}-1} i[m, n] W_N^{mnx/2} + \sum_{m=0}^{M/2-1} \sum_{n=0}^{N/2-1} i[m + \frac{N}{2}, n + \frac{N}{2}] W_N^{mnx} \tag{3}$$

Since $W_N^{MNk/2} = (-1)^k$. This method can be recursively applied until the information is applied to changes of just two focuses. The radix-4 calculation improves $x[k]$ in four groupings, the repeated calculations being finished by 4-point DFTs.

$$Y(k, l) = \sum_{m=0}^{M-1} \sum_{n=0}^{N-1} i[m, n] + (-1)^x \sum_{m=0}^{M/2-1} \sum_{n=0}^{N/2-1} i[m + \frac{N}{2}, n + \frac{N}{2}] W_N^{mnx} \tag{4}$$

Where N is the changing size, $W_N^{mnx} = e^{-j2\pi/N}$, and $j = \sqrt{-1}$. For a solitary processor environment, this sort of butterfly with r equal multipliers would bring about a shrinking in time delay for the total FFT by a factor of $Y(k, l)$. A second part of the adjusted radix–r FFT butterfly, is that they are additionally helpful in equal multiprocessing conditions. The point of reference relations between the motors in the radix–r Experimenting with this algorithm showed the Fourier coefficients in the first row and first column of each 8×8 blocks were sufficient for the encoded image. The Fig. 4 shows the original (Cameraman) image and image of FFT applied. Figure 5 shows the after FFT swapping image.

Fig. 4. (a) Original image (Cameraman) and (b) FFT applied image

Fig. 5. After FFT swapping

4   Results and Discussion

This work is done by MATLAB R2020b using a computer with CPU Intel(R) Core(TM)i5-3320M CPU @ 2.60 GHz, .60 GHz, and 2 GB of RAM. The dataset images with size 256×256 pixel as input sample used as the cameraman. One image is a Cameraman greyscale image. Figure 6 shows the sample of the process of optical encryption.

Fig. 6. Original image (b) encrypted image and (c) decrypted image

4.1 Performance Analysis

The following performance metrics are calculated in this work.

Correlation Coefficient. The correlation coefficient (CC) is a numerical measure of some form of correlation, which refers to a statistical relationship between two variables. The variables may be two columns from a given series of observations, known as a sample, or two components of a quantitative probability distribution with a known distribution. The CC can be expressed as

$$CC(F, f) = \frac{E\{[F - E(F)][f - E(f)]\}}{\sqrt{E\{[F - E(F)]^2\}E\{[f - E(f)]^2\}}} \tag{5}$$

Here F and f represent the plain image and decrypted image.

Mean Square Errors. The MSE is a metric of the consistency of an estimator; it is often non-negative, with values closer to zero being greater. MSE is the difference between the original and decrypted images [12]. Mathematically it can be expressed as

$$MSE = \frac{1}{Px * Px} \sum_{i=1}^{Px} \sum_{j=1}^{Px} |\hat{I}(i, j) - I(i, j)|^2 \tag{6}$$

where $Px * Px$ denotes the number of image pixels, $\hat{I}(i, j)$, $I(i, j)$ signify original image values, decrypted image values, and at that pixel value (i, j).

Peak Signal to Noise Ratio. The PSNR is a technical term that refers to the ratio of the signal's maximum potential power to the power of fully corrupted noise, which influences the precision of its representation.

$$PSNR = 20.\log_{10} MAX_{PX} - 10.\log_{10} MSE \tag{7}$$

Where MAX_{PX} represents a maximum image pixel value.

From Table 1 shows the better comparative performances are CC, PSNR, and MSE on previous works. This work has higher CC, PSNR, and lower MSE with compared to other previous methods. GI-XOR method returns 0.42 CC. Diffractive imaging (DI) returns only a PSNR value like 22.18. Fresnel diffraction (FD) have MSE 0.078. JTC method has 0.89 CC only. The asymmetric encryption scheme with biometric keys gives 0.97 CC which is 0.1CC lower than this proposed method and 0.0045 MSE which is 0.0013 higher than the proposed method. Also, this works provides less computational cost as 1.8 s which is less when compared to previous techniques as shown in Table 1. Figure 7 displays the comparative performance CC with previous methods and Fig. 8 displays the comparative performance of MSE with previous works.

Table 1. Comparative performance of previous works

Method	CC	PSNR	MSE	Time (s)
GI-XOR [8]	0.42	–	–	5
DI [9]	–	22.18	–	3
FD [13]	–	–	0.078	6
AES-BK [14]	0.97	–	0.0045	2
JTC [15]	0.89	–	–	3.5
This work	0.98	25.36	0.0032	1.8

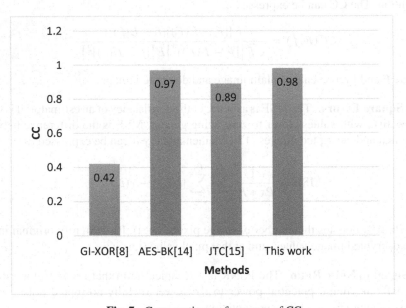

Fig. 7. Comparative performance of CC

5   Conclusion

A highly secure optical encryption technique is proposed in this work for highly secure image transmission applications. Also, text messages are hidden inside the cover image.

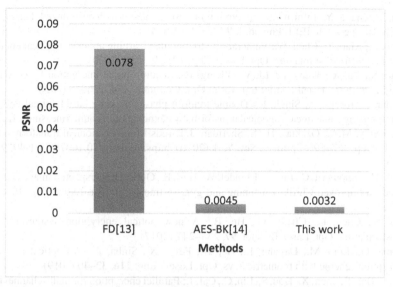

Fig. 8. Comparative performance of MSE

Initially, the text message is converted into a small binary image for the hiding process. LSB replacing technique is used to perform the data hiding process. FFT is used to perform frequency-domain transformation for the block swapping process. To increase the security level 8x8 patch is used. To convert the image into a spatial domain IFFT is used. On the receiver side, the reverse process is performed to the transmission side. Various performance measures are used to evaluate the performance of the proposed technique in different noise conditions. This work provides best performance as an average of 1.8 s for compression process and 0.98, 25.36, 0.0032 of SSIM, PSNR, RMSE respectively.

References

1. Sanpei, T., et al.: Optical encryption for large-sized images. Opt. Commun. **361**, 138–142 (2016)
2. Jiao, S., Gao, Y., Lei, T., Yuan, X.: Known-plaintext attack to optical encryption systems with space and polarization encoding. Opt. Express **28**(6), 8085–8097 (2020)
3. Xiong, Y., Du, J., Quan, C.: Optical encryption and authentication scheme based on phase-shifting interferometry in a joint transform correlator. Opt. Laser Technol. **126**, 106108 (2020)
4. Liu, S., Sheridan, J.T.: Optical encryption by combining image scrambling techniques in fractional Fourier domains. Opt. Commun. **287**, 73–80 (2013)
5. Zhao, S., Wang, L., Liang, W., Cheng, W., Gong, L.: High performance optical encryption based on computational ghost imaging with QR code and compressive sensing technique. Opt. Commun. **353**, 90–95 (2015)
6. Xi, S., et al.: Optical encryption method of multiple-image based on θ modulation and computer-generated hologram. Opt. Commun. **445**, 19–23 (2019)
7. Wang, X., Zhou, G., Dai, C., Chen, J.: Optical image encryption with divergent illumination and asymmetric keys. IEEE Photon. J. **9**(2), 1–8 (2017)

8. Qin, Y., Zhang, Y.: Information encryption in ghost imaging with customized data container and XOR operation. IEEE Photon. J. **9**(2), 1–8 (2017)
9. Zhou, L., Xiao, Y., Chen, W.: Vulnerability to machine learning attacks of optical encryption based on diffractive imaging. Opt. Lasers Eng. **125,** 105858 (2020)
10. Zhao, S., Yu, X.: Wang, L., Li, W., Zheng, B.: Secure optical encryption based on ghost imaging with fractional Fourier transform. Opt. Commun. **474**, 126086 (2020)
11. Kumar, P., Joseph, J., Singh, K.: Double random phase encoding based optical encryption systems using some linear canonical transforms: weaknesses and countermeasures. In: Healy, J.J., Kutay, M.A., Ozaktas, H.M., Sheridan, J.T. (eds.) linear canonical transforms. SSOS, vol. 198, pp. 367–396. Springer, New York (2016). https://doi.org/10.1007/978-1-4939-3028-9_13
12. Chen, H., Tanougast, C., Liu, Z., Blondel, W., Hao, B.: Optical hyperspectral image encryption based on improved Chirikov mapping and gyrator transform. Opt. Lasers Eng. **107**, 62–70 (2018)
13. Yao, S., Chen, L., Chang, G., He, B.: A new optical encryption system for image transformation. Opt. Laser Technol. **97**, 234–241 (2017)
14. Verma, G., Liao, M., Dajiang, L., He, W., Peng, X., Sinha, A.: An optical asymmetric encryption scheme with biometric keys. Opt. Lasers Eng. **116**, 32–40 (2019)
15. Liu, J., Bai, T., Shen, X., Dou, S., Lin, C., Cai, J.: Parallel encryption for multi-channel images based on an optical joint transform correlator. Opt. Commun. **396**, 174–184 (2017)

On Improving Quality of Experience of 4G Mobile Networks – A Slack Based Approach

Gracia S.[1](✉), P. Beaulah Soundarabai[1], and Pethuru Raj[2]

[1] Christ University (Deemed), Bangalore, Karnataka, India
s.gracia@res.christuniversity.in
[2] Reliance Jio Platforms Ltd., Bangalore, Karnataka, India

Abstract. This paper analyses India's four top 4G Mobile network Providers with respect to five key user experience metrics – Video, Games, Voice app, Download speed and Upload speed. Results using Data Envelopment Analysis show Airtel and Vodafone-Idea performing with maximum relative efficiency with respect to these metrics, while BSNL and Jio closely follow them. Further analysis using the Slack Based Measure shows where and by how much BSNL and Jio need to improve to perform at par with Airtel and Vodafone-Idea. On certain variables, for instance Voice app, BSNL and Jio perform well, with no need for improvement. On the contrary, for Upload and Download speed experiences, both BSNL and Jio lag. For Video and Games, there is still scope for improvement, although both these players are reasonable in their performance. Thus, this analysis provides an accurate and optimal benchmark for each variable whose user experience has been evaluated.

Keywords: 4G mobile networks · Data Envelopment Analysis · Slack Based Measure

1 Introduction

The 4G mobile network provides higher bandwidths and supports Multimedia services, with about 100 Mbps of bandwidth achievable. Based entirely on packet switched networks, the 4G network differs from 3G networks which use both circuit and packet switching. On the network security front, 4G has an enhanced security mechanism compared to that of 3G networks. 4G networks are also better scalable with their architecture built in such a way and it achieves global mobility as well [1].

With the onset of Big Data and Cloud-based services, there arose new requirements for Network performance. It has become important for service providers to provide a minimum level of performance, with additional challenges arising from the evolution of new network technologies towards voice and data services. Operators therefore need to have a thorough understanding to manage quality in terms of technical operations (QoS-Quality of Service) and the Quality of Experience at the user level (user perceived QoS or the QoE) [2].

© Springer Nature Switzerland AG 2021
K. R. Venugopal et al. (Eds.): ICInPro 2021, CCIS 1483, pp. 129–141, 2021.
https://doi.org/10.1007/978-3-030-91244-4_11

2 Literature Review

2.1 Video Experience

The Quality of Service (QoS) from the Service Provider's network plays a vital role in the Quality of Experience (QoE) of the end user. It is no exception in the case of Video streaming and video conferencing. There are many real-time video creating apps like YouTube that require a very high bandwidth to work well. This adds a push to the Service Provider that must see to it that the network provides the necessary quality. Different types of streaming (Person to Person- share video in real time) or (Content to Person) need different service provisioning capabilities and quality. For real-time video sharing, a low latency and a high bandwidth for both uplink and downlink is essential, when compared to Content to Person streaming. The QoE expectation for video streaming is high with the need of low jitter and extremely low packet loss. Also, video conferencing needs a high service quality and video, and audio streaming requires median service quality [2]. For multimedia streaming, the QoE is measured by the streaming bit rate, playback smoothness, Peak to signal noise ratio, among other satisfaction factors [3]. To provide better QoS to the users, many new devices and streaming codec have been introduced. However, network traffic is still unpredictable and still causes issues [4]. Since network traffic passes through routers, gateways, and firewalls, it goes through packet loss, delay, jitter and packet reordering many a time. The traffic from routers sometimes waits in queue or network path for a long time that is unsteady due to load balancing [5, 6] and all this causes network disturbances. This has an impact on the streaming of videos and hence the video quality at the end user may not be as expected. Sometimes, the broadcaster sends the data through the network which passes the data packets in a particular sequence, but they arrive at the destination out of sequence. This is called packet reordering and leads to missing data packets. This packet reordering has a huge impact on quality of the video and the perceived QoS is not attained at the end user level [7, 8]. When the user does not experience results as on the Service Level agreements (SLA), they tend to leave the network thus leading to a drop in business revenue to the service provider [9]. QoE as a parameter of network satisfaction is better than network QoS in the sense that feedback received helps improve the network parameters, for a better User QoS (QoE) which is the user perception about the service [10].

2.2 Games Experience

Gaming has become popular in today's context with people playing online games through varied platforms like PC, mobile applications, online servers and cloud services. A QoE assessment was made on user experience of gaming through mobile applications and online. The research measured the satisfaction level of playing through these platforms, also avoiding the usage of game CDS or kiosks. Results showed that application based gaming and big screen displays had better visual quality and was more sought after by players than online platform where the graphics quality was low. Another QoE assessment experiment that was carried out involved the network speed measurement using speedtest.net and upload speed was recorded to be 3.78 MB for PC online gaming over a 4G network, and it was found that at instances where compression was done to save the delay/buffer time, the quality of gaming reduced and so did the QoE of the user [11]. Multiplayer games require very low latencies and there is a need to develop new standards in data transmission and routing to enjoy least latency experience [12].

2.3 Voice App Experience

To carry voice over the internet, the channel used should have a low bandwidth. Low latency packets must be transmitted, so that it gives out low jitter. While determining the channel bandwidth, it must be emphasized that packet transmission is done at the right time period. Voice over the internet would require that the network has a low latency and jitter combination and low bandwidth of 21–320 kbps per call and one way latency of less than 150 ms and a one-way jitter less than 30 ms, to satisfy QoE of the end user by providing optimal network quality as mentioned [2]. QoE of voice applications depends on packet loss rate, delay and encoding rate [13]. There are many approaches that have been proposed to improve user perceived speech quality and bandwidth utilization. A few of these are mentioned here.

Adaptive rate control is a technique that provides optimized voice quality and manages congestion problems by matching the transmission rate to the network capacity. New hybrid algorithms that adjust the rate of transmission based on network parameters and perceptual quality show better quality of speech and improve bandwidth utilization. Rate control mechanisms that are based on only network impairments may fail to provide optimum QoS in terms of user-perceived quality. Another study shows the dynamic adjustment of encoding bit rate with feedback about network congestion through RTCP (Real-time Transport Control Protocol) packets. Other works include rate adaptation using Adaptive Multi-rate Codec and voice packets being marked on priority basis. All these are aimed at minimizing packet loss and delay. One of the concerns on Voice over IP (Internet Protocol) is that it does not use TCP (Transport Control Protocol). Hence the huge volume of VOIP flows that have increased over the past few years may cause network instability and congestion collapse. VOIP uses UDP (User Datagram Protocol) and does not provide any congestion control. It is understood that for network stability, any new protocols and services should be TCP-friendly, consuming comparable bandwidth as with TCP, under the same network conditions. Thus, any adaptive networked multimedia applications like voice and video must also be TCP-friendly [14].

The network's perceived quality is not measured by taking it at any one moment in time, rather over a period, according to a prior work. Hence, regression analysis was done on a voice service for a period of 120 days, and it was found that the user perceived QoS became slightly lower as the user got used to utilizing it more often with more familiarity [15].

2.4 Download Speed Experience

Download speed is defined as the rate at which data is received from the internet. It is a good indicator of browsing experience. As part of a study, this was measured by downloading a large file over a minute to deal with temporary speed changes [16]. File transfer and web browsing are elastic services that are considered to have a utility function that is increasing, concave and continuously differentiable function of throughput.

Previous work based on this assumption shows that this kind of data transfer utility function is logarithmic with respect to rate R [17]. The MOS (Mean Opinion Score) that was proposed in another work [18] was used in this context to measure the quality of user experience [19].

$$MOS = a \cdot \log 10(b \cdot R)$$

Here the maximum and minimum user perceived quality determined the parameters a and b. If the subscriber had signed up for a particular rate service and got the same rate R with no packet loss, then the MOS was set to a maximum, say 4.5. A minimum transmission rate (say, 10 kbps) was also defined and MOS set to 1. MOS as a function of data transmission rate was plotted by varying the data transmission rates [16]. Many studies use a MOS to measure the user perceived QoS or the QoE.

The key to the success of video applications like Ultra HD and high frame rate technology that require higher bandwidth lies in high download rates. ISPs therefore have to continuously monitor the quality of the network at the user level, so that they know the quality a customer using a particular bandwidth will experience. In this respect, a study was done to measure the impact of upload and download speeds on the QoE parameter. The test was done on different upload and download speeds. Multiple speed tests were done to alleviate any influence network connectivity issues may have on the results. The maximum upload and download speeds were measured. There were test cases that were done on single video browsing, multiple video browsing and more interactive behavior-based browsing. Each case was coupled with a unique combination of upload and download speeds. Different KPIs were studied – Page Load time, Initial loading delay, Total stalling time. All these as known affected the QoE measured. Moreover, ISPs are also evaluated publicly which results in competition with other providers on network quality [20].

2.5 Upload Speed Experience

Upload speed is the rate at which data can be sent to the internet. This helps in studying the responsiveness of real-time applications [16]. Upload of large files to servers over the internet through mobile communications has increased a lot. This has its own limitations – bottlenecks in server processing ability or insufficient available bandwidth. There are many measures taken to better the user perceived quality. For example, there was a study that was done by using network resources at the edges flexibly and uploading parallelly, divided files. This improved the upload throughput 10 times more [21].

A proposal was made for an intelligent media distribution architecture of IP Multimedia Subsystem (IMS) for video streaming purposes which involved uploading a multimedia file to a server in the IMS which could later be downloaded as well. This study took into consideration bandwidth, jitter, delay and packet loss that impact the QoE of the user. In addition, it also considered CPU, RAM temperature and the number of users connected to the network, since these also affect the QoE of the end users from an energy efficiency standpoint. Per the article, it was successful in terms of ensuring the upload speed of the multimedia file and guaranteed QoE of the end user. One of the areas of focus was to optimize the upload time of multimedia users. For this, an upload client was defined as an active device that uploaded the videos in to the proposed IMS. It connected the system via SIP (Session Initiation Protocol) and opened the Application Programming Interface (API) and allowed to record videos and to give permissions for sharing them. A manager controlled the upload of the videos and also selected the server to which the uploaded video had to be sent to be downloaded at the other end. For the entire process, the system used optimal transcoding that includes parameters like codec, bitrate and resolution. Evaluating the described multimedia system, with these parameters and also determining the MOS (Mean Opinion Score) showed the impact of the system on the QoE of end users [22].

Apart from prior work, it is to be noted that the download usage of a user is more than the upload usage. In fact, for a median European user, this download-upload share is 88%–12% [23]. Other features of the LTE (Long Term Evolution) technology to be understood is that it uses Single-Carrier Frequency Division Multiple Access (SC-FDMA) to increase coverage and reduce user equipment cost and for energy efficiency on the upload communication side. In this technology, uplink data transmission rates up to 75 Mbps can be achieved. LTE technology thus provides a very high rate of data transmission [24].

3 Methodology

3.1 Data Source

The latest publicly available data for India's 4G Mobile network providers is published by Open Signal at https://www.opensignal.com/reports/2020/09/india/mobile-network-experience [25]. The public reports and insights are made available to Network providers to understand and improve their service and is also available for use by regulators and analysts.

3.2 Time Period

The latest available time period as of now is September 2020, including data over 90 days starting from May 1, 2020.

3.3 Sample Representativeness

The sample size is large with over 100 million devices used for collecting data, running into daily measurements in billions globally. No restrictions were made in terms of apps, type of user or age of device. Measurements were made on smart phones both indoors and outdoors. The end-to-end consumer network experience from the user device to the Content Delivery Networks was measured.

3.4 Technique

Data Envelopment Analysis (DEA). DEA could be seen as a multiple-criteria evaluation methodology, that minimizes inputs and maximizes output. The DEA score gives the overall performance of an organization (Decision Making Units or DMUs). This 'overall performance' is a composite measure out of aggregating individual indicators [42].

Various studies have been made on determining the efficiency of services provided by the Telecommunications and other sectors as follows: Evaluated the efficiency of the wireless communications sector in around 42 countries [26]; proposed a benchmark to the wireless communications sector in the USA [27]; evaluated the performance of the telecommunications companies in Korea using the Data Envelopment Analysis [28]; assessed the performance and rank the telecommunications industry based on the International Telecommunications Union standard [29]; performed an evaluation of mobile subscribers using the DEA technique and the Principal Components Analysis (PCA) techniques. Data sample was taken out of 27 countries and nine indicators were used to show the significance of the approach. It identified strong and weak points of the telecommunication companies and gave the most efficient output and input of each sector [30]; compared the efficiencies of 30 OECD (Organization of Economic Co-operation and Development) members using the Analytical Hierarchy process (AHP) and DEA tools. This study showed that 8 countries proved to be efficient in terms of productivity and revenue [31]; evaluated the efficiency of Telecommunication companies using DEA for mobile operators in Tanzania. Two approaches - a parametric approach using an econometric model and a non-parametric approach using a mathematical model were used in evaluating the efficiency. The DEA technique was used for measuring the efficiency and the best Decision-Making Unit (DMU) was used for comparison with the others. 7 DMUs were used, with a Constant Return to Scale (CRS), non-radial, oriented model on a Slack Based Measure to determine the efficiency. For ranking the DMUs by their efficiency, the Super SBM- oriented model was used to produce a score [32]. A recent study was done on evaluating the efficiency of public sector banks in India during the financial year 2018–2019 using the Data Envelopment Analysis. The mix efficiency, CCR efficiency/SBM efficiency was calculated to reduce the error rate [32]. Another study reviewed the application of different DEA models on airline efficiency. It concluded that in radial models, the standard CCR, BCC models, a combination of standard and other approaches, and extended and modified models were used for airline studies. For non-radial models, the airline studies were made using Slack Based Model (SBM), RAM and EBM models were used [33]. A robot selection problem for a Taiwanese manufacturing company is used for illustrating the use of radial DEA and the SBM based models. The results showed that the efficiency from the radial DEA model overstated the robot's performance while the SBM based model was more reliable in ranking the efficiency and for the robot selection [34]. A new risk analysis model for Failure Mode and effect analysis (FMEA) was proposed based on DEA, for ranking the failure modes per risk priority [35]. DEA was used to evaluate the efficiency of 109 sampled small towns in the Jiangsu province of China. Different types of towns showed differences in efficiency characteristics [36]. A novel, modified SBM based model and least absolute shrinkage and selection operator was used to evaluate the effect of entropy-based variable on the efficiency score for hotel performance [37].

Slack Based Measure (SBM). For an optimal solution that is better than the efficiency scores produced by the DEA technique, a slack based measure (SBM) in DEA was proposed which directly deals with output shortfalls and input excesses of the Decision-making unit. The CCR (Charnes-Cooper-Rhodes) model that was used in earlier studies provided ratio maximization, while this proposed Slack Based Measure could be interpreted as profit maximization. An efficiency measurement tool, this measure is also compatible with other measures of efficiency. This technique considers the ratio efficiency and the slacks. The SBM can be understood as the product of input and output efficiencies [38]. Although the results of this analysis are discussed later, target setting using the SBM approach for a single output measure is shown in Fig. 1.

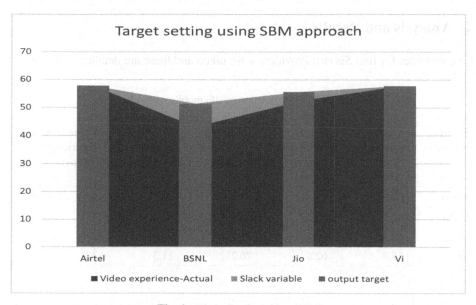

Fig. 1. Target setting using SBM

Advantages of the SBM Based Approach [39]

- It divides the set of observations into two – efficient and inefficient
- SBM model evaluates input excesses and output shortfalls (slacks) and identifies all inefficiencies in the concerned Decision-Making Units (DMUs). A DMU is SBM efficient if the slack is zero.
- Has features like indication of efficiency, monotonicity and unit invariance

Limitations of the SBM Model [40, 41]

- While evaluating efficiency change over time, the non-zero slacks tend to vary during different time periods. This causes the problem of finding which pattern is reasonable

- When these slacks are used as sources of inefficiency, and further statistical analyses are made over these, distortion of results is possible.
- Lack of discrimination in the efficient set of observations

3.5 Software Used

DEA Frontier Solver [42] was used to carry out the analysis and included both DEA efficiency output and the output targets for each metric.

4 Analysis and Results

Five variables for four Service Providers were taken and these are detailed in Table 1.

Table 1. User experience on 4G mobile networks

Service provider	Video experience (0–100 points)	Games experience (0–100 points)	Voice app experience (0–100 points)	Download speed experience (Mbps)	Upload speed experience (Mbps)
Airtel	57.9	56.3	75.6	10.5	2.9
BSNL	43.1	39.1	67.9	4.3	1.7
Jio	52.3	50.1	73.4	6.9	2.3
Vi	57.8	62.8	76.2	11.3	4.0

From the data, we see that Airtel is performing best for Video experience, with Vodafone-Idea very close. For all the other metrics, we see that Vodafone-Idea (Vi) is performing best. BSNL appears to lag the other service providers. The below question though remains – is Vodafone-Idea the undisputed leader; is BSNL performing way below the others?

To answer these questions, we perform the Data Envelopment Analysis that will provide relative efficiencies of the four Service providers. Table 2 shows the results of the analysis.

Table 2. Data envelopment analysis – overall efficiency

Service provider	Efficient input target output
Airtel	100.00%
BSNL	89.11%
Jio	96.33%
Vi	100.00%

The results show why such a mathematical technique is needed, as only by looking at Table 1, we would have interpreted results differently. From Table 1, we would have expected to see Vodafone-Idea to be the undisputed leader. The results from the analytical technique, though, show that both Airtel and Vodafone-Idea fall on the Efficiency frontier. Jio is close behind with 96.33%. Similarly, from Table 1, we would have expected BSNL to be far behind. The results from Table 2, though, shows that BSNL is close to 90% of the efficiency mark.

We are next interested to see what the best possible path for Jio and BSNL are to reach the efficiency frontier. In simple terms, the lines from the origin passing through the current positions of Jio and BSNL and touching the envelope, are the recommended paths for them. A slack based measure in DEA achieves the same.

The results from SBM gives the below results as shown in Fig. 2.

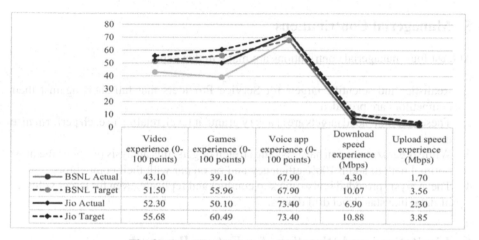

	Video experience (0-100 points)	Games experience (0-100 points)	Voice app experience (0-100 points)	Download speed experience (Mbps)	Upload speed experience (Mbps)
—●— BSNL Actual	43.10	39.10	67.90	4.30	1.70
--●-- BSNL Target	51.50	55.96	67.90	10.07	3.56
—◆— Jio Actual	52.30	50.10	73.40	6.90	2.30
--◆-- Jio Target	55.68	60.49	73.40	10.88	3.85

Fig. 2. Recommendations for BSNL and Jio

Out of the five Experience metrics, we see both BSNL and Jio do not require any improvement in their Voice App Experience. On download and upload speed experiences, BSNL requires improvement of more than 100%, while Jio requires more than 50%. In terms of Video and Games Experience, the improvement required is lower but still significant for BSNL and Jio.

To summarize, if the Service Providers have the following metrics as in Table 3, they would all be performing equally.

One strong point of the technique is that recommendations are not the same as the best provider, rather, the closest path to the efficiency frontier. Let us take BSNL as an example; all recommended targets are still below their competitors, yet still falls on the efficiency frontier.

For Jio too, all targets are below what Vi currently achieves but still falls on the efficiency frontier.

Table 3. SBM recommendations for all providers to achieve 100% efficiency

Service provider	Efficient output target				
	Video experience (0–100 points)	Games experience (0–100 points)	Voice app experience (0–100 points)	Download speed experience (Mbps)	Upload speed experience (Mbps)
Airtel	57.90	56.30	75.60	10.50	2.90
BSNL	51.50	55.96	67.90	10.07	3.56
Jio	55.68	60.49	73.40	10.88	3.85
Vi	57.80	62.80	76.20	11.30	4.00

5 Managerial Contributions

At least four managerial contributions are made in this paper.

1. Realistic, and scientific targets for Service Providers that fall short against their competitors are provided.
2. These recommended targets are at a very granular level, relating to each performance metric.
3. While organizations can have specific stretch targets, this analysis provides the most optimal path to attain 100% efficiency and be on par with their competitors.
4. The analysis provided is also quick to conduct and replicate, even for larger volumes of data, like state level drill downs.

6 Limitations and Directions for Future Research

This research assumed all input parameters to be equal. The reason for this is two-fold. First, a separate study needs to be conducted to identify all significant variables that impact the studied output metrics. Second, even post such a study, organizations may be reluctant to post such material in the public domain. It was therefore decided that few and arbitrary inputs should not be considered while making recommendations for Experience metrics. Future studies though can attempt to analyse data if all the information is made available. The study presents results for India as a whole. What will be more useful to Service Providers could be state or region wise breakdown of such targets. This study has taken into consideration five Experience metrics. Future studies can attempt combining other performance metrics such as Quality of Service (QoS) like bandwidth, latency, packet loss and other related KPIs (Key Performance Indicators).

7 Conclusion

This paper has attempted to scientifically measure performance of Quality of Experience metrics for the top four 4G Service Providers. This is vital in providing recommendations to the Service Providers by analyzing the User perceived QoS. Such performance analysis not only helps the Service Providers to improve their QoS provision but could also help them increase their active customer base.

References

1. Kasera, S., Narang, N.: 3G Networks Architecture, Protocols and Procedures. Tata McGraw-Hill Education, New Delhi (2004)
2. Malisuwan, S., Milindavanij, D., Kaewphanuekrungsi, W.: Quality of service (QoS) and quality of experience (QoE) of the 4G LTE perspective. Int. J. Future Comput. Commun. **5**(3), 158 (2016)
3. Su, G.-M., Su, X., Bai, Y., Wang, M., Vasilakos, A.V., Wang, H.: QoE in video streaming over wireless networks: perspectives and research challenges. Wirel. Netw. **22**(5), 1571–1593 (2016). https://doi.org/10.1007/s11276-015-1028-7
4. Joshi, R., Mandava, M., Saraph, G.P.: End-to-end quality of service (QoS) over internet. IETE Tech. Rev. **25**(4), 216–221 (2008)
5. Zhou, X., Van Mieghem, P.: Reordering of IP packets in internet. In: Barakat, C., Pratt, I. (eds.) PAM 2004. LNCS, vol. 3015, pp. 237–246. Springer, Heidelberg (2004). https://doi.org/10.1007/978-3-540-24668-8_24
6. Asano, A., Nishiyama, H., Kato, N.: The effect of packet reordering and encrypted traffic on streaming content leakage detection. In: 2010 Proceedings of 19th International Conference on Computer Communications and Networks, pp. 1–6. IEEE (2010)
7. Exarchakos, G., Druda, L., Menkovski, V., Liotta, A.: Network analysis on Skype end-to-end video quality. Int. J. Pervasive Comput. Commun. (2015)
8. Laghari, A.A., He, H., Shafiq, M., Khan, A.: Assessing effect of Cloud distance on end user's quality of experience (QoE). In: 2016 2nd IEEE international Conference on Computer and Communications (ICCC), pp. 500–505. IEEE (2016)
9. Laghari, A.A., He, H., Ibrahim, M., Shaikh, S.: Automatic network policy change on the basis of quality of experience (QoE). Procedia Comp. Sci. **107**, 657–659 (2017)
10. Mok, R.K.P., Chan, E.W.W., Chang, R.K.C.: Measuring the quality of experience of HTTP video streaming. In: 12th IFIP/IEEE International Symposium on Integrated Network Management (IM 2011) and Workshops, pp. 485–492. IEEE (2011)
11. Laghari, A.A., Memon, K.A., Soomro, M.B., Laghari, R.A., Kumar, V.: Quality of experience (QoE) assessment of games on workstations and mobile. Entertain. Comput. **34**, 100362 (2020)
12. Panwar, S.: Breaking the millisecond barrier: robots and self-driving cars will need completely reengineered networks. IEEE Spectr. **57**(11), 44–49 (2020)
13. Janssen, J., De Vleeschauwer, D., Buchli, M., Petit, G.H.: Assessing voice quality in packet-based telephony. IEEE Internet Comput. **6**(3), 48–56 (2002)
14. Moura, N.T., Vianna, B.A., Albuquerque, C.V.N., Rebello, V.E.F., Boeres, C.: MOS-based rate adaption for VoIP sources. In: 2007 IEEE International Conference on Communications, pp. 628–633. IEEE (2007)
15. De Pessemier, T., Stevens, I., De Marez, L., Martens, L., Joseph, W.: Analysis of the quality of experience of a commercial voice-over-IP service. Multimedia Tools Appl. **74**(15), 5873–5895 (2015)
16. Budiman, E., Moeis, D., Soekarta, R.: Broadband quality of service experience measuring mobile networks from consumer perceived. In: 2017 3rd International Conference on Science in Information Technology (ICSITech), pp. 423–428. IEEE (2017)

17. Kelly, F.: Charging and rate control for elastic traffic. Eur. Trans. Telecommun. **8**(1), 33–37 (1997)
18. Khan, S., Duhovnikov, S., Steinbach, E., Kellerer, W.: MOS-based multiuser multiapplication cross-layer optimization for mobile multimedia communication. Adv. Multimedia **2007** (2007)
19. Tsibonis, V., Georgiadis, L., Tassiulas, L.: Exploiting wireless channel state information for throughput maximization. : IEEE INFOCOM 2003. Twenty-second Annual Joint Conference of the IEEE Computer and Communications Societies (IEEE Cat. No. 03CH37428), vol. 1, pp. 301–310. IEEE (2003)
20. Robitza, W., Kittur, D.G., Dethof, A.M., Görin, S., Feiten, B., Raake, A.: Measuring YouTube QoE with ITU-T P. 1203 under constrained bandwidth conditions. In: 2018 Tenth International Conference on Quality of Multimedia Experience (QoMEX), pp. 1–6. IEEE (2018)
21. Tokunaga, K., Kawamura, K., Takaya, N.: High-speed uploading architecture using distributed edge servers on multi-RAT heterogeneous networks. In: 2016 IEEE International Symposium on Local and Metropolitan Area Networks (LANMAN), pp. 1–2. IEEE (2016)
22. Cánovas, A., Taha, M., Lloret, J., Tomás, J.: Smart resource allocation for improving QoE in IP multimedia subsystems. J. Netw. Comput. Appl. **104**, 107–116 (2018)
23. Boz, E.: A hybrid approach to quality measurements in mobile networks (2020)
24. Stojanović, I., Koprivica, M., Stojanović, N., Nešković, A.: Analysis of the impact of network architecture on signal quality in LTE technology. Serbian J. Electr. Eng. **17**(1), 95–109 (2020)
25. https://www.opensignal.com/reports/2020/09/india/mobile-network-experience
26. Azadeh, A., Asadzadeh, S.M., Bukhari, A., Izadbakhsh, H.R.: An integrated fuzzy DEA algorithm for efficiency assessment and optimization of wireless communication sectors with ambiguous data. Int. J. Adv. Manuf. Technol. **52**(5–8), 805–819 (2011)
27. Kwon, H.-B., Stoeberl, P.A., Joo, S.-J.: Measuring comparative efficiencies and merger impacts of wireless communication companies. Benchmark. Int. J. (2008)
28. Zhu, J.: Efficiency evaluation with strong ordinal input and output measures. Eur. J. Oper. Res. **146**(3), 477–485 (2003)
29. Azadeh, M.A., Bukhari, A., Izadbakhsh, H.: Total assessment and optimization of telecommunication sectors by multivariate approach. In: 2006 4th IEEE International Conference on Industrial Informatics, pp. 659–663. IEEE (2006)
30. Shin, H.-W., Sohn, S.Y.: Multi-attribute scoring method for mobile telecommunication subscribers. Expert Syst. Appl. **26**(3), 363–368 (2004)
31. Giokas, D.I., Pentzaropoulos, G.C.: Efficiency ranking of the OECD member states in the area of telecommunications: a composite AHP/DEA study. Telecommun. Policy **32**(9–10), 672–685 (2008)
32. Suleiman, M.S., Hemed, N.S., Wei, J.: Evaluation of telecommunication companies using data envelopment analysis: toward efficiency of mobile telephone operator in Tanzania. Int. J. e-Educ. e-Bus. e-Manag. e-Learn. (2017)
33. Vittal, B., Reddy, M.K.: Application of slack based measure of efficiency in data envelopment analysis: review on public sector banks in India. Researchgate.net
34. Cui, Q., Yu, L.-T.: A review of data envelopment analysis in airline efficiency: state of the art and prospects. J. Adv. Transp. **2021** (2021)
35. Kao, C., Liu, S.-T.: Group decision making in data envelopment analysis: a robot selection application. Eur. J. Oper. Res. (2021)
36. Yu, A.-Y., Liu, H.-C., Zhang, L., Chen, Y.: A new data envelopment analysis-based model for failure mode and effect analysis with heterogeneous information. Comput. Industr. Eng. **157**, 107350 (2021)
37. Yin, X., Wang, J., Li, Y., Feng, Z., Wang, Q.: Are small towns really inefficient? A data envelopment analysis of sampled towns in Jiangsu Province, China. Land Use Policy **109**, 105590 (2021)

38. Tan, Y., Jamshidi, A., Hadi-Vencheh, A., Wanke, P.: Hotel performance in the UK: the role of information entropy in a novel slack-based data envelopment analysis. Entropy 23(2), 184 (2021)
39. Tone, K., Toloo, M., Izadikhah, M.: A modified slacks-based measure of efficiency in data envelopment analysis. Eur. J. Oper. Res. 287(2), 560–571 (2020)
40. Avkiran, N.K., Tone, K., Tsutsui, M.: Bridging radial and non-radial measures of efficiency in DEA. Ann. Oper. Res. 164(1), 127–138 (2008)
41. Tone, K.: A slacks-based measure of efficiency in data envelopment analysis. Eur. J. Oper. Res. 130(3), 498–509 (2001)
42. http://www.deafrontier.net/

STVM: Scattered Time Aware Energy Efficient Virtual Machine Migration in Cloud Computing

G. Santhosh Kumar[1]([⊠]) and C. A. Latha[2]

[1] Don Bosco Institute of Technology, Bengaluru, India
[2] RVITM, Bengaluru, India

Abstract. Cloud computing offers services to large number of consumers, applications are hosted through large scale datacenters. Virtualization technology provides virtual servers from the underlying physical resources. Cloud service providers objective is to utilize the computing resources efficiently. Achieving energy efficiency gains more attention in these days, the existing cloud service managements are based on centralized architecture. This paper proposes a decentralized method for virtual machine consolidation and migration. Scattered time aware virtual machine migration method has been designed and implemented, which provides energy efficient virtual machine management solution by handling over-utilized and under-utilized physical machines. The efficacy of algorithm is evaluated through simulation using cloudsim toolkit.

Keywords: Cloud computing · Datacenter · Physical machine (PM) · Virtual machine (VM) · VM migration

1 Introduction

In recent years, cloud computing has been accepted by a number of data centers as a common medium for the management of most operations. Cloud computing naturally increases the utilization and scalability of the physical resources underlying it. Cloud computing offers virtual resources in a packaged way conveniently via the internet as a replacement for individually allocating the computing facilities upon request. In addition, it is significant notice that cloud computing be able to be used to increase infrastructure usage by virtualizing the composition of the service to a higher level. Hence, it is possible to unify the ability of the physical infrastructure to offer better QoS.

Cloud computing has recently become a modern model for dynamic provisioning of various services as a newly established technology [1]. It offers a way for the infrastructure, network and applications to be provided in a pay-as-you-go way as services accessible to consumers [2]. Amazon, Google, and Microsoft are other traditional commercial service providers. Datacenters are typically the fundamental computing infrastructure in cloud computing environments, consisting of several physical nodes and various virtual machines that operate on them [3].

The virtualization technique in datacenters enables a new model so that the physical infrastructure can build personalized virtual environments [4]. An important flexibility

© Springer Nature Switzerland AG 2021
K. R. Venugopal et al. (Eds.): ICInPro 2021, CCIS 1483, pp. 142–151, 2021.
https://doi.org/10.1007/978-3-030-91244-4_12

of Virtualization technique is to give the ability to combine and run one or more number of VM'S on a single physical machine [5].

The service provider must guarantee that the services can be flexibly offered to satisfy different customer criteria that are typically defined by SLAs [6] (service level agreements) to fully leverage the primary resources of cloud computing, thus holding users apart from the physical infrastructure underlying them. However, as time passes, the workloads of various applications or services change dramatically, adaptive management of resource allocation and making rational choices is a challenge. Hence, the future benefits of virtual machine migration include the ability to address the service provider's high-performance problems.

In this way, it was possible to resize the resources allocated to diverse virtual machines and to move virtual machines through various physical machines on demand to achieve different purposes [7].

Green computing, on the other hand, is gaining traction as a result of large data centres' high energy consumption. As a result, researchers are focusing their efforts on improving energy efficiency while retaining high service quality rather than pure performance. As a result, electricity costs become a part of the overall cost of ownership, which must be shielded from the view of the supplier. Further consolidation could be possible as a result of proper virtual machine migration and thus the unlocking of certain underutilized nodes, resulting in even more energy savings [8].

The main goal of this paper is to design a scattered/decentralized approach to virtual machine migration for cloud computing datacentres. The aim of this strategy is to optimize the amount of resource allocation dynamically by migration while also reducing possible energy waste.

2 Literature Study

Authors addressed the issue of virtual machine consolidation and relocation and proposed various resource allocation approaches for cloud computing in order to provide a cost-effective service.

The technique of Dynamic Idle Prediction is implemented in [9]. VM requests are considered in this technique when scheduling/assigning the virtual machines to the host. In order to predict the existence of VM load in the future, this technique uses an artificial intelligence classifier. In order to achieve more effective migration and consolidation-based energy efficient approaches compared to other strategies, this Dynamic Idle Prediction approach dictates the initial VM assignment and VM migration. This strategy is often used to dictate the overprovision of VMs by selecting the idlest VM to be turned off and minimizing missed requests in turn.

The virtual machine optimization in the datacenter for the current allocated is carried out in two steps: In the first step, identify and pick VMs that need to be migrated; in the second step, the selected VMs are placed on hosts using the Modified Best Fit Decreasing (MBFD) algorithm. To determine when and which VMs should be migrated, this algorithm employs double-threshold VM selection policies. The basic idea is to set upper and lower CPU utilisation thresholds for the host and keep total CPU utilisation between these thresholds for all VMs assigned to the host [10].

Hui Xiao et.al., suggested multi-objective VM consolidation strategy using double threshold and ant colony method [11]. As the resource competition becomes more concentrated, specifically resource exploitation crosses a certain level, the physical machine's efficiency becomes worse, and QoS is decreased. In the course of a multi-stage dependent ant colony scheme, to judge the circumstances that lead to virtual machine consolidation and optimization, the double threshold approach is used.

Xijia Zhou et al. suggested a weighted VM selection algorithm and an observational forecast algorithm [12]. When resource utilization exceeds the total capacity allocated to the physical computer, the algorithm determines the migration priority of each overloaded host based on historical exploitation of the host and weight for several reasons of exploitation. The Virtual Machine Manager is in charge of virtual machine migration.

An algorithm called layered VM migration was suggested by Xiong Fu et al. [13]. This algorithm uses the minimum spanning tree theory to segment the data centre into a number of regions based on the physical host's network bandwidth exploitation levels. VM migration in the data center then balances each region's network resources and achieves load balancing.

Nguyen Khac Chien et al. proposed a live migration technique in cloud computing for VM migration [14]. As the resource competition becomes more concentrated, specifically resource manipulation, exceeding a certain threshold, this technique of selecting VMs that need the least amount of migration time and have the least amount of available processors causes the virtual machine to migrate, contributes to less SLA breach occurrence.

3   Scattered Time Aware Virtual Machine Migration Algorithm

The global central resource manager is typically structured and put into operation to handle the resources within a large-scale data center to optimize the system-wide physical resources and construct the right decisions. On the other hand, centralized resource management is exposed to single-point failure, which could lead to the entire system's unmanaged status. Here, we suggest a decentralized mechanism to solve these problems and include assurances in different cases for the availability of VM management behavior.

A multi-phase decentralized algorithm scheduling approach is used in the algorithm. It consists of a VM monitor phase which is responsible for collecting physical machine and virtual machine statistical data, VM Predictor is responsible for identifying and selecting the VM to be migrated, and a VM optimizer phase is responsible for optimizing the use of the virtual machine.

VM Monitor: The virtual machine monitor collects data for both physical and virtual devices. In the next step, the collected data is essentially used as input for the prediction process. Note that a detailed observation of the virtual machine pool's performance, including not only virtualization-related metrics but also server CPU, memory, disc, and network IO performance metrics, is needed to fully characterise VM pool efficiency (Fig. 1).

In order to make VM migration decisions, on each physical machine, virtual machine load data must be collected. The decentralized mechanism obtain the load indexes from

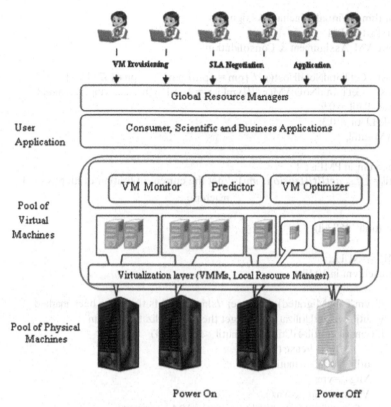

Fig. 1. Scattered time aware virtual machine migration system architecture

other physical machines, each physical machine maintains a workload data. In this work it uses a tuple, defined as follows, to represent the workload index LDI.

ldi = {src,dest,util$_i$}.

Where dest is the unique ID of the target physical machine that will receive the workload index, SRC is the source of the workload index information, and UTIL$_I$ is the source node's current CPU use.

VM Predictor: Since virtual machines host numerous applications with varying workloads, the CPU consumption of the physical node can fluctuate dramatically over time. If the CPU usage of certain virtual machines exceeds a certain threshold, they should be transferred to other physical machines to free up resource power. Find the physical machine "underutilised" if the CPU consumption is too low, below a certain level, and migrate the virtual machines on it to alleviate the physical machine load. In this case, the blank node with no work could be put into a sleeping state to save resources.

VM Optimizer: It's possible to use the VM predictor's prognostic tracking statistics as a helpful feedback for deciding the best consolidating guidance. If at all possible, the strategy should save as much energy as possible while maintaining service quality. In fact, this segment is in charge of deciding the smallest number of PMs that can be used to host the growing number of VMs.

Algorithm – Virtual Machine Assignment
Input: List of PM
Output: VM Assignment & Consolidation

```
PMList   GetTotalNoOfHost(); //{pm = {pm1,pm2, ..., pmn} ⊆ Uhs}
VMList   GetT otalNoOfVM(); //{vmList = {vm1, vm2, ...., vmj , ....vmm}}
double HUtil = 0.9;
double LUtil = 0.3;
MigVM=null;

foreach pm in PMList {
    utilᵢ= $\sum_{j=0}^{n} getUtilization\,(vm)$  //get the current utilization of ith physical
                                                machine
}
while (utili > HUtil)
{
    Mig_vm=NULL;
    foreach vm in VMlist
    {
        if (vm.TobeMigrated) continue; //skip the VMs that have been marked
        vmutil=vm.getUtilization();  //get the CPU utilization of vm
        if (vmutil > utili-HUtil  && vmutil <utili-LUtil)
        {            //case (1)
            utili = utili – vmutil;
            Mig_vm=vm;
            Vm vm=info.getVm();
            info.getTime(vm);  break;    //find a VM to migrate
        }
    }
    if (Mig_vm==null)
    {
        foreach vm in VMlist
        {
            if (vm.TobeMigrated) continue;
            vmutil= vm.getUtilization();
            if (vmutil <utili-LUtil)
                {          //case (2)
                    utilᵢ = utilᵢ – vmutil;
                    Mig_vm=vm;
                    Vm vm=info.getVm();
                    info.getTime(vm);  break;  //identifies VM to migrate
                }
        }
        if (Mig_vm==null) break;    //case (3)
                targetHost=info.getHost(); //identify target physical machine for the
                                        current VM to migrate to
    }
}
```

Performance metrics are used to gauge the algorithm's effectiveness. Table 1 shows a formula for measuring the amount of energy used by physical devices in a data centre. In an Infrastructure as a Service environment, it's important to identify a self-determining workload metric that can be used to estimate the service level agreement given to any Virtual Machine configuration. This paper suggests two criteria for assessing the scope of a service level agreement violation in an IaaS setting.

Violation time of SLA per Active Host (SLATH) is computed in Eq. (1).

$$SLATH = \frac{1}{N} \sum_{i=1}^{N} \frac{T\nabla t}{T\partial t} \tag{1}$$

Where N is the number of physical hosts, and $T\nabla t$ is the total time spent by each Physical Machine that experienced resource overuse, resulting in a service level agreement violation. $T\partial t$ is a figure that shows how long the ith physical host has been active. Performance dilapidation owing to Virtual machine migration is computed in Eq. (2).

$$PDM = \frac{1}{M} \sum_{j=1}^{M} \frac{VMdj}{VMrj} \tag{2}$$

M is the number of virtual machines, and $VMdj$ is the measure of the virtual machine VM_j's performance loss due to migrations. $VMrj$ is a figure that shows the total amount of CPU expertise required by the Virtual Machine VM_j over its lifetime.

As a result, the SLA Violation (SLAV) metric was designed to compensate for performance degradation caused by physical host overloading as well as higher number of virtual machine migrations in Eq. (3).

$$SLAV = SLATH * PDM \tag{3}$$

The simulations in this paper are based on the assumption that during the migration process, an equal computing competency of CPU power is allocated to each VM on the target physical host, and that each migration results in a service level agreement violation. As a consequence, it's important to reduce the number of virtual machine migrations. The total migration time is determined in Eq. (4) by the total memory consumption of the virtual machine and the available network bandwidth. It is not possible to clone the Virtual machine storage since the image and data must be stored on a NAS. As a result of the tests, the migration time has decreased.

$$T_m j = \frac{Ram_j}{Bd_j} \tag{4}$$

Ram_j represents the memory space used by j^{th} VM, and Bd_j represents the available network bandwidth and performance degradation experienced by a ^{jth} VM were described in Eq. (5),

$$U_{dj} = 0.1 * \int_{t0}^{t0+T_{mj}} uj(t)dt \tag{5}$$

Where U_{dj} represents jth VM's total performance degradation, t_0 represents the migration start time, T_{mj} represents the time it took to complete the migration, represents jth VM's CPU exploitation.

The migration time is calculated by dividing the VM's memory (RAM) capacity by the auxiliary network bandwidth available to the host j in (6). Let V_j represent a set of VMs currently assigned to host j.

$$v \in V_j | \forall \in V_j, \frac{Ram_u(v)}{N_j} \leq \frac{Ram_u(a)}{N_j} \tag{6}$$

Where $Ram_u(v)$ is the memory (RAM) space currently in use by VM, and $Ram_u(a)$ is the auxiliary network bandwidth available to host j.

4   Results and Discussion

Experimenting on a real infrastructure in a bulky-scale manner is extremely difficult. The simulation framework chosen was the CloudSim toolkit [15]. The University of Melbourne created and released the Cloud-Sim toolkit, which is an extensible simulator aimed at supporting the simulation and modelling environment for cloud computing. It allows cloud computing organisations to simulate virtualized infrastructure, data centres, virtual servers, and physical hosts. As a result, the CloudSim toolkit makes it easier to enforce different resource allocation policies and estimate policy performance.

Table 1. Power consumed by servers at various load levels in Kwh

Server workload level	HP Pro ML G4	HP Pro ML G5
0%	86	93.7
10%	89.4	97
20%	92.6	101
30%	96	105
40%	99.5	110
50%	102	116
60%	106	121
70%	108	125
80%	112	129
90%	114	133
100%	117	135

A data centre simulation was used for this study, which included 800 homogeneous and heterogeneous physical devices, half of which were HP Pro ML 110 G4 servers and the other half were HP Pro ML 110 G5 servers. The power consumption of the server is shown in Table 1. Each server's network bandwidth is modelled at 1 GBPS, with the HP

Pro ML 110 G4 server having 1860 MIPS per core and the HP Pro ML 110 G5 server having 2660 MIPS per core.

Experiments were carried out using data from PlanetLab's monitoring infrastructure [13]. During the experiment, the workload outline from one of the Physical Machines from the following day is randomly assigned to each Virtual Machine, and the monitoring time is set to five minutes.

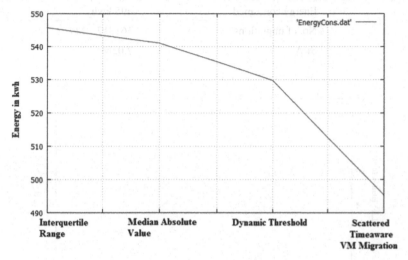

Fig. 2. Energy consumption comparison of STVM with other methods

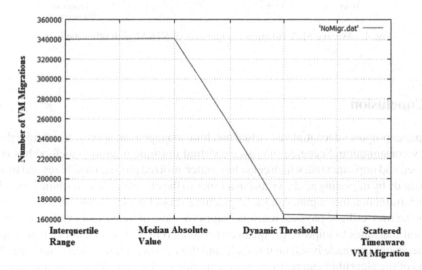

Fig. 3. Comparison of total number of VM migration of STVM with other methods

Table 2 describes the system efficiency measurements, which include 800 hosts and 1033 virtual machines for a one-hour simulation of the proposed system. Figures 2, 3 and 4 depicts the proposed method's inferences and outcomes, as well as a comparison of system performance to the various methods available.

Table 2. System performance metrics

Performance metrics		
1	Energy consumed	495 kwh
2	No. of migrations	161817 VMs
3	SLA	7.02%

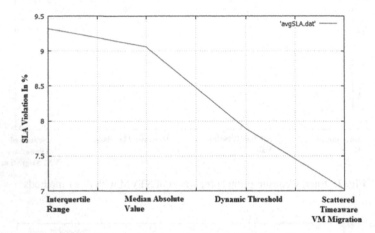

Fig. 4. Average SLA violation comparison of STVM with other methods

5 Conclusion

This paper proposes decentralized virtual machine management approach to diminish the energy consumption. Scattered time aware virtual machine migration method has been designed and implemented, which considers under-utilized physical machines and turn to sleep mode by migrating all the virtual machines to the other physical machine satisfying the constraint that the migrated physical machine should not be over-utilized. Also considers the over-utilized physical machine migrates some of the virtual machines to other physical machines to improve the quality of service (QoS). The virtual machine migration decisions are made based on threshold and the minimum time taken to migrate. The results of the algorithm and performance evaluation of the simulation experiments illustrates that the scattered time aware virtual machine migration method achieves energy efficacy also improves the quality of service.

References

1. Hayes, B.: Cloud computing. Commun. ACM **51**(7), 9–11 (2008)
2. Armbrust, M., et al.: Above the clouds: a Berkeley view of cloud computing. Technical report UCB/EECS-2009-28, EECS Department, University of California, Berkeley (2009)
3. Buyya, R., Yeo, C.S., Venugopal, S., Broberg, J., Brandic, I.: Cloud computing and emerging IT platforms: vision, hype, and reality for delivering computing as the 5th utility. Future Gener. Comput. Syst. **25**(6), 599–616 (2009)
4. Stillwell, M., Schanzenbach, D., Vivien, F., Casanova, H.: Resource allocation using virtual cluster. In: Proceedings of the 9th IEEE/ACM International Symposium on Cluster Computing and the Grid (CCGRID 2009), pp. 260–267, May 2009
5. Wang, X., et al.: Appliance-based autonomic provisioning framework for virtualized outsourcing data center. In: Proceedings of the 4th International Conference on Autonomic Computing (ICAC 2007), p. 29, Fla, USA, June 2007
6. Feng, W.-C., Feng, X., Ge, R.: Green supercomputing comes of age. IT Professional **10**(1), 17–23 (2008)
7. Rosenblum, M., Garfinkel, T.: Virtual machine monitors: current technology and future trends. Computer **38**(5), 39–47 (2005)
8. Sharkh, M.A., Shami, A.: An evergreen cloud: optimizing energy efficiency in heterogeneous cloud computing architectures. Veh. Commun. **9**, 199–210 (2017)
9. Beloglazov, A., Abawajy, J., Buyya, R.: Energy-aware resource allocation heuristics for efficient management of data centers for cloud computing. Futur. Gener. Comput. Syst. **28**(5), 755–768 (2012)
10. Hui, X., Hu, Z., Li, K.: Multi-objective vm consolidation based on thresholds and ant colony system in cloud computing. IEEE Trans. **7**, 53441–53453 (2019). https://doi.org/10.1109/Access.2019.9.2912722
11. Zhou, X., Li, K., Liu, C., Li, K.: An experience-based scheme for energy-SLA balance in cloud data centers. IEEE Access **7**, 23500–23513 (2019). https://doi.org/10.1109/ACCESS.2019.2899101
12. Fu, X., Chen, J., Deng, S., Wang, J., Zhang, L.: Layered virtual machine migration algorithm for network resource balancing in cloud computing. Front. Comput. Sci. **12**(1), 75–85 (2018). https://doi.org/10.1007/s11704-016-6135-9
13. Chien, N.K., Dong, V.S.G., Son N.H., Loc, H.D.: An Efficient Virtual Machine Migration Algorithm Based on Minimization of Migration in Cloud Computing. In: Vinh, P., Barolli, L. (eds.) Nature of Computation and Communication. ICTCC 2016. Lecture Notes of the Institute for Computer Sciences, Social Informatics and Telecommunications Engineering, **168**, 62-71. Springer, Cham (2016). https://doi.org/10.1007/978-3-319-46090-6_7
14. Calheiros, R.N., Ranjan, R., Beloglazov, A., De Rose, C.A.F., Buyya, R.: CloudSim: a toolkit for modeling and simulation of Cloud computing environments and evaluation of resource provisioning algorithms. Softw. Pract. Exp. **41**(1), 23–50 (2011). https://doi.org/10.1002/spe.995
15. Park, K.S., Pai, V.S.: CoMon: a mostly-scalable monitoring system for PlanetLab. ACM SIGOPS Oper. Syst. Rev. **40**(1), 74 (2006)

DRP-DBAS: Dynamic Resource Provisioning for Deadline and Budget Aware Workflow Scheduling in IaaS Clouds

Naela Rizvi and Dharavath Ramesh(✉)

Indian Institute of Technology (ISM), Dhanbad 826004, Jharkhand, India
naela.17dr000330@cse.ism.ac.in, drramesh@iitism.ac.in

Abstract. The cloud being the latest technology, provides immense opportunities to solve large-scale scientific problems. This computing paradigm's success enables the IaaS provider to offer infinite resources based on the pay-per-use model. Although the workflow scheduling problem has been widely investigated, however, most of them are concerned with a single QoS constraint and ignores the consideration of multiple QoS constrained problems. Therefore, this paper proposes a novel strategy of dynamic resource provisioning and scheduling approach for deadline-budget constrained workflows (DRP-DBAS). The algorithm intends to minimize makespan while subject to deadline and budget constraints for the hourly-based cost model of the IaaS cloud. For resource provisioning, the number of instances leased is based on the budget available. For the scheduling, the HEFT algorithm has been extended with the deadline and budget constraint. Further, the new scheduling approach incorporates the clustering mechanism to cluster the pipelined task, which reduces the overall execution time and enhances the algorithm's performance. The DRP-DBAS is compared against the existing BDSD, BDAS, and GRP-HEFT and the obtained result proves the efficacy of the DRP-DBAS algorithm. DRP-DBAS outperforms the other algorithm by achieving a PSR of 56.11%, followed by GRP-HEFT with PSR 44.61%, BDAS with PSR 39.13%, and BDSD with PSR 26.13%.

Keywords: Workflow · Budget · Deadline · Scheduling

1 Introduction

Scientific applications are assumed to be composed of hundreds or even thousands of computational tasks. Such applications are modeled as a Directed Acyclic Graph (DAG), where nodes are the tasks and edges represents the dependencies [1]. With the complex structure and ever-growing data and computing requirements, there is a demand for high computational power. Traditionally, these applications were deployed on high-performance clusters and grid platform

© Springer Nature Switzerland AG 2021
K. R. Venugopal et al. (Eds.): ICInPro 2021, CCIS 1483, pp. 152–165, 2021.
https://doi.org/10.1007/978-3-030-91244-4_13

[2]. However, the existing systems are expensive to maintain and also incompetent to adapt to the escalated resource demand. Cloud offers a highly scalable pool of infinite resources with varying configuration to suit the application's requirements and, thus, motivated such applications for execution. This technology delivers these resources as a service over the internet and offers 'pay-per-use' cost model where the users are charged for the time interval these resources are utilized. Moreover, the flexibility to acquire and release the resources gives more control to the users and empowers them to develop innovative scheduling techniques to efficiently utilize the resources.

Workflow scheduling aims at mapping the tasks to a particular time slot of resources to meet certain performance criterion. Being a well-known NP-hard problem, it has been widely investigated for decades in the distributed computing environment [3]. Although many algorithms have been proposed for grids and clusters, they are primarily concerned with the scheduling phase. This is because they have a fixed number of resources for the execution whose configuration is already known in advance. On the other hand, this scheduling mechanism is not applicable in the cloud, where the appropriate resources are leased from the infinite pool of resources.

Besides this, the algorithms designed for the traditional systems are mainly dedicated to meeting the deadline or minimizing the execution time while ignoring the cost of leased resources. In contrast, the cloud provides heterogeneous resources with varying processing capabilities and costs, where the total cost of cloud services is also a matter of concern. Several scheduling approaches have been developed for the QoS-constrained workflows in the cloud, but most of them focus on either time or cost [4,5]. However, time and cost are the two relevant concerns in the Quality of Service (QoS) constraints that need to be addressed simultaneously. Considering the time and cost simultaneously makes the workflow scheduling problem even more challenging, and hence, up to the present, only a few approaches have been developed [4,6,7]. However, the existing methods consider static VM provisioning, while the resources are dynamically provisioned in the cloud. Further, the existing work [4,7]utilizes the basic cost model, which ignores the idle time intervals [8] and thus causes failure of budgets when the complex cost model is applied. Therefore, to overcome the drawbacks of the existing methodologies, a new approach for budget and deadline constrained workflows have been developed. The proposed method utilizes an 'Hourly-based cost model' for attaining the objective of scheduling the workflows within the defined deadline and budget constraint. For assessing the quality of the proposed DRP-DBAS algorithm, it has been compared with the other state-of-the-art algorithms. Further, various scientific workflows are considered to analyze the performance of the algorithm [9]. Experimental results demonstrate that DRP-DBAS shows a remarkable performance and outperforms state-of-the-art algorithms. The contributions of the paper are illustrated in the following manner.

- A new dynamic resource provisioning and scheduling algorithm for the deadline and budget constrained workflows utilizing the hourly based cost model in the cloud environment has been proposed

- Budget constrained resource provisioning mechanism has been introduced for the selection of the appropriate resources from the available budget
- A novel deadline and budget aware workflow scheduling approach has been developed. The scheduling approach extends the HEFT algorithm with budget and deadline in the scheduling process.
- The scheduling approach includes a clustering operation, which minimizes the volume of transferred data between the directly connected tasks. This mechanism reduces the run time and enhances the performance of the algorithm.

The remainder of the paper has been organized as: Sect. 2 introduces the related work, which includes the review of existing approaches. Section 3 presents the scheduling model and problem formulation. The proposed method has been demonstrated in Sect. 4, followed by the results and comparison of the considered algorithms in Sect. 5. Finally, the conclusion and future work are presented in Sect. 6.

2   Related Work

Generally, there are three types of constrained workflow scheduling: i) budget constrained makespan minimization, ii) deadline constrained cost minimization, iii) budget-deadline constrained scheduling.

2.1   Budget Constrained Workflow Scheduling and Deadline Constrained Workflow Scheduling

Faragardi et al. [8] proposed a new budget constrained resource provisioning and workflow scheduling algorithm for IaaS cloud named: GRP-HEFT. The algorithm aims to minimize the overall execution time while satisfying the budget constraint for the hourly based cost model. Another algorithm named FBCWS (fair budget-constrained workflow scheduling), is designed by Rizvi et al. [12]. The algorithm minimizes the makespan and satisfies the budget constraints. Further, FBCWS offers a fair scheduling policy for every task. However, the algorithm utilizes the basic cost model of the grid, which is not applicable in the cloud. An energy-aware multi-objective reinforcement learning (EnMORL) has been proposed by Qin et al. [13]. The algorithm simultaneously optimizes makespan and energy consumption under the defined budget constraint.

Rodriguez and Buyya [14] use a Particle swarm optimization algorithm for resource provisioning and workflow scheduling in the cloud. The algorithm minimizes the overall execution cost while satisfying the deadline constraints. Wu et al. [15] introduced a meta-heuristic approach L-ACO and a heuristic approach ProLis for minimizing the workflow execution time while satisfying the budget constraint. L-ACO utilizes ant colony optimization to schedule deadline constrained workflows. On the other hand, ProLis distributes the deadline to each task and then follows two steps of scheduling, i.e. task ranking and sequentially allocating tasks to those services which satisfy the sub-deadline and minimizes the cost.

2.2 Budget-Deadline Constrained Workflow Scheduling

An algorithm named: BDSD has been proposed to schedule workflows within the defined deadline and budget constraint. The algorithm works in two phases: task sorting and processor selection. Tasks are sorted according to the sub-deadline (SDL) and scheduled on those processors satisfying the condition. The algorithm behaves proficiently from the other existing algorithms. However, the algorithm has static VM provisioning and a basic cost model of the grid [7]. Arabnejad et al. [6] proposed budget deadline aware scheduling (BDAS). The algorithm has a novel time-cost trade-off function to select heterogeneous resources that satisfy both budget and deadline constraints. The efficacy of the algorithm is proved by performing sensitivity analysis. However, the algorithm uses a fixed number of resources. Zhou et al. [4] proposed a Budget-Deadline constrained workflow scheduling (BDCWS) algorithm. The algorithm has the task optimistic available budget and the optimistic spare budget to select the VMs and control the execution cost. Ghasemzadeh et al. [11] proposed an algorithm having a quadratic time complexity considering the budget and deadline constraints. The algorithm DWBS exhibits better success rates compared to other heuristic approaches.

3 Scheduling Model and Problem Formulation

In this section, the cloud-based workflow scheduling model and the problem formulation have been discussed.

3.1 System Model

A workflow application can be represented as a Directed Acyclic Graph $DAG = \{T, E, d, f\}$. T is the set of nodes that denotes the tasks of the workflow, and E is the set of edges that represents precedence constraints between tasks. The weight $d(t_i)$ assigned with the node $t_i \in T$ is the computational workload, and the weight $f(e_{i,j})$ between the edges $e_{i,j} \in E$ is the amount of communication data between the task t_i and t_j. Each edge $e_{i,j} \in E$ indicates the dependency between the parent task t_i and child task t_j, which means t_j cannot be executed until the task t_i finishes its execution. All the predecessors of t_i can be represented as $pred(t_i) = \{t_j | e_{j,i}\} \in E$, and the immediate successor of t_i can be denoted as $succ(t_i) = \{t_j | e_{i,j}\} \in E$. For the generalization of the workflow with a single entry and single exit, two dummy nodes, t_{entry} and t_{exit} with zero execution and zero communication time are added at the beginning and the end of the workflow. In addition, each workflow is associated with a deadline (D_w) and budget (B_w), where the deadline is the time limit, and the budget is the cost limit within which the workflows are executed.

In the cloud environment, the IaaS provider offers the pool of infinite resources of different processing capabilities (PC_{vm_k}) and cost (C_{vm_k}) to its customer. The maximum number of VMs available for each type of VM is equal to the total number of tasks T in a workflow. Generally, cloud services with higher processing capabilities are assumed to be of higher cost. In this paper, the proposed work considers the commonly used cost model, i.e. the 'pay-per-use' on-demand cost model. In this cost model, users have to pay for the time interval (μ) they have used the resources, and the partial utilization of resources is also charged for the full-time period. Suppose a resource has been used for 80 min, but it is charged for the full 120 min. In the experiments, the hourly based cost model of Amazon EC2 has been used. Amazon EC2 provides free internal data transfer among the EC2 instances; therefore, the proposed model assumes no charges for internal data transfer while calculating the execution cost of the workflow. Further, it is assumed that all the services are present in the same physical region; therefore, the bandwidth (β) is supposed to be roughly equal.

3.2 Problem Definition

A schedule S can be defined as $\{VM_{Leased}, M, TTE_w, TCE_w\}$ where VM_{Leased} is the set containing the VMs leased for the execution; M denote the mapping of the task to VM, TTE_w is the total time of execution, and TCE_w is the total cost of execution. $VM_{Leased} = \{vm_{1,1}, vm_{2,1}, ..., vm_{z,k}\}$ is the set of VMs leased, each $vm_{z,k}$ represents the zth vm of kth type and is associated with lease start time (LST) and lease end time (LET). In contrast, M comprises the tuples $m(t_i, vm_{z,k}) = \{t_i, vm_{z,k}, st(t_i), ft(t_i)\}$ where the tuple $m(t_i, vm_{z,k})$ can be interpreted as a task t_i is scheduled on z^{th} VM of kth type and starts its execution at $st(t_i)$ (start time) and completes the execution at $ft(t_i)$ (finish time). Equation (1) and (2) are defined to show how the and are computed.

$$TTE_w = max\{AFT(t_i), t_i \epsilon T\} \tag{1}$$

$$TCE = \sum_{l=1}^{vm_{Leased}} \frac{LET_l - LST_l}{\mu} * C_{vm_k} \tag{2}$$

Where, $AFT(t_i)$ is the Actual Finish Time of task t_i. Based on the above definitions, the problem can be formulated as: Find a schedule S with minimum TTE_w and minimum TCE_w such that the value of TTE_w and TCE_w is within the specified Deadline and Budget as shown in Eq. (3).

$$\begin{aligned} Minimize\ TTE_w, TCE_w \\ subject\ to\ TTE_w \leq D_w, TCE_w \leq B_w \end{aligned} \tag{3}$$

In this section, some of the basics definitions are discussed, which are important to understand the workflow scheduling mechanism.

Definition 1: The execution time of task t_i on VM of type k is computed using the size Z_{t_i} and the processing capacity of the VM PC_{vm_k} as shown in Eq. 4.

$$ET(t_i, vm_k) = \frac{Z_{t_i}}{PC_{vm_k}} \tag{4}$$

Definition 2: $TT(t_i, t_j)$ is the data transfer time between a parent task t_i and its child task t_j as depicted in Eq. 5.

$$TT(t_i, t_j) = \begin{cases} 0 & \text{when } t_i \text{ and } t_j \text{ are assigned on same } VM \\ \frac{data_{t_i}^{out}}{\beta} & \text{otherwise} \end{cases} \tag{5}$$

Where $data_{t_i}^{out}$ is the output data from t_i.

Definition 3: $MET(t_i)$ is the minimum execution time of task t_i on $vm_k \epsilon VMTypeList$ which is computed as:

$$MET(t_i) = min_{vm_k \epsilon VMTypeList}\{ET(t_i, vm_k)\} \tag{6}$$

Where, $VMTypeList$ is the the type of VMs stored in a List.

Definition 4: $XST(t_i)$ is the expected start time of task t_i, which is the time at which t_i begin its execution after all the predecessors of task t_i finish their execution.

$$XST(t_i) = \begin{cases} 0 & \text{if } t_i = t_j \\ max_{t_j \epsilon pred(t_i)} XST(t_j) + MET(t_j) + TT(t_i, t_j) & \text{otherwise} \end{cases} \tag{7}$$

Definition 5: $XFT(t_i)$ it is the expected finish time of task t_i defined in Eq. 8.

$$XFT(t_i) = XST(t_i) + MET(t_i) \tag{8}$$

Definition 6: $MET(t_i)$, is the minimum time at which the entire workflow finishes its execution.

$$MET_w = XFT(t_{exit}) \tag{9}$$

Definition 7: SDL is the sub-deadline assigned to each task such that all the tasks complete their execution before the defined sub-deadline. The SDL for task t_i is shown in Eq. 10.

$$SDL(t_i) = \begin{cases} D & t_i = t_j \\ min_{t_j \epsilon succ(t_i)} SDL(t_j) - TT(t_i, t_j) - MET(t_j) & \text{otherwise} \end{cases} \tag{10}$$

Definition 8: $EST(t_i, vm_{z,k})$ is defined as the earliest start time of task t_i. It is the time when task t_i start its execution on $vm_{z,k}$, which is formulated as:

$$EST(t_i, vm_{z,k}) = max\{avail(vm_{z,k}), max_{\{t_j \epsilon pred(t_i)}(AFT(t_j) + TT(t_i, t_j)\} \tag{11}$$

where $avail(vm_{z,k})$ is the time at which $vm_{z,k}$ is available for execution and $AFT(t_j)$ denote the actual finish time of task t_j

Definition 9: $EFT(t_i, vm_{z,k})$ is the earliest finish time of task t_i on $vm_{z,k}$, which is computed as:

$$EFT(t_i, vm_{z,k}) = EST(t_i, vm_{z,k}) + ET(t_i, vm_{z,k}) \tag{12}$$

4 Proposed Methodology (DRP-DBAS)

The proposed algorithm is designed for the deadline and budget-constrained workflows in IaaS clouds. The algorithm has i) resource provisioning phase, which is responsible for selecting the type and the number of VMs from the pool of IaaS resources ii) scheduling phase is responsible for assigning workflows to those selected resources provided by the resource provisioning phase.

4.1   Resource Provisioning Algorithm

The proposed approach provides a simple resource provisioning technique which is shown in Algorithm 1. The algorithm's main aim is to determine the type and the number of instances to be used for the scheduling. The algorithm starts with the selection of the different types of VMs, which are stored in $VMTypeList$. $VMTypeList = \{vm_1, vm_2, ..., vm_m\}$ is the types of VMs provided by the cloud provider. Initially, RB (Remaining Budget) is set to the user-defined budget, as shown in **step 2**. Afterward, an array Y has been formed to contain possible VMs that can be leased for execution. **Steps [4–9]** are repeated to specify the number of VMs taken for each type from the available budget. The algorithm takes as much of the most efficient VMs possible from the available budget. If the remaining budget is not enough to purchase the most efficient VM, then the second most efficient VM has opted. These steps are executed until the remaining budget is not enough to buy a single cheapest VM. Lastly, the algorithm returns the array Y in **step-10** containing all the possible instances from the given budget. The Y is then given as input to the workflow scheduling phase, where all the tasks are assigned to these VMs such that the makespan and cost obtained is within the deadline and budget.

Algorithm 1: Resource provisioning algorithm

   Result: possible instances (Y)

1 Set of different types of VMs present in $VMTypeList$;

2 $RB = B_w$;

3 Set of possible VMs from the given budget,$Y = \{\}$;

4 **while** $RB >= Cost(VM_{cheapest})$ **do**

5      $\chi = Select fastest VM(vm_k) from VMTypeList$;

6      $n_1 = \lfloor \frac{RB}{cost(vm_k)} \rfloor$, select n_1 instances of type vm_k and add in Y;

7      $RB = RB - n_1 * vm_k$;

8   Remove vm_k from $VMTypeList$;

9 **end**

10 **return** Y

4.2   Workflow Scheduling Phase

The proposed algorithm extends the existing HEFT [16] algorithm to include deadline and budget constraints. The pseudo-code of the workflow scheduling algorithm has been provided in Algorithm 2. The workflow scheduling approach aims to assign the tasks to the available resources in such a way that the budget and deadline constraints are satisfied. The algorithm starts as: In the first step, an array $VM_{available}$ has been assigned to store the Y (an array containing all the possible VMs) generated from Algorithm 1. The reason for considering another array is $VM_{available}$ is updated each time when a task from the $TaskList$ is assigned. In **step 2**, another empty array named VM_{Leased} has been created to store the leased VMs. Initially, the algorithm contracts 1 VM of each type for 1 hr. Subsequently, leased VMs from the $VM_{available}$ are removed using **step**

4. In **step 5**, an array $TaskList$ has been created to store the number of tasks present in a workflow. **Steps [6–8]** are executed to evaluate MET, XST, and XFT for each task using Eq. (6), (7) and (8). Afterward, **step 9** calculates the minimum execution time MET_w for the workflow from the Eq. (9) which is then compared with the defined deadline. If the value of the deadline is greater than or equal to MET_w then **steps 10** onwards are executed; otherwise, the value of the deadline constraint must be updated. Once the possibility of achieving the deadline is confirmed, then the clustering operation [10] is performed on the DAG to group those parent-child tasks, which is the only child and has a single parent. Now, for each task of new DAG, sub-deadline is computed using the Eq. 10, and the tasks are ranked in ascending order of their SDL (steps [12–13]). For every task in the Sorted Task List (STL) **(step 15)**, compute EST and EFT on each VM present in VM_{Leased} **(Steps [16–18])**. The VM having minimum EFT and which satisfies the condition defined in **step 20** is selected for execution. Otherwise, if there exists another VM in VM_{Leased} whose EFT is not minimum, however, it still satisfies the condition defined in step 20, then that VM will be selected **(steps [20–24])**. Further, there may be a situation when none of the VM from VM_{Leased} is eligible for the selections, then the **steps [26–52]** are executed. If the $VM_{available}$ is not empty, **steps [26–48]** are executed; otherwise, **steps [49–52]** are executed. When there are VMs available for the selection and the already leased VMs are incapable of executing the tasks because the execution time of the VMs goes beyond the full hour, or maybe the deadline condition is violated, then a new fastest VM can opt for the execution. For selecting the new VM (VM_{new}), EST and EFT of task t_i on VM_{new} are obtained **(Step 27)**. EFT of task t_i on VM_{new} is compared with the EFT of the already leased VMs. The VM having minimum EFT and satisfying the sub-deadline assigned to t_i, is selected for the execution **(step 29)**. If the VM is from the VM_{Leased} then **steps [30–32]** are executed; otherwise, the algorithm runs **steps [33–37]**.

When the already leased VM is selected, but the lease time interval is not satisfied, then the proposed approach tries to avoid budget violations by extending the lease time interval of the VM. The algorithm first searches the unused VMs from the $VM_{available}$ with the same type as the VM selected. If there is such VM, it is removed from the $VM_{available}$. Removing this VM keeps the algorithm safe because assigning the task t_i on $VM_{selected}$ increases the cost of execution (since the VM requires an additional 1 or more hr to run), another VM of the same type from $VM_{available}$ is removed to compensate for this cost inflation. When the $VM_{selected}$ is from $VM_{available}$, the selected VM is added into the VM_{Leased} array with the Lease time interval of 1 hr **(steps [34–35])**. **Steps [39–41]** are executed to return the N number of the selected VM such that the desired criterion is satisfied. The variable N is created to store the number of times the conditions are not satisfied. When the $VM_{available}$ has N number of VMs, then N number of VMs from the $VM_{available}$ is removed to compensate for the increased cost **(steps 42–48)** and avoid the budget violation. Otherwise, the budget is updated to increase VMs when no VMs present in $VM_{available}$ can schedule tasks within the defined constraint. Subsequently, the AFT of the task t_i on $VM_{selected}$ is obtained. Further, (steps [49–52]) are executed only when the $VM_{available}$ is empty, i.e., there is no more VMs available, and the already leased VMs violate the budget constraints. All these mentioned steps are repeated for each task of a DAG, and then the makespan and the execution cost of the workflow is obtained in (steps [55–56]).

5   Result and Discussions

The section begins with the experimental settings for evaluations, and then the comparison of the proposed with the other state-of-the-art algorithms is demonstrated.

5.1   Experimental Settings

The DRP-DBAS algorithm's performance is demonstrated through rigorous simulation considering five realistic scientific workflows, including Cybershake, Montage, SIPHT, Epigenomic, and Inspiral [9]. For the simulation, three different sizes of workflows, i.e., small (comprising 50 tasks), medium (consists of 400 tasks), and large (consists of 1000 tasks), are considered.

The experiments were conducted on eight different virtual machines [9], each with varying processing and cost capabilities. The average bandwidth among the VMs is assumed to be 20 Mbps, roughly equal to the average bandwidth between VMs in Amazon EC2. The time interval for billing is set as one hour, which is the same as Amazon EC2. Further, the experiments are performed on PC with core i7, 1.80 GHz, 8 GB RAM, Window 10, and CloudSim toolkit. DRP-DBAS algorithm has been compared with the other state-of-the-art algorithms, including BDA [6], BDSD [7], and GRP-HEFT [8]. The experiments were conducted using four different deadlines and four budget constraints. The deadline constraints are evaluated such that their values lie between the slowest and the fastest runtime. On the other hand, for the budget constraint evaluation, the values lie between the cost of execution on the cheapest and the most expensive VMs.

5.2   Performance Metric

Planning Successful Rate (PSR) is computed to determine the percentage of successful schedule [11]. The PSR can be evaluated to analyze the performance of each algorithm. In addition to this, the ratio of defined deadline and makespan obtained (NM) and the ratio of defined budget and cost obtained (NC) is also computed from [11].

5.3   Performance Comparison

The algorithms are analyzed in terms of the percentage of successful schedules (PSR) obtained. The PSR of different workflows has been plotted in Fig. 1. For the Cybershake workflow (Fig. 1(a)), at the strict deadline constraint ($D_w = 1, B_w = 1$), only DRP-DBAS generates a successful schedule. However, when the budget constraint is relaxed to $B_w = 3, 4$, BDAS also generates a successful schedule. BDSD and GRP-HEFT show 100% failure at the strict deadline. The result of Montage workflow has been shown in Fig. 1(b), from the analysis, it is proved that all the algorithm fails at the strict constraint ($D_w = 1, B_w = 1, 2$). However, the DRP-DBAS improves its performance when the budget constraint is relaxed. BDSD has the worst performance and is outperformed by the rest of the algorithms. When the constraints are relaxed, all the algorithms exhibit 100% success. The PSR of Epigenomic has been shown in Fig. 1(c). The DRP-DBAS delivers remarkable performance by generating viable schedules, even at strict constraints. GRP-HEFT shows 100% failure at the strict deadline ($D_w = 1$) and improves its performance when the deadline constraints are relaxed. BDSD again indicates the worst performance because of the violation of both budget and deadline constraints. For the SIPHT workflow, DRP-DBAS outperforms the other algorithms. The success rate of DRP-DBAS at the constraint ($D_w = 1, B_w = 1$) is nearly 40%, while the rest of the algorithms exhibit 0% success. Similarly, for the Inspiral (LIGO), all the algorithms indicate 100% failure at the constraint ($D_w = 1, B_w = 1, 2, 3, D_w = 2, 3, 4, B_w = 1$). However, when the constraint is relaxed to ($D_w = 1, B_w = 4$), only DRP-DBAS produces a viable schedule.

Figure 2 and Fig. 3 represents time efficiency (Budget constraint Vs Normalized Makespan) and cost efficiency (Deadline constraint Vs. Normalized Cost). To have better representation, the box plot graph has been plotted here to examine algorithms' behavior in terms of meeting/failing deadline over the constant budget or meeting/failing budget over the constant deadline. Further, the metric NM and NC have been used for the analysis. A value greater than NM or NC means the algorithm generates a schedule of makespan/cost lower than the defined deadline/budget. However, the value of $NM/NC < 1$ means the schedule generated has a higher makespan/cost than the defined deadline/budget.

From Fig. 2, it has been shown that for all the range of budget, DRP-DBAS generate schedules of lower makespan and satisfies the deadline for most of the time. The cost-efficiency of the considered algorithms has been demonstrated in Fig. 3. BDSD exhibits the worst performance by obtaining the highest cost of execution. GRP-HEFT produces schedules of minimum cost for most of the range of deadline constraints. However, GRP-HEFT and DRP-DBAS violate some of the budget constraints at all the deadline constraints. Further, each algorithm's overall success rate for all value of budget and deadline constraint on all types of workflows are shown in Table 1.

Overall, it can be concluded that DRP-DBAS behaves outstandingly for the strict constraint. DRP-DBAS satisfies most of the deadline constraints and generates a schedule of minimum makespan. However, most of the failure occurs from the budget violation. BDSD generates most of the failed schedules because of the budget violation; this may be due to the adoption of the basic cost model.

Algorithm 2: Deadline and Budget Aware Workflow Scheduling (DBAS)

Result: TTE_w, TCE_w

1 $VM_{available} = Y$;
2 $VM_{Leased} = \{vm_{1,i}, vm_{1,j}, \ldots vm_{1,k}\}$,, Lease 1 VM of each type present in the list Y;
3 $LT(VM_{Leased}) \leftarrow 1$ hr;
4 $VM_{available} = VM_{available} - VM_{Leased}$;
5 $TaskList \leftarrow NumberofTasksinaDAG$;
6 **while** *(TaskList is not empty)* **do**
7 | Compute $MET(t_i), XST(t_i), XFT(t_i)$ for each task;
8 **end**
9 Compute MET_w;
10 **if** $MET_w \leq D$ **then**
11 | $[TT_{new}, ET_{new}, G_{new}(T_{new}, E_{new})] = Clustering(TT, ET, G(T, E))$;
12 | Compute $SDL$of each task;
13 | $STL \leftarrow$ Sort task according to their SDL;
14 | **for** *each task* $t_i \epsilon STL$ **do**
15 | | Select the first task (t_i) from the list;
16 | | **for** *each instance in* VM_{Leased} **do**
17 | | | Compute EST, EFT of t_i on every VM present in
18 | | **end**
19 | | $EFT(t_i, vm_{z,k})$, Select that VM having minimum EFT
20 | | **if** $EFT(t_i, vm_{z,k} \leq SDL(t_i) \&\& EFT(t_i, vm_{z,k} \leq LT(vm_{z,k})$ **then**
21 | | | $AFT(t_i) = EFT(t_i, vm_{z,k})$
22 | | **else if**
23 | | **then**
24 | | | Select another VM from VM_{Leased} satisfying the condition defined in 20
25 | | **else**
26 | | | **if** $VM_{available}! = empty$ **then**
27 | | | | $SelecttheFastestVM(VM_{new})fromtheVM_{available}$ Compute $EST(t_i, VM_{new}), EFT(t_i, VM_{new})$;
28 | | | | Compare $EFT(t_i)$ on the already leased VMs and new VMs from $VM_{available}$;
29 | | | | $VM_{selected} \leftarrow min(EFT(t_i))$, Select the VM having minimum EFT and satisfying SDL;
30 | | | | **if** $VM_{Selected}$ *is from* VM_{Leased} **then**
31 | | | | | Go to line no 39
32 | | | | **end**
33 | | | | **if** $VM_{Selected}$ *is from* $VM_{available}$ **then**
34 | | | | | $LT(VM_{selected}) = 1hr$;
35 | | | | | $VM_{Leased} = VM_{new}$;
36 | | | | | Remove VM_{new} From $VM_{available}$;
37 | | | | **end**
38 | | | | $N = 0$;
39 | | | | **while** $(EFT(t_i, VM_{Selected}) <$ $SDL(t_i) \&\& (EFT(t_i, VM_{Selected}) > LT(VM_{1,k})$ **do**
40 | | | | | $N = N + 1$;
41 | | | | **end**
42 | | | | **if** *(N number of VMs of the same type of* $VM_{Selected}$ *are present in* $VM_{available}$) **then**
43 | | | | | $LT(VM_{Selected}) = LT(VM_{Selected}) + Nhr$;
44 | | | | | Remove N number of VMs of the same type of $VM_{Selected}$ from $VM_{available}$;
45 | | | | | $AFT = EFT(t_i, VM_{Selected})$;
46 | | | | **else**
47 | | | | | Select another VM from $VM_{available}$ satisfying the condition 39-41 or Go to line no 50 if NO capable VMs are present
48 | | | | **end**
49 | | | **else**
50 | | | | Update the budget constraint;
51 | | | | Go to line no 25
52 | | | **end**
53 | | **end**
54 | **end**
55 | $TTE_w = max(AFT)$;
56 | $TCE_w = Cost(VM_{Leased})$;
57 **end**

BDAS performs better than BDSD and GRP-HEFT but lags behind the DRP-DBAS. On the other hand, GRP-HEFT produces a minimum cost schedule and satisfies most of the budget constraints but fails to fulfill the deadline constraint.

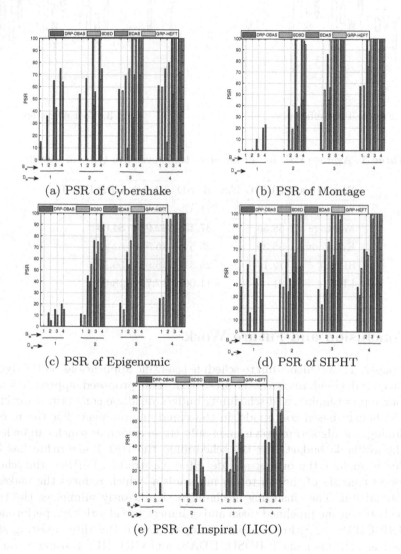

(a) PSR of Cybershake

(b) PSR of Montage

(c) PSR of Epigenomic

(d) PSR of SIPHT

(e) PSR of Inspiral (LIGO)

Fig. 1. PSR

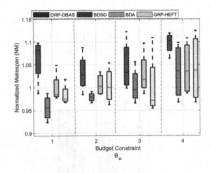

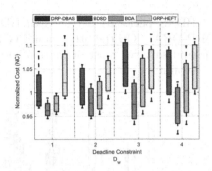

Fig. 2. Time efficiency   **Fig. 3.** Cost efficiency

Table 1. Average PSR for all the algorithms over five types of workflows

Workflows	DRP-DBAS	BDSD	BDAS	GRP-HEFT
Cybershake	73.43	21.37	39.56	42
Montage	58.56	37.43	42.94	54.06
Epigenomic	55.38	35.5	46.5	52
SIPHT	72.18	25.31	48.93	56.31
LIGO	21	11.06	17.75	18.69

6   Conclusion and Future Work

In this paper, a novel algorithm to schedule scientific applications on the dynamically acquired cloud resources is presented. The proposed approach focuses on resource provisioning, and scheduling budget-deadline constrained workflows under an hourly based cost model in the cloud environment. For the resource provisioning, the algorithm provisioned only those resources which can be leased from the available budget. For the scheduling, the HEFT algorithm has been extended to include the budget and deadline constraints. Further, the scheduling process consists of the clustering mechanism, which reduces the makespan of the algorithm. The clustering mechanism significantly minimizes the transfer time between the pipelined tasks and enhances the algorithm's performance. The DRP-DBAS algorithm has been compared with the three existing state-of-the-art algorithms named: BDSD, BDAS, and GRP-HEFT using a range of metrics. In terms of PSR, DRP-DBAS outperforms the rest of the algorithm by achieving an average PSR of 56.11%, while BDSD is the worst performer by achieving 26.134% success.

For the future, it is intended to extend the DRP-DBAS algorithm for a multi-cloud environment. Further, the resource model can be extended such that the data transfer cost between data centers are considered, and the VMs can be deployed in different regions.

References

1. Van Der Aalst, W., Van Hee, K.M., van Hee, K.: Workflow Management: Models, Methods, and Systems. MIT Press, Cambridge (2004)
2. Gupta, A., Garg, R.: Workflow scheduling in heterogeneous computing systems: a survey. In: 2017 International Conference on Computing and Communication Technologies for Smart Nation (IC3TSN), pp. 319–326. IEEE, October 2017
3. Michael, L.P.: Scheduling: Theory, Algorithms, and System (2008)
4. Zhou, N., Lin, W., Feng, W., Shi, F., Pang, X.: Budget-deadline constrained approach for scientific workflows scheduling in a cloud environment. Cluster Comput. 1–15 (2020)
5. Mboula, J.E.N., Kamla, V.C., Djamegni, C.T.: Cost-time trade-off efficient workflow scheduling in cloud. Simul. Model. Pract. Theory **103**, 102107 (2020)
6. Arabnejad, V., Bubendorfer, K., Ng, B.: Budget and deadline aware e-science workflow scheduling in clouds. IEEE Trans. Parallel Distrib. Syst. **30**(1), 29–44 (2018)
7. Sun, T., Xiao, C., Xu, X.: A scheduling algorithm using sub-deadline for workflow applications under budget and deadline constrained. Clust. Comput. **22**(3), 5987–5996 (2018). https://doi.org/10.1007/s10586-018-1751-9
8. Faragardi, H.R., Sedghpour, M.R.S., Fazliahmadi, S., Fahringer, T., Rasouli, N.: GRP-HEFT: a budget-constrained resource provisioning scheme for workflow scheduling in IaaS clouds. IEEE Trans. Parallel Distrib. Syst. **31**(6), 1239–1254 (2019)
9. Juve, G., Chervenak, A., Deelman, E., Bharathi, S., Mehta, G., Vahi, K.: Characterizing and profiling scientific workflows. Futur. Gener. Comput. Syst. **29**(3), 682–692 (2013)
10. Ahmad, W., Alam, B., Ahuja, S., Malik, S.: A dynamic VM provisioning and de-provisioning based cost-efficient deadline-aware scheduling algorithm for Big Data workflow applications in a cloud environment. Clust. Comput. **24**(1), 249–278 (2020). https://doi.org/10.1007/s10586-020-03100-7
11. Ghasemzadeh, M., Arabnejad, H., Barbosa, J.G.: Deadline-budget constrained scheduling algorithm for scientific workflows in a cloud environment. In: 20th International Conference on Principles of Distributed Systems (OPODIS 2016). Schloss Dagstuhl-Leibniz-Zentrum fuer Informatik (2017)
12. Rizvi, N., Ramesh, D.: Fair budget constrained workflow scheduling approach for heterogeneous clouds. Cluster Comput. **23**(4), 3185–3201 (2020). https://doi.org/10.1007/s10586-020-03079-1
13. Qin, Y., Wang, H., Yi, S., Li, X., Zhai, L.: An energy-aware scheduling algorithm for budget-constrained scientific workflows based on multi-objective reinforcement learning. J. Supercomput. **76**(1), 455–480 (2019). https://doi.org/10.1007/s11227-019-03033-y
14. Rodriguez, M.A., Buyya, R.: Deadline based resource provisioning and scheduling algorithm for scientific workflows on clouds. IEEE Trans. Cloud Comput. **2**(2), 222–235 (2014)
15. Wu, Q., Ishikawa, F., Zhu, Q., Xia, Y., Wen, J.: Deadline-constrained cost optimization approaches for workflow scheduling in clouds. IEEE Trans. Parallel Distrib. Syst. **28**(12), 3401–3412 (2017)
16. Topcuoglu, H., Hariri, S., Wu, M.Y.: Performance-effective and low-complexity task scheduling for heterogeneous computing. IEEE Trans. Parallel Distrib. Syst. **13**(3), 260–274 (2002)

Predicting the Movement of Cryptocurrency "Bitcoin" Using Random Forest

Shivani Inder[(⊠)] and Sandhir Sharma

Chitkara Business School, Chitkara University, Rajpura, Punjab, India
{Shivani.chopra,sandhir}@chitkara.edu.in

Abstract. Predicting cryptocurrency is a challenging and interesting job for traders, investors and researchers because of the cost and complexity involved. The current study focuses on predicting the direction of the cryptocurrency 'Bitcoin' for trading window of 3 days, 5 days, and 10 days ahead of the current day. The predictive model proposed is built using ensemble learning via random forest. The model uses feature matrix drawn on technical indicators as input for learning and predicting. The proposed model is empirically evaluated and achieves good levels of accuracy 92%, 85% and 87% for trading windows of 3 days, 5 days and 10 days ahead respectively.

Keywords: Bitcoin · Cryptocurrency · Random forest · Market Prediction

1 Introduction

Crypto-currency markets present a unique extension to the financial markets. Since its birth in 2009, the cryptocurrency market has witnessed phenomenal growth. It has been of peculiar interest to portfolio managers, market regulators, and policy makers.

Concerning the crypto currency markets, efficiency of cryptocurrency markets is a crucial aspect that needs to be assessed. Under the ambit of classical asset pricing theories, financial returns have been assumed to be linear function of an array of underlying factors i.e. risk, momentum, market capitalization, valuation ratios etc. (Fama and French 1993). With increase in the intensity of competition in financial markets, analysis of price performance and behavior of financial instruments play an instrumental role in understanding the financial markets (Parot et al. 2019). Analyzing the behavior and efficiency of cryptocurrency market would update the investor's knowledge in improving the decisions of investors while managing portfolios and its risk. Additionally, it provides a provision of international investment opportunities for the investors.

Predicting the information efficiency of cryptocurrency markets is a tricky job as the exchange rates hold few statistical properties like Dynamic, non-parametric, chaotic and noisy. Such properties make them noisy and non-stationary in nature. Random forest presents an interesting and alternative paradigm to investigate the dynamics of the stock dynamics and information deriving the movement of stock. Additionally, it offers an opportunity to extend the theoretical model for asset pricing to machine learning algorithm. The current study aims to predict the target signal for the movement of exchange

K. R. Venugopal et al. (Eds.): ICInPro 2021, CCIS 1483, pp. 166–180, 2021.
https://doi.org/10.1007/978-3-030-91244-4_14

rate of cryptocurrency 'Bitcoin' by using random forest. The rest of the paper has been organized as follows. Section 2 discusses the related literature. Section 3 elaborates on the research methodology and computation of technical indicators. Section 4 summarizes the data set and features. Section 5 presents followed by conclusion of the study in Sect. 6.

2 Related Literature

'Efficient market hypothesis' (Malkiel and Fama 1970) underpins the efficiency theory for financial markets. The hypothesis suggests that the current market price of any financial asset contains all the information available in the market and prices are reflective of all the relevant information. In other words, the hypothesis suggests that it is not possible to predict the market on the basis of historical information and book profits. The notion of examining the factors influencing the prices and predicting the stock market movement is assumed to be against the basic presumption of 'Efficient Market Hypothesis'. This hypothesis has been highly debated.

Urquhart (2016) analyzed informational efficiency of Bitcoin and found the market for cryptocurrency to be inefficient. Phillip and Gorse (2017) used the information on social media to predict the price bubbles in cryptocurrency markets by using Markov model. Phillip et al. (2017) studied 224 cryptocurrencies to find the presence of long memory, leverage, volatility and heavy tails. Ciaian et al. (2017) found interdependence among 16 altcoins and Bitcoin in the short run, whereas it was missing in the long run. Jiang and Liang (2017) used convolutional neural netowrks for analyzing the cryptocurrency portfolio management and supported the use of deep reinforcement learning. Alessandreti e al. (2018) used machine learning to investigate the efficiency of cryptocurrency market. They suggested that combining machine learning algorithms with simple trading strategies can help traders in guessing the evolution of cryptocurrency market in the short term. Analyzing four cryptocurrencies for the persistence in price behavior from 2013 to 2017, Caporale et al. (2018) suggested inefficiency for Litecoin market. Persistence has been considered as indicator of predictability and reflects on the evidence of market inefficiency and the possibility of earning abnormal profits in the cryptocurrency market. Whereas, Zargar and Kumar (2019) used variance ratio tests to provide evidence on the presence of informational efficiency in Bitcoin market at higher frequency levels. On examining 456 cryptocurrencies, Wei (2018) found Bitcoin returns showing signs of efficiency, whereas a number of crypto-currencies exhibited signs of autocorrelation and non-independence. The results obtained by Kristoufek (2018) indicated that markets were efficient only during cooling downs after bubble like price surges. He also found evidence in favour of market inefficiency for USD and CNY Bitcoin markets from 2010 to 2017. Nan and Kaizoji (2019) proposed a bitcoin based exchange rate of USD/EUR and tested for cointegration with FX series. They found the exchange rate is unbiased in the long run and is unbiased in short run. Whereas, Tran and Leirvik (2020) used adjusted market inefficiency magnitude measure to find prices as significantly inefficient.

A number of studies have been carried out to predict the market through machine learning or deep learning methods. Gencay (1997) examined the forecasting of spot foreign exchange rate returns by drawing information form past buy/sell signals. They used feedforward network regressions and nearest neighbors regression. Patel et al. (2015) employed two stage fusion model by employing different machine learning techniques to predict future values of CNX Nifty and S&P BSE Sensex. They used ten technical indicators as inputs for Support Vector Regression, Artificial Neural Network and Random Forest. Tanaka et al. (2016) utilized random forest to identify bank failures by employing large set of financial statement data. Long et al. (2017) analyzed capital operations on performance of listed companies under different market conditions. The accuracy of random forest was found to be highest in comparison to other machine learning methods. Similarly, Selvin et al. (2017) supported the capacity of deep neural network architecture to make predictions of stock prices and capturing hidden dynamics. They suggested the use of networks like convolutional neural network for relying on current information. Whereas, Zhong and Enke (2017) supported the use of fuzzy robust principal component analysis and Kernel principal component analysis for forecasting the stock prices on daily basis. For selecting a company for investment, Jeevan et al. (2018) used collaborative and content based machine learning model and recurrent neural networks to predict the stock prices. Lohrmann and Luukka (2019) support the use of random forest classifier to predict future state of stock markets. Mercadier and Lardy (2019) appreciated the transparency property of the random forest regression in confirming the impact of equity-to-credit estimate the impact of variables in prediction of credit default swap. Naik and Mohan (2019) and Nikou et al. (2019) favoured deep learning in predicting the stock price movements. Baba and Sevil (2020) used random forest to examine the price performance and underpricing in IPOs issued on Borsa Istanbul. Comparing the accuracy, they find that random forest outperforms other regression models. However, literature suggests inconclusiveness on the efficiency of the cryptocurrency markets. Also, the use of machine learning has been supported for prediction purposes. Traditional methods for assessing the efficiency have suffered from limitation related to medium to long-run predictions like filtering of outliers out of sample, including substantial set of factors, inculcating linearity in the econometric models. Such limitations affect the robustness and reliability of the results. Enhancing the predictability and information set for predicting the market, the study presents a fresh and gritty perspective on examination of the cryptocurrencies.

3   Research Methodology

The first step for analysis is smoothening of the data. For the purposes of the prediction of the movement of the closing prices of the USD/Bitcoin exchange rate, the direction of the movement of closing prices is captured in the series. For exponential smoothing of the exchange rates, we have adopted the 'Risk metrics approach' which smoothens the timeseries recursively.

$$P_o = C_o \tag{1}$$

$$for \ t > 0; P_t = \alpha * C_t + (1 - \alpha) * C_{t-1}$$

Where α is the smoothing factor with $\alpha \in [0, 1]$. The smoothed exchange rate is used for all the further calculations of technical indicators, which are subsequently used for organizing the feature matrix. The 'target' t_i signal indicates the movement of cryptocurrency with reference to 'd' days before the reference day 'i', and is calculated as follows

$$t_i = \gamma \, (P_{i+d} - P_i) \tag{2}$$

$$\text{Where } t_i = \begin{cases} 1 \rightarrow if \ \gamma \ is \ positive \\ 0 \rightarrow if \ \gamma \ is \ either \ negative \ or \ nochange \end{cases}$$

Employing t_i as a variable of choice for prediction offers few statistical benefits. First, the response to t_i will be rule based. Second, it derived from trading book which includes the complete market responses. Third, t_i indicates the generation of excess returns in statistical manner (Avellaneda and Lee 2010, p. 761). The choice of 'alpha' influences the degree of smoothening of the variation or noise. The smoothed time series is used to identify the movement of the price behavior. The study aims to predict the movement of cryptocurrency 'Bitcoin' by drawing knowledge from the set of technical indicators.

3.1 Feature Extraction

We have calculated the following indicators as important parameters for forecasting the direction of the exchange rate for bitcoin. Technical indicators are indicative tools which are used by traders to identify the bullish or bearish signals in the market. We have used the following indicators for the study:

Relative Strength Index (RSI): RSI is a momentum indicator which is used to identify whether the currency is over-purchased or over-sold. If the price is pushed upwards or is moving in upward direction, then the currency is said to be in demand and it is overbought. This indicates overvaluation in the market and the exchange rate can be suggested to go down.

$$RSI = 100 - \frac{100}{1 + RS} \tag{3}$$

$$\text{Where } RS = \frac{Average \ gain \ over \ past \ 14 \ days}{Average \ loss \ over \ past \ 14 \ days} \tag{4}$$

$$\text{For } RSI = \begin{cases} below \ 30 \rightarrow oversold \\ above \ 70 \rightarrow overbought \end{cases}$$

Stochastic Oscillator: It is a technical indicator which captures the speed or momentum of price. Momentum is expected to change before the change in price. It gauges the ratio obtained through the formula given below:

$$SO = \%K = 100 * \frac{(C_o - L_{14})}{(H_{14} - L_{14})} \tag{5}$$

Where,

C_o →
Current closing price, L_{14} → Lowest closing price over past14days, H_{14} →
Highest closing price over the past 14 days.

Williams Percentage range (W%R): It is also a momentum indicator. It is another momentum indicator. It measures the ratio of closing value to highest price for past period of 14 days.

$$W\%R = \frac{(H_{14} - C_o)}{(H_{14} - L_{14})} * (-100) \tag{6}$$

Where
C_o →
Current closing price, L_{14} → Lowest closing price over past 14 days, H_{14} →
Highest closing price over the past 14 days

$$\text{For } W\%R = \begin{cases} above(-20) \to Sell\ signal \\ below(-80) \to Buy\ signal \end{cases} \tag{7}$$

Moving average convergence divergence (MACD): MACD is a momentum indicator that makes a comparison of moving average of prices. The exponential moving average for 12 days is differenced from 12 days exponentially moving average. It considers values drawn on 9 –day EMA as signal line.

$$MACD = EMA_{12}(C_o) - EMA_{26}(C_o) \tag{8}$$

Where
MACD → Moving average convergence divergence, C_o →
closing value of exchange rate, $EMA_n \to n$ day Exponential moving average.

$$\text{When } MACD = \begin{cases} Above\ signal\ line \to Sell\ signal \\ Below\ signal\ line \to Buy\ signal \end{cases} \tag{9}$$

Price rate of change: This indicator comprehends the percentage change in price over different windows of time period.

$$PROC_t = \frac{(C_t - C_{t-n})}{C_{t-n}} \tag{10}$$

Where $PROC_t \to$ Price rate of change time 't'; $C_t \to$ Closing price at time 't'.

For the study, we have used PROC calculated over period of 3 days, 5 days and 10 days.

On Balance Volume: It uses volume shifts to examine the change in exchange rates. It considers cumulative volume and the effect of volume with increase or decrease in the price.

$$OBV_t = \begin{cases} OBV_{t-1} + Vol_t\ if\ C_t > C_{t-1} \\ OBV_{t-1} - Vol_t\ if\ C_t < C_{t-1} \\ OBV_{t-1}\ if\ C_t = C_{t-1} \end{cases} \tag{11}$$

Where,
OBV_t → On balance volume at time t, Vol_t →
trading volume at time t, and $C_t \to$ closing price at timet.

3.2 Random Forest

Bootstrap aggregation is a technique which aims to reduce the variance for the predicted values. In regression the same regression equation is fit number of times to bootstrap the sample and average the result.

Random forests (Breiman 2001), as a modified form of bagging, draw on huge collection of de-correlated trees and then comprehend them through average. Random forests attempt to reduce the variance of the data. They capture complex structures and intricacies in the data and provide a solution with lower levels of bias. Random forest trains the data on the basis of recursive partitioning of the feature matrix by deploying a tree structure. The algorithm divides each child node till the point we reach a pure node. Splitting is carried out by choosing a criteria which achieve the maximum purity of child nodes. With maximum purity, pure nodes are obtained. The pure nodes do not need further splitting and are converted into leaf nodes. For reaching any prediction, the decision tree is traced across all the nodes till the leaf node of decision tree. With ensemble learning, random forests reduce over-fitting of data. The feature space captures available knowledge and forward it into the decision trees.

For the purpose of current study, we have constructed random forest model by using the algorithm given by Breiman (2001). For constructing each decision tree, we divide the original data into subsets randomly. Then, from the training data set, decision trees are grown to the maximum depth of random forest. At the same time, technical indicators from the feature space are used on each split at random. After repeating the procedure n times, we generated 90, 95 and 41 number of trees for $t_i \rightarrow 3, 5$ and 10 days respectively. The results, thus, obtained are discussed in following sections.

4 Data Set and Features

The results obtained are based on data for cryptocurrency 'Bitcoin' and the data has been collected from 'coinmarketcap.com'. The daily closing values for the exchange rate has been collected from 2015 to 31 July 2020. Bitcoin has been selected for two reasons. First, Bitcoin is a pioneer in the cryptocurrency market. Second, it holds the largest market share in the cryptocurrency markets. We would like to emphasize that the Bitcoin is still considered as a representative for the cryptocurrency markets with a debatable legal status.

The basic input data include the closing values, and the volume values for Bitcoin from 1 Jan 2015 to 31 July 2020. Based on the closing values of the exchange rate and volume, the remaining technical indicators are calculated. Subsequently, these indicators act as feature set for learning algorithm. The data set contains all numeric values for features and all the feature values are continuous. The feature values as well as the input data are non-linear in nature. This sets suitable ground for the use of tree-based classifiers for the purposes of further exploration.

The steps followed in the current paper are summarized as follows

$$\begin{pmatrix} Collection\ of\ raw\ data \\ (Exchange\ rate\ Volume\ valueson\ daily\ basis) \end{pmatrix} \rightarrow (Smoothing\ of\ series) \rightarrow \begin{pmatrix} Feauture\ extraction \\ Feature\ matrix \end{pmatrix} \rightarrow \begin{pmatrix} Ensemble\ learning \\ Random\ forest \end{pmatrix}$$

$$\rightarrow \begin{pmatrix} Target\ prediction \\ (over\ trading\ windows\ of\ 3,\ 5\ and\ 10\ days) \end{pmatrix} (1)$$

(12)

5 Results and Discussion

The aim of the model proposed is to predict the direction of the target signal by obtaining information from the technical indicators. This can assist in making a trading decision of buy or sell w.r.t. cryptocurrency 'Bitcoin'. A positive value of 1 for the predicted value indicates that the exchange rate is expected to rise. On the other hand, a prediction of '0' suggests that the rate would fall for next trading data and a suggestion for 'sell' can be made. Thus, the 'target' predicted value attains either '1' or '0'. The 'target' prediction imbibes the stocks' variance over the considered time period. Lower variance observed in the data suggests that we can proceed with the Machine Learning classifiers. Additionally, the lower variance indicates that the proposed model is stable and is relevant from economic point of view. Observing the value of exchange rate, by capturing exchange rate information from past 90 days, the robustness of the proposed model should be evaluated. Being a binary classifier, the results are analyzed for accuracy, precision, recall and specificity.

For assessing the efficacy of the proposed model, we present the results for predicting the 'target' signals over three short term trading windows (t_i): 3 days, 5 days and 10 days ahead. The entire experiment was implemented for three different short term trading windows ahead of the current trading day separately.

5.1 Results of Random Forests

In the current work, we have used the closing values of the exchange rate of 'Bitcoin' cryptocurrency as basic input. On surveying the existing literature, the employment of random forest as processing algorithm with inputs from technical indicators is scant. The current methodology is expected to perform better than the linear classifiers or other statistical analysis which considers linearity of inputs as basic assumption. In the current study, the underlying nature of the input features has not been disturbed. The proposed learning model is expected to perform better in terms of predicting the movement of cryptocurrency 'bitcoin' at large scale.

For the purposes of current study, we have capture exchange rate for cryptocurrency 'Bitcoin' from Jan1, 2015 to July 31, 2020. This provides us with 2039 data points. After calculating all the technical indicators (capturing information for approximately 90 past days) (Refer to methodology section) and cleaning data, we have 1947 datapoints available for further processing. As data split preferences, we have used 20% of all datapoints as holdout test data and 20% of sample datapoints for the purposes of training and validation. This provides us with 1246 datapoints to train the algorithm, 312 datapoints for validation and 389 for testing purposes (See Fig. 1). As part of algorithmic settings, we have used 50% data as training data used per tree and the algorithm optimized at 90, 95 and 41 trees for 'target' signals for 3days, 5 days and 10 days.

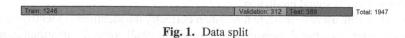

Fig. 1. Data split

The results obtained for the test data for different trading days windows are given in Table 1. In general, the accuracy decreases with increase in the number of trading days to the 'target' signal. Additionally, a relevant observation is that F1-score also decreases with increase in the number of trading days to the 'target' signal.

Table 1. Results obtained for proposed model by using Random Forests

Particulars	$t_i \rightarrow$ 3 days	$t_i \rightarrow$ 5 days	$t_i \rightarrow$ 10 days
Trees	90	95	41
Predictors per split	02	02	02
n(Train)	1246	1246	1246
n(Validation)	312	312	312
N(Test)	389	389	389
Validation accuracy	0.978	0.843	0.897
Test accuracy	0.923	0.851	0.879
OOB accuracy	0.949	0.863	0.892
Recall	0.936	0.851	0.879
Precision	0.936	0.858	0.879
F1-score	0.936	0.851	0.879
AUC	0.983	0.923	0.956

'Accuracy' indicates the extent to which the sample is classified correctly. Referring to Table 1, we find the validation accuracy and test accuracy improves as the time to predict the signal is reduced. In other words, the accuracy of the model is highest i.e. greater than 90% in case of predicting the 'target' signal for 3 days trading window, whereas it reduces with increase in number of trading days.

OOB accuracy refers to the 'Out Of Bag' accuracy. Brieman (1996) provides empirical evidence to indicate that out-of-bag estimate provides the same accuracy as using test set of same size. Subsequently, out-of-bag error estimate eliminates the requirement for using a set aside test set. Brieman (1996) in his original implementation of random forest has used 75% of the total data points for training. As we construct the forest, each tree is tested on the sample data points not included while building the tree. Thus, out of bag error estimate measures an internal error estimate of the random forest while the random forest is being constructed. Observing the OOB accuracy (refer Table 1), we can see that the random forest constructed for 3 days target signal performs better that 5 days and 10 days. The figure no. 1 shows the out-of-classification accuracy plots obtained. The plots indicate that variation in training set and the validation set is least in case 5days of target signal prediction.

'Recall' is a measure to indicate how much correctly the identifier is identifying positive labels. And, 'precision' measures the percentage of all correctly identified samples as 'positive labels'. The results indicate that recall and precision is approximately 93.6% for 3days target signal prediction, whereas it get reduced to around 85% and 89% in case of 5 days and 10 days target signal prediction.

Additionally, confusion matrix for each random forest has been summarized in Table 2.

Table 2. Confusion matrix

Predicted						
		$t_i \to$ 3 days		$t_i \to$ 5 days		$t_i \to$ 10 days
		0	1	0	1	0	1
Observed	0	158 (0.41)	21 (0.05)	152 (0.39)	18 (0.05)	149 (0.38)	22 (0.06)
	1	9 (0.02)	201 (0.52)	40 (0.1)	179 (0.46)	25 (0.06)	193 (0.50)

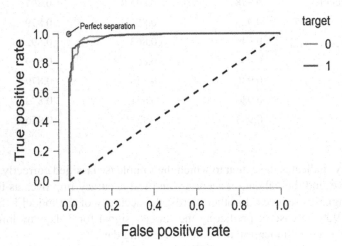

Fig. 2. ROC curves plot (for $t_i \to$ 3 days)

Receiver operating Characteristic or ROC curves have been obtained from random forest for examining the effect of duration of number of trading days. ROC is a graphical method which is instrumental in evaluating the performance of binary classifier. We observe that as we increase the trading days in the trading windows for predicting the 'target' signal, there is decrease in classification accuracy achieved via random forest (See Fig. 2). One plausible reason for decrease in accuracy is that with increase in number of trading days, the technical indicators may lose information with the increase in number of days.

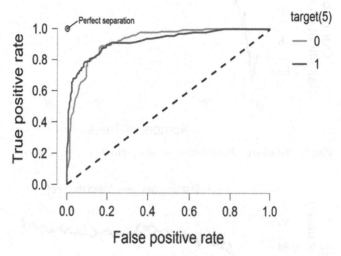

Fig. 3. ROC curves plot (for $t_i \to$ 5 days)

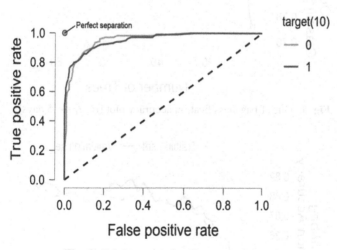

Fig. 4. ROC curves plot (for $t_i \to$ 10 days)

Discovering knowledge from the proposed model should provide new frontiers which can be used as basis for designing the trading strategies. The knowledge should be built on the classification accuracy based on the random forest output. We achieved this through the 'Mean decrease in accuracy' and 'Mean decrease in node purity'. Mean decrease in

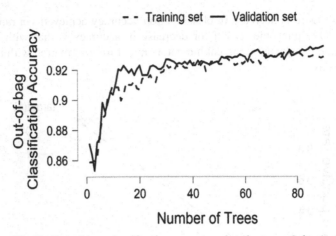

Fig. 5. Out-of-bag classification accuracy plot (for $t_i \rightarrow$ 3 days)

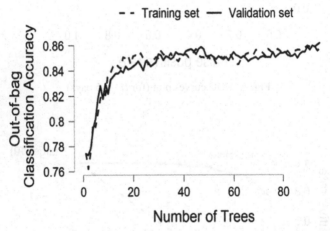

Fig. 6. Out-of-bag classification accuracy plot (for $t_i \rightarrow$ 5 days)

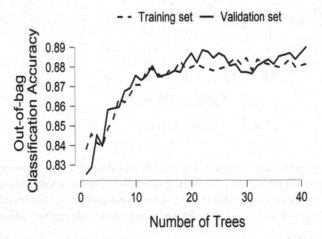

Fig. 7. Out-of-bag classification accuracy plot (for $t_i \rightarrow$ 10 days)

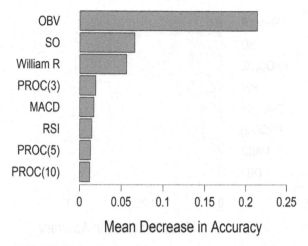

Fig. 8. Mean decrease in accuracy* (for $t_i \rightarrow$ 3 days)

accuracy provides an estimate of loss in prediction capacity when that particular variable is removed or omitted from the training set. Observing the Figure no. 3, we can find that OBV shared redundant information and if we permute this it will not obstruct the model, for predicting the cryptocurrency over a trading window of 3 days. Whereas, in case of trading window of 5 days, we can permute the either William R or SO features to obtain. Similarly for predicting over trading window of 10 days, either PROC (3) or PROC (5) can be permuted (Figs. 3, 4, 5, 6, 7, 8, 9 and 10).

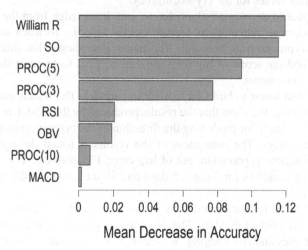

Fig. 9. Mean decrease in accuracy* (for $t_i \rightarrow$ 5 days)

*where William R is Williams Percentage Range, SO is Stochastic Oscillator, RSI is Relative Strength Indicator, OBV is On Balance Volume, MACD is Moving average convergence divergence, PROC(i) is Price rate of change is for 'i' days.

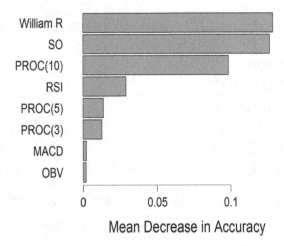

Fig. 10. Mean decrease in accuracy* (for $t_i \rightarrow 10$ days)

6 Conclusion

The use of machine learning techniques for prediction of stock market requires care in execution and detail in avoidance of technical problems. Subsequently, cryptocurrency markets do not depend on set of circumstances belonging to a single country or region, rather they depend on the macro-economic variables and events happening at global level. The current study assumes that all the macro-economic or global events are reflected in the exchange rates values for the cryptocurrency.

The paper tries to develop an algorithm to gain knowledge from the past data and use a set of technical indicators as feature matrix in order to learn and predict the movement of cryptocurrency 'bitcoin'. The paper also shows that machine learning tries to understand the scope of big data and can be used to forecast the direction of cryptocurrency movements.

We use random forest to build on a predictive model. The results produced by the model are impressive. We show that the results produced by the model are robust with a greater level of accuracy for predicting the direction of the cryptocurrency for a trading window for three days. The robustness of the results is tested through a battery of parameters like accuracy, precision, out of bag error, F1 score etc. The comparison of model for trading windows for 3 days, 5 days and 10 days testifies the efficacy for the output of model.

The model can be further extended by including a number of indicators which can capture for the volatility of the data. The proposed model can be used for the design of the trading strategies and risk hedging. We recommend that future research can explore deep learning practices for forecasting the direction of cryptocurrency in the long run.

References

Alessandretti, L., ElBahrawy, A., Aiello, L.M., Baronchelli, A.: Anticipating cryptocurrency prices using machine learning (2018). arXiv: arXiv:1805

Andriyashin, A., Härdle, W.K., Timofeev, R.V.: Recursive portfolio selection with decision trees (2008)

Avellaneda, M., Lee, J.H.: Statistical arbitrage in the US equities market. Quant. Finance **10**(7), 761–782 (2010)

Baba, B., Sevil, G.: Predicting IPO initial returns using random forest. Borsa Istanbul Rev. **20**(1), 13–23 (2020)

Breiman, L.: Random forests. Mach. Learn. **45**(1), 5–32 (2001)

Caporale, G.M., Gil-Alana, L., Plastun, A.: Persistence in the cryptocurrency market. Res. Int. Bus. Finance **46**, 141–148 (2018)

Chavez-Dreyfuss, G., Connor, M.: All the rage a year ago, bitcoin sputters as adoption stalls. Reuters News (2014). www.reuters.com. Accessed 9 Sep 2020

Ciaian, P., Rajcaniova, M., Kancs, D.: The economics of bitcoin price formation. Appl. Econ. **48**(19), 1799–1815 (2016)

Fama, E.F., French, K.R.: Common risk factors in the return on stocks and bonds. J. Finance **33**, 3–56 (1993)

Freifeld, K., Chavez-Dreyfuss, G.: New York regulator issues final virtual currency rules. Technology News (2015). www.reuters.com. Accessed 9 Sep 2020

Gencay, R.: Linear, non-linear and essential foreign exchange rate prediction with simple technical trading rules. J. Int. Econ. **47**(1), 91–107 (1999)

Hu, Y., Valera, H.G.A., Oxley, L.: Market efficiency of the top market-cap cryptocurrencies: Further evidence from a panel framework. Finance Res. Lett. **31**, 138–145 (2019)

Jeevan, B., Naresh, E., Vijaya Kumar, B.P., Kambli, P.: Share price prediction using machine learning technique. In: 2018 IEEE 3rd International Conference on Circuits, Control, Communication and Computing, I4C 2018. Institute of Electrical and Electronics Engineers Inc. (2018)

Jiang, Z., Liang, J.: Cryptocurrency portfolio management with deep reinforcement learning. In: 2017 Intelligent Systems Conference (IntelliSys), pp. 905–913. IEEE (2017)

Krauss, C., Do, X.A., Huck, N.: Deep neural networks, gradient-boosted trees, random forests: Statistical arbitrage on the S&P 500. Eur. J. Oper. Res. **259**(2), 689–702 (2017)

Kristoufek, L.: On Bitcoin markets (IN) efficiency and its evolution. Physica A **503**, 257–262 (2018)

Le Tran, V., Leirvik, T.: Efficiency in the markets of crypto-currencies. Finance Res. Lett. **35**, 101382 (2020)

Lintner, J.: The valuation of risk assets and the selection of risky investments in stock portfolios and capital budgets. In: Stochastic optimization models in finance, pp. 131–155. Academic Press (1975)

Lohrmann, C., Luukka, P.: Classification of intraday S&P500 returns with a random forest. Int. J. Forecast. **35**(1), 390–407 (2019)

Long, W., Song, L., Cui, L.: Relationship between capital operation and market value management of listed companies based on random forest algorithm. Procedia Comput. Sci. **108**, 1271–1280 (2017)

Malkiel, B.G., Fama, E.F.: Efficient capital markets: a review of theory and empirical work. J. Finance **25**(2), 383–417 (1970). https://doi.org/10.1111/j.1540-6261.1970.tb00518.x

Mercadier, M., Lardy, J.-P.: Credit spread approximation and improvement using random forest regression. Eur. J. Oper. Res. **277**(1), 351–365 (2019)

Nakamoto, S.: A peer-to-peer electronic cash system. Bitcoin (2008). https://bitcoin.org/bitcoin.pdf

Nan, Z., Kaizoji, T.: Market efficiency of the bitcoin exchange rate: weak and semi-strong form tests with the spot, futures and forward foreign exchange rates. Int. Rev. Financ. Anal. **64**, 273–281 (2019)

Nikou, M., Mansourfar, G., Bagherzadeh, J.: Stock price prediction using DEEP learning algorithm and its comparison with machine learning algorithms. Intell. Syst. Acc. Finance Manage. **26**(4), 164–174 (2019)

Parot, A., Michell, K., Kristjanpoller, W.D.: Using artificial neural networks to forecast exchange rate, including VAR–VECM residual analysis and prediction linear combination. Intell. Syst. Acc. Finance Manage. **26**(1), 3–15 (2019)

Patel, J., Shah, S., Thakkar, P., Kotecha, K.: Predicting stock market index using fusion of machine learning techniques. Expert Syst. Appl. **42**(4), 2162–2172 (2015)

Phillip, A., Chan, J., Peiris, S.: A new look at cryptocurrencies. Econ. Lett. **163**, 6–9 (2017)

Phillips, R.C., Gorse, D.: Predicting cryptocurrency price bubbles using social media data and epidemic modelling. In: 2017 IEEE Symposium Series on Computational Intelligence (SSCI), pp. 1–7. IEEE (2017)

Ross, S.A.: The arbitrage theory of capital asset pricing. In: MacLean, L.C., Ziemba, W.T. (eds.) Handbook of the Fundamentals of Financial Decision Making: In 2 Parts, pp. 11–30. WORLD SCIENTIFIC (2013). https://doi.org/10.1142/9789814417358_0001

Schwartz, R.A., Whitcomb, D.K.: Evidence on the presence and causes of serial correlation in market model residuals. J. Finan. Quant. Anal. **12**(2), 291–313 (1977)

Selvin, S., Vinayakumar, R., Gopalakrishnan, E.A., Menon, V.K., Soman, K.P.: Stock price prediction using LSTM, RNN and CNN-sliding window model. In: 2017 International Conference on Advances in Computing, Communications and Informatics (2017)

Sharpe, W.F.: Capital asset prices: a theory of market equilibrium under conditions of risk. J. Finan. **19**(3), 425–442 (1964)

Tanaka, K., Kinkyo, T., Hamori, S.: Random forests-based early warning system for bank failures. Econ. Lett. **148**, 118–121 (2016)

Urquhart, A.: The inefficiency of Bitcoin. Econ. Lett. **148**, 80–82 (2016)

Vidal-Tomás, D., Ibañez, A.: Semi-strong efficiency of Bitcoin. Finan. Res. Lett. (2018). https://doi.org/10.1016/j.frl.2018.03.013

Wang, J.J., Wang, J.Z., Zhang, Z.G., Guo, S.P.: Stock index forecasting based on a hybrid model. Omega **40**(6), 758–766 (2012)

Wei, W.C.: Liquidity and market efficiency in cryptocurrencies. Econ. Lett. **168**, 21–24 (2018)

Zargar, F.N., Kumar, D.: Informational inefficiency of Bitcoin: a study based on high-frequency data. Res. Int. Bus. Finan. **47**, 344–353 (2019)

Zhong, X., Enke, D.: Forecasting daily stock market return using dimensionality reduction. Expert Syst. App. **67**, 126–139 (2017)

Zhu, M., Philpotts, D., Stevenson, M.J.: The benefits of tree-based models for stock selection. J. Asset Manage. **13**(6), 437–448 (2012)

Evaluating the Performance of POX and RYU SDN Controllers Using Mininet

Nafees M. Kazi[1]($\boxtimes$), Shekhar R. Suralkar[1], and Umesh S. Bhadade[2]

[1] SSBTs COET, Bambhori, Jalgaon, India
[2] KBC North Maharashtra University, Jalgaon, India

Abstract. Software Defined Networks (SDN) has attracted the researchers and industry due to their flexibility and programmability. SDN has been differentiated from traditional networks in terms of separation of the control plane and forwarding functions. The forwarding decisions are sent by the controller to switches and routers. The switches are responsible only for logical forwarding of the packets. Hence performance of any SDN network depends on the performance of the controller. Lot of SDN controllers are available. In this paper we have evaluated the performance of two well-known python base SDN controllers POX and RYU. Mininet is used as the simulation tool. The performance is evaluated for linear topology, tree topology and datacenter topology with varying scales. We have used D-ITG for performance evaluation. Iperf is also used for measuring the maximum available bandwidth. RYU controller performs better in terms of average delay, jitter, bitrate and throughput. The selection of the controller depends on the application requirements.

Keywords: POX · RYU · SDN · Mininet · D-TIG

1 Introduction

The aim of Software-defined networking (SDN) is to make networks agile and flexible. The network control is in\improved in SDN. SDN has been successful in satisfying the changing needs if the enterprises and service providers. The network engineer or administrate is able to change the traffic from control plane without touching individual switches in the network. Switches are directed by the centralized SDN controller for delivering the services as per requirement. In the traditional networks, individual network devices were making the traffic decisions based on their routing tables.

SDN architecture.

Figure 1 shows the SDN architecture. As depicted in the figure, SDN architecture is divided into three layers: the application layer, the control layer and the infrastructure layer.

The application layer contains the typical network applications or functions organizations use. The applications are intrusion detection systems, load balancing or firewalls. In the traditional networks a specialized module is used for firewall or load balance. In SDN the separate appliance is replaced by the centralized controller which is responsible for management of the data plane.

© Springer Nature Switzerland AG 2021
K. R. Venugopal et al. (Eds.): ICInPro 2021, CCIS 1483, pp. 181–191, 2021.
https://doi.org/10.1007/978-3-030-91244-4_15

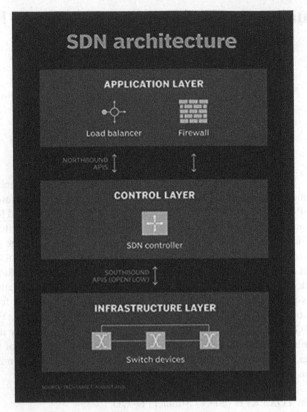

Fig. 1. SDN architecture

The three layers of SDN architecture are communicating through northbound and southbound APIs.

The controller is the brain of the SDN. It resides in the control plane. The controller is responsible for management of policies and network flow. The controller lies in the server. The physical switches in the network lies in the infrastructure layer. The communication between these three layers occurs using respective northbound and southbound application programming interfaces (APIs). Northbound interface is used for communication of applications to the controller. The southbound API are used for communication between controller and switches. OpenFlow protocol is and example of southbound protocol.

- How SDN works

Initially the major focus of SDN was the separation of control plane and data plane. The control plane is responsible for decision making about packet forwarding through the network. The data plane is responsible for logical packet forwarding.

In the classic scenario of SDN, on arrival of the packet, the rules are built on the switch's proprietary firmware were telling the switch regarding packet forwarding. The centralized controller was sending these rules to the switch.

The switch is known as data plane device. It requests the controller for the guidance and provides with the traffic. Every packet having the same destination is treated as the same by the switch and sent along the same path.

In adaptive or dynamic mode of SDN, route request is sent to the controller by the switch for the packet which is not having any specific route. Adaptive routing is different from this. In adaptive routing the route request is dine through routers as well as algorithms which are topology specific. It is not sent through controller.

- Benefits of SDN

SDN offers several advantages over traditional networks as below:

Administrators are able to change the rules including prioritization or blocking specific packets.

The role of the network administrator is to deal with the centralized controller and distribution of packets to the switches. He is not needed to configure the multiple devices connected to the network.

Due to this, the controller can monitor the traffic and security can be deployed.

The controller is able to reroute or drop the packet if found suspicious.

Using SDN, previously dedicated hardware can be virtualized. Hence the operational cost can be reduced.

Software-defined wide area network (SD-WAN) technology is another important aspect of SDN.

- Challenges with SDN

Although security can be achieved through the centralize controller, it is also a cause of concern. The centralized controller becomes the single point of failure.

There is no specific definition of SDN. Different vendors design different approaches according to their requirements.

2 Related Work

Researchers have found the need to change the network architecture from root. SDN is the new changing environment for networks where control plane is separated from forwarding plane. In this paper authors have presented the introduction of SDN along with the architecture, application, different controllers, comparison and limitations of the SDN networks [1].

SDN has introduced the programmability and flexibility into the network by decoupling of control plane from data plane. SDN controller is the important network entity which allows to setting the policies of the network. In the paper [2] authors have compared the performance f new and different SDN controllers.

In [3] authors have compared SDN networks based on security, openflow, mininet, pox controller, floodlight controller and virtual switches. Different virtual security functions are implemented. Hence network security is increased by avoiding loops and broadcasting. Availability of the network is increased by avoiding loops and eliminating different attacks such as congestion driven attacks, distributed denial of service attacks etc.

In SDN the traditional network is split into centralized control plane and programmable data plane. The controller provides flow to switches and optimizes the network performance. Hence the performance of the controller directly affects the performance of SDN controller. Also the benchmarking tool to evaluate the performance must be reliable. In [4] authors have presented a comprehensive qualitative comparison of different SDN controllers. Also quantitative performance is measures. 34 controllers are categorized and compared three benchmarking tools are used to compare 9 controllers.

SDN applications are mostly depend on controllers. Authors in [5] focus on the problem of choosing the right controller. A new method for controller's benchmarking is proposed. The method is divided in two steps. First the controllers are classified according to their features. In the second step the controller performance is evaluated using CBENCH tool where on it three controllers form the first step are considered. The methods is tested on 10 controllers. Authors found the open daylight controller to respond better for all constraints and requirements.

In SDN the architecture has moved from traditional fully distributed model to a more centralized model. This approach is characterized by the separation of control plane and data plane. A more flexible network is created by using the controller to manage the flow. The controller is responsible for making the forwarding decisions and switches only performs the logical forwarding [6].

In SDN the traditional network is distributed in control plane and forwarding plane. This approach is characterized by the separation of control plane and data plane. The control plane contains the controller and the data plane performs the forwarding including switches and routers. The success and failure of the SDN network depends on the controller. In [7] authors have discussed the basic concepts of SDN and compared existing controllers. Different parameters used for comparing controllers are cost, efficiency, ability to support various protocols etc. The controllers are compared based on programming languages.

The advent of smart grid technology has promoted the upgradation of substation automation technology. It has lead to the addition of processing and communication capabilities for controlling and protecting devices. Hence a research is done for supporting such scenario. SDN is able to provide the tools for fulfilling the needs at low costs. In [8] authors provide the survey and comparison of different SDN controllers available.

One of the essential part of the SDN is its controller. Lot of research has been done on numerous controllers as well as comparison is done. The paper [9] tests the newly arrived SDN controllers including ONOS, LibFluid based controllers etc. The benchmarking tool used for evaluation is CBENCH. Results depicts the MUL, Beacon to be the best controllers. But selection of the controller depends on the application and user requirements.

Software Defined Networks depends in the centralized controller. The routing intelligence is decoupled from forwarding functions. It is necessary to understand the performance of the controller before its use. But as lot of controllers are available for research, it becomes difficult to choose the appropriate one. In [10] authors have compared the performance of ONOS, RYU, FloodLight and Open Daylight controllers. The performance parameters used are latency and throughput. Cbench is used as benchmarking tool. Also controllers are compared based on their features.

SDN is able to handle the data forwarding and control plane separately. Programmability has given the central importance in SDN. Most of the researchers are attracting towards SDN due to proprietary and open source. In [11] authors have compared these two strategies of SDN. They have identified the operating principles of both strategies, strengths and weakness.

3   Experimental Setup

During this experiment we have used Mininet as software simulation tool. Different network topologies used are linear, tree and datacenter topology with varying scales. Different steps followed during simulation are as follows.

- Start the mininet
- Create different network topologies
- Create linear topology with 5 switches and varying number of hosts: We have increased the number of hosts from 25, 50, 75,100 per switch.

The command to create the linear topology with fixed 5 switches and 25 hosts per switch is.

- Sudo mn –topo linear, 25, 5 –controller remote
  Create linear topology with varying number of switches: We have increased the number of switches from 8,16,24,32,40,48,56,64,72,80.
  The command to create the linear topology the 16 switches and 5 hosts is.
- Sudo mn –topo linear,5,16 –controller remote
  Create the tree topology with varying depth: We have created the tree topology with varying depth using following command.
- sudo mn –topo tree,depth = 6 –controller remote
  We have evaluated the tree topology performance for depth = 1,2,3,4,5,6,7.
  Create a datacenter topology: We have created the custom datacenter topology. In this topology, the switches and hosts are divided into racks. The centralized switch controls the flows in the racks along with the controller. The rank of the datacenter topology defined the number of switches and hosts in each rack. We have evaluated the datacenter topology for 4,5,6,7,10,12,14,16,18 and 20 ranks.
  The python script 'datacenter.py' is placed in mininet/custom folder. The command to run custom python script is.
- Sudo mn –custom datacenter.py –topo mytopo –controller remote

Start the RYU Remote Controller. RYU is the python based SDN controller. In RYU, various software components are provided with well-defined API. It becomes easy for the developers to create different network management and control applications. Different protocols are supported by RYU for management of network devices including OpenFlow.

In this experiment we have used simple_switch.py application of RYU controller. The application is in RYU/app folder.

The command to start the RYU controller is.

- cd /home/ubuntu/ryu &&./bin/ryu-manager --verbose ryu/app/simple_switch.py

POX Controller of SDN. POX provides the framework for communication along with the switches using OpenFlow protocol. POX is an open source python based controller. Developers can create their own SDN controller using Python and the components provided by POX. POX can be used as the basic SDN controller.

The command to run the pox controller is

- cd /home/ubuntu/pox &&./pox.py log.level --DEBUG forwarding.tutorial_l2_hub

4 Performance Evaluation

To evaluate the performance of the network we have used the D-ITG traffic generator. Different performance parameters considered are.

- Average delay
- Average jitter
- Average bitrate
- Throughput

Throughput of the network is the maximum available bandwidth in the network. We have used the iperf for measuring the throughput of the network.

Table 1 depicts the performance of the linear topology with 5 switches and different number of hosts per switches using RYU controller.

Table 2 depicts the performance of the linear topology with 5 switches and different number of hosts per switches using POX controller.

As shown in Tables 1 and 2, the performance of the RYU controller is better than POX controller for linear topology with 5 switches. The average delay, Jitter and Bitrate are low for the RYU controller. The RYU controller gives better throughput as compared to POX controller.

Table 3 depicts the performance of RYU controller for different number of switches using linear topology. During this experiment the number of switches are increased from 8, 16, 24, 32, 40, 48, 56, 64, 72, 80. Number of hosts connected to each switch are kept constant as 5.

Table 1. Topology with 5 switches for RYU controller

Number of hosts per switch	Linear topology			
	Average delay	Average jitter	Average bitrate	Throughput
25	0.00003	0.000078	7.928062	6.22
50	0.000137	0.000069	8.003869	6.41
75	0.000073	0.000068	7.983175	6.61
100	0.000046	0.000093	7.996148	6.42

Table 2. Topology with 5 switches for POX controller

Number of hosts per switch	Linear topology			
	Average delay	Average jitter	Average bitrate	Throughput
25	0.043475	0.025617	7.08795	5.45 Mb/s
50	0.046752	0.023906	7.962638	4.57 Mb/s
75	0.046215	0.028191	7.887414	3.21 Mb/s
100	0.048998	0.029015	7.995198	3.01 Mb/s

Table 3. Linear topology with 5 hosts per switch for RYU controller

Number of switches	Linear topology			
	Average delay	Average jitter	Average bitrate	Throughput
8	0.00313	0.000074	7.982405	4.47
16	0.000417	0.000101	7.97266	2.09
24	0.000446	0.000162	7.751627	1.45
32	0.000611	0.000156	7.890272	1.64
40	0.00072	0.000192	7.075748	808 MB/s
48	0.000782	0.000162	7.476058	738 MB/s
56	0.000957	0.000338	7.200725	416 MB/s
64	0.000919	0.000215	7.694275	518 MB/s
72	0.000987	0.000304	7.653575	447 MB/s
80	0.001092	0.000212	7.993848	243 MB/s

Table 4 shows the performance of the network using linear topology with varying number of switches for the POX controller.

Table 4. Linear topology with 5 hosts per switch for POX controller

Number of switches	Linear topology			
	Average delay	Average jitter	Average bitrate	Throughput
8	0.04548	0.034342	7.804887	5.47 Mb/s
16	0.080046	0.03112	7.966169	1.67 Mb/s
24	0.218235	0.028962	8.012513	808 Kb/s
32	0.279502	0.029206	7.940093	558 Kb/s
40	0.463598	0.035118	7.790673	418 Kb/s
48	0.538863	0.39861	7.965852	526 Kb/s
56	0.719632	0.045195	8.01967	524 Kb/s
64	0.865373	0.047615	7.786185	319 Kb/s
72	1.092198	0.042286	8.051757	244 Kb/s
80	3.703019	0.101509	7.437361	192 Kb/s

As shown in Tables 3 and 4, the average delay, average jitter and average bitrate is low for RYU controller. For the linear topology with varying number of switches the RYU controller outperforms the POX controller.

A tree topology with varying number of depth (increasing number of switches and hosts) is used. Tree topology with 2 layers is shown in the Fig. 2.

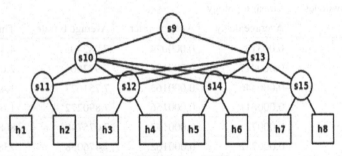

Fig. 2. Tree topology with level = 2

The tree topology is evaluated for varying depth from 1, 2, 3, 4, 5, 6, 7. Table 5 depicts the performance of the network using RYU controller.

Table 6 shows the performance of the POX controller for tree topology.

As shown in Tables 5 and 6 the low average delay, low average jitter, higher bitrate and higher throughput proves the RYU controller to be better than POX controller.

In datacenter topology with 4 rank, we have 4 racks with 4 hosts and single top of the rack switch. These switches are connected to centralized switch. We have evaluated the performance of custom datacenter topology for varying ranks from 4, 5, 6, 7, 10, 12, 14, 16, 18, 20. Table 7 shows the performance of the network for the RYU controller.

Table 5. Tree topology performance using RYU controller

Number layers	Tree topology			
	Average delay	Average jitter	Average bitrate	Throughput
1	0.000258	0.000078	7.986302	7.24
2	0.000271	0.000071	7.823257	7.51
3	0.000308	0.000121	7.995707	6.15
4	0.000339	0.0001	7.451616	5.06
5	0.000293	0.00018	7.982877	4.44
6	0.000372	0.000113	7.826859	4
7	0.000263	0.000098	7.980832	1.88

Table 6. Tree topology performance using pox controller

Number layers	Tree topology			
	Average delay	Average jitter	Average bitrate	Throughput
1	0.0.27389	0.023222	7.930075	18.4 Mb/s
2	0.35395	0.025335	7.896354	8.01 Mb/s
3	0.046666	0.02767	7.79832	4.70 Mb/s
4	0.064173	0.016972	7.892138	2.20 Mb/s
5	0.093521	0.015434	7.991486	1.19 Mb/s
6	0.259626	0.027456	7.958231	526 Kb/s
7	10.178	0.269027	4.047482	237 Kb/s

Table 7. Datacenter topology using RYU controller

Number of racks	Datacenter topology			
	Average delay	Average jitter	Average bitrate	Throughput
4	0.000317	0.000097	7.959375	5.94
5	0.000304	0.00009	7.996362	7.67
6	0.000324	0.000135	7.914234	6.84
7	0.000286	0.000085	7.883428	7.01
10	0.000305	0.000103	7.962812	7.32
12	0.00034	0.000162	7.705806	7.09
14	0.000298	0.000089	7.958661	7.56
16	0.000283	0.00008	7.987428	6.4
18	0.000313	0.000102	7.945782	6.18
20	0.000323	0.000131	7.713832	7.34

Table 8 shows the performance of the datacenter topology using POX controller.

Table 8. Datacenter topology using POX controller

Number of racks	Datacenter topology			
	Average delay	Average jitter	Average bitrate	Throughput
4	0.037798	0.024756	7.890169	7.18 Mb/s
5	0.03711	0.024965	7.914407	6.13 Mb/s
6	0.037606	0.024422	7.956066	5.78 Mb/s
7	0.03917	0.025092	7.934689	4.25 Mb/s
10	0.044474	0.025186	7.911541	4.15 Mb/s
12	0.039727	0.027088	7.927641	3.35 Mb/s
14	0.039834	0.024134	7.938844	2.72 Mb/s
16	0.04914	0.02787	7.869888	2.55 Mb/s
18	0.044209	0.034153	7.877594	2.17 Mb/s
20	0.045272	0.023001	7.954815	1.83 Mb/s

As shown in Tables 7 and 8 the performance of RYU controller is better than the performance of POX controller for the datacenter topology for all scales under consideration.

The performance of RYU controller is better for all network topology under consideration in this study. The performance of the RYU controller does not affect with the increase in number of switches or hosts in the network. While the performance of POX controller degrades with increase in number of switches and hosts.

5 Conclusion

Software Defined Networks are differentiated from the traditional networks due to the separation of control plane from the forwarding functions. The controller in the control plane is responsible for making decisions about how to forward the packets. All instructions are given to the switches regarding packet routing. Decision regarding packet drop or packet forwarding in case of any suspicious packets is done by the controllers. Hence controllers are known as the brain of the SDN. Although lot of SDN controllers are available today, in this research we have used two python based SDN controllers namely, POX and RYU. We have evaluated the performance of the controllers in terms of average delay, average jitter and average bitrate along with throughput. D-ITG tool and iperf is used for measuring the performance of the network. The experiment is carried out using different scenarios. In the first scenario we have created the linear topology with 5 switches and varying number of hosts per switch. In second scenario we have used linear topology with 5 hosts per switch and varying the number of switches. In third case we have used tree topology with increasing depth. In the last scenario we have created custom datacenter topology with increasing number of switches and hosts. After evaluating the performance, we have concluded that for all topology under consideration, average delay, average jitter is, low for RYU controller as compared to POC controller. Hence RYU controller outperforms the POX controller. Average bitrate and throughput is more for RYU controller as compared to POX controller. Performance of the RYU controller does not vary with increasing number of hosts and switches in the network. Performance of POX controller degrades with increasing number of switches and hosts. Although the performance of POX controller is low as compared to RYU controller, the selection of appropriate controller will depends on the requirement and specification of the application.

Acknowledgment. Authors would like to thanks to the management of Shram Sadhana Bombay Trust's College of Engineering & Technology, Bambhori, Jalgaon for providing the infrastructure for carrying out this research work.

References

1. Goswami, B.: Software defined network, controller comparison. Int. J. Innov. Res. Comput. Commun. Eng. An ISO 3297: 2007 Certified Organization **5**(2), 211–217 (2017)
2. El Khalfi, C., El Qadi, A., Bennis, H.: A comparative study of software defined networks controllers. In: ICCWCS 2017, November 14–16, 2017, Larache, Morocco © 2017 Association for Computing Machinery. ACM ISBN 978–1–4503–5306–9/17/11 (2017)
3. Dobrev, D., Avresky, D.: Comparison of SDN controllers for constructing security functions, comparison of SDN controllers for constructing security functions. In: 2019 IEEE 18th International Symposium on Network Computing and Applications (NCA) (2019)
4. Zhu, L., Karim, M.M., Sharif, K., Li, F., Du, X., Guizani, M.: SDN controllers: benchmarking & performance evaluation. IEEE J. Sel. Areas Commun.
5. Belkadi, O., Laaziz, Y.: A systematic and generic method for choosing a SDN controller. Int. J. Comput. Netw. Commun. Secur. **5**(11), 239–247 (2017)
6. Sudarsana Raju, V.R.: SDN Controllers Comparison. In: Proceedings of Science Globe International Conference, 10th June 2018, Bengaluru, India (2018)
7. Semenovykh, A.A., Laponina, O.R.: Comparative analysis of SDN controllers. Int. J. Open. Inf. Technol. **6**(7), 50–56 (2018)
8. Quincozes, S.E., et al.: Survey and Comparison of SDN Controllers for Teleprotection and Control Power Systems. 978–3–903176–23–2 IFIP (2019)
9. Salman, O., Elhajj, I.H., Kayssi, A., Chehab, A.: SDN controllers: a comparative study. In: 2016 18th Mediterranean Electrotechnical Conference (MELECON)
10. Mamushiane, L., Lysko, A., Dlamini, S.: A Comparative Evaluation of the Performance of Popular SDN Controllers. CSIR Pretoria, South Africa
11. Razaa, M.H., Sivakumarb, S.C., Nafarieha, A., Robertsona, B.: A comparison of software defined network (SDN) implementation strategies. In: 2nd International Workshop on Survivable and Robust Optical Networks (IWSRON) Procedia Computer Science, vol. 32, pp. 1050 – 1055 (2014)

Data Science

Design and Development of Data Warehousing for Bookstore Using Pentaho BI Tools

C. S. Anagha[1,2]([✉]) and Siddhaling Urolagin[1,2]

[1] Department of Computer Science, APP Center for AI Research (APPCAIR),
Birla Institute of Technology and Science, Pilani, India
[2] Dubai International Academic City, Dubai, United Arab Emirates

Abstract. The bookstore data warehouse is proposed to analyze the book sell-ing patterns to increase the turnover by means of best sellers and other factors which affect a bookstore. This analysis will help bookstore owners to purchase books according to their customers' interests, language preferences, book types, and other factors. This paper tries to overcome the obstacles the bookstore faces in collecting data for analysis. The data will be in different formats and may contain false values or empty cells. So, analytical query response time will not be efficient. The data needs to be cleaned and integrated before analysis. The data needs to be changed to a standard format which will speed up the query response time. Data warehouse design is very dynamic. A data warehouse can be used for analysis, decision making, and for future planning of a bookstore. The schema design and storage of data is done using the MySQL relational database management system. The Pentaho Server, MySQL workbench, and Data Integration tools are used for data integration and transformation into a standard form. The multi-dimensional database modelling approach used is STAR Schema, which uses dimension tables and a fact table. The data was collected from different bookstores online. The collected information is cleaned and transformed at the bookstore data ware-house. The Pentaho Server is used for querying the cube using the MDX query for transforming the data into tabular and chart representation. Based upon the data transformed and descriptions, the decision-maker or stakeholders can decide on upcoming strategies for successfully achieving target turnover.

Keywords: Data extraction · Data analysis · Data warehouse · Pentaho

1 Introduction

If one can understand the products on a deeper level, it will be easier to order and maintain a stock of those items. It is challenging to analyze the best sellers in a bookstore. A bookstore has several branches, opened in different parts of the world. Analyzing the best sellers is not an easy task in a bookstore. The number of books is increasing day by day, as well as the number of authors. In different places, according to the language, people, etc., the best sellers may vary. Nowadays, many people prefer audio books or kindles rather than printed versions. Since there are different types of single book available on the market, people will choose from these according to their interests. Selling methods

K. R. Venugopal et al. (Eds.): ICInPro 2021, CCIS 1483, pp. 195–210, 2021.
https://doi.org/10.1007/978-3-030-91244-4_16

have also changed. Online ordering of books and homedelivery services also need to be considered along with employee performance for the better performance of a bookstore. The statistical analysis will help a bookstore owner to make better decisions to ensure better turnover and service for the bookstore. This analysis will help to predict the collection of books to be stored in the bookstore. The data from heterogeneous sources is collected and pushed using services to the data warehouse. This study focuses on more aspects which affect a bookstore. This study utilizes data integration tools and the Pentaho Schema workbench. A recent study says 39% read printed books, 7% read digital books and 29% read both printed and digital books. Most surveys come up with the finding that printed book format readers are always higher in number compared to e-books and audio books. A study shows the Holy Bible is the best-selling book in the world. Some readers like a particular author, while some readers like to read a particular type of book. Some publishers' books sell more than others. This is all information that concerns a bookstore owner for decision making.

We identified the need for a data warehouse for a bookstore. The objective of this paper mainly focuses on Datawarehouse architecture for bookstores, where it will be very useful for doing analysis to find the selling pattern. We developed a data warehouse model for the bookstore using Pentaho tools. This model is useful for bookstores to analyze their selling patterns and make decisions to improve their sales and thus make more profit. It is hard to see a Datawarehouse model developed particularly for a bookstore. Our Datawarehouse model discusses more dimensions, and it can be adapted to any bookstore, which will be helpful for any sellers. In the following section, we will describe the background of the bookstore warehouse. Data warehouse architecture is discussed in Sect. 3, schema design in Sect. 4, Extract Transformation and Loading in Sect. 5, Experimental Results in Sect. 6, followed by conclusion in Sect. 7.

2   Literature Survey

A data warehouse is the main repository of an organization's historical data. The data warehouse is optimized for reporting and analysis [1]. A growing number of large enterprises select a data warehouse to help with their decision-making analysis. With a data warehouse, enterprises can understand the information of customers, business conditions, sales channels, and make timely and effective decisions, thereby reducing operating costs, improving customer satisfaction, increasing operating profits and expanding market share [2]. OLAP (Online Analytical Processing) tools offer the possibility of archiving, management, analysis, and multidimensional modelling [3]. Before applying the Business Intelligence technique, data is processed and normalized to avoid data redundancy [4]. The Multidimensional Data Model can be used for the creation of multiple data marts and the design of an ETL process for populating the data marts from the data source [5]. The use of dynamic ETL process using metadata ETL is required when the ETL process is dealing with the operational system and to address the increasing requirements for reporting from users [6]. Business Intelligence tools (BI tools) can be used to analyze large amounts of data [7]. A top-down approach makes it possible to elicit and consolidate user requirements and expectations [8]. The star is composed of two kinds of basic tables: fact tables and dimension tables. The fact table includes operational transactions, or the analysis values wanted, and dimension tables include the description information related to these transactions and values. The star schema exists widely in database application systems [9]. Pentaho data integration can give much better results [10].

In paper [11] we discuss the constraint of a bookstore in processing data, since data increases with increasing time. A bookstore has problems analyzing bestsellers because of the increasing number of authors and book publishers who work together. This paper found a solution for GIS Bookstore by implementing business intelligence that plays a role in the processing of raw data into well-structured data information. The results are collected and processed in a data warehouse scheme. Even though the model needs to be improved for the current situation, with online book purchase, implementation changes are needed to improve the analysis of the old model. This model analyses only 7 dimensions. There are more dimensions which affect the sale of books in bookstores. So, this model cannot be considered as a common model which is applicable to any bookstores. We need to develop a common model for bookstores. In paper [12], they developed a snowflake schema for the library system. which is used to decide on the purchase of books for the library according to students' needs. But this cannot be applicable in the case of a bookstore. This model includes dimensions for selection and management of books. Here, data dimensions can be broken down in more detail. But the snowflake schema requires normalization of tables.

3 Data Warehouse Architecture for Bookstore Data

The Data Warehouse Architecture consists of different parts, as shown in Fig. 1. Extraction Transformation Loading (ETL), Data Warehouse, Data Marts, and Business Intelligence (BI) tools. The data is collected from heterogeneous sources like the operational system, ERP, CRM, and flat files. ETL tools are used for extracting, transforming, and loading data. The data is first stored in the staging data warehouse. Once confirmed and verified, the data is pushed from staging to the production Data Warehouse. The data warehouse consists of metadata, summary data, and raw data. This data will be in the form of a DataMart. By using OLAP tools, the data is presented and visualized. The same data is used for the mining process as well as for future analysis, prediction, and automation.

The Bookstore Datawarehouse shown in Fig. 2 consists of two major parts. One is Online Transactional Processing (OLTP) and the other is Online Analytical Processing (OLAP). This paper deals extra with the OLAP part. The source data is taken from various internal and external sources. Various bookstores and publishers have their own transaction sources, which is known as Online Transactional Processing (OLTP). In this study, we are generating data from bookstore-based applications and spreadsheets. The data from various sources is integrated. This process is well known as Extract Transformation Loading (ETL). The ETL process is practical by using tools. We are using MySQL and Pentaho Tools (Pentaho Server, Schema Workbench, and Pentaho Data Integration Tools). On the Bookstore Data Warehouse design and implementation, we apply the top-down approach to designing the data warehouse.

The data includes information related to bookstores, such as bookstore information, book information, publisher information, customer information, employee information,

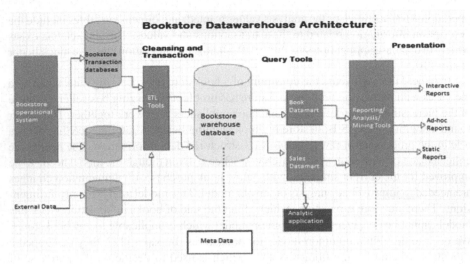

Fig. 1. Datawarehouse architecture

etc. This data is stacked and approved to ensure the extracted information is correct. This is done in the staging area. Here, ETL operations will take place which will utilize ETL devices. Then the information is pushed to the information distribution center. Various deletion, aggregation, and summarization techniques will be done on extracted data, and then loaded back into the data warehouse. After all the rundown is finished, some more changes will be made to the data to make the information structure characterized by the data marts. The source data from different sources is extracted and pushed to the bookstore data warehouse. The data warehouse is built on a relational database management system (RDBMS), and the star schema style is used to develop data warehouses and dimensional data marts. The Cube schema is generated using the Pentaho Schema workbench tool. For querying the data from the bookstore data warehouse database, the MDX (Multidimensional Expression) query is used. For transforming and showing the bookstore statistics in tabular and chart format, the Pentaho Data Integration or JPivot View on the Pentaho BI tool is used.

4   Schema Design

The Star schema has become a common term used to connote a dimensional model. Database designers have long used the term star schema to describe dimensional models because the resulting structure looks like a star and the logical diagram looks like the physical schema. The star model is the basic structure for a dimensional model. It typically has one large central table (called the fact table) and a set of smaller tables (called the dimension tables) arranged in a radial pattern around the fact table. It is more powerful for dealing with straight-forward questions. Usually, the truth tables in a star model are in the third normalized form (3NF), while dimensional tables are in the de-normalized form. The star model is the most used model nowadays and is suggested by Oracle.

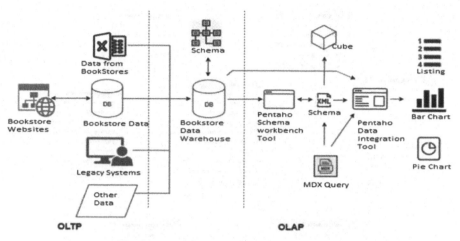

Fig. 2. Bookstore data warehouse architecture

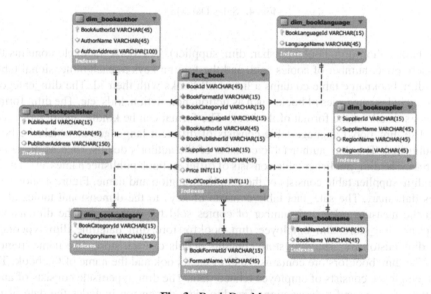

Fig. 3. Book DataMart

In this study, the multidimensional data is illustrated by the Star Schema. The multi-dimensional database consists of dimensions and fact tables. A fact table has two sorts of sections: foreign keys to measurement tables and measures those numeric attributes of a fact, representing the performance or behavior of the bookstore relative to the dimensions. A fact table can contain a fact's data in detail or gathered level. Dimensions are the parameters over which we want to perform Online Analytical Processing (OLAP). For example, time, location, customers, salespeople, etc... Figure 3 shows one of the data marts in the bookstore data warehouse. It consists of a fact table (fact_book) and dimension tables (dim_bookname, dim_bookcategory, dim_bookformat, dim_booklanguage,

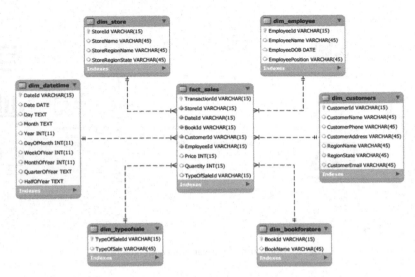

Fig. 4. Sales DataMart

dim_bookauthor, dim_bookpublisher, dim_supplier). The book_facttable contains the measure price, number of copies sold, and the foreign keys to each dimensional table. The dim_bookname table contains a list of all books with their id. The dim_category table includes categories of books such as fiction, non-fiction, kids, etc. The dim_format includes details of the format of the book. The format can be kindle, audio, paperback etc. The dim_bookLanguage table consists of different languages in which the book is published. The dim_author table consists of the author's details, such as name and address. The dim_publisher table consists of different book publisher names and details. The dim_supplier table consists of the supplier's region and name. Figure 4 shows the sales data mart. The sale_fact table consists of keys to the dimensional tables along with the measure price and number of copies sold for each book. The dimensional tables are dim_store, dim_employee, dim_bookforstore, dim_datetime, dim_typeofsale and dim_customers. The dim_store contains details of each store, like name, region, etc. The dim_bookforstore contains the id of each book and the name of the book. The dim_employee consists of employee related data. The dim_typeofsale consists of an id for the sale type and a description. The dim_date dimension includes the date of the transaction (the year, month, week etc.). The dim_customer stores the details about the customer, such as name, region, email etc.

5   Extract Transformation and Loading (ETL)

ETL means Extract Transformation and Loading. Extraction is the process by which data is grabbed from heterogeneous sources and saved in another repository to analyse it later. The transformation of data changes the form of the data from the form in which it is stored to another form which suits the database where the data is going to be stored. It can be done using data filtration-selection of relevant data, by joining data from different sources, by removing duplicate data, etc. For data transformation, specific rules and query tables can be utilized. Data is extracted from different sources and stored on excel sheets. The csv file for each dimensional table as well as fact tables are created. Then the csv files are imported into the Bookstore Schema. Sample data for the Book_Fact table is shown in Table 1. Sample data for Sales_Fact is shown in Table 2.

Tables 3 and 4 show sample data for dimensional tables Dim_BookCategory from Book DataMart and Dim_Date from Sales DataMart. The Mondrian Schema was created using the Pentaho workbench tool. The Mondrian Schema is a logical model that consists of cubes, different levels of hierarchies, and members.

Table 1. Sample data for book fact data

BookId	Book-FormatId	BookCat-egoryId	BookLan-guageId	BookAu-thorId	Book-PublisherId	Sup-plierId	Book-NameId	Book-PublishedDate	Price	NoOfCopies-Sold
B1	F40	BC53	L14	AU65	P8	SU2	PZNJBW	06/04/1914	1344	93470014
B10	F40	BC144	L33	AU261	P50	SU7	VPPNOR	15/05/1932	1545	74678639
B100	F15	BC52	L25	AU199	P37	SU4	SINUQV	19/04/1989	384	95219250
B1000	F27	BC3	L28	AU122	P51	SU5	MXBUCO	26/03/2017	1314	75782165
B1001	F6	BC200	L7	AU147	P53	SU1	BQGKKE	24/05/1901	1779	37631794
B1002	F13	BC173	L13	AU131	P38	SU3	KJEXGK	08/08/2017	1564	14974535
B1003	F41	BC51	L24	AU245	P58	SU2	MQRYSI	14/01/2011	157	11015760
B1004	F5	BC230	L27	AU212	P49	SU6	FBDVLV	07/01/1974	259	18519626
B1005	F1	BC148	L33	AU95	P18	SU3	SXXFED	10/02/1922	1932	19190162
B1006	F35	BC33	L23	AU190	P56	SU1	OYICFW	14/12/1912	884	15665714

Table 2. Sample data for sales fact data

TransactionId	StoreId	DateId	BookId	CustomerId	EmployeeId	Price	Quantity	TypeOfSaleId
T1	ST3	534	B1663	348	EM14	1582	12	Online
T10	ST1	249	B753	462	EM5	331	10	Online
T100	ST4	47	B2138	204	EM7	1548	12	Online
T1000	ST2	1685	B1911	347	EM19	99	19	Online
T1001	ST3	121	B1858	406	EM9	1393	18	WalkIn
T1002	ST1	1471	B326	227	EM9	1819	9	Online
T1003	ST4	1204	B1011	112	EM11	1004	14	Online
T1004	ST5	1336	B697	172	EM16	843	3	WalkIn
T1005	ST5	840	B1436	335	EM14	1883	12	Online
T1006	ST4	1743	B2106	166	EM6	1551	10	Online

Table 3. Sample data for book category dimension data

Book categoryId	Category name
BC1	Prolegomena. Fundamentals of knowledge and culture. Propaedeutics
BC10	Computer communication
BC100	Astronomy. Astrophysics. Space research. Geodesy
BC101	Physics
BC102	Mechanics
BC103	Optics
BC104	Heat. Thermodynamics. Statistical physics
BC105	Electricity. Magnetism. Electromagnetism
BC106	Condensed matter physics. Solid state physics
BC107	Physical nature of matter

Table 4. Sample data for date dimension data

DateId	Date	Day	Month	Year	Day of month	Week of year	Month of year	Quarter of year	Half of year
1	01/01/2001	Monday	January	2001	1	1	1	Q1	H1
10	10/01/2001	Wednesday	January	2001	10	2	1	Q1	H1
100	10/04/2001	Tuesday	April	2001	10	15	4	Q2	H1
1000	27/09/2003	Saturday	September	2003	27	39	9	Q3	H2
1001	28/09/2003	Sunday	September	2003	28	40	9	Q3	H2
1002	29/09/2003	Monday	September	2003	29	40	9	Q3	H2
1003	30/09/2003	Tuesday	September	2003	30	40	9	Q3	H2
1004	01/10/2003	Wednesday	October	2003	1	40	10	Q4	H2
1005	02/10/2003	Thursday	October	2003	2	40	10	Q4	H2
1006	03/10/2003	Friday	October	2003	3	40	10	Q4	H2

6   Experimental Results

We have created about 2500 records in the Book Datamart and about 3000 records in the Sales Datamart. The sample data is randomly generated in Excel documentation (CSV format) and converted to star schema as shown in Figs. 3 and 4. The Mondrian schema and cubes are created using the Pentaho workbench tool as shown in Fig. 5. The created schema is validated and then published to the Pentaho repository. The JPivot view on the Pentaho BI server is used for the bookstore data analysis. The statistics are viewed in tabular or various kinds of chart formats. The MDX Query is used for querying the

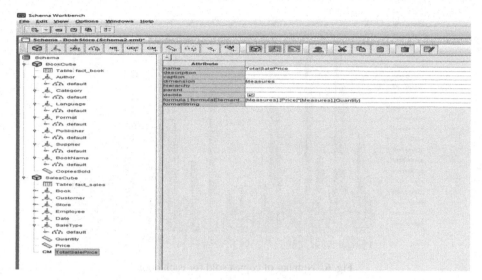

Fig. 5. The bookstore Mondrian schema

Bookstore Datawarehouse database. Some test MDX queries and display results are shown below.

Number of Copies Sold for Each Book. Figure 6 shows the number of copies sold for each book. This report helps to find the best sellers. From the figures, we can find copies of some books sold for 80000000, some are sold for less than 2000000.

> MDX Query: select NON EMPTY {[Measures].[CopiesSold]} ON COLUMNS, NON EMPTY [BookName].[All BookNames].Children ON ROWS from [BookCube].

Books Sold According to Author. The report in Fig. 7 indicates which author's books are more in demand. The author's name is shown in the x axis and the copies of books sold in the y axis. This report will help the analyser to include more books by a particular author.

> MDX Query: select NON EMPTY {[Measures].[CopiesSold]} ON COLUMNS, NON EMPTY [Author].[All Authors].Children ON ROWS from [BookCube].

Books Sold According To Language. Figure 8 shows a pie chart in which each slice represents a language. It can help to find out the sales of books according to their customers' preferred language. This helps to incorporate books in customers' preferred languages.

> MDX Query: select NON EMPTY {[Measures].[CopiesSold]} ON COLUMNS, NON EMPTY [Language].[All Languages].Children ON ROWS from [BookCube].

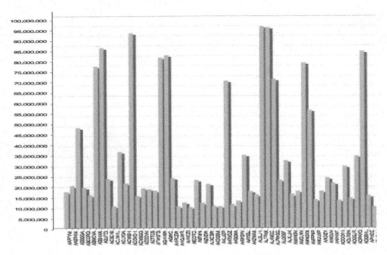

Fig. 6. Number of copies sold for each book

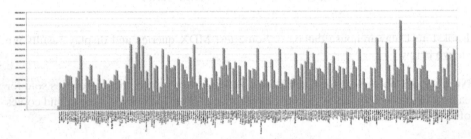

Fig. 7. Number of Books sold according to authors

Sales of a Selected Store in the Year 2001 Quarter Q1 Month of February (Slicing the Cube). Figure 9 shows the results of sales of a particular store in a particular year in a particular month. Here the sales of a single branch, AK Bookstore, are represented for the year 2001 in the quarter Q1 of the month of February. For each day of the month, the sales are represented using a bar chart.

MDX Query: select NON EMPTY Crossjoin({[Store].[AK Bookstore]}, [Date].[2001].[Q1].[February].Children) ON COLUMNS, NON EMPTY {[Measures].[Quantity]} ON ROWS from [SalesCube].

Total Sales of Each Book for the Company. Figure 10 shows the total sales of each book for the company. The result is in a tabular form. It helps to find out the total sale price of each book for the company. The total sale price is calculated by multiplying the sale quantity by the price.

MDX Query: select NON EMPTY {[Measures].[Quantity], [Measures].[Price], [Measures].[TotalSalePrice]} ON COLUMNS, NON EMPTY [Book].[All Books].Children ON ROWS from [SalesCube].

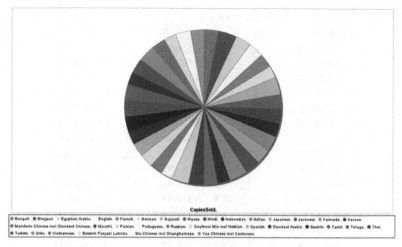

CopiesSold.

◆ Bengali. ◆ Bhojpuri. Egyptian Arabic. English. ◆ French. German. ◆ Gujarati. ◆ Hausa. ◆ Hindi. ◆ Indonesian. Italian. Japanese. ◆ Javanese. Kannada. ◆ Korean.
◆ Mandarin Chinese incl Standard Chinese. ◆ Marathi. Persian. Portuguese. ◆ Russian. Southern Min incl Hokkien. Spanish. ◆ Standard Arabic. ◆ Swahili. ◆ Tamil. ◆ Telugu. ◆ Thai.
◆ Turkish. Urdu. ◆ Vietnamese. Western Punjabi Lahnda. Wu Chinese incl Shanghainese. Yue Chinese incl Cantonese.

Fig. 8. Number of books sold according to the language

	* Store				
	AK Bookstore	AK Bookstore	AK Bookstore	AK Bookstore	AK Books
	* Date	* Date	* Date	* Date	* Date
Measures	6	7	9	10	18
Quantity	6	23	10	20	

Slicer:

```
25
20
15
10
 5
 0
              Quantity
```

Slicer:
■ AK Bookstore.2001.Q1.February.6. ■ AK Bookstore.2001.Q1.February.7.
  AK Bookstore.2001.Q1.February.9.   AK Bookstore.2001.Q1.February.10.
■ AK Bookstore.2001.Q1.February.18.   AK Bookstore.2001.Q1.February.19.
  AK Bookstore.2001.Q1.February.20. ■ AK Bookstore.2001.Q1.February.22.
■ AK Bookstore.2001.Q1.February.23. ■ AK Bookstore.2001.Q1.February.25.
■ AK Bookstore.2001.Q1.February.27.   AK Bookstore.2001.Q1.February.28.

Fig. 9. Sales of a selected store

Performance of a Selected Employee For Each Year (Slicing the Cube). An employee's performance can be calculated by the total sales he made to the company. The bar chart in Fig. 11 shows the sales made by a single employee for each year. It is basically slicing the cube with the dimensions of sales time and employee name (Fig. 12).

MDX   Query: select
NON EMPTY {[Measures].[TotalSalePrice]} ON COLUMNS, NON EMPTY

Book	Measures Quantity	Price	TotalSalePrice
AAPFYM	19	$1,481	$28,139
AASRKM	12	$1,704	$20,448
ABCQRQ	7	$1,328	$9,296
ABMCVA	29	$4,900	$142,100
ABUFEI	12	$1,548	$18,576
ACBCHI	49	$5,840	$286,160
ACJMHL	11	$3,468	$38,148
ACLYOA	11	$1,545	$16,995
ACQVDH	15	$751	$11,265
AETTES	35	$4,144	$145,040
AGHHIW	14	$1,284	$17,976
AGIIYC	1	$1,308	$1,308
AHQJWR	15	$504	$7,560
AIITMN	13	$39	$507
AIZVZW	48	$2,922	$140,256
AJXCBW	13	$1,204	$15,652
AKDQSM	5	$631	$3,155
AKLUZP	5	$2,480	$12,400
AKRXDZ	18	$1,676	$30,168
AKSQXA	3	$1,146	$3,438
AKWSYV	20	$2,232	$44,640
AKYSSJ	24	$198	$4,752
AKZNXM	6	$866	$5,196
ALFPAX	8	$844	$6,752
ALKMEC	17	$2,324	$39,508
ALQOSF	33	$5,619	$185,427
ALXLUK	22	$2,674	$58,828
AMANSV	45	$4,098	$184,410
AMNGYX	4	$1,760	$7,040
AMPPNP	17	$162	$2,754
ANDIZY	26	$234	$6,084
ANYKNF	13	$418	$5,434

Fig. 10. Total sales of each book for the company

Crossjoin({[Employee].[Abrahams]}, [Date].[All Dates].Children) ON ROWS from [SalesCube].

Performance of a Selected Employee for a Selected Year (Dicing the Cube). The chart in Fig. 13 shows the performance of a particular employee for a particular period. It is basically dicing the cube.

MDX Query: select NON EMPTY {[Measures].[TotalSalePrice]} ON COLUMNS, NON EMPTY ([Employee].[Abrahams], [Date].[2001])} ON ROWS from [SalesCube].

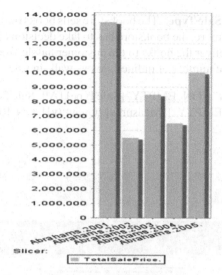

Fig. 11. Performance of a selected employee

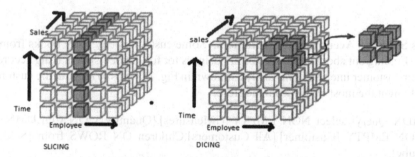

Fig. 12. Slicing and dicing of a cube

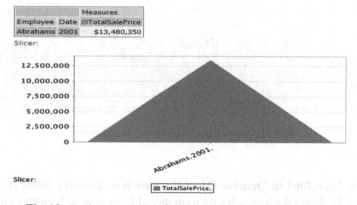

Fig. 13. Performance of a selected employee for a selected year

Sales According to the Sale Type. The bookstore sells books online as well as in person. If online book sales are more, the bookstore has to take decisions about the management of their employees to deliver the books to the customer hassle-free. Figure 14 shows the results of finding out the number of online sales vs walk-in sales.

MDX Query: select NON EMPTY [SaleType].[All SaleTypes].Children ON COLUMNS, NON EMPTY {[Measures].[Quantity]} ON ROWS from [Sales-Cube].

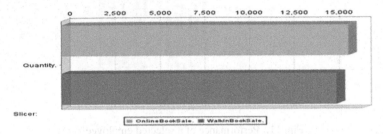

Fig. 14. Online and walk-in sale statistics

Sales Statistics According to Customers. Some customers buy more books from the store. Finding out about them is a part of business for further decision making according to their customer interests. This analysis shown in Fig. 15 helps to find out the customers who bought the most books from the store.

MDX Query: select NON EMPTY {[Measures].[Quantity]} ON COLUMNS, NON EMPTY [Customer].[All Customers].Children ON ROWS from [Sales-Cube].

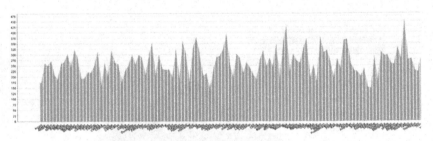

Fig. 15. Sales statistic according to customers

Sales in the Year 2001 in Quarter Q1 of the month of January (Drill down). The result in Fig. 16 shows the total sales for each day of the month of January in quarter Q1 of the year 2001 for each branch of the company. This will help to find out the monthly income for the company from all the stores.

MDX Query: select NON EMPTY {[Measures].[Quantity], [Measures].[Price], [Measures].[TotalSalePrice]} ON COLUMNS, NON EMPTY Crossjoin([Store].[All Stores].Children, Date].[2001].[Q1].[January].Children) ON ROWS from [Sales-Cube].

Store	Date	⊙Quantity	⊙Price	⊙TotalSalePrice
AK Bookstore	2	25	$2,559	$63,975
	3	26	$999	$25,974
	4	9	$323	$2,907
	15	12	$661	$7,932
	25	20	$1,958	$39,160
	29	13	$314	$4,082
	31	5	$1,439	$7,195
Anchor Bookstore	1	8	$248	$1,984
	7	1	$1,459	$1,459
	8	9	$718	$6,462
	11	18	$719	$12,942
	15	20	$1,121	$22,420
	23	23	$2,186	$50,278
	26	6	$641	$3,846
	27	20	$1,876	$37,520
	30	4	$1,160	$4,640
Follet Books	10	1	$1,331	$1,331
	13	3	$1,607	$4,821
	17	5	$332	$1,660
	21	6	$537	$3,222
	24	9	$366	$3,294
	28	11	$1,160	$12,760
Greenleaf Books	2	18	$456	$8,208
	8	19	$439	$8,341
	17	16	$1,120	$17,920
	25	5	$1,659	$8,295
	27	10	$1,541	$15,410
Hachette Book Group	6	23	$1,628	$37,444
	26	9	$1,598	$14,382

Fig. 16. Sales in the year 2001 in quarter Q1 in the month January

7 Conclusion and Future Work

Using Data warehouse and Pentaho tools, the data can be grouped into the newly developed data source. The Analysis of data can be done by using tabular and chart forms. This result can be generated more orderly and easily. The data can be made ready to process by Pentaho schema workbench. Using Pentaho Business Intelligence (BI) Server the dashboard can be prepared. This Panels helps to make decisions for the Analyzer. The Datawarehouse could analyze the bookstore data successfully. It could produce analysis results for number of books sold according to author, format of the book, written language, category, publisher, and supplier. The system enables to find out the best sellers in a bookstore. The Sales data mart enables the analyzer to understand the current sale of the books as well as past sale records. The employee performance could monitor using this system. The online transactions and walk-in transactions also analyzed successfully. The Data warehouse extracted and cleaned data, which can be used for datamining related activities. This data mart can be used by business organizations to plan things accordingly for the future.

References

1. Rifaie, M., Kianmehr, K., Alhajj, R., Ridley, M.J.: Data warehouse architecture and design. In: 2008 IEEE International Conference on Information Reuse and Integration, Las Vegas, NV, USA, 2008, pp. 58–63 (2008).https://doi.org/10.1109/IRI.2008.4583005
2. Song, Q., Liu, L.: Research on the design of a new data warehouse system. In: 2009 2nd IEEE International Conference on Computer Science and Information Technology, Beijing, 2009, pp. 462–465 (2009). https://doi.org/10.1109/ICCSIT.2009.5234503
3. Boutkhoum, O., Hanine, M., Tikniouine, A., Agouti, T.: Integration approach of multicriteria analysis to OLAP systems: Multidimensional model. In: 2013 ACS International Conference on Computer Systems and Applications (AICCSA), Ifrane, 2013, pp. 1–4 (2013). https://doi.org/10.1109/AICCSA.2013.6616464
4. Wang, J., Kourik, J.L.: Data warehouse snowflake design and performance considerations in business analytics. J. Adv. Inf. Technol. **6**(4), 212–216 (2015). https://doi.org/10.12720/jait.6.4.212-216
5. Seah, B.K., Selan, N.E.: Design and implementation of data warehouse with data model using survey-based services data. In: Fourth edition of the International Conference on the Innovative Computing Technology (INTECH 2014), Luton, 2014, pp. 58–64 (2014). https://doi.org/10.1109/INTECH.2014.6927748
6. Wijaya, R., Pudjoatmodjo, B.: An overview and implementation of extraction-transformation-loading (ETL) process in data warehouse (Case study: Department of agriculture). In: 2015 3rd International Conference on Information and Communication Technology (ICoICT), Nusa Dua, 2015, pp. 70–74 (2015). https://doi.org/10.1109/ICoICT.2015.7231399
7. Moscoso-Zea, O., Andres-Sampedro, Luján-Mora, S.: Datawarehouse design for educational data mining. In: 2016 15th International Conference on Information Technology Based Higher Education and Training (ITHET), Istanbul, 2016, pp. 1–6 (2016). https://doi.org/10.1109/ITHET.2016.7760754.
8. https://dl-acm-org-bpdc.new.knimbus.com/journal/tosem
9. Zhang, C., Zhuang, L.: Association rules expand technology research based on attributes. In: 2013 IEEE 4th International Conference on Software Engineering and Service Science, Beijing, pp. 133–135 (2013). https://doi.org/10.1109/ICSESS.2013.6615272
10. Dwiandriani, F., Kusumasari, T.F., Hasibuan, M.A.: Fingerprint clustering algorithm for data profiling using pentaho data integration. In: 2017 2nd International conferences on Information Technology, Information Systems and Electrical Engineering (ICITISEE), Yogyakarta, pp. 359–363 (2017). https://doi.org/10.1109/ICITISEE.2017.8285528
11. Harvy, G.A., Matitaputty, A.S., Girsang, S.M., Isa, S.M.: The use of book store GIS data warehouse in implementing the analysis of most book selling. In: 2019 7th International Conference on Cyber and IT Service Management (CITSM), Jakarta, Indonesia, 2019, pp. 1–5 (2019). https://doi.org/10.1109/CITSM47753.2019.8965404
12. Dahlan, A., Wibowo, F.W.: Design of library data warehouse using snowflake scheme method: case study: library database of campus XYZ. In: 2016 7th International Conference on Intelligent Systems, Modelling and Simulation (ISMS), pp. 318–322. IEEE (2016).

Prediction of Essential Protein Using Machine Learning Technique

Md. Inzamam-Ul-Hossain[✉] and Md. Rafiqul Islam

Computer Science and Engineering Discipline, Khulna University,
Khulna, Bangladesh
https://ku.ac.bd/

Abstract. For the survival and reproduction of organisms, essential proteins are crucial. Identification of the essential protein is important for cell working and drug design. Essential proteins are predicted from many protein-protein interactions (PPI) networks that are developed using high-throughput techniques. Computational methods are used by many existing proposed techniques to identify essential proteins. Many of them considered topological features for essential protein prediction. Some of the research works consider both topological and biological features to identify essential proteins. In this paper, we have proposed a method using machine learning techniques to accomplish the purpose. Here the Saccharomyces Cerevisiae dataset is considered for essential protein prediction. Three classifiers such as XGBoost, Random Forest, and decision tree have been used to predict the essential proteins. We also apply ensemble methods combined with the three classifiers XGBoost, Random Forest, and decision tree for essential protein prediction. The ensemble method gives the highest accuracy to identify essential proteins compared with the other existing methods. On the other hand, XGBoost gives the highest F1 score.

Keywords: Biological information · Ensemble method · Essential proteins · PPI networks · Topological features

1 Introduction

Proteins are the molecular compound and the product of genes [1]. For the activities of the life cells, the role of proteins is very important [2]. Proteins involved in several biological processed and by interacting with other proteins or DNA, carry out cellular functions [3]. Proteins can be classified as essential and non-essential proteins. Essential proteins are found in protein complexes and lack of this type of protein in the cell can cause infertility or even cell death [2,4]. The role of the essential proteins is vital to understand the structure of the organism, designing potential drugs, synthetic biology, to the diagnosis and treatment of disease [2]. To prevent and treat disease, vaccines and new drugs are created with the help of essential proteins. It is also a significant target to detect infections

© Springer Nature Switzerland AG 2021
K. R. Venugopal et al. (Eds.): ICInPro 2021, CCIS 1483, pp. 211–223, 2021.
https://doi.org/10.1007/978-3-030-91244-4_17

or to improve the diagnostic tools [5]. Essential proteins have more tendency to be conserved in biological evaluation compared to non-essential proteins [5]. The deletion of the essential proteins causes the loss of the protein complexes function and deactivates the organisms to function [1]. Because of the importance of the essential proteins, the identification of the essential protein is not only important to understand the key biological process of the organism at the molecular level or the minimum requirement of this type of proteins for the life of the cell but also for human drug and disease target [2,4].

In previous studies, experimental and computational methods were used to identify essential proteins [2]. The former type uses biological experiments such as RNA interference, single-gene knockout, and conditional knockout [1]. The problems with this type of methods are that they requires lots of time, inefficient and costly [4]. Also, these methods are the application for complex organisms, e.g., humans [6]. The later type of method to identify essential proteins is the computational method which costs less than the experimental method [2].

At present, a variety of genome-related datasets are available due to the high throughput of the experimental techniques such as cellular localization data Protein-Protein Interaction (PPI) data, gene expression data, and protein sequence data [6]. High throughput techniques like Yeast two-hybrid system (Y2H), Mass Spectrometry (MS), and tandem Affinity Purification (AP) are the modern technologies developed to generate protein-protein interaction (PPI) data [3]. The computational methods use these types of data to identify essential proteins. Through the analysis of the biological information, many features were discovered to identify essential proteins [6]. The computational method can be classified as supervised and unsupervised machine learning techniques [4,6]. In unsupervised techniques, essential proteins are identified from some essentiality-related data such as cellular localization, gene expression, and PPI networks data, etc.

The centrality-lethality rule was used to identify essential proteins using measures such as Degree Centrality (DC), Betweenness Centrality (BC), Eigenvector Centrality (EC), Subgraph Centrality (SC), Closeness Centrality (CC), Information Centrality (IC), Local Average Connectives (LAC) and Edge Cluster Coefficient Centrality (NC) [1]. The quality of the PPI networks is closely related to these centrality methods. There are many false positives and false negatives interactions and unstable interactions that exist in PPI networks [6,7]. With the availability of genome data, many researchers considered topological features with biological features to identify essential proteins. Peng et al. [8] proposed a method using the PPI and orthology networks to identify essential proteins. In this research, the observation is taken into consideration of the conserved property of the essential proteins compared to the non-essential proteins. Koschutzki et al. [9] proposed a network-motif-based centrality method using the functional substructures and the network centrality to predict the essential proteins. Essential proteins can be identified from comprehensive dynamic networks. For example, Li et al. [10] combined gene expression profile and subcellular localization information to construct TS-PIN dynamic networks to identify essential proteins. In supervised techniques, machine learning algorithms such as Random Tree, RBF network, SVM, etc. are used to predict essential proteins. The prediction of the supervised technique is higher than the unsupervised technique [4]. Deng et al. [11] combined the Naïve Bayes classifier, CN2 rule, C4.5 decision tree, and logistic regression model to predict the essential proteins.

Even though modern technologies identify PPI networks, there are limits to effectively identify essential proteins. In the proposed method, the PPI network of Saccharomyces Cerevisiae (S. Cerevisiae) is used to get seven topological features such as NC, LAC, IC, EC, BC, CC, and DC; biological features such as subcellular Localization (SL) and gene expression data to predict essential proteins. The PPI network data is collected from DIP [12]. SL feature collected from compartments [13]. Log Fold Change and P values from three experiments (ada2, adr1, and aft2) of the gene expression collected from the research proposed by Jüri Reimand [14]. The features are combined and added to the essential proteins list that was collected from [15]. Three classifiers such as Decision Tree, XGBoost, and Random Forest are used to predict the essential proteins. Besides three classifiers, an ensemble method combines the earlier mentioned three classifiers. In our proposed method, the highest accuracy has been achieved to predict the essential proteins from the ensemble method compared with the existing methods.

2   Related Work

Chiou-Yi Hor et al. [16] proposed a modified backward feature selection method and based on selected features, develop the Support Vector Machine (SVM) predictor. From the DIP [12] database, S. cerevisiae and E. coli. datasets were collected. In this method, protein properties (cell cycle and metabolic process), sequence properties (average amino acid and amino acid occurrence), topological properties (betweenness centrality and bit string of double screening scheme related to physical interactions), and other properties (essential index and phyletic retention) were used as features. In this proposed method, the modified sequential backward feature selection was used to address imbalanced data. In this model, at first, the features set was used as the benchmark that was used in literature (Hwang's feature set for S. cerevisiae dataset and Gustafson's feature set for the E. coli dataset). After that, to gain the highest average performance, SVM was applied to the dataset. For the highest performance of the SVM, the parameters of the SVM are recorded as the reference parameter. To get feature subsets, this method applied the backward feature selection procedures. 10-fold cross-validation was used on the feature subsets to observe the significance of the performance. In this paper, only SVM was used for essential protein prediction but other recent classifiers were not applied those can give better results. Although the accuracy is higher than the existing methods, it is possible to further increase the prediction rate.

Wel Liu et al. [7] proposed an improved particle swarm optimization method to identify essential protein named EPPSO (Essential Protein Particle Swarm

Optimization). Both topological features and biological attribute information are used to construct a weighted network and then define update rules of the velocity vector and the position of the particle-based on the network. An index is measured to identify the top-p essential proteins overall essentiality. EPPSO initialized the particles by the degree-based method which selects p nodes with the highest degree in PPI network G. Then, a swarm with high performance in both essentiality and diversity can be obtained. The EPPSO used the yeast dataset collected from the DIP database [12]. The yeast gene expression data collected from GEO [17], domain information collected from the DOMINE database [18], phylogenetic profile data collected from the InParanoid database [19] databases. Total 1011 essential protein collected from MIPS [20], SGD [21], DEG [22] and SGDP [23]. The proposed method EPPSO showed significant results if considered the top 5% and 10% candidates and got accuracy 73.3% and 61.6% respectively. This paper only shows the top selected essential protein prediction rate than all essential proteins that exist in the dataset. Although the prediction accuracy is higher than other existing methods, the accuracy rate can be increased to correctly identifying the more essential proteins.

Wei Liu [1] proposed an essential protein discovery method called ETB-UPPI on uncertain networks. The topological features were integrated with the biological information to identify essential protein. At first, the Simrank problem in an uncertain network converted into a Simrank calculation of a deterministic network. Then, to calculate the protein similarity in a PPI network, the Simrank method was used. Finally, the feature selection technique and integrated multivariate biological data were used to measure the essentiality of the protein. The multivariate data used in this method are subcellular localization information, the Pearson correlation coefficient, and gene ontology (GO) similarity. The proposed method used PPI datasets which are collected from four different sources as DIP [12], MIPS [20], Gavin [24], and Krogan [25]. The list of essential proteins is collected from four different sources such as MIPS [20], SGD [21], DEG [22], and SGDP [23]. This method showed improved accuracy than the other exists essential protein prediction methods. The accuracy rate of this method can be improved to get higher prediction rate to identify essential protein.

3   Proposed Method

Our proposed method is divided into two steps. At first, we prepare the dataset by calculating the values of features, combine them and finally integrate the essential protein status. After this step, the dataset is used for classification using different classifiers to get the performance of each classifier. These two steps are discussed in the following sub-sections.

3.1   Dataset Representation

In the proposed method, we collected the PPI data from DIP [12]. There are 5093 proteins and 24743 interactions among the proteins are available in the

DIP dataset. From this dataset, we got EC, DC, IC, LAC, BC, CC, and NC from cytospace molecular interacting networks visualization software [26]. After collecting the topological features, the subcellular localization feature of S. Cerevisiae is collected from compartments [13] and gene expression data from the existing method proposed by JüriReimand [14]. Here three log fold change values of all proteins from three knockout experiments names ada2, adr1, and aft2 are used. We also used P values of the three experiments of ada2, adr1, and aft2 as three features for the proposed method. The essential protein list is collected from [15] which contains 1167 essential proteins. The description of the dataset of our proposed method is given in Table 1.

In Table 1, all features along with their range are given. Here both topological features (2–8) and biological features (1, 9–14) of S. Cerevisiae species are used. Each feature is at first normalized. After that, each feature is combined to generate a single dataset. To combine the features, a procedure is used. This procedure is applied to all features to make a single dataset. Here with the following steps, the DC feature is combined with the SL feature.

i   Consider a protein of the DC feature and find the same protein in the SL feature.
ii If the protein name matches, then append the SL value of the protein at the end of the DC value of that protein.
iii If the protein name of the DC feature does not exist in the SL feature, append zero (0) as the protein SL value of the protein.
iv Repeat step I to III for all proteins in the DC feature.

After applying the process to all single features, the final dataset is prepared. After that, the essential protein feature is added as the target feature. For this, the following procedure is applied:

i   Compare the protein name of the feature's protein with the essential protein list.
ii If the protein name exists in the essential protein list, then append 'E' after all the protein features. Otherwise, append 'NE' at the end of the protein feature values.

At the time of the feature combined, some missing values appeared in SL, LFC_ada2, LFC_adr1, LFC_aft2, P_ada2, P_adr1, P_aft2. The reason behind this problem is that SL values are collected from the compartment [13] and the topological features are calculated from the DIP [12] dataset. DIP dataset has different versions. So, some proteins in DIP may not be considered in the compartment for the SL feature. Similarly, biological features are collected except for SL from [13]. This paper also used the DIP dataset but it was published in 2010. So, they used the older version of DIP data that is used in the proposed method. The missing value handling technique is applied to replace all missing values. Here, in the proposed method, all missing values are replaced with the mean value of the feature that has missing values. In Fig. 1(a), there is a missing value for protein YIL002W-A for ada2. So, this value is replaced with the mean of the other three values. The mean of the other three values is -0.05059. The missing value is replaced with 0 as shown in Fig. 1(b). Finally, the dataset is normalized to get the final dataset for essential protein prediction. Figure 2, shows the process of the dataset representation.

Table 1. Description of the dataset of the proposed method

Sl No.	Feature name	Description	Range
1	SL	Subcellular Localization	[0, 56]
2	DC	Degree Centrality	[1, 1951]
3	EC	Eigenvector Centrality	[0.0, 0.16175519]
4	IC	Information Centrality	[1.561319, 7.08321]
5	LAC	Local Average Connectivity	[0, 27.31915]
6	BC	Betweenness Centrality	[0, 4856163]
7	CC	Closeness Centrality	[0.000177, 0.129082]
8	NC	Edge clustering coefficient centrality	[0, 814.4247]
9	LFC_ada2	Log fold change values of each protein from ada2 experiment	[−1.14237, 1.195077]
10	LFC_adr1	Log fold change values of each protein from adr1 experiment	[−0.96182, 0.751473]
11	LFC_aft2	Log fold change values of each protein from aft2 experiment	[−1.40095, 1.040725]
12	P_ada2	P value of all proteins based on ada2 experiment	[−0.99926, 0.994302]
13	P_adr1	P value of all proteins based on adr1 experiment	[−0.98093, 0.987936]
14	P_aft2	P value of all proteins based on aft2 experiment	[−0.94769, 0.982318]
15	Essentiality	Values for essential or non-essential	Essential or Non-essential

XGBoost is a supervised learning classifier that is considered as one of the state-of-the-art classifiers for the prediction of the structured dataset. It has faster speed as it implements using the design of gradient boosted decision tree and can use multicore for computation [27]. This classifier can improve the accuracy and can control the overfitting by averaging decision trees. XGBoost can be used in both regression and classification-related problems. In our proposed method, $max\_depth = 4$ is used as a parameter in XGBoost to generate the tree with maximum depth as four. Another parameter $scale\_pos\_weight = 3$ is also applied. By setting the values to the parameters, the highest accuracy is achieved than the other combination of these two parameters.

Random Forest is an ensemble learning method that can be used in classification and regression. It is a boosting system and has some advantages such as reduces the risk of overfitting, offers a high accuracy level, works well on large datasets, produces high accuracy by estimating missing values [28]. It searches for the best features among all features used in classification. It randomized the decision tree and then aggregate prediction by averaging. It has better performance than the decision tree but shows lower performance than the gradient boosted tree. The parameters such as n_ estimators, max_depth, max_features, etc. can be tuned to achieve better performance for prediction. In the proposed method, two parameters named max_ depth and n_ estimator were tuned for better performance and the highest performance was achieved for the values of 6 and 100 respectively. The max_ depth parameter enables Random Forest to generate a tree with a maximum depth of 6. Another parameter n_ estimator helps to generate 100 trees for prediction. Among these 100 trees generated by random forest, the maximum voting of prediction is selected as a performance.

Ensemble learning is a method of combining different models to get better performance than the individual models used in an ensemble. It is also considered as the collection of k models to create an improved model. This method uses the same training dataset on different models and based on the outcome of the models, the final decision can be taken. The outcome of the ensemble method is the weighted average of all models or the majority vote based on each model. In our proposed method, three classifiers named XGBoost, Random Forest, and decision tree were used individually to predict the essential protein. Here, majority voting is used to get the final result from the ensemble method. For all individual and ensemble methods, 10-fold cross-validation is used to get the test accuracy.

Protein	ada2
YGR144W	0.01539
YPL259C	-0.19722
YIL131C	-0.06479
YIL002W-A	0

(a)

Protein	ada2
YGR144W	0.01539
YPL259C	-0.19722
YIL131C	-0.06479
YIL002W-A	-0.05059

(b)

Fig. 1. Proposed dataset (a) Feature with missing value, (b) Feature with replacing missing value by mean value

Figure 3 shows the proposed method where at first the features and the target field were set on the dataset for classification of the essential protein. After that, all the classifiers were implemented in python and then each classifier is tuned as discussed earlier in this section. The ensemble method was also implemented by combining three classifiers. After implementation, confusion matrix of each classifier is obtained as output. This confusion matrix is then used to calculate the performance of each classifier. Figure 4 shows the confusion matrix for XGBoost.

In Fig. 4, NE represents non-essential and E represents essential proteins. Here 3440 non-essential proteins are correctly classified and 678 essential proteins are correctly classified.

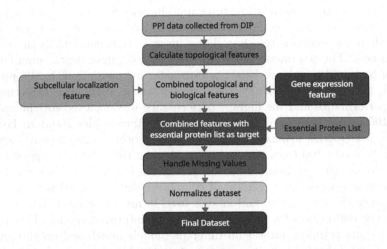

Fig. 2. Dataset representation of the proposed method

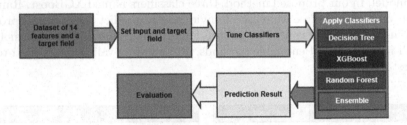

Fig. 3. Proposed method of the essential protein prediction

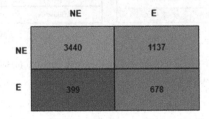

Fig. 4. Confusion matrix of XGBoost

4   Experimental Results

The proposed method was implemented using python programming language and Pycharm as the IDE. The programs were executed in Intel Core i5 @2.50

GHz and 8 GB RAM. The outcome of the proposed method is the confusion matrix which is used to measure the performance of each classifier. In the confusion matrix, TP represents the predicted essential protein which is also represented as essential proteins as in the dataset, TN represents the predicted non-essential protein which is also defined as non-essential in the dataset, FP represents the predicted essential protein which is non-essential in the dataset and FN represents the predicted the non-essential protein which is essential in the dataset.

Table 2. Performance Metrics for XGBoost for Different Parameter Values of max_ depth and scale_ pos_ weight to Predict Essential Protein

max_depth, scale_pos_weight	Number of features	Accuracy	Recall or sensitivity or true positive rate	Specificity	Precision or positive predictive value	Negative predictive value	Error rate	False positive rate	F1 score
3, 3	10	75.49	53.85	80.58	39.48	88.13	24.51	19.42	45.56
3, 4	6	72.57	61.19	75.25	61.19	89.17	27.43	24.75	61.192
5, 4	8	76.34	45.87	83.51	39.55	86.77	22.10	16.49	42.48
6, 3	11	76.91	41.13	85.32	39.73	86.03	23.10	14.68	40.42
6, 4	9	75.92	45.59	83.05	38.76	86.84	24.09	16.95	41.90

To evaluate the performance of the proposed algorithm, the seven measurements that are used are as follows:

Accuracy (ACC): The ratio of correct results in all identified results. To measure the accuracy, Eq. (1) is used.

$$ACC = \frac{TP + TN}{TP + TN + FN + FP} \qquad (1)$$

Precision (PR): The ratio of the correctly detected essential protein in all number of proteins predicted as essential. The equation of PR is as follows:

$$PR = \frac{TP}{FP + TP} \qquad (2)$$

Recall/Sensitivity (SN): Measure the ratio of correctly identified essential proteins in all essential proteins. To compute the sensitivity, the following equation is used:

$$SN = \frac{TP}{FN + TP} \qquad (3)$$

Specificity (SP): Measurement of the ratio of correctly detected non-essential proteins in all the non-essential proteins. It is calculated using the following equation:

$$SP = \frac{TN}{FP + TN} \qquad (4)$$

Positive Predictive Value (PPV): Ratio of correctly detected essential proteins in all the proteins that are predicted as essential. The equation of PPV is shown in Eq. (5).

$$PPV = \frac{TP}{FP + TP} \tag{5}$$

Negative Predictive Value (NPV): Ratio of correctly predicted non-essential proteins in all the non-essential proteins that are predicted as non-essential proteins. It can be calculated using Eq. (6).

$$NPV = \frac{TN}{FN + TN} \tag{6}$$

F-measure (F): The harmonic value of precision and recall. Equation (7) is used for F-measure.

$$F = \frac{2 * (precision * recall)}{precision + recall} \tag{7}$$

All measurements are considered to observe the performance of each classifier used in our proposed method. Different values to the parameters named max_ depth and scale_ pos_ weight are used for XGBoost and got the performance for each variation of the parameter values which is shown in Table 2. In Table 2, it is observed that the maximum accuracy gained is 76.91 for max_ depth= 6 and scale_ pos_ weight = 3 using 11 features. On the other hand, other performance except the accuracy is better for max_ depth = 3 and scale_ pos_ weight= 4 by using only 6 features. After calculating the performance of the XGBoost using different values of its parameters, the performance of the other two classifiers named random forest and decision tree and ensemble method are calculated which are shown in Table 3. In Table 3, it is observed that the highest accuracy is achieved from the ensemble method using 8 features and the second-highest accuracy is achieved from the random forest by using 9 features. Besides accuracy, specificity, negative predictive value, error rate, and the false-positive rate are also good for the ensemble method. But other performance measures such as precision, f1 score are good for the random forest. Table 4 shows the comparison of the XGBoost of our proposed method and the other existing methods.

Table 3. Performance matrics for the different classifiers to predict essential protein

Classifier	Number of features	Accuracy	Recall or sensitivity or true positive rate	Specificity	Precision or positive predictive value	Negative predictive value	Error rate	False positive rate	F1 score
Decision tree	7	79.43	15.97	94.56	40.38	82.71	20.50	05.54	22.89
Random Forest	9	80.39	18.76	94.69	45.39	83.20	19.77	05.31	26.55
Ensemble	8	81.22	13.74	96.57	48.53	82.63	19.20	03.43	21.42

Table 4. Comparison of results of the proposed method and existing methods

Method name	Number of features	Accuracy	Recall or Sensitivity or True positive rate	Specificity	Precision or Positive Predictive Value (PPV)	Negative Predictive Value (NPV)	False Positive Rate (FPR)	F1 Score
Proposed method (ensemble)	8	81.22	0.137	0.966	0.485	0.826	0.304	0.214
Proposed method (XGBoost)	6	72.57	0.612	0.753	0.612	0.892	0.247	0.612
[1]	–	76.40	0.532	0.833	0.487	0.857	–	0.508
[7]	5	70.20	0.498	0.842	0.500	–	–	0.503
[16]	16	0.433	–	0.748	–	–	0.549	–
[29]	–	73.40	0.468	0.824	0.475	0.820	–	0.471
[30]	10	79.27	.4576	0.839	0.458	0.865	0.134	0.547

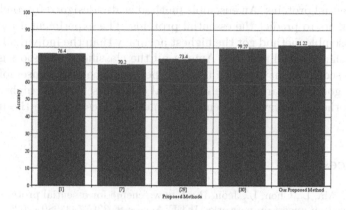

Fig. 5. Comparison of the proposed methods with the existing method based on the accuracy

In Table 4, it is observed that the highest accuracy and specificity are achieved from the ensemble method of the proposed method. In this method, the correct classification rate of the non-essential proteins is high compared to the other existing methods. But the correct classification rate of essential proteins compared to the actual essential proteins in the dataset is low. XGBoost has the highest f1 score of our proposed method along with the sensitivity. Thus, the essential protein prediction rate has the highest value compared with the other existing methods. Thus, in terms of overall performance (accuracy), the ensemble method shows the highest result. The true positive rate and precision for XGBoost are higher than the other classifiers that were used in the proposed method. These values help to get the highest F1 score for XGBoost compared to the other classifiers of our proposed method. This indicates that the correctly predicted essential protein rate is high for XGBoost. So, in terms of essential protein prediction rate, XGBoost shows the best result using 6 features. In Table 4, the dash (-) indicates that the value is not mentioned in the existing method. Table 5 shows the comparison between the existing methods and the proposed method(ensemble) in terms of accuracy. Here, the ensemble achieved the highest accuracy than the existing methods.

5   Conclusion

In recent past research, many computational approaches use topological features, biological information, or the combination of those to predict the essential proteins. But still, it is a big issue to design an efficient method that can integrate different features to predict the essential proteins. We have proposed a machine learning approach to predict essential proteins. Here three classifiers are applied individually such as XGBoost, random forest, and decision tree to predict essential proteins. An ensemble method is also designed by combining the three classifiers to predict the essential proteins. The experimental results show that the ensemble method got the highest accuracy than the individual classifiers to predict the essential proteins. Among all the classifiers, XGBoost gained the highest F1 score. We have shown that the performance of the ensemble method is not only good compared to the individual classifiers, but also higher than the existing methods. However, the ensemble method does not give the highest F1 score.

References

1. Liu, W., Ma, L., Chen, L., Jeon, B.: A new scheme for essential protein identification based on uncertain networks. IEEE Access **8**, 33977–33989 (2020)
2. Zhang, Z., Luo, Y., Hu, S., Li, X., Wang, L., Zhao, B.: A novel method to predict essential proteins based on tensor and HITS algorithm. Human Genom. **14**, 1–12 (2020). https://doi.org/10.1186/s40246-020-00263-7
3. Lei, X., Wang, S., Wu, F.: Identification of essential proteins based on improved HITS algorithm. Genes **10**(2), 177 (2019)
4. Zhong, J., et al.: A novel essential protein identification method based on PPI networks and gene expression data (2020)
5. Qin, C., Sun, Y., Dong, Y.: A new method for identifying essential proteins based on network topology properties and protein complexes. PloS One **11**(8), e0161042 (2016)
6. Zhong, J., Wang, J., Peng, W., Zhang, Z., Li, M.: A feature selection method for prediction essential protein. Tsinghua Sci. Technol. **20**(5), 491–499 (2015)
7. Liu, W., Wang, J., Chen, L., Chen, B.L.: Prediction of protein essentiality by the improved particle swarm optimization. Soft. Comput. **22**(20), 6657–6669 (2017). https://doi.org/10.1007/s00500-017-2964-1
8. Peng, W., Wang, J., Wang, W., Liu, Q., Wu, F.X., Pan, Y.: Iteration method for predicting essential proteins based on orthology and protein-protein interaction networks. BMC Syst. Biol. **6**(1), 1–17 (2012)
9. Koschützki, D., Schwöbbermeyer, H., Schreiber, F.: Ranking of network elements based on functional substructures. J. Theoret. Biol. **248**(3), 471–479 (2007)
10. Li, M., Ni, P., Chen, X., Wang, J., Wu, F.X., Pan, Y.: Construction of refined protein interaction network for predicting essential proteins. IEEE/ACM Transactions Comput. Biol. Bioinform. **16**(4), 1386–1397 (2017)

11. Deng, J., et al.: Investigating the predictability of essential genes across distantly related organisms using an integrative approach. Nucleic Acids Res. **39**(3), 795–807 (2011)
12. DIP Database. www.dip-mbi.ucla.edu/. Accessed 13 Mar 2019
13. Subcellular Localization Database. www.compartments.jensenlab.org/Downloads. Accessed 23 Apr 2020
14. Reimand, J., Vaquerizas, J.M., Todd, A.E., Vilo, J., Luscombe, N.M.: Comprehensive reanalysis of transcription factor knockout expression data in Saccharomyces cerevisiae reveals many new targets. Nucleic Acids Res. **38**(14), 4768–4777 (2010)
15. Chen, W.-H., Minguez, P., Lercher, M.J., Bork, P.: OGEE: an online gene essentiality database. Nucleic Acids Res. **40**(D1), D901–D906 (2012)
16. Hor, C.-Y., Yang, C.-B., Yang, Z.-J., Tseng, C.T.: Prediction of protein essentiality by the support vector machine with statistical tests. Evol. Bioinform. **9**, EBO-S11975 (2013)
17. Gene Expression Omnibus. www.ncbi.nlm.nih.gov/geo/. Accessed 13 Jan 2020
18. DOMINE: Database of Protein Domain Interactions. www.manticore.niehs.nih.gov/cgi-bin/Domine. Accessed 12 Feb 2020
19. InParanoid—ortholog groups with inparalogs. www.inparanoid.sbc.su.se/cgi-bin/index.cgi. Accessed 19 Feb 2020
20. Mewes, H.-W.: MIPS: analysis and annotation of proteins from whole genomes in 2005. Nucleic Acids Res. **34**(suppl-1), 169–172 (2006)
21. Cherry, J.M.: SGD: saccharomyces genome database. Nucleic Acids Res. **26**(1), 73–79 (1998)
22. Zhang, R., Lin, Y.: DEG 50, a database of essential genes in both prokaryotes and eukaryotes. Nucleic Acids Res. **37**(suppl-1), D455–D458 (2009)
23. Saccharomyces Genome Deletion Project. www.sequence.stanford.edu/group/yeast_deletion_project/deletions3.html. Accessed 07 Mar 2020
24. Gavin, A.-C., et al.: Proteome survey reveals modularity of the yeast cell machinery. Nature **440**(7084), 631–636 (2006)
25. Krogan, N.J., et al.: Global landscape of protein complexes in the yeast Saccharomyces cerevisiae. Nature **440**(7084), 637–643 (2006)
26. Cytoscape. www.cytoscape.org/what_is_cytoscape.html. Accessed 21 Oct 2020
27. GPU Accelerated XGBoost. www.xgboost.ai/2016/12/14/GPU-accelerated-xgboost.html. Accessed 21 Sept 2020
28. Random Forest Algorithm. https://www.www.simplilearn.com/tutorials/machine-learning-tutorial/random-forest-algorithm. Accessed 07 Nov 2020
29. Wang, J., Li, M., Wang, H., Pan, Y.: Identification of essential proteins based on edge clustering coefficient. IEEE/ACM Trans. Comput. Biol. Bioinform. **9**(4), 1070–1080 (2011)
30. Zhong, J., Wang, J., Peng, W., Zhang, Z., Pan, Y.: Prediction of essential proteins based on gene expression programming. BMC Genom. **14**(4), 1–8 (2013)

Diabetes Prediction Using Boosting Algorithms: Performance Comparison

Gururaj N. Kulkarni[1], Sateesh Ambesange[2(✉)], A. Preethi[3], and A. Vijayalaxmi[3]

[1] SBS Arts, Commerce and Science for Women, Vijayapur, India
[2] Pragyan School of AI, Bangalore, India
[3] Department of ECE, CMR Institute of Technology, Bangalore, India
{preethi.a,vijayalaxmi.a}@cmrit.ac.in

Abstract. Early detection and control of diabetes can help to prevent associated long term health risks to heart, lungs, kidneys, neural system etc. In this work we have developed a boosting ensemble machine learning (ML) model to predict diabetes based on Pima Indian Diabetes Dataset (PIDD). The data set is preprocessed to enhance the learning ability of the model. We have used various data preprocessing techniques like standardization, outlier removal, data balancing and dimension reduction. The performance of various machine learning algorithms like Logistic Regression (LR), Random Forest (RF) classifier, AdaBoost and Extreme Gradient Boost (XGBoost) are compared to select the best model. The performance metric used for comparison consists of Accuracy, Recall, Precision, F1-Score and Area under ROC Curve (AUC-ROC). Since the application is medical diagnosis, the cost associated with false negative is of utmost importance, thus Recall value played significant role in selecting the best model. Among the basic ML, LR and RF, based models; RF with power transformer achieved highest prediction accuracy and recall value of 0.968 and 0.924 respectively. The boosting ensemble ML models predicted diabetes with better performance metric, which was further improved by hyper-parameter tuning. The AdaBoost based model achieved an accuracy of 0.966 and recall value of 0.97. The best model to predict diabetes based on PIDD, as per this work is hyper-parameter tuned XGBoost model with accuracy and recall value of 1.

Keywords: Diabetes prediction · ML · Boosting algorithm · PIDD · Data balancing · Hyper parameter tuning

1 Introduction

Fatality owing to diabetes every year asper world health organization's report is over 1.6million [1]. A person is said to be diabetic when the blood sugar level in the body is above the normal limit. Cells absorbs glucose from blood with the help of a hormone called insulin produced by Pancreas gland and converts that to energy required for the normal functioning of human body. If the body is not producing sufficient insulin (cause for Type-1 diabetes) or the cells fails to use the insulin produced (cause for Type-2 diabetes) the glucose level in the blood stream increases. The accumulation of blood

© Springer Nature Switzerland AG 2021
K. R. Venugopal et al. (Eds.): ICInPro 2021, CCIS 1483, pp. 224–236, 2021.
https://doi.org/10.1007/978-3-030-91244-4_18

glucose will cause various long term health issues like heart failure, nervous system break down, kidney failure, oral health issues, glaucoma etc.

There is no permanent cure for diabetes, early detection of diabetes or pre diabetes condition opens up the possibility for efficient control of the disease. A medical practitioner's experience and knowledge are in general used for early diagnosis with the help of medical records and family history. Such diagnosis can become inaccurate if some hidden pattern goes unnoticed. Automatic detection of pre-diabetic condition or early diabetic symptoms can remove the ambiguity in analyzing hidden patterns and can provide possibilities of early diagnosis and control of diabetes.

Recent advances in ML domain have improved accuracy of medical diagnosis by facilitating hidden pattern recognition and prediction without ambiguity. In this work eight different ML models to predict diabetes have been developed by employing different ML algorithms combined with data preprocessing and parameter tuning. The database used to train and test the developed models is PIDD. The performances of these models are tested and compared using accuracy, recall and F1 score.

The paper is organized as follows; Sect. 2 reviews the related works for diabetes prediction using ML. Algorithm overview with various techniques used are described in Sect. 3. Discussions or results are available in Sect. 4 and Conclusive remarks on the developed Models and future scope of the present work is discussed in Sect. 5.

2 Literature Review

A few related works available on prediction of Diabetes using ML techniques are discussed in this section [2–16]. Tigga et al. [2] compared the performance of six different ML classification methods, support vector machine (SVM), LR, K-nearest neighbor, Naive Bayes (NB) and Random forest, to predict risk of type-2 diabetes. The study concluded RF Classifier as the most accurate method with an accuracy of 94%. The trained models were tested to self-assess the risk of diabetes by answering simple queries based on health statistics, life style and family back ground.

Meng et al. [3] conducted a study on diabetes or pre-diabetes prediction using three predictive models, artificial neural networks (ANNs), LR, and decision tree. The data set comprising anthropometric measurements, lifestyle, family diabetes history and demographic indicators, was collected through questionnaire from a sample set of normal and diabetic individuals. The study reported the Decision tree based model performed best with 77.87% accuracy.

Aishwarya et al. [4] presented a work to predict Diabetes Mellitus by making use of various ML algorithms. Among various classification algorithms LR gave the highest accuracy, 96%. The best model asper this work is based on AdaBoost classifier which predicted diabetes with highest accuracy of 98.8%. T Alam et al. [5] implemented artificial neural network (ANN), RF and K-means clustering techniques to predict diabetes. In this work with an accuracy of 75.7% the ANN technique proved to be the best and further the work gave a conclusive indication of a strong association of diabetes with glucose level and body mass index.

Vijayan et al. [6] conducted a comparative study of the performance of various base classifiers for AdaBoost algorithm in the prediction of diabetes. Authors used SVM, NB,

Decision tree and Decision Stump as base classifiers for AdaBoost algorithm. According to the study the maximum accuracy of 81.72% was achieved with Decision stump as base classifier. Hasan et al. [7] proposed a weighted ensemble of AdaBoost and XgBoost for diabetes prediction. The data preprocessed to remove outliers and missing values improved robustness and precision. AUC-ROC was the performance metric chosen and further hyper-parameter tuning using the grid search technique is employed to maximize the performance. The work reported an AUC of 0.95 for the proposed model. Mingqi Li et al. [8] proposed an improved XGBoost algorithm to predict diabetics by generating new variables by cross combining different features of the data set. The new features were derived to simulate a clinical practitioner's inquisitiveness in analyzing diabetes data. The proposed Data feature stitching with XGBoost achieved accuracy of 80.2%.

Rahman et al. [9] used boosting ensemble technique with random committee classifier on data, which includes clinical and personal information of patients, to predict the risk of diabetes. The proposed model reported an accuracy of 81%. Deepti et al. [10] conducted a comparative study of diabetes prediction using various ML algorithms like SVM, Naive Bayes and Decision Tree. This study conducted based on PIDD reported maximum accuracy of 76.30% for Naive Bayes. Xu et al. [11] proposed a hybrid model to predict diabetes with an accuracy of 93.75%. The proposed hybrid model used improved RF algorithm for feature selection and XGBoost as classifier.

The data pre-processing is crucial in improving the overall performance of prediction algorithms. Hanskunatai et al. [12] proposed the hybrid model with data balancing which performed better in terms of F-measure compared to model without balancing. The work incorporated combination of under sampling technique and SMOTE- over- sampling algorithm with Decision Tree and Naive Bayes prediction algorithms. Nonso et al. [13] quantified the improvement offered by data preprocessing on PIMA database by comparing the performance of various ML algorithms with and without preprocessing. The SMOTEd + IQRd preprocessing offered an improvement of 3–5% in accuracy compared to the model trained by original data. Paing et al. [14] compared the performance of RF algorithm with various sampling methods such as Random under sampling, random oversampling, Tomek link and SMOTE and the results indicated SMOTE performed better among these four methods. Wang et al. [15] proposed a new ML algorithm to augment the minority class data in an imbalanced classification problem. The algorithm incorporated Relief F based feature selection and AdaBoost for augmenting minority samples. The proposed algorithm, AdaBoost-SVM_OBFS, performance is compared with AdaBoost-SVM to demonstrate the classification effectiveness. Khanam et al. in [16] compared the performance of diabetes prediction by seven ML algorithms on PIDD dataset. The study concluded that the best models for prediction are based on LR and SVM.

3   Algorithm Overview

3.1   Implementation Environment

For implementation of the ML algorithms in this work Anaconda 4 (version 1.9.7), python 3.7 environment is employed. The execution environment for the code is Jupyter Notebook (version 6.0.1).

3.2 Dataset Description

This work compares the performance of various ML models to predict diabetes based on PIDD dataset. Figure 1 illustrates the work flow, each model differs from the others based on the classification method, sampling, feature engineering and tuning. Data set provided by the National Institute of Diabetes and Digestive and Kidney Diseases (NIDDK) is used in this work. The NIDDK aims at improving access to data and bio specimens for the wider science community. The data set for diabetes prediction consisted of measured values, like Insulin level, body mass index Skin thickness, and survey based information, like age, number of pregnancies, Diabetes pedigree function.

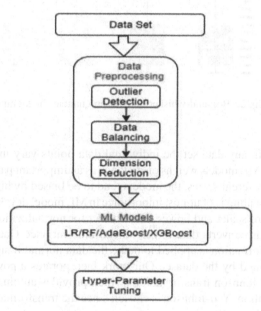

Fig. 1. Model development work flow

3.3 Data Preprocessing

Outlier Detection. An outlier, a data point which differs significantly from the set of observations either due to variability or measurement error, causes significant errors in statistical analyses. The outliers due to measurement errors need to be removed before processing the data set. Among the several methods available to classify outliers, box plot is a popular choice. Box plot is a visual indicator of spread, skewness, quartiles and outliers in a data set. Box length is an indicator of data variability and analysis of data using box plot is based on direct parameters like minimum, maximum, quartile, median and derived parameters like whiskers, interquartile range. The approach of identifying outliers as the data points outside the whiskers is termed as Turkey Fences method. This approach creates a "fence" boundary at a distance of 1.5 time's interquartile range beyond the 1st and 3rd quartiles in this work we have employed turkey outlier detection technique. Figure 2 depicts outlier for the single feature, pregnancies. Once the outliers are treated the data set goes for standardization.

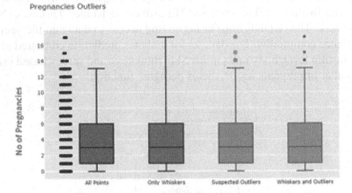

Fig. 2. Box-and-whisker plot for pregnancies histogram

Standardization. In any data set the individual data points vary in magnitude range, realization of many ML models weighs on distance as an important parameter. If features are represented in different scales, the models tend to be biased by higher weighted features. Thus feature scaling is of utmost importance in ML model development. This work incorporates standard scalar and Power Transformer for normalization and standardization. Standard scalar converts data to zero mean unit variance Gaussian distribution. Some type of transformation is applied to make the data normal if significant skewness or kurtosis is exhibited by the dataset. Our work incorporates a power transformation method called Yeo-Johnson transformation, which provides automatic transformation to Gaussian distribution. Yeo-Johnson is chosen since the transformation provides more symmetric distribution.

Data Balancing. Higher accuracy models are generated if the data set available is balanced. The data sets available might not be balanced, as depicted in Fig. 3; the data points might be supporting one class. Sampling techniques are employed to derive balanced data. To emphasis minority classes, over sampling techniques can be employed, which relies on adding copies of instances. The newly generated samples are termed as synthetic data. Over emphasized classes can be balanced by deleting the instances through under sampling. The data balancing can be done by hybrid methods incorporating both under and over sampling. This work make use of Instance Hardness Threshold (IHT), an under sampling technique.

Dimensionality Reduction. The dimension refers to the number of inputs or the features given to an algorithm. If large number of features are used to train an ML model, the chances of over fitting is more and the model will be completely depended on training data. If fewer features are used the model makes less assumptions and it will be a simpler model. Improved accuracy, lesser training time, removal of redundant features and

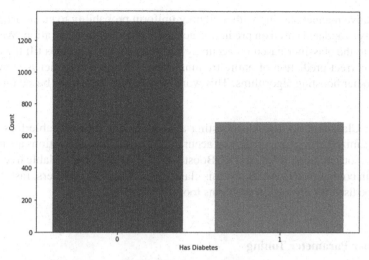

Fig. 3. Histogram of data

saving of storage space are some advantages of dimensionality reduction. In this work we have used Principle Component Analysis (PCA) for dimensionality reduction along with boosting algorithms. In this technique, maximum variance features are chosen as principle components, by projecting the data in the increasing variance direction.

3.4 ML Algorithms

LR. LR classifier is based on probability concept and is a predictive algorithm to solve binary classification problems. This supervised learning technique derives the connection between dependent and independent variable by probability evaluation with the help of sigmoid functions.

RF Classifier. RF is a flexible and easy to use supervised learning algorithm. Decision trees are created using randomly selected data points, prediction from each tree is evaluated and the best solution is chosen by means of voting. This algorithm is a good indicator of important features and can be used to solve classification and regression problems.

Boosting Algorithms. To create improved models ensemble approach is a popular technique. Boosting, one of the ensemble approach, is a method of converting a weak classifier to strong classifier model. In boosting weak learner, which individually not capable giving satisfactory results, is combined with family of week learners to make them strong and outperform other algorithms. AdaBoost, Gradient Boosting and Extreme Gradient Boosting (XGBoost) are the different boosting mechanisms.

AdaBoost Classifier. In this approach a weak learner like decision tree or support vector classifier (SVC), is given with a sub set of data. The data points are accurately predicted in an iterative manner starting with assigning uniform probability to each instance. The probability is updated based on prediction accuracy obtained from iteration. Weights are assigned to the classifier based on accuracy and the process continues till the classifier achieves correct prediction of entire training data. AdaBoost classifier is slower compared to other boosting algorithms. This work implements AdaBoost based on decision tree.

XGBoost Classifier. In gradient boosting algorithms any differentiable loss function can be optimized to improve prediction accuracy. The difficult observations are identified by the previous iteration residuals. XGBoost is one of the efficient scalable tree boosting method. In comparison to AdaBoost this classifier is resistant to outliers, faster, flexible and can be used for regression problems too.

3.5 Hyper Parameter Tuning

The learning process can be controlled by the value of a set of optimal parameters of the algorithm called hyper parameters, thus the hyper parameter tuning is a significant part of ML. Tuning has a direct effect on the behavior of training algorithm and impacts the performance of the obtained model. In our work we have employed grid search for tuning. This technique constructs all possible architectures for the model based on every possible combination of hyper parameters. These architectures are evaluated based on performance matrix and the architecture which gives best result is chosen.

3.6 Models

In this work we have implemented eight ML models to predict diabetes based on different data preprocessing techniques and four different ML algorithms namely LR, RF, AdaBoost and XGBoost as depicted in Table 1.

In Model 1, scaling is done with standard scalar and prediction is done using LR. Further in Model 2, data cleaning and outlier detection was incorporated to improve the performance. The scaling technique is changed to Power Transformer in Model 3 which improved results, henceforth Power Transformer is used in the later models. Keeping the preprocessing techniques same, LR is replaced with RF classifier to arrive at Model 4.

From Model 5 – Model 8 we have used Boosting algorithms with data balancing, dimensionality reduction. The data is balanced using IHT and dimensionality reduction is done through PCA. The Boosting algorithm used in Model 5 and Model 6 is AdaBoost whereas XGBoost is used in Model 7 and Model 8. Further the hyper-parameter tuning is introduced in Model 6 and Model 8.

4 Results

Performances of the developed models are compared using their achieved Accuracy, Precision, Recall, F1 score and AUC-ROC. Accuracy can be a judgement metric for simple classification scenarios whereas Precision and recall gives better insight into the performance of classification algorithms. Precision is the best evaluation metric if the cost associated with false positive is high. Whereas recall is the best metric when the cost associated with false negative is of highest importance. F1 ratio provides a balanced evaluation between precision and recall when the dataset is imbalanced, with true negative as the majority class. Since our work concentrates on prediction of diabetes, the cost associated with false negative is high; the best model selection is majorly based on Recall and F1-score.

Table 2 illustrates the performance of developed ML models based on the selected parameters. Normalised data, using standard scalar, without any other data pre-processing was used to train LR based Model 1. The model predicted diabetes with

Table 1. ML models

Models	Data cleaning	Outlier removal	Scaling	Dimension reduction	Sampling	Algorithm	Hyper parameter tuning
Model 1	-	-	Standard scalar	-	-	LR	-
Model 2	Yes	Turkey outlier Detection	Standard scalar	-	-	LR	-
Model 3	Yes	Turkey outlier detection	Power transformer	-	-	LR	-
Model 4	Yes	Turkey outlier detection	Power transformer	-	-	RF Classifier	-
Model 5	Yes	Turkey outlier detection	Power transformer	PCA	IHT	AdaBoost classifier	-
Model 6	Yes	Turkey outlier detection	Power transformer	PCA	IHT	AdaBoost classifier	Randomized search CV
Model 7	Yes	Turkey outlier detection	Power transformer	PCA	IHT	XGB classifier	-
Model 8	Yes	Turkey outlier detection	Power transformer	PCA	IHT	XGB classifier	Randomized search CV

an accuracy of 0.783 and a recall value of 0.531.Slight improvement in performance was obtained with Model 2, where we have incorporated outlier detection and data cleaning. To further improve the performance standard scalar was replaced with power transformer in Model 3 which resulted in an accuracy of 0.815 and a recall value of 0.664. RF based Model 4 performed better than LR based models and achieved a prediction accuracy of 0.968 and a recall value of 0.924.

Using balanced data and Boosting algorithms great improvement in performance was obtained in Model 5–8, which was further improved by hyper parameter tuning. Model 5, even though the accuracy is lower, is a better model than Model 4 based on recall value. The Boosting algorithm based models performed equally well with dimensionality reduction using only five features. The Model 7 without dimensionality reduction achieved an F1-score of 0.978 compared to the present score of the model 0.973. The hyper-parameter tuning in AdaBoost did not improve recall value but the performance was better based on accuracy and f1score.

Among these models, Model 8 incorporating XGBoost algorithm with hyper-parameter tuning evolved as the best model with an accuracy and recall value of 1. Table 3 depicts the classification report, confusion matrix and ROC_AUC for the best models based on LR, RF, AdaBoost and XGBoost.

Table 2. Results

Models	Accuracy	Balanced accuracy	Precision	Average precision	F1 Score	Recall	ROC- AUC
Model 1	0.783	0.716	0.716	0.530	0.610	0.531	0.834
Model 2	0.788	0.730	0.709	0.542	0.632	0.570	0.839
Model 3	0.815	0.776	0.744	0.604	0.702	0.664	0.892
Model 4	0.968	0.958	0.985	0.937	0.953	0.924	0.995
Model 5	0.945	0.946	0.922	0.909	0.945	0.970	0.971
Model 6	0.966	0.966	0.956	0.941	0.963	0.970	0.990
Model 7	0.973	0.973	0.960	0.953	0.973	0.986	0.997
Model 8	1	1	1	1	1	1	0.999

Table 3. Results: Reports for best models

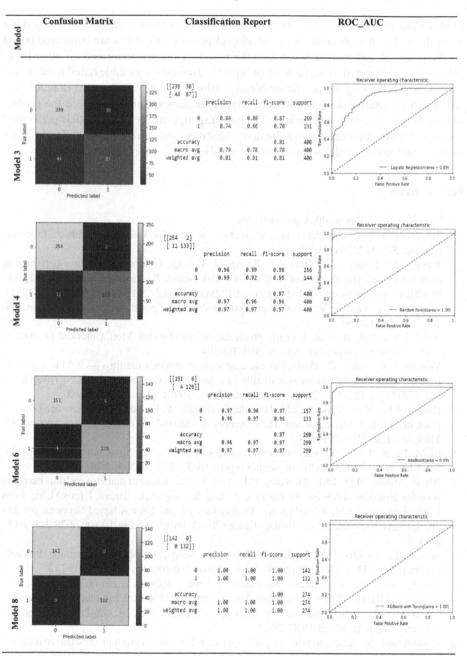

5   Conclusion

In this work diabetes is predicted using basic ML algorithms and boosting ensemble algorithms. The performance of the developed prediction models are compared based on Accuracy, Recall and F1-score. The RF based model outperformed all other basic ML models. Compared to the basic algorithms the Boosting ensemble based models performed better. Among the two ensemble algorithms, major improvement in prediction was given by XGBoost over AdaBoost. Hyperparameter tuning enhanced the prediction metrics of ensemble algorithms. The boosting algorithms based models performed equally well with dimensionality reduction. The study can be further extended to different data sets and boosting algorithms to arrive at robust model for diabetes prediction.

References

1. https://www.who.int/health-topics/diabetes
2. Tigga, N.P., Garg, S.: Prediction of type 2 diabetes using ML classification methods. Procedia. Comput. Sci. **167**, 706–716 (2020)
3. Meng, X.H., Huang, Y.X., Rao, D.P., Zhang, Q., Liu, Q.: Comparison of three data mining models for predicting diabetes or prediabetes by risk factors. Kaohsiung J. Med. Sci. **29**(2), 93–99 (2013). https://doi.org/10.1016/j.kjms.2012.08.016. Epub 2012 Oct 16 PMID: 23347811
4. Mujumdar, A., Vaidehi, V.: Diabetes prediction using ML algorithms. Procedia Comput. Sci. **165**, 292–299 (2019)
5. Alam, T.M., et al.: A model for early prediction of diabetes. Inf. Med. Unlocked **16**, 100204 (2019). https://doi.org/10.1016/j.imu.2019.100204
6. Vijayan, V.V., Anjali, C.: Prediction and diagnosis of diabetes mellitus — A M.L approach. In: 2015 IEEE Recent Advances in Intelligent Computational Systems (RAICS) Trivandrum, India 2015, pp. 122–127 (2015). https://doi.org/10.1109/RAICS.2015.7488400
7. Hasan, M.K., Alam, M.A., Das, D., Hossain, E., Hasan, M.: Diabetes prediction using ensembling of different ML classifiers. IEEE Access **8**, 76516–76531 (2020). https://doi.org/10.1109/ACCESS.2020.2989857
8. Li, M., Fu, X., Li, D.: Diabetes prediction based on XGBoost algorithm. IOP Conf. Ser. Mater. Sci. Eng. **768**, 072093 (2020). https://doi.org/10.1088/1757-899X/768/7/072093
9. Ali, R., Siddiq, M.H., Idris, M., Kang, B.H., Lee, S.: Prediction of diabetes mellitus based on boosting ensemble modeling. In: Hervásx, R., Lee, S., Nugent, C., Bravo, J. (eds.) Ubiquitous Computing and Ambient Intelligence. Personalisation and User Adapted Services, pp. 25–28. Springer International Publishing, Cham (2014). https://doi.org/10.1007/978-3-319-13102-3_6
10. Sisodia, D., Sisodia, D.S.: Prediction of diabetes using classification algorithms. Procedia Comput. Sci. **132**, 1578–1585 (2018). https://doi.org/10.1016/j.procs.2018.05.122
11. Xu, Z., Wang, Z.: A risk prediction model for type 2 diabetes based on weighted feature selection of RFand XGBoost ensemble classifier. In: 2019 Eleventh International Conference on Advanced Computational Intelligence (ICACI), Guilin, China, pp. 278–283 (2019). https://doi.org/10.1109/ICACI.2019.8778622
12. Hanskunatai, A.: A new hybrid sampling approach for classification of imbalanced datasets. In: 2018 3rd International Conference on Computer and Communication Systems (ICCCS), Nagoya, pp. 67–71 (2018). https://doi.org/10.1109/CCOMS.2018.8463228
13. Nnamoko, N., Korkontzelos, I.: Efficient treatment of outliers and class imbalance for diabetes prediction. Artif. Intell. Med. **104**, 101815 (2020). https://doi.org/10.1016/j.artmed.2020.101815

14. Paing, M.P., Pintavirooj, C., Tungjitkusolmun, S., Choomchuay, S., Hamamoto, K.: Comparison of sampling methods for imbalanced data classification in random forest. In: 2018 11th Biomedical Engineering International Conference (BMEiCON) Chiang Mai 2018, pp. 1–5 (2018). https://doi.org/10.1109/BMEiCON.2018.8609946

15. Wang, Q.: Imbalanced classification based on over-sampling and feature selection. In: 2020 IEEE 5th International Conference on Cloud Computing and Big Data Analytics (ICCCBDA), Chengdu, China, pp. 325–330 (2020). https://doi.org/10.1109/ICCCBDA49378.2020.9095693

16. Khanam, J.J., Foo, S.Y.: A comparison of ML algorithms for diabetes prediction. ICT Express. (2021). ISSN 2405–9595. https://doi.org/10.1016/j.icte.2021.02.004

Novel Classification Modelling for Bipolar Disorder Using Non-verbal Attributes for Classification

K. A. Yashaswini[✉] and Aditya Kishore Saxena

Department of Computer Science and Engineering, School of Engineering, Presidency University, Bengaluru, India
{yashaswini,adityasaxena}@presidencyuniversity.in

Abstract. Identification of a subject suffering from Bipolar Disorder (BD) is still an open end problem in present time even after they undergo effective prescribed stages of treatments. Review of literature towards existing investigation in BD shows presence of different variants of approaches in order to diagnose BD. However, the prominent research challenge is to consider the non-verbal behavioral signals and perform investigation in order to accurately identify the actual state of diagnosis by explicitly differentiating the normal user (control group) from the subjects suffering from different variant of BD. Implemented on Python, the proposed system make use of Bayesian Neural Network to showcase it as a suitable machine learning approach for correctly classifying the type of BD using statistical inference for better modelling purpose.

Keywords: Bipolar disorder · Unipolar disorder · Depression · Manic · Deep neural network · Bayesian neural network

1 Introduction

Bipolar Disorder (BD) is considered as one of the prominent psychiatric condition which is represented by various forms and types of symptoms associated with depression [1]. There are two classified taxonomies of BD which is Bipolar Disorder (BD-I) and Bipolar Disorder (BD-2) characterized by elevated mood and depressive episodes. However, BD II separates from BD-I with respect to hypomanic episodes [2–4]. The most challenging aspect of the BD is that it would exist in latent fashion even after the treatment has been successfully completed. Hence, precise diagnose of BD still persists currently. At present, there are various attributes responsible for operational recovery of BD; however, there is lesser number of investigations in this regards. There has been various research works being carried out towards facilitating precise diagnosis of BD; however, there is less number of research towards connecting interpersonal accuracy with the standard taxonomies of BD. There are different studies carried out towards unipolar disorder in existing times [5–8], however, they are mostly associated with the recognition of the facial feature. Existing approach towards BD has witnessed usage of different variants of technical approaches however, they lack using non-verbal factors towards investigating

© Springer Nature Switzerland AG 2021
K. R. Venugopal et al. (Eds.): ICInPro 2021, CCIS 1483, pp. 236–247, 2021.
https://doi.org/10.1007/978-3-030-91244-4_19

the precise cases of the BD types [9, 10]. This is quite essential as the symptomatic representation of the BD patient as well as the normal patient could be quite confusing which leads to an impediment towards an effective diagnosis. It is still and uncertainty to understand the relationship significant between the control user and the actual BD user with different types i.e. BD-I, BD-II, and Unipolar disorder (UD). Therefore, this paper present a novel framework of classification of the bipolar disorder considering the non-verbal attributes has been used. Further, machine learning approach has been used in order to achieve this research aim. The paper organization is as follows: Sect. 2 discusses about the existing studies on BD followed by highlights of research problem in Sect. 3. Proposed methodology is discussed in Sect. 4 while system design is carried out in Sect. 5. Result discussion in Sect. 6 followed by conclusion in Sect. 7.

2 Related Work

In present times, there has been various study carried out towards investigating Bipolar Disorder (BD). An identification algorithm of BD has been developed by Fitriati et al. [11] by applying back propagation algorithm over the data collected by questionnaire. Although, the study has simplified usage of training scheme but it lacks reliability assessment towards it aggregated verbal data. A neuro-imaging method was presented by Evgin et al. [12] for performing classification of BD using deep learning approach. This model offers faster classification performance but it suffers from extensive dependency of training operation leading to unscalable performance. Usage of electric impulse and optical spectroscopy is reported in study of Qaseem et al. [13] to develop a predictive model. The mechanism is capable of processing and predicting larger size of data, however, the resultant of the model is specific to its domain of data only. A scheme using affective structure model of latent type is presented by Huang et al. [14] to identify unipolar and bipolar disorder. Although the model offers better detection performance but the accuracy completely depends upon pre-defined threshold score; thereby not applicable for complex form of data. A study to identify BD from the data extracted from questionnaire is carried out by Palmius et al. [15] where a linear regression model has been used for predictive purpose. The advantage and limitation of this work is equivalent to that carried out by Fitriati et al. [11].

Study towards mood dynamics has been carried out by Villasanti and Passino [16]; however, the study doesn't discuss about the dominant attributes leading to mood dynamics. An investigated clinical characteristic of BD is presented by Demir et al. [16] using tractogram dataset in order to generate a dispersion map. However, the study is highly specific to this dataset and doesn't offer flexibility of the scheme on different set of information. The work carried out by Liu et al. [17] have used manifold learning analysis in order to perform stochastic analysis for investigating different forms of psychiatric disorder. This model contributes towards an efficient construction of learning scheme; however, the scheme is highly iterative irrespective of the size of the data leading to extensive resource usage. A classification-based approach is presented by Alimardani et al. [18] for schizophrenia and BD using statistical measures. The beneficial aspect of this study is its simplified inferential modelling of its outcome using statistics while it also suffers due to similar reason as granularity in its accuracy cannot be concretely defined towards diagnosis of BD. Castro et al. [19] have carried out a study where association rules are used for investigating BD and associated disease. This model uses a conventional mining approach which is only suitable in restricted use cases of BD and it doesn't offer flexibility in its extensive diagnosis process. A study is carried out where a speech signals are considered to be subjected to convolution neural network as discussed by Li et al. [20]. This study offers a better learning scheme and network model; however, speech signal and its possible variation with their connectivity with the BD is still not established. This limits the model even to perform accurate diagnosis of mental disorder of various forms. Support Vector Machine is used by Gao et al. [21] for kernel-based system in order to classify bipolar disorder. The only advantage of this model is its usage of supervised learning scheme to make classification simpler; however, the model doesn't address considering various essential attributes that contributes towards latent BD. Su et al. [22] have used wavelet decomposition as well as Long Short Term Memory (LSTM) in order to classify mood disorder. Adoption of wavelet decomposition as well as LSTM assists in classification but the study doesn't connect any form of clinical rules while performing decomposition which could eventually results in outliers while subjected to training. A similar approach for standard database is presented by Su et al. [24] along with applying hierarchical spectral clustering over facial data as well as speech data. However, this model is only applicable for well-defined image dataset for making classification and cannot use any form of non-verbal information. A simulation study using experimental approach is carried out by Constantinou et al. [24] have carried out a. Similarly, there are various classes of research-based approach towards BD investigation viz. data science approach (Lee et al. [25]), generative model using deep learning (Matsubara et al. [26]), covariate based (Cigdem et al. [27]) single sliding window based approach (Yang et al. [28]), etc. Therefore, existing approaches offers different methods in order to ensure a better form of investigation of BD. A closer look into all the above mentioned techniques showcase that majority of the approaches towards investigating BD is either based upon specific form of dataset or commonly uses learning techniques for well-defined verbal dataset. The challenging factors of this model are that they are heavily computational complex process owing to involvement of higher iterative learning techniques. The next section discusses about research problems concluded from this studies.

3 Problem Description

From the prior section, it can be seen that there are different variants of approaches towards investigating BD; however, there are significant loopholes. The primary problem is there are less studies carried out towards non-verbal analysis of behavior, which could offer more insight towards proper classification of BD. The secondary problem is that machine learning has not being implemented in computationally efficient way over the data, which lead to lesser applicability in practical scenario. The tertiary problem is associated with that existing approaches doesn't seem to emphasize on interpersonal accuracy in order to offer more granularity in classification of bipolar disorder. The next section discusses about a novel approach where a unique machine learning model has been used to address the above mentioned problem.

4 Proposed Methodology

The proposed system constructs a computational learning model in order to perform diagnosis of the Bipolar Disorder (BD) on the basis of the non-verbal behavior attributes. The study hypothesize that extraction of this essential non-verbal cognitive patterns will assists more in understanding the function of the interpersonal and social behavior of the subject. The prime aim of the proposed study is to perform a comprehensive assessment of interpersonal precision towards realizing the types of BD. The study make use of a dataset [29] which consists of reports of 119 subjects with reported BD where 70 subjects have BD-I and 49 individual have BD-II. This data has been subjected to the comparative assessment with 39 subjects reported with Unipolar Depression (UD) as well as 119 normal individual who doesn't have BD termed as control group.

For this purpose, the proposed study makes use of MiniPONs which was actually developed for evaluating an ability to perform decoding of nonverbal signal [30]. Such form of non-verbal signals can be obtained from tone of voice, body language, facial expression, etc. This tool is used for assessing the difference within an individual with respect to interpersonal sensitivity along with identifying the differences communication channel. The study hypothesize that control group to offer better outcome with much predominant outcome compared to other groups with performance parameters tested in MiniPONs. The study also hypothesize that different groups of BD (BD-I, BD-II, and UD) doesn't offer better outcome in contrast to control group with respect to identification of inference of voice intonation, body language, facial gestures, etc.

The proposed study make us of machine learning approach using multiple variants to evolve up with Bayesian Neural Network to be the preferred model in order to carry out an effective classification process. Apart from this, the study also formulates a hypothesis that outcomes of unipolar depression are not much significantly different from BD groups (BD-I and BD-II). The main idea of the proposed model is to highlight that existing database lacks of accuracy in interpersonal facts and it offers an insight that enhanced information of deficiencies associated with accuracy of interpersonal facts in DB will further help in constructing better training scheme to enhance these subjects to offer better influence over their psychological functionality. The flow of proposed system is shown in Fig. 1

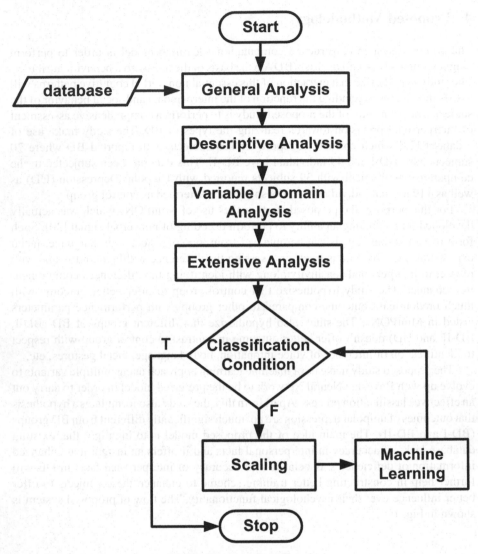

Fig. 1. Proposed flow of study

5   Implementation

An explicit environment for implementing proposed model is created from the base environment of anaconda which is activated in the anaconda ecosystem. A professionally managed browser based kernel Jupyter notebook is used for coding and data analysis process. The system makes use of an explicit tool MiniPONs where the dataset associated with BD is read. The system read the data and it specifies the separator as semi-colon and saves in a raw data. The next part of the implementation is about performing *general analysis* of the dataset. The study considers Number of samples are from 0 to 276 with total 12 columns of elements. The description of datatype is given in a table below where 10 variables are of integer type and 2 variables are categorical. This operation leads to outcome of 12 elements viz. group, type, age, right answer, audio prosody, combined channel, face video, body video, positive valence, negative valence, dominant, and submissive. Out of 277 entries, group and type is basically object and rest all attributes are numerical values. The study considers numerical values for feature engineering and doesn't consider group/type which are basically categorical data type.

This operation is followed by implementing descriptive statistics using mean, standard deviation, min, max, and quantile in order to observe this. Variable analysis is also carried out in order to offer more insight towards the domain of the dataset. In this part of analysis, the higher score will eventually means enhanced performance. Further the domain-based analysis is carried out in order to extract the clinical trials, and associated information of samples of BD and UD. The study is then carried out towards analyzing the effect of variables on depression levels. The idea is to find out which type is most dominant with the score of ages among the subjects with BD-I, BD-II, control, and UD. Further, the study obtains the right answer for each groups to cross-check the reliability of the answer furnished by the participants. This analysis is further combined for each group with respect to the type.

The next part of implementation focuses about assessing the individual score on the basis of the voice response for each group. The study is also carried out for analyzing score of face video as well as average body language score for each group. The analysis is further continued for assessing the reliability for each reactions of the BD patient with respect to positive and negative valence score for each group. The next part of the study is about the computation of the dominant and submission score for the subjects in dataset. Basically, dominance and submissive score highlights the reactiveness and non-reactiveness to the test environment, which is unlikely for control group.

The next part of the implementation is about performing correlation analysis of each factor and variables. In this assessment, different factors e.g. age, right answer, audio prosody, combined channel, face video score, body video score, positive/negative valence, dominant/submissive score are studied with respect to 4 significant groups of the study i.e. BD I, BD II, control, and UD. Irrespective of any type and group, BD-I, BD-II, UD do not have any significant differences in their scores and hence they are termed as not normal whereas the UD is termed as normal. And accordingly, an additional column for the class inference is added with class-I not normal and class-II as normal.

The raw data was not suitable for supervised learning (multi-class classification) which is being deduced to two binary classes. The next part of the implementation is to carry out analysis of density based on inference using Kernel Density Estimation (KDE) by highlighting "not normal" and "normal". This is the initialized point of applying machine learning method.

For this purpose, certain level of preprocessing of the data is required which is carried out using scaling mechanism. If data are not scaled then the function approximators will consume more computational resources (min max scaler for scaling is used). This is followed by effective visualization of a scaled data pertaining to KDE to observe that the peak separation is still more visible. The proposed system initiates its machine learning model by apply logistic regression, Gaussian Naïve Bayes, and K-neighbor classifier by doing appropriate encoding of inference as normal and nor normal class as 0 and 1. Training data and testing data are prepared in the next set of action followed by training the model and testing the model. The outcome of the study has been investigated using confusion matrix for logistic regression. The accuracy assessment is carried out using ROC curve.

While doing this assessment, it has been observed that KNN, Gaussian Naïve Bayes, and Logistic regression works well with the feature engineering. In order to improvise the performance, a deep learning model is presented followed by benchmarking the three methods of machine learning. The observation carried out in this part of the investigation shows that KNN performance is not satisfactory and so the proposed study considers logistic regression as well as Gaussian Naïve Bayes. The study further enhances the machine learning model using Deep learning. However, until recently there was no method to build up probabilistic method to design artificial neural network as referred in [31]. The proposed study built a Bayesian Neural Network (BNN) which allows probabilistic model to be used in neural network. The proposed study aims for forwarding the essential characteristics of function that are sampled from the Gaussian process before the functions were sampled from BNN. The study offers emphasis to the priors associated with BNN connected with the function space and not over the space parameters. The study of proposed machine learning is carried out over different activation function as well as kernels using various test environment. The proposed study is associated with constructing a bridge between the higher scalable Bayesian based deep learning with non-parametric Bayesian approach. Hence, the proposed system targets towards building a better form of Bayesian models for an effective classification of the signals by capturing the required prior beliefs that can be useful for posterior inference.

In the complete implementation of the proposed design, the proposed system is found to offer simplified and progressive scheme in its design methodology with better diagnosis of BD. The study offers much better classification performance too using proposed machine learning. The next section discusses results obtained.

6 Results Discussion

This section discusses about the outcome obtained from the proposed system. The implementation of the proposed study is carried out from standard Kaggle dataset associated with bipolar disorder [31]. The dataset consist of 277 samples arranged in columnar wise with both categorical and integer form. The ecosystem of the implementation is as follows: an open source distribution to manage python projects e.g. Anaconda is used as an IDE. Since, the implementation model uses parallel computing; therefore, a parallel computing platform and programming model e.g. Computed Aided Device Architecture (CUDA) that uses a general purpose computing through Graphical Processing Unit (GPU) provided by artificial intelligence flagship company NVIDIA along with CPU resources. This set up overcomes the sequential computing constraints of CPU alone. Additionally, a deep neural network library which is GPU accelerated e.g. cuDNN is used, which finetunes the routines involved in the different layers of deep learning including convolution, normalization, pooling, and activation layers. The most importantly, the machine learning model requires extensive support of complex matrix manipulation, which is being provided by TensorFlow, a generalized architecture of the computational implementation platform for the BiP – ML. The analysis of the normal (orange) and not-normal (blue) curve is shown in Fig. 2 as follow.

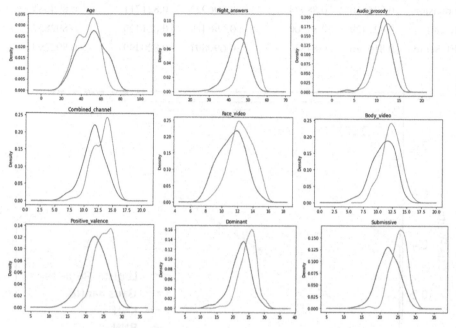

Fig. 2. Analysis of normal and not-normal curve w.r.t. each attributes

Since the separation of the pick points are clearly visible therefore, at the first glance we may feel that KNN could be a suitable algorithm here, however if we look at the number of samples, we realize that this cannot be done since KNN is suitable for bigger datasets. If the dataset is in the order of 10000 then KNN is more suited. However since the dataset contains only 277 samples, we have to prefer a probabilistic model like naive Bayes method. Keeping the logistic regression as baseline we will test both K-Nearest Neighboring and Gaussian Naïve Bayes on this particular dataset. (Blue is not normal and orange is normal). The study make use of min-max scaling process as if data are not scaled then the function approximator will consume more computational resources. This process is followed by applying logistic regression, Gaussian NB, and K-neighbor classifier by doing appropriate encoding of inference as normal and nor normal class as 0 and 1. The next step is preparation of training data and testing data. In order to assess machine learning model, the proposed system chooses 2 performance parameters viz. ROC curve and Accuracy. The obtained numerical data for comparative analysis is shown in Table 1.

Finally it is observed that when we make use of a regular neural network, the results are at par with Logistic regression however when we add a probabilistic Bayesian layer to the same model, we see that results improve significantly. From the outcome shown in Fig. 3, it can be seen that proposed system offers better outcome in contrast to existing

Table 1. Comparative performance analysis

	Logistic regression	Gaussian naive bayes	K-nearest neighbor	Artificial neural network	Bayes neural network
Precision	0.825510	0.88074	0.696925	0.821711	0.894081
Recall	0.821429	0.87500	0.696429	0.821429	0.892857
F1-Score	0.819076	0.87536	0.690891	0.820499	0.892299

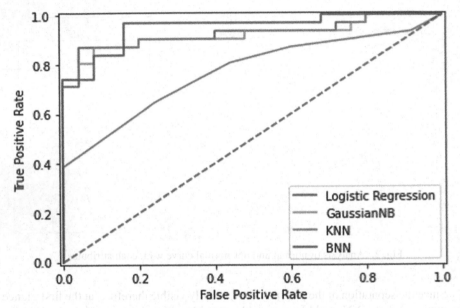

Fig. 3. ROC curve of comparative analysis

approaches of machine learning proving that it can be used efficiently in order to perform classification of normal and abnormal classes of BD. Apart from this, the complete analysis is also carried out by increasing the size of the dataset to 10–50% to find that there is no significant changes in the accuracy score. This exhibit that proposed system offers consistency in its identification of BD.

7 Conclusion

There has been archives of research work being carried out towards facilitating an effective diagnosis of bipolar disorder. After reviewing the existing approaches in BD, it has been found that there are various machine learning approaches towards precise classification of BD. As a part of novelty of this paper, the proposed system implements a Bayesian Neural Network in order to perform optimized outcome towards classification process. The study shows that inclusion of non-verbal attributes significantly assists in classification process of BD. The analysis of the proposed study is carried out in Python to show that Bayesian Neural Network offers better outcome in contrast to existing machine learning approaches.

References

1. Carvalho, A.F., Vieta, E., Quevedo, J.: Neurobiology of Bipolar Disorder Road to Novel Therapeutics. Elsevier Science, Amsterdam (2020). ISBN: 9780128191835
2. Gruber, J.: The Oxford Handbook of Positive Emotion and Psychopathology. Oxford University Press, Oxford (2019). ISBN: 9780190653200
3. Bhalla, I.P., Tampi, R.R., Srihari, V.H.: 50 Studies Every Psychiatrist Should Know. Oxford University Press Incorporated, Oxford (2018). ISBN: 9780190625085
4. Geddes, J.R.: New Oxford Textbook of Psychiatry. Oxford University Press, Oxford (2020). ISBN: 9780198713005
5. Pampouchidou, A., et al.: Automatic assessment of depression based on visual cues: a systematic review. IEEE Trans. Affect. Comput. **10**(4), 445–470 (2019). https://doi.org/10.1109/TAFFC.2017.2724035
6. Liu, Z.: Detection of depression in speech. In: International Conference on Affective Computing and Intelligent Interaction (ACII), Xi'an, China, pp. 743–747 (2015). https://doi.org/10.1109/ACII.2015.7344652
7. Zhao, Z., et al.: Automatic assessment of depression from speech via a hierarchical attention transfer network and attention autoencoders. IEEE J. Sel. Topics Signal Process. **14**(2), 423–434 (2020). https://doi.org/10.1109/JSTSP.2019.2955012
8. Umar, A., Qamar, U.: Detection and diagnosis of psychological disorders through decision rule set formation. In: IEEE 17th International Conference on Software Engineering Research, Management and Applications (SERA), Honolulu, HI, USA, pp. 33–37 (2019). https://doi.org/10.1109/SERA.2019.8886786
9. Valenza, G., et al.: Wearable monitoring for mood recognition in bipolar disorder based on history-dependent long-term heart rate variability analysis. IEEE J. Biomed. Health Inf. **18**(5), 1625–1635 (2014). https://doi.org/10.1109/JBHI.2013.2290382
10. Pan, Z., Gui, C., Zhang, J., Zhu, J., Cui, D.: Detecting manic state of bipolar disorder based on support vector machine and Gaussian mixture model using spontaneous speech. PMC-Psych. Investig. **15**(7), 695–700 (2018)
11. Fitriati, D., Maspiyanti, F., Devianty, F.A.: Early detection application of bipolar disorders using back propagation algorithm. In: 6th International Conference on Electrical Engineering, Computer Science and Informatics (EECSI), Bandung, Indonesia, pp. 40–44 (2019). https://doi.org/10.23919/EECSI48112.2019.8977102
12. Evgin, H. B.: Classification of fNIRS data using deep learning for bipolar disorder detection. In: 27th Signal Processing and Communications Applications Conference (SIU), Sivas, Turkey, pp. 1–4 (2019). https://doi.org/10.1109/SIU.2019.8806435
13. Qassem, M., Constantinou, L., IasonasI, F., Triantis, M.H., Palazidou, E., Kyriacou, P.A.: A method for rapid, reliable, and low-volume measurement of lithium in blood for use in bipolar disorder treatment management. IEEE Trans. Biomed. Eng. **66**(1), 130–137 (2019). https://doi.org/10.1109/TBME.2018.2836148
14. Huang, K.-Y., Wu, C.-H., Su, M.-H., Kuo, Y.-T.: Detecting unipolar and bipolar depressive disorders from elicited speech responses using latent affective structure model. IEEE Trans. Affect. Comput. **11**(3), 393–404 (2020). https://doi.org/10.1109/TAFFC.2018.2803178
15. Villasanti, H.G., Passino, K.M.: Modeling and analysis of mood dynamics in the bipolar spectrum. IEEE Trans. Comput. Soc. Syst. **7**(6), 1335–1344 (2020). https://doi.org/10.1109/TCSS.2020.3028205

16. Demir, A., Ozkan, M., Ulug, A.M.: A macro-structural dispersion characteristic of brain white matter and its application to bipolar disorder. IEEE Trans. Biomed. Eng. **68**(2), 428–435 (2021). https://doi.org/10.1109/TBME.2020.3002688

17. Liu, W., Li, D., Han, H.: Manifold learning analysis for allele-skewed dna modification snps for psychiatric disorders. IEEE Access **8**, 33023–33038 (2020). https://doi.org/10.1109/ACC ESS.2020.2974292

18. Alimardani, F., Cho, J.-H., Boostani, R., Hwang, H.-J.: Classification of bipolar disorder and schizophrenia using steady-state visual evoked potential based features. IEEE Access **6**, 40379–40388 (2018). https://doi.org/10.1109/ACCESS.2018.2854555

19. Castro, G.: Applying association rules to study bipolar disorder and premenstrual dysphoric disorder comorbidity. In: IEEE Canadian Conference on Electrical & Computer Engineering (CCECE), Quebec, QC, Canada, pp. 1–4 (2018). https://doi.org/10.1109/CCECE.2018.844 7747

20. Li, Y., Yang, L., Chen, H., Jiang, D., Sahli, H.: Audio visual multimodal classification of bipolar disorder episodes. In: 8th International Conference on Affective Computing and Intelligent Interaction Workshops and Demos (ACIIW), Cambridge, United Kingdom, pp. 115–120 (2019). https://doi.org/10.1109/ACIIW.2019.8925023

21. Gao, S.: Discriminating bipolar disorder from major depression based on kernel SVM using functional independent components. In: IEEE 27th International Workshop on Machine Learning for Signal Processing (MLSP), Tokyo, Japan, pp. 1–6 (2017). https://doi.org/10. 1109/MLSP.2017.8168110

22. Su, M., Wu, C., Huang, K., Hong, Q., and Wang, H.: Exploring microscopic fluctuation of facial expression for mood disorder classification. In: International Conference on Orange Technologies (ICOT), Singapore, pp. 65–69 (2017). https://doi.org/10.1109/ICOT.2017.833 6090

23. Su, M.-H., Wu, C.-H., Huang, K.-Y., Yang, T.-H.: Cell-coupled long short-term memory with L -skip fusion mechanism for mood disorder detection through elicited audiovisual features. IEEE Trans. Neural Netw. Learn. Syst. **31**(1), 124–135 (2020). https://doi.org/10. 1109/TNNLS.2019.2899884

24. Constantinou, L., Kyriacou, P.A., Triantis, I.F.: Towards an optimized tetrapolar electrical impedance lithium detection probe for bipolar disorder: a simulation study.In: IEEE SENSORS, Glasgow, UK, pp. 1–3 (2017). https://doi.org/10.1109/ICSENS.2017.8234225

25. Lee, C.-Y., Zeng, J.-H., Lee, S.-Y., Lu, R.-B., Kuo, P.-H.: SNP data science for classification of bipolar disorder I and bipolar disorder II. IEEE/ACM Trans. Comput. Biol. Bioinf. **99**, 1 (2020). https://doi.org/10.1109/TCBB.2020.2988024

26. Matsubara, T., Tashiro, T., Uehara, K.: Deep neural generative model of functional MRI images for psychiatric disorder diagnosis. IEEE Trans. Biomed. Eng. **66**(10), 2768–2779 (2019). https://doi.org/10.1109/TBME.2019.2895663

27. Cigdem, O.: Effects of covariates on classification of bipolar disorder using structural MRI. In: Scientific Meeting on Electrical-Electronics & Biomedical Engineering and Computer Science (EBBT), Istanbul, Turkey, pp. 1–4 (2019). https://doi.org/10.1109/EBBT.2019.874 1586

28. Yang, Y., Qian, C., and Huafu, C.: Altered inter-hemispheric functional connectivity dynamics in bipolar disorder. In: 17th International Computer Conference on Wavelet Active Media Technology and Information Processing (ICCWAMTIP), Chengdu, China, pp. 334–337 (2020). https://doi.org/10.1109/ICCWAMTIP51612.2020.9317369

29. Theory of mind in remitted bipolar disorder. https://www.kaggle.com/mercheovejero/theory-of-mind-in-remitted-bipolar-disorder. Accessed 4 Aug 2020

30. Swiss Center for Affective Sciences. https://www.unige.ch/cisa/emotional-competence/home/research-tools/minipons/. Accessed 4 Aug 2020
31. Flam-S. D., Requeima, J., Duvenaud, D.: Mapping Gaussian process priors to Bayesian neural networks. In: 31st Conference on Neural Information Processing Systems (NIPS 2017) (2017)

Towards More Accurate Touchless Fingerprint Classification Using Deep Learning and SVM

K. C. Deepika[1(✉)] and G. Shivakumar[2]

[1] Department of E&C, MCE, VTU, Hassan, Belagavi, India
`kcd@mcehassan.ac.in`
[2] Department of E&I, MCE, VTU, Hassan, Belagavi, India
`gs@mcehassan.ac.in`

Abstract. Automated fingerprint classification is one of the mostly used human identity verification system. The touchless fingerprint identification and classification system offers a higher user convenience and hygiene compared to conventional touch-based identification. Recently, convolutional neural networks (CNN) are trained to achieve better performance on fingerprint classification. For classification, deep learning models utilizes the SoftMax layer for prediction which limits cross entropy loss. In this proposed method SoftMax layer is replaced by multi class Support Vector Machine which reduces the margin-based loss as well. There are various combinations of deep learning models and support vector machines. In this paper experiments on AlexNet with multi class SVM and AlexNet with SoftMax have been demonstrated on PolyU 3D benchmark fingerprint Database to improve the recognition accuracy on test data set and to minimize the training time involved. The results show that the deep learning model with SVM achieve better results both in terms of validation accuracy and time in training than the modified transfer learning network with SoftMax as classifier.

Keywords: Fingerprint · Touchless · Deep learning · Deep convolutional neural network · Multiclass SVM · Classification

1 Introduction

With the vast development in information technology and network in recent years, the society has put forward a higher requirement for the security of information systems. Biometric recognition technology has gradually become one of the important methods to enhance the security and stability of information systems. Biometrics identification uses human physiology and behavioral features for automatic identification. Biometrics are unique personal attributes, which have the characteristics of stability, diversity, and individual differences. Several biometrics have been used including fingerprint, iris, signature [24] and finger vein [13]. Compared with other biological features, fingerprint has many unique advantages and each person's fingerprint has different characteristics [1].

© Springer Nature Switzerland AG 2021
K. R. Venugopal et al. (Eds.): ICInPro 2021, CCIS 1483, pp. 248–257, 2021.
https://doi.org/10.1007/978-3-030-91244-4_20

In general, the classical fingerprint system employs contact-based fingerprint acqui-sition gadgets like solid state or optical sensors. Nonetheless, these traditional method-ologies regularly go through adversaries because of equipment conditions and neigh-borhood ecological deformities. Unlike classical fingerprint identification approaches, the touchless fingerprint classification seems to be promising and more capable as it can diminish the issue of contact of fingertip with the surface of acquisition [2]. Also, since the fingerprint acquisition is touchless or contactless with the acquisition system it can likewise stay away from likelihood of transmission of skin illnesses and pandemic sicknesses like COVID19. Touchless strategy likewise decreases procurement time and client preparing requests [3]. Touchless fingerprint system which functions based on 3D mechanism is typically more accurate [4].

Deep learning using convolutional neural networks have achieved better perfor-mances in feature extraction and the recognition of touchless fingerprint [5, 6]. Most of the conventional deep learning methods use the SoftMax activation function/layer for classification [25]. In this paper a multi class Support Vector Machine is utilized as swap to SoftMax for classification. The proposed method is a two-stage process where con-volutional neural network is followed by SVM. A feature learner dependent on the deep CNN engineering is proposed to extract the fingertip features with no earlier information on the data. Then, at that point, classification is performed by the help of support Vec-tor Machines to augment the classification capacity of convolutional neural network. Experiments are performed on the benchmark PolyU 3D fingerprint database [23] to validate the method and the result show that the proposed combined CNN+SVM system can outperform both individual SVMs and CNNs.

2   Previous Work

This segment is chiefly about conversation of a portion of the key late written works relating to fingerprint identification and classification. A.K. Jain and Lin Hong are the couple of researchers who are working more on touchless 3D fingerprint classification. To improve classification accuracy, A. K. Jain et al. [9] suggested double stage classifiers. Ding et al. [10] applied Gabor filter with dictionary learning idea to perform finger impression identification and classification, however, the traditional ROI segmentation and wavelet-based method limits its heartiness over complex info nature. Tan et al. [11] inferred an information driven model with feature learning method dependent on Genetic Programming, which was essentially used to choose features, while Bayesian classifier model is utilized to classify the images. A criticism-based line-identifier was designed by Shah et al. [12] to obtain better ROI location, which was subsequently followed by classification utilizing SVM, K-Nearest Neighbor and ANN. Of all the classifiers, researchers found that ANN can empower higher precision. Halici et al. [13] applied a self-coordinating element map with ANN classifier to perform unique finger impression classification. Researchers changed SOFM learning and ANN with "certainty" idea to manage contorted locales of unique finger impression pictures to empower higher exactness. Considering skin mutilation over fingerprint, Si et al. [14] at first rectified skin distortion with finger impression, which was subsequently classified utilizing SVM.

Not at all like conventional wavelet-based methodologies, deep learning ideas have acquired far reaching consideration for finger analysis and classification. Thinking about its robustness, a couple explores have been completed by utilizing deep learning techniques for fingerprint recognition and classification. Zhang et al. [16] fostered a DCNN based fingerprint liveness identification called Slim-ResCNN to recognize true fingerprints over the phony ones. Yuan et al. [17] expressed a superior Deep CNN with picture scale adjustment to safeguard picture resolution and texture data, which was prepared utilizing versatile learning rate to carry out exact fingerprint classification. Tertychnyi et al. [19] applied DCNN idea to distinguish the bad quality fingerprints with good exactness and reliability. Wang et al. [20] has proposed a fingerprint recognition strategy dependent on deep learning methods by picking direction field as the element. As of late, Labati et al. [21] proposed ANN concept for the assessment of picture quality of the unwrapped 3D fingertip. Acquiring diverse set of elements authors have performed fingerprint picture quality appraisal for better classification precision and accuracy.

Seeing above expressed written works, it is very well discovered that irrefutably various explores have been done towards fingerprint classification; notwithstanding, larger part of the current methodologies utilizes 2D feature set with conventional wavelet analysis or ridge information. Tragically, these methodologies can't be appropriate for Touchless based environment. The conventional finger impression identification models are equipment or sensor subordinate, while Touchless arrangement can ease such reliance and thus can be more powerful for ongoing security purposes and most of the existing deep learning models [22] for touchless fingerprint identification use the Soft-Max activation for classification. In this paper the Support Vector Machine is utilized as a replacement for SoftMax activation layer for fingerprint classification.

3   Methodology

The proposed method is mainly based on deep convolutional neural network model with its two different tasks for touchless 3D fingerprint classification as shown in Fig. 1.

The methodology is explained in the following steps:

Step1: The 3D fingerprint image dataset is given to the Deep convolutional neural network.
Step2: The features are extracted/learned in many abstraction levels.
Step3(i): Modification of CNN layers to adopt to the new dataset and number of output classes.
Step3(ii): Learned features are used to train multi class SVM.
Step 4: Achieve fingerprint classification in both the tasks.

3.1   Deep Convolutional Neural Network

Nowadays solutions for image identification and classification generally use deep learning techniques. A convolutional neural network is a feed forward network able to learn the deep features of image in many abstraction levels. It can extract deep features from

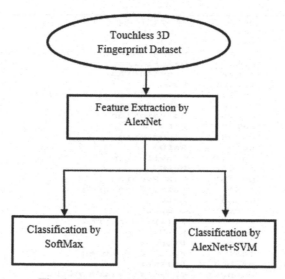

Fig. 1. Framework of the proposed system.

the input image in its earlier layers and classify the images with its final few layers. The architecture is trained by using back propagation method. The raw fingerprint image is provided in the form of array of pixels. The CNN performs a sequence of convolutions, activations, and pooling operations on the hidden layer as a step of feature extraction/learning. After learning the features, the architecture performs classification.

AlexNet AlexNet is a pretrained network which is already trained on a million number of images. AlexNet is a 25 layered architecture as listed in Fig. 2. It consists of five convolutional layers in which the 1st, 2nd and 5th convolutional layers are followed by max-pooling layers. This part is responsible for feature learning/extraction. The second part of the AlexNet includes three fully connected layers with an output layer of thousand SoftMax neurons for classification.

3.2 SoftMax Layer

In classification tasks using deep convolutional neural network techniques, it is more common to use the SoftMax layer for classification. For example, for 10 number of classes, the SoftMax layer consists of 10 number of nodes signified by p_i, where $i = 1, ..10$. p_i indicates a Discrete PDF* so, ($\sum_1^{10} p_i = 1$)*.

In case h be the activation of the penultimate layer, w is the weight which is associating the penultimate layer to the SoftMax layer. The input into SoftMax layer given by,

$$a_i = \sum_k h_k w_{ki} \qquad (1)$$

$$\text{We have, } p_i = \frac{exp(a_i)}{\sum_j^{10} exp(a_j)} \qquad (2)$$

```
1    'data'       Image Input
2    'conv1'      Convolution
3    'relu1'      ReLU
4    'norm1'      Cross Channel Normalization
5    'pool1'      Max Pooling
6    'conv2'      Convolution
7    'relu2'      ReLU
8    'norm2'      Cross Channel Normalization
9    'pool2'      Max Pooling
10   'conv3'      Convolution
11   'relu3'      ReLU
12   'conv4'      Convolution
13   'relu4'      ReLU
14   'conv5'      Convolution
15   'relu5'      ReLU
16   'pool5'      Max Pooling
17   'fc6'        Fully Connected
18   'relu6'      ReLU
19   'drop6'      Dropout
20   'fc7'        Fully Connected
21   'relu7'      ReLU
22   'drop7'      Dropout
23   'fc8'        Fully Connected
24   'prob'       Softmax
25   'output'     Classification Output
```

Fig. 2. AlexNet layers.

The labelled class $\hat{i}$ can be

$$\hat{i} = arg_i max(p_i)$$
$$= arg_i max(a_i) \tag{3}$$

3.3 Multi Class SVM

Most of the learning methods for classification using convolutional layers and fully connected layer use SoftMax layer with an objective to learn the lower-level parameters. In this paper a multi class SVM is also used for classification. A simple way to build multi class SVM is to extend SVMs for multi class problems by using One versus Rest approach. For a problem of n number of classes, n linear SVMs are trained separately by the features extracted from the penultimate layer independently, where the information from rest of the classes is considered as adverse class.

$$\text{The output of n-SVM } a_k(x) = w^T x \tag{4}$$

$$\text{The identified class would be } arg_k max a_k(x) \tag{5}$$

The prediction of class utilizing SVM is equivalent to utilizing a SoftMax. The distinction among SoftMax and multiclass SVM is in their destinations parametrized by all the weight frameworks w.

4 Implementation

AlexNet is implemented in two different tasks as discussed below.

4.1 AlexNet with SoftMax Classifier

In this implementation images are taken from PolyU 3D fingerprint dataset. The AlexNet is loaded with the dataset for feature extraction. The last three layers of the AlexNet that are fully connected layer fc8, SoftMax layer named as prob and classification output layer named as output are designed for thousand number of classes, These layers should be tuned for the new classification problem. So, all the layers of AlexNet excluding the last three layers are extracted from the pretrained network and the last three layers are replaced with new layers for adopting to a new task and dataset. The dataset is split into training and testing phase with 70–30 ratio. All images of dataset are resized with input size of AlexNet network (227 × 227 × 3) by augmented image datastore which automatically resize the input image and do some more augmentation operations to prevent the network from overfitting. Then training of modified network is carried out by deciding the number of epochs, iterations, learning rate, weight learn rate factor, bias learn rate factor and other options, so that there is a fast-learning rate in the initial layers. By keeping the features from the earlier layers of the network, a slow learning rate is fixed for the final layers. The network that consists of transferred and new layers as shown in Fig. 3, by default, train the network using GPU if available, otherwise it uses CPU. This is followed by classification by SoftMax layer which classifies the validation images. Classification accuracy will be calculated on the validation set where accuracy is the portion of the total classes that the network predicts correctly.

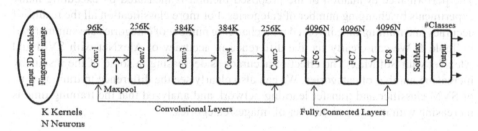

Fig. 3. Block diagram of modified AlexNet+SoftMax.

4.2 AlexNet with SVM as Classifier

The Deep Convolutional Neural Network consists of a hierarchical representation of input raw images. Deeper layers contain a higher-level deep feature which are configured using the low-level features obtained from the lower layers. In the proposed method the AlexNet model is used for only feature extraction or learning from the pretrained convolutional neural network and the extracted features are used to train the image classifier for classification. So, the last three layers of pretrained model that are fc8, SoftMax and output layer are removed from pretrained model and a multiclass SVM is fitted after the penultimate layer fc7 to classify the fingerprint images as shown in Fig. 4. Feature extraction is the simplest and fastest way to use the representational powers of deep networks. We can train a multi class SVM on the extracted features because unlike training, feature extraction requires a single pass through the data to learn the features. To get the feature representation of training and test images, activations on fully connected layer fc7 are used. The class labels are learned from the training and test dataset. The features that are extracted from the training image dataset are considered as predictor variables. Then the test images are classified using the trained multi class SVM model. Classification accuracy and the training time are noted for performance evaluation.

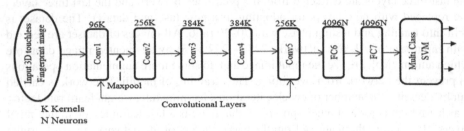

Fig. 4. Block diagram of modified AlexNet+SVM.

5 Results and Discussion

The performance evaluation of the proposed method is measured by executing many experiments by changing number of categories. For more classification all the results of accuracy and training time involved with changing number of categories are summarized in Table 1. We can clearly see the difference in accuracy of AlexNet with SVM and AlexNet with SoftMax. The effect of accuracy is decreasing in some of the results by increasing number of categories. We can also clearly see the difference in training time of SVM classifier and transfer learning network and analyzed that the training time is increasing with increase in number of images/categories.

Table 1. Comparative results of classification.

No of categories	Accuracy in %		Training time	
	AlexNet+SoftMax	AlexNet+SVM	AlexNet+SoftMax (in Mins)	AlexNet+SVM (in Mins.)
10	97.5	100	30	0.11
20	93.75	98.8	60	0.28
30	94.17	99.17	90	0.49
40	91.25	99.38	105	0.79
50	93.5	99.5	120	1.08

In transfer learning while training its newly assigned layers, the training progress, training details with validation accuracy and validation loss of the network for the assigned categories for example for category 50 is shown in Fig. 5. The Fig. 5 shows that the percentage of accuracy in the initial epoch is less but with the ascending epochs the percentage of accuracy gradually increases. Figure 5 also shows that validation loss is high at epoch1 and becomes less with jerks and becomes almost smooth when accuracy approximates to 100%.

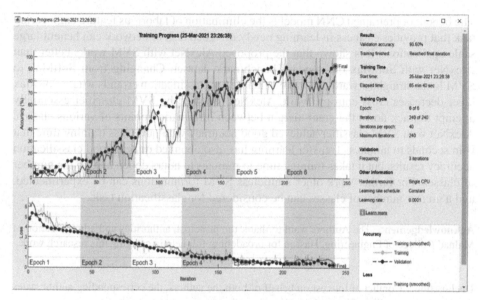

Fig. 5. Training progress showing all the training details.

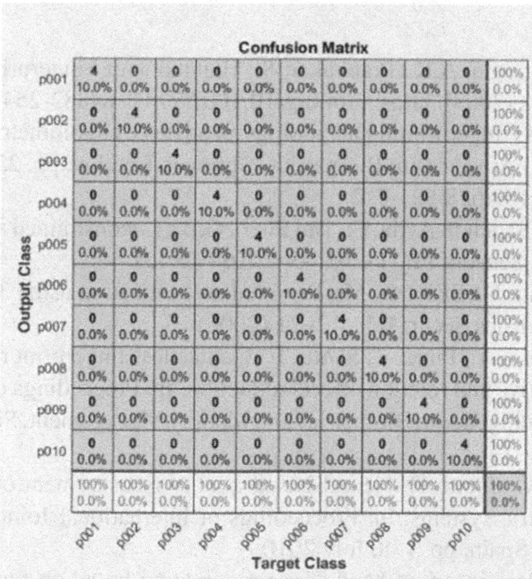

Fig. 6. Confusion matrix showing the classification of 10 categories by SVM.

Once the SVM classifier is trained with the features obtained from AlexNet, the classification of test dataset which corresponds to 30% of the total images is plotted as Confusion matrix as shown in Fig. 6.

6 Conclusion

The benefit of pretrained CNN model is the elimination of laborious feature engineering task that provides easiness in learning newly assigned task to network can benefit large scale applications. It is shown that the proposed AlexNet with SVM works better than AlexNet with SoftMax on standard benchmark dataset. Changing from SoftMax to SVM looks straightforward and more valuable for classification errands where SoftMax layer decreases cross-entropy, while AlexNet followed by SVM classifier essentially attempt to track down the great margin between information points of various classes. AlexNet with SVM classifier achieved good accuracy with quite less training time that is in seconds to minutes. Transfer learning has also obtained quite a good classification accuracy results, but it takes training time in minutes to hours depending on the number of classes. For further work other multiclass SVM formulations can be experimented, and a greater number of classes can be considered for classification task.

Acknowledgement. The Authors want to thank the principal, authorities and administration of Malnad College of Engineering, Hassan for broadening full help in carrying this research work.

Declaration. Conflict of Interest on behalf of both the authors, the corresponding author states that there is no conflict of interest.

References

1. Maltoni, D., Maio, D., Jain, A.K., Prabhakar, S.: Handbook of Fingerprint Recognition, 2nd edn. Springer, London (2009). https://doi.org/10.1007/978-1-84882-254-2
2. Parziale, G.: Touchless fingerprinting technology in advances in biometrics. In: Ratha, N.K., Govindaraju, V. (eds.) Advances in Biometrics. Springer, London, pp. 25–48 (2008). https://doi.org/10.1007/978-1-84628-921-7_2
3. de-Santos-Sierra, A., Sánchez-Avila, C., Del Pozo, G.B.: Unconstrained and contactless hand geometry biometrics. Sensors 11(11), 10143–10164 (2011)
4. Malassiotis, S., Aifanti, N., Strintzis, M.: Personal authentication using 3-D finger geometry. IEEE Trans. Inf. Forensics Secur. 1(1), 12–21 (2006)
5. Labati, R.D., Genovese, A., Piuri, V., Scotti, F.: Contactless fingerprint recognition: a neural approach for perspective and rotation effects reduction. In: Proceedings of IEEE Symposium on Computational Intelligence in Biometrics and Identity Management, Singapore, pp. 22–30, April 2013
6. Labati, R.D., Piuri, V., Scotti, F.: Neural-based quality measurement of fingerprint images in contactless biometric systems. In: Proceedings of International Joint Conference Neural Network, Barcelona, Spain, pp. 1–8, July 2010
7. Kang, W., Wu, Q.: Pose-invariant hand shape recognition based on finger geometry. IEEE Trans. Syst. Man Cybern. Syst. 44(11), 1510–1521 (2014)

8. Khan, M.M.U., Sadi, M.S.: An efficient approach to extract singular points for fingerprint recognition. In: 2012 7th International Conference on Electrical and Computer Engineering, Dhaka, pp. 13–16 (2012)
9. Jain, A.K., Prabhakar, S., Hong, L.: A multichannel approach to fingerprint classification. IEEE Trans. Pattern Anal. Mach. Intell. 21(4), 348–359 (1999)
10. Ding, S., Bian, W., Liao, H., Sun, T., Xue, Y.: Combining Gabor filtering and classification dictionaries learning for fingerprint enhancement. IET Biomet. 6(6), 438–447 (2017)
11. Tan, X., Bhanu, B., Lin, Y.: Fingerprint classification based on learned features. IEEE Trans. Syst. Man Cybern. Part C (Appl. Rev.) 35(3), 287–300 (2005)
12. Shah, S., Sastry, P.S.: Fingerprint classification using a feedback-based line detector. IEEE Trans. Syst. Man Cybern. Part B (Cybern.) 34(1), 85–94 (2004)
13. Halici, U., Ongun, G.: Fingerprint classification through self-organizing feature maps modified to treat uncertainties. In: Proceedings of the IEEE, vol. 84, no. 10, pp. 1497–1512 (1996)
14. Si, X., Feng, J., Zhou, J., Luo, Y.: Detection and rectification of distorted fingerprints. IEEE Trans. Pattern Anal. Mach. Intell. 37(3), 555–568 (2015)
15. Deepika, K.C., Shivakumar, G.: Touchless 3D fingerprint classification: a systematic survey. In: 2018 Third International Conference on Electrical, Electronics, Communication Computer Technologies and Optimization Techniques (ICEECCOT). GSSS, Mysore, December 2018
16. Zhang, Y., Shi, D., Zhan, X., Cao, D., Zhu, K., Li, Z.: Slim-ResCNN: a deep residual convolutional neural network for fingerprint liveness detection. IEEE Access 7, 91476–91487 (2019)
17. Yuan, C., Xia, Z., Jiang, L., Cao, Y., Jonathan Wu, Q.M., Sun, X.: Fingerprint liveness detection using an improved CNN with image scale equalization. IEEE Access 7, 26953–26966 (2019)
18. Deepika, K.C., Shivakumar, G.: Touchless 3D fingerprint identification using SSIM template matching technique. Dogo Rangsang Res. J. (UGC Care Group I J.) 10(08), 13 (2020). ISSN 2347–7180
19. Tertychnyi, P., Ozcinar, C., Anbarjafari, G.: Low-quality fingerprint classification using deep neural network. IET Biomet. 7(6), 550–556 (2018)
20. Wang, R., Han, C., Guo, T.: A novel fingerprint classification method based on deep learning. In: 2016 23rd International Conference on Pattern Recognition (ICPR), Cancun, pp. 931–936 (2016)
21. Labati, R.D., Genovese, A., Piuri, V., Scotti, F.: Quality measurement of unwrapped three-dimensional fingerprints: a neural networks approach. In: The 2012 International Joint Conference on Neural Networks (IJCNN), Brisbane, QLD, pp. 1–8 (2012)
22. Lin, C., Kumar, A.: Contactless and partial 3D fingerprint recognition using multi-view deep representation. Pattern Recogn. 83, 314–327 (2018). https://doi.org/10.1016/j.patcog.2018.05.004
23. The Hong Kong Polytechnic University 3D Fingerprint Images Database (2016). http://www.comp.polyu.edu.hk/~csajaykr/3Dfingerprint.htm.
24. Lauer, F., Suen, C.Y., Bloch, G.: A trainable feature extractor for handwritten digit recognition. Pattern Recogn. 40(6), 1816–1824 (2007). https://doi.org/10.1016/j.patcog.2006.10.011ff. ffhal-00018426f
25. Deepika, K.C., Shivakumar, G.: A Robust Deep Features Enabled Touchless 3D-Fingerprint Classification System. SN Computer Science 2(4), 1–8 (2021). https://doi.org/10.1007/s42 979-021-00657-x

Real Conversation with Human-Machine 24/7 COVID-19 Chatbot Based on Knowledge Graph Contextual Search

Tanuja Patgar[1], Ripal Patel[1(✉)], and S. Girija[2]

[1] Dr. Ambedkar Institute of Technology, Bangalore 560056, India
[2] Visvesvaraya Technological University, Belgaum, Karnataka, India

Abstract. The outbreak of the COVID-19 pandemic has changed the whole world scenario and made researchers innovate on the corona virus. Researchers are working on information that includes symptoms, Infection spreading, preventive measures, health and travel advisories, and help lines for further assistance. During this pandemic scenario, the health assistant Chatbot is a very useful conversation tool for COVID-19, which provides preliminary medical advice and preventive measure suggestions. The paper proposes an Artificial Intelligence-based Re-Co Chatbot to provide information about the corona virus and also assist with customer queries. The goal is to build a 24/7 COVID Chatbot capable of answering user questions and to emphasize and stress the concept of contextual semantic search and Knowledge Graph to serve as the FAQ for Corona information. Natural Language Processing (NLP) is used to process the user question and the SpaCy library is used for text processing. Once the question is processed, entities (the subject of question) and relations (predicate of the question) are recognized and extracted. The Chatbot is designed for about 100 question-answers pairs in the CSV file and will create about 575 relationships in the Knowledge Graph.

Keywords: Knowledge Graph · Re-Co Chatbot · Flask API · NLP · Contextual Search

1 Introduction

The idea of building human-machine interaction from major tech companies such as Apple Siri, Amazon Alexa, IBM Watson is growing excitement around conversational agents. Nowadays many start-up companies are working on the idea of advances in artificial intelligence combined with a natural form of human-machine interactions. Though, innovation on Chatbot makes market booming strategy a long way but facing criticism as well. The accidental failure is such a lack of knowledge on user experiences and human-machine conversations in the real world. The user can be satisfied when a real conversation between machine-human Chatbot fills the gap of the marketing wave. Artificial intelligence software that can mimic a human conversation is generally termed as Chatbot. It

© Springer Nature Switzerland AG 2021
K. R. Venugopal et al. (Eds.): ICInPro 2021, CCIS 1483, pp. 258–272, 2021.
https://doi.org/10.1007/978-3-030-91244-4_21

just acts as a virtual assistant with natural language as means for messaging applications, websites, mobile apps, or through the telephone. Much research has been carried out in Chatbot technology and achieved the optimal interaction result where the user is unable to compare with an actual human being. The tech companies such as Google, Amazon, Microsoft, Facebook are fetching new Chatbot that can interact with the user in text, speech, and human action formats. Chatbots are just apps where users connect conversation with text or speech. As technology grows further, they are designed the conversation with a complex pattern for able to understand artificial intelligence.

Chatbots are generally classified into:

- **Scripted Chatbots:** The short and long text conversation Chatbot is designed based on creating single and multiple entities. If the Chatbot will acknowledge for single entity and replies with a specific response it is termed a short text conversation. The conversations are based on multiple entities with managing multiple responses that are termed as long text Chatbot. It has several significant features like recognition, unique phrases with pre-defined responses, user-friendly machine-human learning capabilities.
- **Virtual Assistant:** Artificial intelligence-based Chatbot is designed like a human toy that can have the capacity to understand, answering questions, and finally maintaining an end-to-end conversation.
- **FAQ Chatbot:** The Chatbots are designed for questions and answer entities with user's single or multiple response data. It is selected as a higher intelligence Chatbot with a simple pattern.
- **Virtual Agent:** Most intelligent Chatbots with complex dialogs, processing and allowed security protocols as a human does.

There are several Chatbot development platforms of characteristics that should be considered when designing Chatbot. The Chatbot characteristics include–Intent Recognition - It offers rel conversation with user and even helps to reduce the user's frustration. Dialog Management - The Chatbot is designed with complex Q&A and enables meaningful conversations with user data. Humanization–Chatbot acts more human like, and then users will be engaged in conversation. Interaction Channels–the platform can be chosen with access to various interactive medium. Task Automation - The ability to perform different task for user. Ease of Implementation - The platform provides flexible software development environment.

For many years Knowledge Graphs (K-Graph) have been looking into our daily lives. Many machine learning algorithms have been used to create K-Graphs with the domain knowledge. It is a structural mapping of all domain data. In the algorithm the K-graph is top level entity and it can be structured or unstructured. It is a single platform where all data and interlocking relationships behind that data are virtually stored. Google search is one such enhanced technology, and LinkedIn uses K-Graphs to boost its search, business, and consumer analytics. The K-Graph contains various nodes containing possible questions, entities, and answers.

2   Literature Survey

A Chatbot with a deep learning algorithm has been proposed in [1]. The presented work proposed by MILABOT is efficient enough to text and speech. The presented system is a combination of different neural network models and their variations. Reinforcement learning is applied and the model is trained to get the best performance out of the presented model. The framework is inspired by ensemble machine learning systems. Different independent and small models are used and combined intelligently. A new Chatbot model has been proposed with a self-feeding mechanism in [2]. Part of the conversations is feed to train the system periodically and improving its dialogue performance. The self-feeding process is divided into three parts: train, deploy and retrain. Self-feeding has been done in this work in two ways: first is generating similar responses or queries from satisfied customers or users and critical assessment of unsatisfied users. However, there are still various ways to improve this model.

Chatbot has changed the scenario of Human-Computer Interaction (HCI) and this is one of the reasons and requirements to incorporate social behavior into Chatbot. The extensive survey of social characteristics for Chatbot is presented in [3]. The author has analyzed research done in various fields to understand. The social mechanism can be useful for interaction and recognizing the challenges and strategies to design Chatbot. Mostly, all Chatbots are designed based on the conversational dataset. Lin et al. proposed a multi-lingual extension of Xpersona [4]. The author made a personalized dataset with 6 different types of language conversations. Both multi-lingual and cross-lingual trained datasets have been used to design the Chatbot.

An open domain Chatbot has been presented in [5]. Till now, all the Chatbots models are trained with various sizes of data and with variations in the number of parameters. In this work, the author had mentioned different ingredients to get high performance out of Chatbot. The main input fed to the Chatbot models is conversation but in this work apart from the conversation, choice of generation strategy and appropriate training data also fed to Chatbot henceforth it can converse like an expert. Chatbot is an outstanding low-cost and effective tool for self-disclosure features, which can do small talk with humans. The work presented in [6], has designed a Chatbot with self-disclosure features, which can do small talks with humans. The advantages of self-disclosure features are conversation feed more genuine and more enjoyable. Many proposed Chatbots are not capable enough to deal with complex interactions and withstand company needs. In [7], the author presented Xatkit, which tackles requirements like, deep understanding of the targeted platform and reduction in development and maintenance cost. A set of domain-specific languages is used to define Chatbot, which makes Chatbot platform-independent. Due to diverse modifications in knowledge improvement, education, and learning method, Sinha et al. [8] introduced an approach where document is converted into knowledge which is later used to feed Chatbot. A virtual Chatbot is implemented in this work, where user queries are input to Chatbot in terms of electronic documents. The documents are converted into editable text form and later used to train Chatbot. The proposed system is to develop an automatic system which can provide an answer to the question asked by the user, solely for educational purpose. Another application of Chatbot is presented in [9] as an expert recommendation task. The main purpose of this Chatbot is to serve the phare community. In [10], the author has proposed an approach to improve group chat discussion. Often group chats discussion has consequences are like unrecognized messages, uneven members' contribution, and longtime reach consensus. To overcome these problems, the author proposed a group feed bot. The functions of this Chatbot agent are discussion time management, member participation encouragement, and member opinions organization.

The thriving usage of social media requires customer care support 24x7. Hence, Xu et al. [11] proposed a Chatbot agent to support customer service for social media. The system implementation has been done using deep learning algorithms and tested and evaluation is done for Twitter conversations. Mental health care digitization requirement is inexpensive and universally available and this can be possible with conversational assistance in terms of the Chatbot. Lee et al. [12] has presented a self-compassion Chatbot, which can find the availability of care giving and care receiving requirements. Recently, Dhyani et al. [13] has proposed a deep learning framework for Chatbot. The deep learning framework comprises of Bidirectional Recurrent Neural Network (BRNN) containing attention layer resulting more appropriate real conversation.

3   Proposed Re-Co FAQ Chatbot Framework

We proposed a Re-Co (Real Conversation) Chatbot framework and the block diagram of the proposed work is shown in Fig. 1. It is FAQ based Chatbot, designed for questions and answer entities with user's single or multiple response data. The user types in the desired question in the Messenger platform built as front-end. The question is received by Python Flask API endpoint named as '/search' and the question is text processed. Natural Language Processing (NLP) is used to process the user question and the SpaCy library is used for text processing. Once the question is processed, entities (the subject of question) and relations (predicate of the question) are recognized and extracted.

To achieve the summarized objective, the Chatbot model is divided into several parts. It started with building K-Graph (KG) using a set of questions and answers. Flask API is created to receive user questions, search for the answer in K-Graph and return the response. Facebook Messenger is used as a front-end tool. It covers two algorithms – one for building K-Graph and another for searching K-Graph and returning an optimal response.

The extracted entities and relations are sent to K-Graph. It contains various nodes with possible questions, entities, and answers. The process will search K-Graph for entities, relations and will retrieve the optimal end node and retrieve the answer as well. Flask API will return this answer as a response to the user. Since the context of the question (entity and relation), and not the entire question is searched from K-Graph is called Contextual Search.

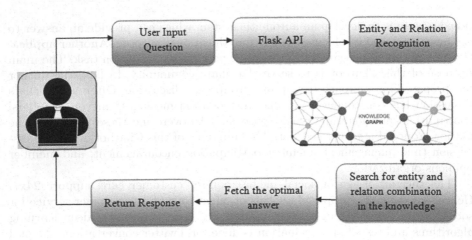

Fig. 1. Re-Co FAQ Chatbot framework

4   Development of Open Front-End Tools and Visualizations

In designing the FAQ Chatbot, the procedure and methodology are portioned into several parts as shown in Fig. 2. It started with building a K-Graph using a set of user questions and answers. Flask API is created to receive user questions, search for an answer and return the response. The front-end tool used is Facebook Messenger. The research work demands many new technologies to be used. Some of them are NLP (Text processing), Contextual search using K-Graph. The Re-Co FAQs bots were designed with 100 distinct answers. The job of the underlying system is to decide based on queries, and make the conversation effective.

The research on natural language processing has been increasingly considered as one of the future technology. It is a computerized approach to analyzing text and being a very thrust area of research and development. The literature distinguishes the main application of NLP for speech synthesis and speech recognition as a major contribution in the Artificial Intelligence field. In the speech synthesis method, the text data is converted to speech. It uses high-level modules for speech synthesis. The sentence segmentation deals with punctuation marks on a simple decision tree. Similarly, an automatic speech recognition system makes use of NLP techniques based on grammar. It uses context-free grammars for representing the syntax of the language presents. Spontaneous dealing through the spotlighting addition of automatic summarization including indexing, which extracts the gist of speech transcriptions to deal with Information retrieval and dialog system issues.

The most explanatory method for presenting what happens within NLP is employing the 'levels of language' approach. This is also referred to as the synchronic model of language and is distinguished from the earlier sequential model, which hypothesizes that levels of human language processing follow one another

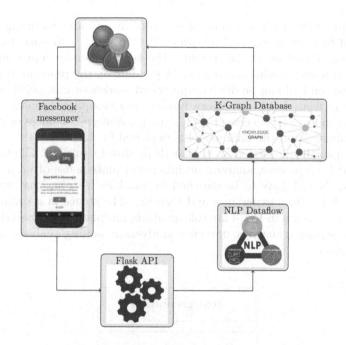

Fig. 2. 5-Frame data-flow visualization chart

in a strictly sequential manner. Psycholinguistic research suggests that language processing is much more dynamic, as the levels can interact in a variety of orders. Introspection reveals that we frequently use the information we gain from what is typically thought of as a higher level of processing to assist in a lower level of analysis. For example, the pragmatic knowledge that document reading is about biology will be used when a particular word that has several possible senses is encountered, and the word will be interpreted as biology sense. Of necessity, the following description of levels will be presented sequentially. The key point here is that meaning is conveyed by each level of language and that since humans have been shown to use all levels of language to gain understanding, the more capable an NLP system is, the more levels of language it will utilize. Typically, any NLP-based problem can be solved by a methodical work-flow that has a sequence of steps. The major steps are depicted in the following Fig. 3.

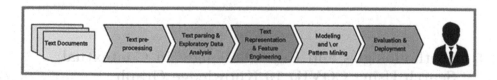

Fig. 3. NLP data-flow

It usually starts with a corpus of text documents and following standard processes of text wrangling and pre-processing, parsing, and basic exploratory data analysis. Based on initial insights, the text is usually represented using relevant feature engineering techniques. Depending on the problem at hand, the focus is either on building predictive supervised models or unsupervised models, which usually focus more on pattern mining and grouping. Finally, the model is evaluated and overall success criteria with relevant stakeholders or customers are also evaluated and the final model is deployed for future usage.

NLP is a small sub-section of AI that deals with linguistics. The final objective of NLP is to process, analyze, decipher and make sense of natural human language. It doesn't have to be enriched by machine learning that enables the Chatbot to learn from experience and training. The stages in a comprehensive NLP system as shown in Fig. 4 are tokenization, morphological analysis, syntactic analysis, semantic analysis, discourse analysis, reasoning analysis, and finally text generation.

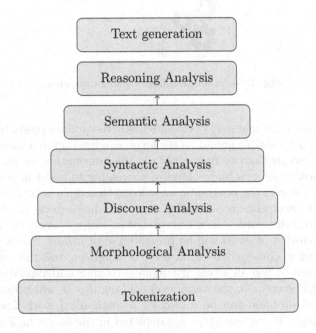

Fig. 4. Text generation flow analysis

5   Methodologies and Key Challenges in Contextual Search Using COVID 19 Knowledge Graph

The algorithm for building K-G Graph, Neo4j graph database is used to store the required information. A CSV file is similar to an excel sheet is created with all questions and corresponding answers. The question and answer pair is processed and K-G graph is set up after processing. The code is written in Python for NLP.

When the user makes a query, process it, and return a response. A Facebook Messenger platform has been built, in which the user types the question and receives a response. Once the user types question, it is received at the Flask API endpoint. Then the question is processed using SpaCy and entities, relations are extracted. The API then searches if the entity node and relation combination are present in the K-G graph. The end node which is the answer is returned to the user (Fig. 5).

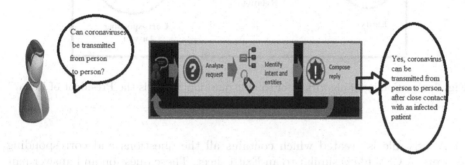

Fig. 5. Process of query and response in Flask API

A Flask API '/search' is written in Python language which does the process of receiving the query, searching K-G graph for an answer, and returning the answer. The below steps depict the algorithm for building a K-Graph:

– Step 1: When the code is executed, initializations by loading a SpaCy model.
– Step 2: The CSV file has to be read first by the code. Pandas library is used to read CSV file.
– Step 3: Each question is read from CSV file and it is processed using SpaCy library. By processing, the corresponding entities and relations are extracted. Entity is the subject of question and relation is predicate of question.
– Step 4: Once all the questions are processed, a data-frame is formed. It contains arrays of all the questions and their corresponding entities, relations and answers.
– Step 5: Now create a relationship between the extracted entity and its corresponding answer and commit it in the K-G graph.
– Step 6: To achieve the above goal, certain conditions are to be satisfied. First, entity nodes are to be created. The code checks if the entity node is already present. If it is not present a new node is created. Else existing node only is used.
– Step 7: The relationship between the entity node and answer node is to be created.

- Step 8: After all the relationship and nodes are created, K-G graph is committed into Neo4j graph database.

Neo4j is a graph database that is used to store the K-G graph. The algorithm is explained by taking a sample question ("Who is the President of India?") as an example as shown in Fig. 6.

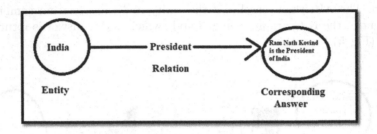

Fig. 6. Functioning of algorithm for sample question "Who is the President of India?"

A CSV file is created which contains all the questions and corresponding answers. A CSV file is similar to an Excel sheet. These question and answer pair is processed and the K-Graph is set up after the processing. The code is written in Python for Natural Language Processing. Below steps defined the algorithm for Searching the Query in Knowledge Graph.

- Step 1: Flask API receives the user text (Question).
- Step 2: If the user has typed only one word, then user text only will be the entity as well as relation.
- Step 3: Else, if the user types in an actual question, the entities and relations are extracted using SpaCy.
- Step 4: K-G graph is built using the algorithm which contains 2 types of relationship. ● Question – 'answer is' – Answer tried, Entity – relation – Answer tried. First, the code searches if whole question is present in Knowledge Graph. If it is present, step 8 is executed. Else, next step is executed.
- Step 5: Since the question is not present, code will now search if the entity-relation combination is present in the K-G graph. If it is present, step 8 is executed. Else, next step is executed.
- Step 6: Both question as and combination is not present, code will now search for the entity node in K-G graph. If it is present, step 8 is executed. Else, next step is executed.
- Step 7: Else, the relation is searched in K-G graph. If it is present, step 8 is executed. Else, 'no answer found' is returned.
- Step 8: The end node which is the answer is returned.
- Step 9: Flask API will return the answer to the user.

6 Result Analysis

The goal is to build a Chatbot capable of answering user questions and to emphasize and stress the concept of contextual semantic search and K-G graph. The Chatbot is just an application of the contextual search, which shows the power of K-G graph by using Natural Language Processing. To meet the goal, the following are objectives - • A code is written in Python language which will build a knowledge graph. • A Flask API is built which will in turn receives user requests, search from the knowledge graph, and returns the response. • Facebook Messenger is a user interface. A Chatbot typically has 3 things in it 1) Intent (Intention of the query asked by user) 2) Entities (Named entities in Query like Location names, People names, date, etc.) #Named Entity Recognition 3) Action or Response (the result to throwback to the user).

To achieve the goal, research work is sectioned into several parts. The work is started with building K-G graph using a set of questions and answers. Then Flask API is created to receive user questions, search for the answer in K-G graph and return the response. The front-end tool used is Facebook Messenger. The extracted entities and relations are sent to K-G graph. It contains various nodes with possible questions, entities, and answers. The process now will search the K-G graph for entities, relations and will retrieve the optimal end node, which in turn contains the answer. The Flask API will return this answer as a response to the user. Since the context of the question (entity and relation), and not the entire question, is searched from K-G graph, the process is called Contextual Search.

6.1 Case 1 List of Promising Data Sets Used to Build COVID-19 Knowledge Graphs

The CSV file is a set of predefined questions and their corresponding answers as shown in Fig. 6. The code takes in this file, reads the file then processes the questions using Natural Language Processing (NLP) to extract the entities required for building K-G graph. Each question is read from the CSV file and is processed using SpaCy library with corresponding entities and relations are extracted. The entity is the subject of the question and relation is the predicate of the question.

6.2 Case 2 K-Graph Construction

Figure 7 shows the constructed Knowledge Graph of all the questions present in CSV file. In K-G graph one can see the blue nodes are the entities(subject) of question and brown nodes are corresponding answers. The blue node is connected to brown node with the help of a relationship.

There are about 100 questions-answers pairs in the CSV file. It will create about 575 relationships in the knowledge graph.

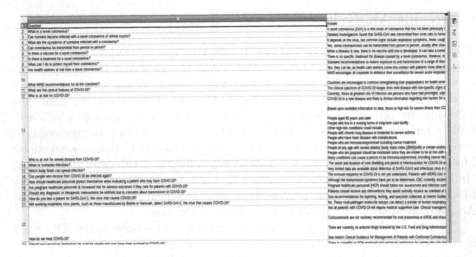

Fig. 7. CSV file containing the Question and Answer

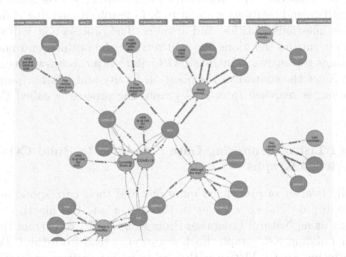

Fig. 8. Knowledge Graph of all the questions present in CSV file

Dataset 1 - Full Question is Given: Figure 8 shows the use case where the user types the whole question and retrieves the answer. Once the user types in the question, it is received at the Flask API endpoint. The question is processed using spaCy and entities, relations are extracted. The API then searches if the entity node and relation combination is present in the Knowledge Graph, end node which is the answer is returned to the user.

Dataset 2 - Only an Entity or a Relation or a Combination (Entity + Relation) is Given: The Fig. 9 shows the use case where the user types in either only entity or relation or a combination of entity and relation and

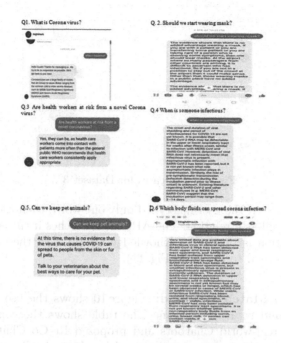

Fig. 9. Use cases for Dataset 1

Fig. 10. Use cases for Dataset 2

Fig. 11. Use cases for Dataset 3

retrieves the answer. If the user has typed only one word, then use text only will be the entity as well as a relation. The API then searches if the entity node and relation combination is present in Knowledge Graph, and the end node is the answer which returned to the user (Fig. 11).

Dataset 3 - No Match is Found: Figure 10 shows the use case where no match is found and returns the same The table shows the comparison chart for the different real-world Chatbots and proposed Re-Co Chatbot. The proposed Chatbot is 24/7 customer support with a high degree of accuracy and high-throughput. Increase customer satisfaction by effectively solving their daily issues. It is keyword recognition-based and contextual-based Chatbots that utilize customizable keywords and AI to determine how to serve an appropriate response to the user.

Table 1 Comparative for Re-Co FAQ based chat bot with real world bot.

Table 1. Comparative analysis of Re-Co FAQ based chat bot with real world bot

Bot name	Features	Channels	Language
Agent bot	Natural language processing	Voice or Messenger channel	English/Spanish
Twyla	Machine learning	Web, Facebook, Telegram	English
Semantic machine	Deep learning	Text/Voice	Language independent
Chatter bot	Matching response (python)	Console API speech recognition	Language independent
Octane.ai	Real time analytic	Facebook messenger	Language independent
Gubshup	Natural language processing	Facebook, Skype, Telegram	English/Hindi
Re-Co bot (our model)	Natural language processing	Facebook messenger	English

7   Conclusion

This paper has highlighted the relationship between semantic information of the context and conceptual graph and designs FAQ-based contextual search Re-Co Chatbot. The work is started with building a K-G graph using a set of questions and answers. Then Flask API is created to receive user questions, search for the answer in K-G graph, and return the response. The front-end tool used is Facebook Messenger.

The paper introduced a new architecture for use in retrieval-based Chatbots. The model in turn shows the ability and power of a K-G graph. Natural Language Processing is still an emerging field, and it is also stressed in this paper. The model mimics the Google search engine; the only difference is it retrieves the answers instead of sites. For the proposed work, the Re-Co Chatbot is designed for about 100 question-answers pairs in the CSV file and will create about 575 relationships in the K-G graph.

References

1. Serban, I.V., et al.: A deep reinforcement learning Chatbot. arXiv preprint arXiv:1709.02349 (2017)
2. Hancock, B., Bordes, A., Mazare, P.E., Weston, J.: Learning from dialogue after deployment: feed yourself, Chatbot! arXiv preprint arXiv:1901.05415 (2019)
3. Chaves, A.P., Gerosa, M.A.: How should my Chatbot interact? A survey on human-Chatbot interaction design. arXiv preprint arXiv:1904.02743 (2019)
4. Lin, Z., et al.: XPersona: evaluating multilingual personalized Chatbot. arXiv preprint arXiv:2003.07568 (2020)
5. Roller, S., et al.: Recipes for building an open-domain Chatbot. arXiv preprint arXiv:2004.13637 (2020)
6. Lee, Y.C., Yamashita, N., Huang, Y., Fu, W.: "I hear you, I feel you": encouraging deep self-disclosure through a Chatbot. In: Proceedings of the 2020 CHI Conference on Human Factors in Computing Systems, pp. 1–12, April 2020
7. Daniel, G., Cabot, J., Deruelle, L., Derras, M.: Xatkit: a multimodal low-code Chatbot development framework. IEEE Access 8, 15332–15346 (2020)
8. Sinha, S., Basak, S., Dey, Y., Mondal, A.: An educational chatbot for answering queries. In: Mandal, J.K., Bhattacharya, D. (eds.) Emerging Technology in Modelling and Graphics. AISC, vol. 937, pp. 55–60. Springer, Singapore (2020). https://doi.org/10.1007/978-981-13-7403-6_7
9. Cerezo, J., Kubelka, J., Robbes, R., Bergel, A.: Building an expert recommender Chatbot. In: 2019 IEEE/ACM 1st International Workshop on Bots in Software Engineering (BotSE), pp. 59–63. IEEE, May 2019
10. Kim, S., Eun, J., Oh, C., Suh, B., Lee, J.: Bot in the bunch: facilitating group chat discussion by improving efficiency and participation with a Chatbot. In: Proceedings of the 2020 CHI Conference on Human Factors in Computing Systems, pp. 1–13, April 2020
11. Xu, A., Liu, Z., Guo, Y., Sinha, V., Akkiraju, R.: A new Chatbot for customer service on social media. In: Proceedings of the 2017 CHI Conference on Human Factors in Computing Systems, pp. 3506–3510, May 2017

12. Lee, M., Ackermans, S., van As, N., Chang, H., Lucas, E., IJsselsteijn, W.: Caring for Vincent: a Chatbot for self-compassion. In Proceedings of the 2019 CHI Conference on Human Factors in Computing Systems, pp. 1–13, May 2019
13. Dhyani, M., Kumar, R.: An intelligent Chatbot using deep learning with Bidirectional RNN and attention model. Mater. Today Proc. **34**, 817–824 (2021)

Positive Correlation Based Efficient High Utility Pattern Mining Approach

Dharavath Ramesh$^{(\boxtimes)}$ ⓘ, Krishan Kumar Sethi ⓘ, and Aman Rathore

Department of Computer Science and Engineering, Indian Institute of Technology (ISM), Dhanbad 826004, Jharkhand, India
{ramesh.d.in, kksethi}@ieee.org

Abstract. The problem of high utility pattern (HUP) mining is an interesting task and comes with various applications. It has been argued in various past studies that the patterns with strong mutual correlation are more useful for decision making. A more powerful tool that takes the inherent correlation into account is more desirable. This paper presents an Efficient Correlated High Utility Miner (ECHUM) that considers the positive correlation among the items to find correlated high utility itemsets. The ECHUM uses a list structure is to avoid multiple scanning of the dataset. The search space is significantly reduced by using two upper-bounds named *sub-tree utility* and the *local utility*. Also, several database projection methods and transaction merging methods are used to reduce the complexity. Several pruning strategies are adopted to make the algorithm fast and memory efficient. A variety of experiments conducted shows that ECHUM is 2–3 times faster, and the memory usage is also 3–4 times lesser than the existing algorithm.

Keywords: High utility pattern mining · Correlation · Pruning strategy · Database projection

1 Introduction

High Utility Itemset Mining (HUIM) appears to be one of the most effective methods for pattern detection, which considers the quantity and profit of an itemset to discover the set of high utility itemsets (HUIs). Several Algorithms, like Two-Phase [1], tree-based [2–4], and one-phase [5–8] were introduced to find HUIs. The two-phase algorithms search potential candidate itemsets in the first phase and HUIs in the second phase. All two-phase algorithms suffer from scalability issues. The tree-based algorithms use a tree structure to produce the candidate itemsets and HUIs. The one-phase algorithms utilize a vertical list structure to search candidates and HUIs in a single phase. Several other one-phase algorithms like FHM [6], HUP miner [7], and EFIM [8] were introduced with several efficient pruning strategies. The EFIM algorithm introduces various novel ideas to improve time efficiency and memory optimization compared to the HUI miner [5]. It presents database projection and transaction merging to reduces the merging complexity and size of the database. EFIM also uses two upper-bounds sub-tree utility and local

© Springer Nature Switzerland AG 2021
K. R. Venugopal et al. (Eds.): ICInPro 2021, CCIS 1483, pp. 273–286, 2021.
https://doi.org/10.1007/978-3-030-91244-4_22

utility for efficient pruning. A fast utility counting method was introduced to calculate the utilities. This new counting method uses an array to store the utilities.

The main focus of the existing HUI mining algorithms [9, 10] is to search for the set of high utility items in a transactional dataset. However, they do not consider the mutual relationship among items of the discovered patterns. For example, a high-profit item like gold may appear with some low-profit items like cotton resulting in a high utility itemset. Still, cotton and gold are unrelated and could have occurred by chance. Hence, it is important to develop a framework that considers the correlation factor and produces high utility itemsets with inherent correlation. Many correlation measures have been commonly used for pattern mining, e.g., support [11], confidence [12], frequency affinity [13], all-confidence [14], and coherence [14]. HUIPM [13] and FDHUP [15] algorithms were introduced in utility-based pattern mining to discover HUIs using frequency affinity. Both algorithms use co-occurrence to calculate the correlation measure, which makes them ineffective. A projection-based algorithm called CoHUIM [16] was introduced, which considers the inherent correlation instead of the co-occurrence frequency. It discovers a set of high utility itemsets correlated using the correlation factor called Kulc [17]. The Kulc factor follows null transaction unvarying property to ensure that the correlation measure should not be affected by the number of null transactions in the database. However, the projection-based CoHUIM was found inefficient as it produces a huge number of candidate itemsets resulting in an extra memory consumption and efficiency decline. Correlated high-Utility Pattern Miner (CoUPM) [18] is a one-phase algorithm that uses the utility-list structure to store important details of the itemsets. This algorithm also uses the correlation factor Kulc to discover correlated patterns efficiently.

We observed that CoUPM [18] is an extension of the HUI-miner [5], which is computationally expensive for the small value of minimum utility. We propose an algorithm ECHUM, which exploits efficient EFIM [8]. Our choice is based on the fact that EFIM takes three times lesser time and takes eight times lesser memory than the HUI-miner. We have used the Kulc measure to find the inherent correlation in the discovered patterns. We adopted three major techniques to search the Correlated High Utility Itemsets (CoHUI). Firstly, database projection and transaction merging are used to reduce the size of the dataset and merging complexity. Secondly, sub-tree utility and local utility upper-bounds are used to prune the unpromising candidate itemsets. Thirdly, a fast counting method is adopted that uses an array to store the utilities to lower the upper bounds. We performed comprehensive experiments to compare ECHUM with the existing CoUPM algorithm. It has been observed that the ECHUM is faster than CoUPM.

The remaining paper is structured as follows. Section 2 discusses various preliminaries. The proposed algorithm is described in Sect. 3. The performance evaluation of the proposed algorithm has been carried out in Sect. 4. We conclude the paper in Sect. 5.

2 Preliminaries

Let D be a dataset consists of a collection of items I. An itemset X with k different items is called k-itemset. If X is a subset of transaction Tq then X is said to be contained in that transaction. An example of transaction and profit tables are shown in Tables 1 and 2.

Table 1. Transaction table

TID	Transactions	Count	TU
T_1	$A_2\ A_3\ B_2$	<1,1,2>	11
T_2	$A_1\ A_2\ B_1$	<3,1,2>	21
T_3	C_1	<1>	1
T_4	$B_1\ B_2$	<2,4>	26
T_5	$A_2\ B_1$	<3,1>	11
T_6	$A_1\ A_3\ B_2\ C_1$	<1,2,1,3>	12
T_7	$B_1\ B_2\ C_1$	<3,3,4>	31
T_8	$A_2\ B_1\ C_1$	<2,2,2>	16
T_9	$A_1\ B_1\ B_2$	<2,2,1>	20
T_{10}	$A_1\ C_1$	<4,5>	17

Table 2. Profit table

Stocks	A_1	A_2	A_3	B_1	B_2	C_1
Utility	3	2	1	5	4	1

Definition 1. Utility of item i in a transaction T, i.e., $U(i, T)$ is the multiplication of its profit $p(i)$ and quantity in that transaction $q(i, T)$.

$$U(i, T) = p(i) \times q(i, T) \tag{1}$$

The utility value for an itemset X in a transaction T and dataset D is calculated as follows.

$$U(X, T) = \sum_{i \in X \wedge X \subseteq T} U(i, T) \tag{2}$$

$$U(X) = \sum_{X \in T \wedge T \subseteq D} U(X, T) \tag{3}$$

E.g., $U(B_2, T_4) = 4 \times 4 = 16$; $U(A_1B_2,T_6) = 1 \times 3 + 1 \times 4 = 7$; $U(A_1B_2) = (1 \times 3 + 1 \times 4) + (2 \times 3 + 1 \times 4) = 17$.

Definition 2. Transactional utility associated with any transaction T is represented by $tu(T)$ and calculated as follows.

$$TU(T) = \sum_{i \in T} u(i, T) \tag{4}$$

E.g., $tu(T_4)$ is $(2 \times 5 + 4 \times 4) = 26$.

Definition 3. The transaction weighted utility of an itemset X is represented by $TWU(X)$ and calculated as follows.

$$TWU(X) = \sum_{X \in T} TU(T) \tag{5}$$

E.g., $TWU(C_1) = 1 + 12 + 31 + 16 + 17 = 77$. The TWU values for running dataset are depicted in Table 3.

Table 3. Transactional weighted utility

Times	A_1	B_2	A_3	B_1	B_2	C_1
TWU	70	59	23	125	100	77

Definition 4 (High Utility Itemset). If the utility of an itemset X is greater than or equal to the minimum utility threshold *minutil*, it is said to be a high utility itemset.

$$HUI = \{X \mid u(X) \geq minutil\} \tag{6}$$

Definition 5 (Remaining Utility). Let T/X be the items appearing after X in a transaction T. The remaining utility is denoted as $ru(X)$ and calculated as follows.

$$ru(X, T) = \sum_{i \in T/X} u(i, T) \tag{7}$$

E.g., $ru(B_1) = (4 \times 4) + (3 \times 4 + 4 \times 1) = 32$.

Property 1. If the value of Transactional Weighed Utility of X, i.e., $TWU(X)$ is smaller than *minutil* then X and all its supersets are pruned.

Definition 6. The upper limit of remaining utility for itemset X, i.e., $reu(X)$, is calculated as follows.

$$reu(X) = u(X) + re(X) \tag{8}$$

E.g., $reu(B_1) = u(B_1) + ru(B_1) = 60 + 32 = 92$.

Property 2. If the remaining utility upper bound for X is smaller than the *minutil* then X and all its supersets are pruned.

Definition 7. The correlation value among the items of an itemset determines the strength of the inherent correlation. We use *Kulc* measure to find the correlation for an itemset X in ECHUM.

$$Kulc(X) = 1/k \sum_{i \in X} SUP(X)/SUP(i) \tag{9}$$

The value of *Kulc* lies between 0 and 1 and helps determine the positive correlation among the items.

2.1 Pruning Strategies

2.1.1 Exploring the Search Space

ECHUM explores the search space recursively in a depth-first search manner. In the search space, items are arranged in increasing order of their TWU. This order is selected because it reduces the number of candidates generated.

Definition 8. Extensions of itemset α is denoted by $E(\alpha)$ and defined as $E(\alpha) = \{y \mid y \in I^\wedge \wedge y > x \forall x \in a$. Let extension of itemset α be Z where $Z = \alpha \cup W$ if $W \in 2^{E(\alpha)}$ and $W \neq \emptyset$. Also, Z is a single itemset extension if $Z = \alpha \cup x$.

E.g., The single extensions of itemset $\{A_3\}$ are $\{A_3, B_2\}$, $\{A_3, A_1\}$, $\{A_3, C_1\}$ etc. The other extensions are $\{A_3, B_2, B_1\}$, $\{A_3, A_1, C_1\}$ etc.

2.1.2 2.1.2. Cost reduction Using Projections

Definition 9. For calculating the utility value for an itemset α, the itemsets that are not the extensions of α, i.e., $E(\alpha)$, are disregarded. The database consisting of such itemsets is called a projected database. The projection of a transaction T is denoted by $\alpha - T$, which is equal to $\{I \mid i \in T^\wedge i \in E(\alpha)\}$. The projection of the database is denoted as $\alpha - D$, which is equivalent to $\{\alpha - T \mid T \in D \wedge \alpha - T \neq \emptyset\}$.

E.g., if $\alpha = \{B_2\}$ then the projected database will consist of 3 transactions i.e. $T_1\{A_2, A_2, B_2\}$, $T_4\{B_1, B_2\}$ and $T_7\{B_1, B_2, C_1\}$.

Database projections are used in ECHUM to reduce the database size. Database projection is carried out efficiently using *pseudo-projection* by sorting the database in total order. The method of pseudo projection is explained in EFIM [8].

2.1.3 Cost Reduction by Transaction Merging

Identical transactions are merged to form a single transaction. A transaction is a duplicate of another transaction if the same items are present in both the transactions, but each item's quantity can be different.

Definition 10. Transaction merging is applied to each of the projected transaction databases $\alpha - T$, and all the duplicate transactions are combined to form a single transaction in the projected database. Consider a transactional database D; if the transactions are sorted in lexicographical order and are read from end to start, then the order is known as total order and is denoted by $>.T$.

Transaction merging is used to merge a set of transactions efficiently, and after applying this method to the projected database, the number of transactions is drastically reduced. The smaller the transaction length, the greater the probability of matching transactions. The database is sorted according to the $>.T$ order to merge identical transactions efficiently. The transactions are combined efficiently using the property introduced in EFIM [8] by first sorting the transactions lexicographically and then introducing various measures to compare the transactions.

Property 3. In a transactional database sorted according to the $>.T$ property, the duplicate transactions will appear consecutively. All the duplicate transactions can be removed directly from the database by comparing the current transaction with its next transaction.

2.1.4 Pruning Using Upper Bounds

Definition 11. Consider an itemset α associated with a single extended itemset y. The *sub-tree* utility of the element y is denoted as $su(\alpha,y)$.

$$su(\alpha, y) = \sum_{T \in g(\alpha \cup \{y\})} [u(\alpha, T) + u(y, T) + \sum_{i \in T \wedge i \in E(\alpha \cup \{y\})} u(i, T)] \qquad (10)$$

E.g., if $\alpha = \{A_3\}$ and $z = \{B_2\}$ then $su(\alpha,z) = (1 + 8) + (2 + 4 + 3) = 18$.

Property 4. Consider an itemset α, the utility value of the single extensions of α is smaller than or equal to the *sub-tree utility*. Mathematically, $su(\alpha, y) \geq u(Y)$, where Y is known to be only the single extended itemset of α.

Pruning Method 1. If the *sub-tree utility* of α is smaller than *minutil* then the single itemset extensions of α cannot be a HUI, and therefore the extensions are pruned.

Definition 12. Consider an itemset α, the extension of α denoted by $y \in E(\alpha)$, the *local utility* of y concerning α is indicated by $lu(\alpha, y)$.

$$lu(\alpha, y) = \sum_{T \in g(\alpha \cup \{y\})} [u(\alpha, T) + ru(\alpha, T)] \qquad (11)$$

Property 5. For an itemset α, the *local utility* value of α is always greater than or equal to the utility of the extensions of α. Mathematically, $lu(\alpha, y) \geq u(Y)$, where Y is any extension of α.

Pruning Method 2. For itemset α, if the *local utility* value of α is less than the *minutil* then the single itemset extensions of α cannot be a high-utility itemset. Therefore, the extensions are pruned.

Property 6. For any itemset α with an extension Y, the following property always holds.

$$TWU(\alpha) \geq lu(\alpha, y) \geq su(\alpha, y) \qquad (12)$$

Definition 13. The *primary items* of an itemset α are the collection of all items, which are the extensions of α whose *sub-tree* utility is greater than *minutil*. The *secondary items* of α are the collection of those items: the extensions of α, and *local utility* is greater than *minutil*. Mathematically, $Primary(\alpha) = \{y|y \in E(\alpha) \wedge su(\alpha,y) > minutil\}$ and $Secondary(\alpha) = \{y|y \in E(\alpha) \wedge lu(\alpha,y) > minutil\}$. By Property 6, it can be said that *Primary items* are the subset of *Secondary items*.

2.1.5 Pruning Using Correlation Properties

Property 7. In a transactional database where are the items in a transaction are arranged according to their support values, the *Kulc* measure uses the following *sorted downward closure property.*

$$Kulc(i_1, i_2 \ldots \ldots \ldots, i_k) \leq Kulc(i_1, i_2 \ldots \ldots \ldots, i_k, i_{k-1}) \qquad (13)$$

Pruning Method 3. Consider an itemset X for which the value of *Kulc(X)* is smaller than any of the *Kulc* values of its subsets, then the collection of every superset of X are not CoHUI's. Let X_k be an itemset and subset X_{k-1} be its subset; Mathematically, if $Kulc(X_k) \leq Kulc(X_{k-1})$, then all supersets of X_k are not in CoHUI's.

2.2 Improving Efficiency Using Fast Utility Counting

In the previous section, two utility-based pruning methods are used to eliminate ineffective itemsets. To compute the utility upper bounds, EFIM introduced a technique that uses an array data structure to store utilities known as Fast Utility Counting (FUC). The novel array structure is called utility-bin [19]. Let a dataset D containing $|I|$ items, and $U[y]$ denotes the *utility bin*, where y is any item in itemset I. The array is initialized with 0. Following is the process for the calculation of upper limits by using the *utility bin* array.

TWU Upper-Bound. The TWU upper bound is calculated using the utility bin array. Initially, values in the array are set to 0. Then by proceeding transaction-wise in the database, the values of each item of array $U[z]$ are modified by the total utility $TU(T)$ of that transaction. i.e., $U[y] = U[y] + TU(T)$, where y is the item contained in transaction T. Finally, all the values of the utility array will be equal to the TWU value.

Sub-tree Utility upper-bound associated with any itemset α is calculated using the *utility bin* in the following manner. We proceed with the dataset transaction-wise, such that the sub-tree utility value modifies each item's utility in U[p] for that transaction.

$$U[p] = U[p] + u(p, T_i) + u(\alpha, T_i) + \sum_{i \in T \wedge i > .z} u(i, T) \qquad (14)$$

Where p is the extension itemset in transaction T_i. Finally, all the utility array values will be equal to the subtree utility value, i.e., *su(p, α)*.

Local-Utility upper-bound associated with any itemset α, the dataset is processed transaction-wise. Utility values of each item of array $U[p]$ are modified by the "local-utility" of that transaction.

$$U[p] = U[p] + \sum_{i \in T \wedge i > .z} u(i, T) + reu(\alpha, T) + u(\alpha, T) \qquad (15)$$

Where *p* is the extension itemset in transaction T. Finally, all the utility array values will be equal to the local utility value, i.e., *lu(y, α)*.

2.3 Problem Statement

An itemset X is called a CoHUI if the value of $u(X)$ is more than *minutil*, and *Kulc(X)* is also greater than the minimum correlation measure *minCor*.

$$\text{CoHUI} = u(X) \geq minutil \text{ AND } Kulc(X) \geq minCor \tag{16}$$

3 ECHUM Algorithm Design

The ECHUM algorithm is the combination of the two algorithms EFIM and CoHUM. The updated utility-list is used in this method, which considers the utility-list formation of EFIM and the support value of each item used in CoHUM. The algorithm takes input as the transactional database, utility table, minimum utility *minutil*, and minimum correlation *minCor*. Initially, an itemset X and the support values are initialized to $\emptyset$. The *local-utility* associated with every item is calculated using the utility-bin array by examining the database. The *secondary itemsets* are then generated with the help of the *local-utility* associated with every item. Now, the *secondary itemsets* are arranged based on the increasing order of their *TWU* values. The scanning of the database is then done again to withdraw the collection of items that are not present on the *secondary itemsets*, and the empty transactions are deleted. Finally, the sorting of D takes place once again according to their $>.T$ order. The sub-tree utility is denoted by $su(X, i)$ for every item present in the secondary itemset andcalculated using the utility bin array. The primary itemset list is generated using these values of the sub-tree utility array. The *primary itemset, secondary itemset, minutil, minCor,* and database D are then passed to the Explore function. Algorithm 1 depicts the pseudo-code of ECHUM.

Algorithm 1: ECHUM Algorithm

Input: Transactional database D, utility table P, *minutil* and *minCor*
Output: The collection of CoHUI's

1. The itemset X is initialized with value $\emptyset$;
2. Initially, the database D is examined, and the local utility of each element is calculated.
3. Using the local-utility for every item the Secondary items are generated by
   $Secondary(X) = \{ i \mid i \in I \wedge lu(X,i) \geq minutil \}$
4. The items in the *Secondary(X)* are sorted according to the *TWU* values.
5. The scanning of the database is then again applied to D to extract the absent transactions in *Secondary(X)*.
6. The transactions in D are again sorted according to the TWU values.
7. Again the database D is examined and the calculation of sub-tree utility of every element for the items that are present in *Secondary(X)*.
8. Using the local-utility associated with every item the Secondary items are generated by
   $Primary(X) = \{i \mid i \in Secondary(X) \wedge su(X,i) \geq minutil \}$
9. Explore($X, D, Primary(X), Secondary(X), minutil, minCor$)

Algorithm 2: Explore Procedure
Input: An itemset X, the projection of the original database X-D, $Primary(X)$, $Secondary(X)$, min-util and $minCor$
Output: The collection of all CoHUI's, which are the extension of X.

1. For each item i present in Primary itemsets do
2. $Y = X \cup \{i\}$
3.   Update the support value and the corresponding $Kulc$ value of Y by increasing its count;
4.   Scan the projected database X-D for the calculation of the utility $u(Y)$
        and for the creation of the projection of the new database Y-D
5.   If the following holds: $u(Y) > minutil$ and $Kulc(Y) > minCor$ then output Y;
6. Calculate $su(Y,y)$ and $lu(Y,y)$ for each item $y \in Secondary(X)$ by examining the extended database β-D once.
7. $Primary(Y) = \{y \mid y \in Secondary(X) \mid su(Y,y) \geq minutil$ and $Kulc(Yy) > minCor \}$
8. $Secondary(Y) = \{y \mid y \in Secondary(X) \mid lu(Y,y) \geq minutil$ and $Kulc(Yy) > minCor \}$
9.   Explore(Y, Y-D, $Primary(Y)$, $Secondary(Y)$, $minutil$, $minCor$);
10. End

The Explore Method takes input as itemset X, projected database $X - D$, the items that primary items to α: $Primary(X)$, and secondary items to X: $Secondary(X)$, user-defined $minutil$ and $minCor$ threshold. For every item i in $Primary(X)$, an itemset Y is created, which is the extension of X. The support values are updated in the utility list of Y by incrementing its count. The projected database $X - D$ is then examined to find the utility of Y, and the projection of the database $Y - D$ is created. If the utility of Y is larger than the $minutil$ and the $Kulc$ value is greater than the $minCor$ then the itemset Y is considered as CoHUI. Then the *sub-tree utility* $su(Y, y)$ and the *local utility* $lu(Y, y)$ are calculated for every item y present in $Secondary(X)$ by scanning the projected database $Y - D$. Finally, the itemsets $Primary(Y)$ and $Secondary(Y)$ are generated from the itemset $Secondary(X)$ by applying Properties 4, 5, and 6. The explore method is recursively called again to search the extensions of X. The final output is accumulated in the set of CoHUIs.

4 Experimental Analysis

The ECHUM algorithm is compared with the CoUPM. We performed the analysis of execution time and memory usage by varying the $minutil$ and $minCor$. The experiments were carried out on an Ubuntu machine with a 2.7 GHz Intel i5 6th generation processor. The RAM allotted to the system was 8 GB, and the program code was written in Java Language. A total of 4 datasets were used to test the algorithms. The datasets were collected from SPMF library [20]. The dataset properties are given in Table 4.

4.1 Runtime Comparison

The runtime comparisons are shown in Figs. 1 and 2. In Fig. 1, the $minutil$ for *foodmart*, *chess, mushroom,* and *retail* are kept fixed at 0.007%, 20%, 8%, and 0.12%, respectively. The $minCor$ value is varied from 0.01 to 0.06 in *foodmart* dataset, from 0.74 to 0.79 in *chess* dataset, from 0.4 to 0.5 in *mushroom* dataset, and from 0.1 to 0.2 in *retail* dataset. For almost all the values, the running time for ECHUM is faster than the existing CoUPM algorithm. As the $minCor$ value is increased, the run time required for each of

Table 4. Dataset properties

Dataset	Transactions	No. of distinct items	Average transaction length
Chess	3196	75	36
Mushroom	8124	120	23
Retail	88162	16470	10.3
Food mart	4141	1559	4.42

the algorithms is persistent. In Fig. 2, the *minCor* for foodmart, chess, mushroom, and retail are kept fixed at 0.03, 0.76, 0.42, and 0.1, respectively. The *minutil* value is varied from 0.006% to 0.011% in the food mart dataset, from 17 to 22 in the chess dataset, from 8 to 13 in the mushroom dataset, and from 0.01 to 0.015 in the retail dataset. For almost all the values, the running time for ECHUM is faster than the existing algorithm CoUPM. As the *minutil* value is increased, the run time required for each of the algorithms is persistent.

When the *minCor* is set fixed and the *minutil* is increased, the time taken by both the CoUPM and the ECHUM Algorithm decreases non-linearly. For the dataset, *foodmart* the time taken by both the algorithms is almost similar but is less for the ECHUM algorithm. For the datasets, *chess* and *mushroom,* we have a drastic time difference in both algorithms with ECHUM the faster. For *retail* dataset, the time difference is the most. The reason is that CoUPM generates more unpromising candidates due to a huge number of transactions and many distinct items. More candidates require extra memory for CoUPM. It can be seen that the ECHUM algorithm outperforms in all the cases as the number of candidates generated is very low because of efficient pruning.

4.2 Memory Comparison

The Memory Consumptions by the two algorithms ECHUM and CoUPM are shown in Figs. 3 and 4. In the first graph, the *minutil* for *foodmart, chess, mushroom,* and *retail* are kept fixed at 0.007%, 20%, 8%, and 0.12%, respectively. The *minCor* value is varied from 0.01 to 0.06 in *foodmart* dataset, from 0.74 to 0.79 in *chess* dataset, from 0.4 to 0.5 in *mushroom* dataset, and from 0.1 to 0.2 in *retail* dataset. For some datasets, the memory consumed by ECHUM is more, while for some datasets, the memory consumed is less. The overall memory consumed is lesser for the ECHUM algorithm than the CoUPM Algorithm.

For the first graph in Fig. 3, the ECHUM algorithm's memory is a little more than the CoUPM algorithm. The average memory consumed is around 100 MB for ECHUM, while it is around 90 MB for CoUPM. In the *chess* dataset, the ECHUM algorithm took less than half the memory consumed by the CoUPM algorithm for all cases. The average memory consumed by the CoUPM algorithm for the *chess* dataset is 230 MB, while for the ECHUM algorithm, it is 120 MB. The ECHUM algorithm performs best in the *mushroom* dataset in terms of memory consumption. It is almost 5 times efficient than the CoUPM algorithm. For the *retail* dataset, the memory taken is similar for both the algorithms, but it is less for ECHUM.

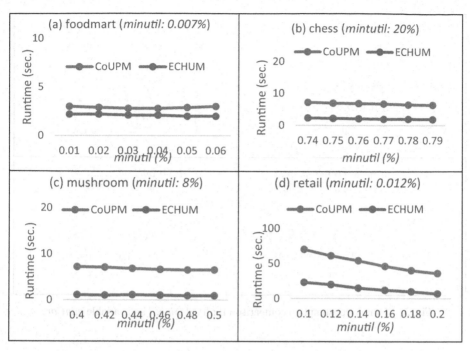

Fig. 1. Runtime comparison for fixed *minutil* and variable *minCor*

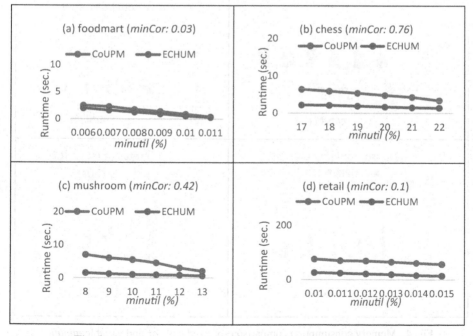

Fig. 2. Runtime comparison for fixed *minCor* and variable *minUtil*

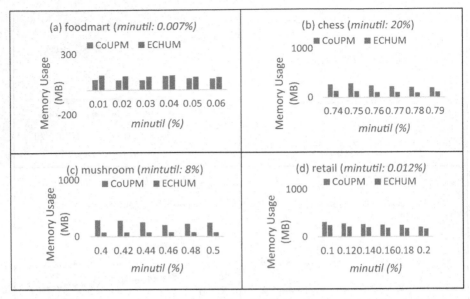

Fig. 3. Memory consumption comparison for fixed *minutil* and variable *minCor*

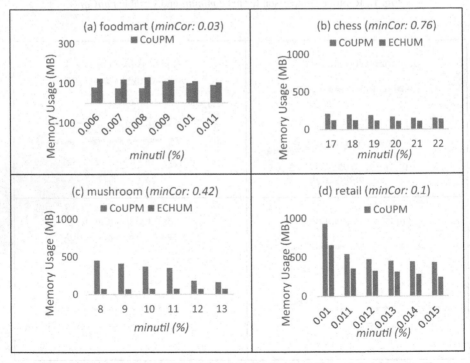

Fig. 4. Memory consumption comparison for fixed *minCor* and variable *minutil*

In Fig. 4, the minCor for *food mart*, *chess*, *mushroom*, and *retail* are kept fixed at 0.03, 0.76, 0.42, and 0.1, respectively. The *minutil* value is varied from 0.006% to 0.011% in food mart dataset, from 17 to 22 in *chess* dataset, from 8 to 13 in *mushroom* dataset, and from 0.01 to 0.015 in *retail* dataset. The first graph in Fig. 4 represents the *food mart* dataset in which for all the *minutil* values, the memory consumed by ECHUM is more than the CoUPM algorithm. The average memory consumed by the CoUPM algorithm for the *food mart* dataset is 90 MB, while for ECHUM, it is 115 MB. For the *chess* dataset, the ECHUM algorithm is faster than CoUPM in all cases. The ECHUM algorithm works best for the *mushroom* dataset as the average memory consumed is almost 5 times less than the CoUPM algorithm. For the *retail* dataset, the memory consumed by both algorithms is high, which is because of the large transactions, but the memory consumed by ECHUM is less than the CoUPM algorithm. Overall the memory consumed by the ECHUM algorithm is less than the CoUPM algorithm.

It is evident from the runtime and memory consumption graphs that ECHUM is almost 200–300% faster than CoUPM and takes nearly 50% less memory than CoUPM. The pruning methods help to reduce the search space, and also, the use of a better utility-list structure minimizes the memory usage.

5   Conclusion and Future Scope

This paper presents an efficient algorithm ECHUM to reduce the time and space required to search the CoHUIs. The idea involved in this paper is to combine the two algorithms EFIM and CoHUM, such that the overall complexity is reduced. The ECHUM uses two upper utility bounds named *sub-tree utility* and the *local utility* and a fast utility counting measure, which improves efficiency. The ECHUM also uses transaction and projection merging, which reduces the dataset size and enhances efficiency. More optimizations will be done for future work in this algorithm by developing a much tighter upper bound to reduce the search space.

References

1. Liu, Y., Liao, W., Choudhary, A.: A two-phase algorithm for fast discovery of high utility itemsets. In: Ho, T.B., Cheung, D., Liu, H. (eds.) PAKDD 2005. LNCS (LNAI), vol. 3518, pp. 689–695. Springer, Heidelberg (2005). https://doi.org/10.1007/11430919_79
2. Ahmed, C.F., Tanbeer, S.K., Jeong, B.S., Lee, Y.K.: Efficient tree structures for high utility pattern mining in incremental databases. IEEE Trans. Knowl. Data Eng. **21**(12), 1708–1721 (2009)
3. Tseng, V.S., Wu, C.W., Shie, B.E., Yu, P.S.: UP-growth: an efficient algorithm for high utility itemset mining. In: Proceedings of the 16th ACM SIGKDD International Conference on Knowledge Discovery and Data Mining, pp. 253–262 (2010)
4. Tseng, V.S., Shie, B.E., Wu, C.W., Philip, S.Y.: Efficient algorithms for mining high utility itemsets from transactional databases. IEEE Trans. Knowl. Data Eng. **25**(8), 1772–1786 (2012)
5. Liu, M., Qu, J.: Mining high utility itemsets without candidate generation. In: Proceedings of the 21st ACM International Conference on Information and Knowledge Management, pp. 55–64 (2012)

6. Fournier-Viger, P., Wu, C..-W.., Zida, S., Tseng, V.S.: FHM: faster high-utility itemset mining using estimated utility co-occurrence pruning. In: Andreasen, T., Christiansen, H., Cubero, J.-C., Raś, Z.W. (eds.) ISMIS 2014. LNCS (LNAI), vol. 8502, pp. 83–92. Springer, Cham (2014). https://doi.org/10.1007/978-3-319-08326-1_9

7. Krishnamoorthy, S.: Pruning strategies for mining high utility itemsets. Expert Syst. Appl. **42**(5), 2371–2381 (2015)

8. Zida, S., Fournier-Viger, P., Lin, J.C.-W., Wu, C.-W., Tseng, V.S.: EFIM: a highly efficient algorithm for high-utility itemset mining. In: Sidorov, G., Galicia-Haro, S.N. (eds.) MICAI 2015. LNCS (LNAI), vol. 9413, pp. 530–546. Springer, Cham (2015). https://doi.org/10.1007/978-3-319-27060-9_44

9. Gan, W., Lin, J.C.W., Fournier-Viger, P., Chao, H.C., Tseng, V.S., Philip, S.Y.: A survey of utility-oriented pattern mining. IEEE Trans. Knowl. Data Eng. **33**(4), 1306–1327 (2019)

10. Sethi, K.K., Ramesh, D., Sreenu, M.: Parallel high average-utility itemset mining using better search space division approach. In: Fahrnberger, G., Gopinathan, S., Parida, L. (eds.) ICDCIT 2019. LNCS, vol. 11319, pp. 108–124. Springer, Cham (2019). https://doi.org/10.1007/978-3-030-05366-6_9

11. Agrawal, R., Srikant, R.: Fast algorithms for mining association rules. In: Proceeding of 20th International Conference on Very Large Databases. VLDB, vol. 1215, pp. 487–499 (1994)

12. Kim, W.-Y., Lee, Y.-K., Han, J.: CCMine: efficient mining of confidence-closed correlated patterns. In: Dai, H., Srikant, R., Zhang, C. (eds.) PAKDD 2004. LNCS (LNAI), vol. 3056, pp. 569–579. Springer, Heidelberg (2004). https://doi.org/10.1007/978-3-540-24775-3_68

13. Ahmed, C.F., Tanbeer, S.K., Jeong, B.S., Choi, H.J.: A framework for mining interesting high utility patterns with a strong frequency affinity. Inf. Sci. **181**(21), 4878–4894 (2011)

14. Omiecinski, E.R.: Alternative interest measures for mining associations in databases. IEEE Trans. Knowl. Data Eng. **15**(1), 57–69 (2003)

15. Lin, J.-W., Gan, W., Fournier-Viger, P., Hong, T.-P., Chao, H.-C.: FDHUP: fast algorithm for mining discriminative high utility patterns. Knowl. Inf. Syst. **51**(3), 873–909 (2016). https://doi.org/10.1007/s10115-016-0991-3

16. Gan, W., Lin, J.C.W., Fournier-Viger, P., Chao, H.C., Fujita, H.: Extracting non-redundant correlated purchase behaviors by utility measure. Knowl.-Based Syst. **143**, 30–41 (2018)

17. Gan, W., Lin, J.C.W., Chao, H.C., Fujita, H., Philip, S.Y.: Correlated utility-based pattern mining. Inf. Sci. **504**, 470–486 (2019)

18. Kulczynski, S.: Die pflanzenassoziationen der pieninen, Imprimerie del'Universit (1928)

19. Song, W., Liu, Y., Li, J.: BAHUI: fast and memory efficient mining of high utility itemsets based on bitmap. Int. J. Data Warehousing Min. (IJDWM) **10**(1), 1–15 (2014)

20. Fournier-Viger, P., Gomariz, A., Gueniche, T., Soltani, A., Wu, C.W., Tseng, V.S.: SPMF: a Java open-source pattern mining library. J. Mach. Learn. Res. **15**(1), 3389–3393 (2014)

Mental Health Analysis of Indians During Pandemic

Pritul Dave[1](✉) (iD), Kushal Master[1] (iD), Prince Makwana[1] (iD), Parth Goel[1] (iD),
and Amit Ganatra[2] (iD)

[1] Computer Science and Engineering Department, Devang Patel Institute of Advance
Technology and Research, Charotar University of Science and Technology (CHARUSAT),
CHARUSAT Campus, Changa 388421, Gujarat, India
`{17dcs008,17dcs028,17dcs027}@charusat.edu.in,`
`parthgoel.ce@charusat.ac.in`
[2] Computer Engineering Department, Devang Patel Institute of Advance Technology and
Research, Charotar University of Science and Technology (CHARUSAT), CHARUSAT
Campus, Changa 388421, Gujarat, India
`amitganatra.ce@charusat.ac.in`

Abstract. The Covid-19 pandemic has severely affected many countries around
the globe in terms of physically as well as mentally. During the initial months
of the pandemic have reported India's deficient cases, but eventually the cases
were proliferated as the time progress. The government's decision to impose a
lockdown without warning has a wide-ranging impact, affecting everyone from
low-wage workers to huge corporations. As a result, there is a negative impact on
people's mental health and emotions. The people had suffered from depressions,
anxiety, fatigue and so forth. Many wide varieties of the people had expressed
their thoughts, viewpoints, and their mental conditions in the form of tweets over
the Twitter, a social media platform. Hence, in this paper, we have statistically
analysed the data of tweeted tweets to elicit the meaningful insights. The data was
analysed using the unsupervised clustering strategy–K-means and LDA–and the
results were reinforced and validated using the pre-trained supervised classification approach–Text to Text transformer. The anticipated data depicted that the fear
was the most common state of mind at the end of the lockdown, followed by joy,
anger, and sadness. Furthermore, the deduced insights will be highly beneficial in
decision-making process when such an epidemic or pandemic situation re-surges.

Keywords: Sentimental analysis · K-nearest neighbour · Latent
Dirichlet allocation · Machine learning · Supervised learning · Unsupervised
learning · Emotion analysis · Twitter · COVID-19 · Pandemic

1 Introduction

The COVID-19 pandemic is the most defining global health emergency in contemporary
history, dating back to World War Two. With the exception for, Antarctica, the virus has
spread around the world since its discovery in Asia in 2019.

The effect was felt across the worldwide. At initial stages of pandemic in March
2020, many people were feared and started expressing their views on social media. It

© Springer Nature Switzerland AG 2021
K. R. Venugopal et al. (Eds.): ICInPro 2021, CCIS 1483, pp. 287–297, 2021.
https://doi.org/10.1007/978-3-030-91244-4_23

imposed the nationwide lockdown in many countries and due to which there was no travel to different countries. So, social media was the only platform where people meet new people and get an idea about the situation of COVID-19 in world. In order to shield people from the awful outbreak, government announced the lockdown, intending to halt the infection's spread. A long-term lockdown has now resulted in psychosocial challenges for the most disadvantaged members of society, including fatigue, anxiety, anger, boredom, depression, and even suicidal ideation and attempts.

We've now passed the disgraceful milestone of over 4,31,240 deaths, and the human family is suffering under an almost insufferable burden of misfortune. In any case, the pandemic isn't just a health emergency; it's also a major financial disaster. Focusing on all the countries it comes into touch with, it has the potential to have obliterating physical, economical, and political consequences that can leave deep and long-lasting scars.

People are becoming more active on social media as their use of the internet grows. The social media usage increased dramatically between 2020 and 2021. Social media has evolved into a type of communal gathering where people may voice their opinions. It has strengthened into a transparent medium for people all around the world to communicate their opinions. As a result, a trend is established to identify and evaluate people's sentiments, also known as opinion mining. In this paper, the real-time twitter data is analysed. The Unsupervised Learning methods, K Means Clustering and Latent Dirichlet allocation, is incorporated to analyse the sentiments of the large corpus of text in real-time when there is unlabelled data. The obtained results are fortified by validating it using supervised approach, Text-to-Text transformer.

From our analysis of the tweets, we concluded that during the period of lockdown and during the end of lockdown, people's mental health was highly suffered. During the unlocking face, the fear was one of the most common and pivotal mindsets of Indians. The people were in fear because of risk of going outside, colleges or someone from the family tested positive, some of losing the employments, flouting of rules of masks, while students are in a panic of giving examinations at centre. This extrapolated analysis will be aid for the Government as well as for an individual in ameliorating the decision-making process.

2 Literature Review

In this section, the survey addressing to different techniques of sentiment analysis which are related to our thesis on mental health analysis is performed. Paulraj and Neelamegam [3] had improved K Means Clustering's performance on the high dimensional dataset. They have proposed K Means Clustering for feature extraction. K Means clustering uses Euclidean distance to measure the distance between the maximum and minimum distances between any two sample points. [9] used the topic modeling approach with sentiment analysis for various tweets and observe the trend on themes and topics related to covid-19-related tweets. [10] used LDA to identify the underlying topics of tweets and observed changes in tweets over time. The [11] has used the LDA for the sentiment analysis of COVID-19 from Twitter data. They differentiate the tweets in the 11 clusters through LDA and find the percentage of the emotion in each cluster. They have considered anger, anticipation, disgust, fear, joy, sadness, surprise, and trust. From many emotions, fear has maximum weight age, around 45% of the total tweets from each topic.

Christy et al. and Ahmad Farhan Hidayatullah et al. [4, 5] shows the idea behind Latent Dirichlet Allocation (LDA) an unsupervised learning approach that finds different clusters using Topic Modelling. Many researches and studies started on COVID-19 drug development, COVID-19 virus, how to reduce chaining of infection in people, etc. Psychologists studied the sentiments or the mood of people using social media tweets on COVID-19. The [12] had done the sentiment analysis of COVID-19 Vaccination Drive and using LDA differentiated the Twitter data into 16 different clusters and find out the most common word from each of the topics. The [13] had proposed a Naïve Bayes classifier approach for performing the sentimental analysis on corona virus outbreak using twitter data. They had classified into positive and negative sentiments. The obtained results are 30% positive tweets, negative tweets are 16% and 56% neutral. However, the author had not categorized the emotions. [14] had proposed an approach for sentimental analysis over the large-scale data based on unsupervised approach. They propose an algorithm which calculate the sentiment score more commonly the polarity scores by parallel processing of big data. The results were compared with senti-WordNet. [15] performed an opinion mining based on Latent Dirichlet Allocation, in which they manually labelled the topics and corresponding words. [16] had performed the sentimental analysis of COVID-19 tweets using Bidirectional Encoder Representations from Transformers (BERT) by collecting the tweeter tweets from 23rd March, 2020 and 15th July, 2020 and the text has been labelled as fear, sad, anger, and joy. Among which, the highest sentiment of emotion found was of fear and followed by sad, anger and joy.

3   Proposed Work

The analysis of the mental state of the Indian is carried out using an unsupervised approach, and the results were further validated by the supervised approach. In this paper, for unsupervised learning, the two approaches are elicited–K-Means Clustering approach and Latent Dirichlet Allocation approach. The results of the both approaches are compared with the supervised approach - Text to Text transformer. Finally, the meaningful insights were made based on the different emotional sentiments.

3.1   K-Means Clustering

K-Means is widely used as a clustering algorithm in research because of its simplicity and acute performance [17]. K-Means algorithm uses an iterative approach and partitions the dataset into K defined distinct non-overlapping clusters represented as centroid points. It computes the Euclidean distance between the data-points and centroids and keeps data points as close as to centroid.

The mean-square-error (MSE) cost-function which is depicted in Eq. 1, is minimized to solve problem of clustering N data-points $x_1, .., x_N$ into k disjoint subsets of C_i, i = 1, ..., k, each containing n_i data points, $0 < n_i < N$:

$$MSE = \sum_{i=1}^{k} \sum_{xt \in Ci} ||xt - ci||^2 \qquad (1)$$

where x_t is a vector representing the tth data point in the cluster Ci which is the geometric centroid of the cluster C_i. An input data point x_t is put into cluster 1 if it satisfies the following condition:

$$I(x_t, i) = \begin{cases} 1 \; if \; i = \text{argmin}\left(||xt - ci||^2\right) j = 1, \ldots, k \\ o \; otherwise \end{cases} \tag{2}$$

(Figure 1) The Elbow approach provides an efficient method for determining the number of clusters [7]. This method is based on visualizing the cost function and detecting the breakpoints for various clusters. If adding more clusters does not considerably reduce variance, then adding more clusters should be halted. This method provides insight into clusters, which serve as a necessary data mining tool prior to clustering.

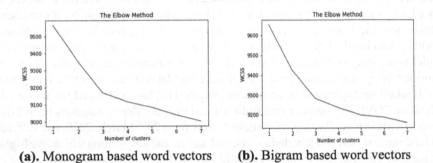

(a). Monogram based word vectors   **(b).** Bigram based word vectors

Fig. 1. Elbow based clustering of K-Means monogram and bigram

3.2 Latent Dirichlet Allocation (LDA)

The Latent Dirichlet Distribution defines a probability density for a vector valued input having the same characteristics as our multinomial parameter fie. It has support (the set of points where it has non-zero values) over where K is the number of variables.

$$\alpha = (\alpha_1 \cdots \alpha_k) \quad \alpha_i \geq 0 \quad i \in \{1 \cdots k\}$$
$$A \; draw \; from \; Dirichlet \; Distribution \; is \; written \; as \tag{3}$$
$$\phi \sim \text{Dirichlet}_k(\alpha)$$

LDA is also widely used for dimensionality reduction [2]. It is a method for clustering large corpus of text from documents. It is also used in topic modelling. In this method, the word-vector was formed, and they are iterated based on Gibb's sampling [9]. Afterwards, the probability of that word is sampled according to the clusters formed.

The LDA finds the term frequency of words in each cluster from the Corpus, created by given documents. Corpus is a bag of words which is created by randomly distributed documents according to Dirichlet distribution. They both describe LDA levels. The first level is corpus level parameters, which are represented by symbol alpha and beta. These corpus-level parameters are assumed to be sampled once in the process of generating a corpus. Secondly, document-level variables (s), tested once of each document—finally, word-level variables symbolized by "z" and "w". Word-level variables are sampled once for each word in each document.

So, the LDA uses three levels of implementation, which gives the far better insights of the document. The unsupervised method is useful to obtain the number of clusters, but it is difficult to tell which cluster representing what, hence we have done analysis using supervised approach as well.

3.3 Text-to-Text Transformer

The first approach used for supervised sentimental analysis is based on a word vector where the word is mapped to the specific entity label. But the problem that researchers caveated in that in this approach the word usage can vary from sentence to sentence. We have observed the same problem in analyzing COVID-19 Data where the "Corona Positive" falls in a category of positive sentiment, but in the real sense, it should be harmful. Hence the transfer learning-based approaches are used. Most commonly, the NLP-based transfer learning approach is over unsupervised learning approach. The Text-to-Text Transformer is trained over the hundreds of gigabytes of scrapped text from the internet whose prominent name is "Colossal Clean Crawl Corpus".

The BERT is taken as a base model for creating the encoder part as BERT is Encoder [6]. They adopted the concept of the "masked language modelling" technique as their based model which is similar to the BERTBASE.

The approach is crafted in Fig. 2, where first the token is mapped to embeddings then passed into BERT based Encoder having self-attention layer followed by Feedforward Network. The concept of Self-Attention [9] has significantly improved the Transfer Learning technique where each sequence of text is replaced by the weighted average of the rest of the arrangement. Finally, a decoder part is based on standard self-attention followed by a Fully Connected Network and a SoftMax activation classifying the labels.

In our paper we are using this Text-to-Text transformer and validating the results for the unsupervised models.

4 Results

The dataset fetched from the twitter. The tweets in the dataset are dated from 13th September 2020 to 22nd September 2020 having 2 million tweets and has four attributes– text, location, date, and time. In primary data analysis, the dataset contains 32% of null values and repeated, leading to redundancy. We observed that after removing null values and redundant data, data has only 9730 tweets. We consider 9730 tweets for our analysis. The results from the unsupervised approach are further compared with pre-trained Supervised based Text-to-Text transformer [1].

4.1 Experimental Results from K-means and LDA

The monogram-based word vectors to find the number of clusters from tweets is illus- trated in the Fig. 3. It is observed that an elbow is seen at 3 and 4 clusters. Figure 4

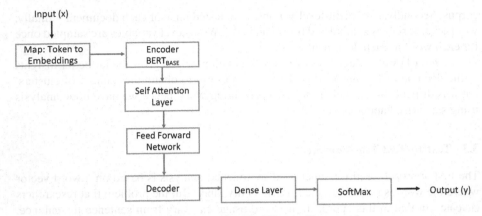

Fig. 2. An architecture of unified Text-to-Text transformer

shows the bigram-based word vectors to find the number of clusters from tweets. It is observed that an elbow is seen at 2, 3 and 5 clusters.

We can conclude from an elbow analysis that the tweets are grouped into 3 or 4 clusters if we consider the monogram-based word vector representation [8] considered that the social media data is increasing and to group these large text data K-means clustering of unsupervised learning is an effective way for preliminary data mining technique. The most frequently appearing keyword in cluster#0 is "test", cluster#1 is "cases", cluster#2 is "covid" for monogram based clustering while for bigram are "covid", "cases", "virus", "covid" respectively related to cluster#0, cluster#1 and cluster#3.

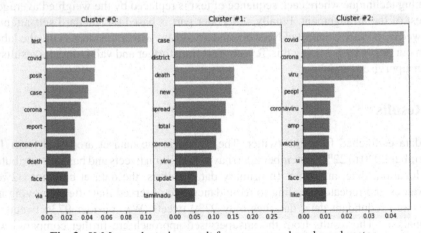

Fig. 3. K-Means clustering result for monogram-based word vector

Figure 5 shows the clustering result using LDA approach. Here clusters (cluster #0 to cluster #4) represents the different sentiments. This figure gives an idea that which words in word-vector contribute to which cluster. The frequent appearing words

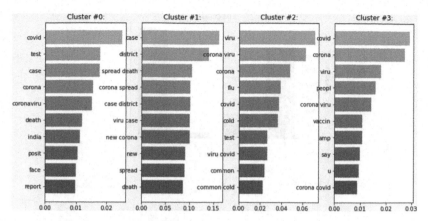

Fig. 4. K-Means clustering result for bigram-based word vector

are "covid", "covid", "corona", "covid" from the cluster#0, cluster#1, cluster#2 and cluster#3 respectively.

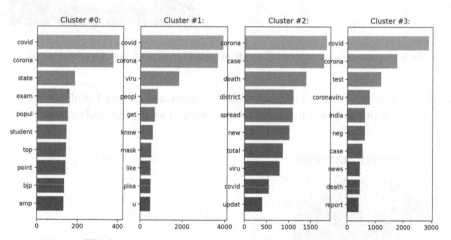

Fig. 5. Latent Dirichlet allocation (LDA) clustering result

4.2 Results and Discussion on Mental Emotions

The obtained results from the Latent Dirichlet Allocation and K-Means clustering Monogram as well as Bigram are depicted in the Fig. 6. From Fig. 7. It was concluded that fear was the highest tweet followed by sadness, fear, and anger. On comparing the results of K-Means as well as LDA with the Supervised Text-to-Text transformer approach, the K-Means Bi-gram is resembling the most. The direct trained unsupervised approach can be used for predicting the state of mental health, by superseding the labels of the supervised approach. On comparing, the clustering frequency of cluster 4 is pairing with

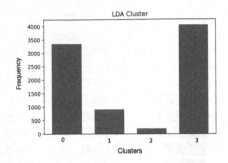

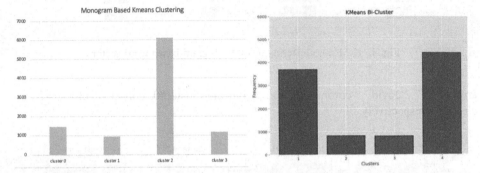

Fig. 6. Results from unsupervised K-Means and LDA (Latent Dirichlet Allocation)

the Fear, cluster 1 with Joy, cluster 2 with the Anger and cluster 1 with the sadness. People are fear due to stress, trauma, anxiety, losing of loved ones and depression.

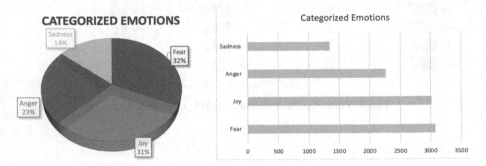

Fig. 7. Results from supervised Text-to-Text transformer

Also, people are fear due to unemployment. Spreading of misinformation is also one of the causes for fear. Some of the analysis for Joy are like people enjoying movie, tweeting about music, reading articles, making memes, trying different dishes at home, enjoyed by hearing news that cases of COVID-19 are decreasing daily, starting of offices & shops, tweeting about vaccines, realizing about health consciousness, celebrating a huge change in behavior, sharing negative result of COVID test, doing social works etc.

Fig. 8. Frequently appearing words in tweets

While some tweets are depending on aspects of person like some enjoy for having strict fine on no mask while some don't which our model has considered as positive. The results are also compared with the unsupervised approach which shows tantamount between supervised and unsupervised approach. The cluster 4 surrogates with Fear, Cluster 1 with joy, Cluster 2 with anger and Cluster 3 with Sadness.

The frequently appearing words in the complete dataset is analyzed and depicted as a word cloud in Fig. 8. In which some of the words are "COVID", "CORONA", "EXAM TIME", "MASK" is some of the frequent appearing words. As illustrated in the Fig. 9 the

Fig. 9. Word cloud of different classes of emotions

sadness tweets are mainly associated with Mortality, criticizing government, GDP rates, sharing horrible experiences, some talking about racism, opening of Schools, protest, rebel etc. Each word cloud shows that which words are related to the sentiments and which words are used majorly and frequently to represent that sentiment.

Some of the frequent words appearing in anger cluster are "death", "mask", "vaccine", "people" etc. while for the fear cluster are "death", "common cold", "flu", "lockdown" etc. Same way, in sadness cluster are "case", "unemployment", "death toll", "exam", "vaccine" etc. The joy cluster includes "confirm case", "tears of joy", "cool things", "awesome stuff" etc.

The tweets which are misclassified are mainly like writing Hindi in English, writing short-forms such as "pls" for "please", sarcasm, irony, idioms etc.

5   Conclusion

In this work, the emotion analysis of the Twitter data through supervised and unsupervised techniques is proposed. The unsupervised learning approach of K-Means and Latent Dirichlet allocation techniques are proposed whose results are compared with the supervised approach text-to-text transformer. The proposed approach is an aid for decision making process. The results of tweets are analyzed during the unlock phase of lockdown, hence it is more useful for the Indian Government. The above analysis concludes that performing an unsupervised approach over the twitter emotion dataset gives the correspondingly nearby same results as that of the supervised approach, having 4 clusters corresponding to fear, joy, sadness, and anger. The combined result of Anger, Fear, Sadness is more than Joy, indicating that people are not feeling optimistic. Angriness is produced due to mismanagement; Fear is produced due to the Covid-19 virus; sadness is due to loss of Jobs & Wages & loss of beloved ones; joy is there because people at lockdown are enjoying and making fun while residing home. The proposed approach can be recast to the pandemic or epidemic that resurges in the future. As this paper is analyzing the emotions of Indians during the pandemic hence it is christened as mental health analysis of Indians during pandemic. Furthermore, the classification can be improved by training a model over multiple languages as a future scope and to expand the scope of analyzing mental health throughout the world.

Acknowledgements. The authors would like to thank CHARUSAT for providing a platform to conduct the research work.

References

1. Raffel, C., et al.: Exploring the limits of transfer learning with a unified text-to-text transformer (2020). arXiv, abs/1910.10683
2. Blei, D.M., Ng, A., Jordan, M.I.: Latent Dirichlet allocation. J. Mach. Learn. Res. **3**, 993–1022 (2003)
3. Prabhu, P.: Improving the performance of k-means clustering for high dimensional dataset (2011)

4. Christy, A., Gandhi, M.: An enhanced method for topic modeling using concept-latent. In: 2019 International Conference on Machine Learning, Big Data, Cloud and Parallel Computing (COMITCon). IEEE (2019). https://doi.org/10.1109/COMITCon.2019.8862179

5. Hidayatullah, A.F., Ma'arif, M.R.: Road traffic topic modeling on Twitter using Latent Dirichlet allocation. In: 2017 International Conference on Sustainable Information Engineering and Technology (SIET). IEEE (2017). https://doi.org/10.1109/SIET.2017.8304107

6. Devlin, J, Chang, M.W., Lee, K., Toutanova, K.: Bert: pre-training of deep bidirectional transformers for language understanding. arXiv preprint arXiv:1810.04805 (2018)

7. Bholowalia, P., Kumar, A.: EBK-means: a clustering technique based on elbow method and k-means in WSN. Int. J. Comput. Appl. **105**, 17–24 (2014)

8. Gelfand, A.E.: Gibbs Sampling (2006). https://doi.org/10.1002/0471667196.ess0302.pub2

9. Shen, T., Zhou, T., Long, G., Jiang, J., Pan, S., Zhang, C.: DiSAN: directional self-attention network for RNN/CNN-free language understanding. In: AAAI (2018)

10. Chandrasekaran, R., Mehta, V., Valkunde, T., Moustakas, E.: Topics, trends, and sentiments of tweets about the COVID-19 pandemic: temporal infoveillance. Study J. Med. Internet Res. **22**(10), e22624 (2020). https://doi.org/10.2196/preprints.22624

11. Xue, J., Chen, J., Chen, C., Zheng, C., Li, S., Zhu, T.: Public discourse and sentiment during the COVID 19 pandemic: using Latent Dirichlet allocation for topic modeling on Twitter. Plos ONE **15** (2020). https://doi.org/10.1371/journal.pone.0239441

12. Lyu, J.C., Han, E.L., Luli, G.K.: COVID-19 vaccine–related discussion on Twitter: topic modeling and sentiment analysis (Preprint) (2021). https://doi.org/10.2196/preprints.24435

13. Zahoor, S., Rohilla, R.: Twitter sentiment analysis using machine learning algorithms: a case study. In: 2020 International Conference on Advances in Computing, Communication & Materials (ICACCM). (2020). https://doi.org/10.1109/ICACCM50413.2020.9213011

14. Pandarachalil, R., Sendhilkumar, S., Mahalakshmi, G.S.: Twitter sentiment analysis for large-scale data: an unsupervised approach. Cogn. Comput. **7**(2), 254–262 (2014). https://doi.org/10.1007/s12559-014-9310-z

15. Sv, P., Lathabhavan, R., Ittamalla, R.: What concerns Indian general public on second wave of COVID-19? A report on social media opinions. Diabetes Metab. Syndr. **15**, 829–830 (2021). https://doi.org/10.1016/j.dsx.2021.04.001

16. Chintalapudi, N., Battineni, G., Amenta, F.: Sentimental analysis of COVID-19 tweets using deep learning models. Infect. Dis. Rep. **13**, 329–339 (2021). https://doi.org/10.3390/idr130 20032

17. Korovkinas, K., Danėnas, P., Garšva, G.: SVM and k-means hybrid method for textual data sentiment analysis. Baltic J. Mod. Comput. **7** (2019).https://doi.org/10.22364/bjmc.2019.7.1.04

Prediction of Breast Cancer Analysis Using Machine Learning Algorithms and XGBoost Technique

Bonda Likitha[1], Jyothsna Nakka[1(✉)], Jyotsana Verma[2], and Nenavath Srinivas Naik[2]

[1] Department of Electronics and Communication Engineering, IIIT Naya Raipur, Raipur, India
{likitha18101,jyothsna18101}@iiitnr.edu.in
[2] Department of Computer Science and Engineering, IIIT Naya Raipur, Raipur, India
{jyotsana18100,srinu}@iiitnr.edu.in

Abstract. Breast Cancer is one of the most prevalent malignancies amongst men and women. Currently, it has become the common health issue all over the world with its drastic increase in death rate every year. Early detection of breast cancer provides high treatment efficiency and better healing chances. The main contribution of this paper is to find the model which is most accurate for predicting the type of tumour cell (Benign or Malignant). ANOVA f-test Feature Selection is applied to the Wisconsin Breast Cancer dataset to select the subsets of input features that are most relevant to the target variable. We compared various machine learning algorithms like Support Vector Machine (SVM), K-Nearest Neighbors (KNN), Decision Tree, Logistic Regression, Gaussian Naive Bayes, Random Forest and XG boost Classifier algorithms. We obtained the highest accuracy of 98.25% in XGboost classifier as it uses ensemble techniques and is a very powerful classifier.

Keywords: Breast cancer · Prediction · Diagnosis · Machine learning · XGboost classifier · Analysis

1 Introduction

Globally, Breast Cancer has remained the second most common disease which causes death among women [1]. Breast cancer classification of tumors accurately helps in curing the disease at the early stage itself. Breast cancer tumors are mainly classified into malignant (Cancerous) and Benign (Non-Cancerous). To discriminate amid these tumors, doctors require a reliable and safe diagnostic system. However sometimes, even the specialists find it challenging to identify the tumors correctly. So the early prediction of the disease is the need of the hour to reduce the risk of death in this case. Breast cancer malignancy is the most prevalent disease among women; it has consistently high mortality and frequency rates. We collected data from Wisconsin breast cancer dataset and applied ANOVA

© Springer Nature Switzerland AG 2021
K. R. Venugopal et al. (Eds.): ICInPro 2021, CCIS 1483, pp. 298–313, 2021.
https://doi.org/10.1007/978-3-030-91244-4_24

f-test Feature Selection method to decrease the high data dimensionality of the feature space before the classification process. It also helped in selecting the subsets of input features that are more relevant to the target variable so that we can get better results. After computing all the models, we compared them based on eight parameters such as Accuracy, Precision, Recall, F1-Score, Sensitivity, Specificity, False Negative Rate and False Positive Rate (Fig. 1).

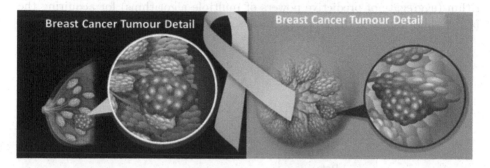

Fig. 1. Breast cancer

2 Related Works

This research paper has gathered the information from various papers who researched the prediction of breast cancer on different datasets, including Wisconsin breast cancer dataset. Paper by Anusha [7], compared Support Vector Machine (SVM), Decision Tree (CART), Naive Bayes (NB) and k Nearest Neighbours (kNN) based on accuracy, Another paper by Naveen [6], compared ensemble machine learning models which gave 100% accuracy in KNN and decision tree on Coimbra breast cancer split train-test dataset in a ratio of 90:10. Fabiano Teixeira [8] evaluated different classification methods: Multilayer Perceptron, Decision Tree, Random Forest, Support Vector Machine and Deep Neural Network and got a good performance in accuracy level of 92%. Gilbert Gutabaga Hungilo [12] in his paper compared AdaBoost, Random Forest, and XGBoost-whose result indicates that the random forest is the best predictive model and has the following performance measure, accuracy 97%, sensitivity 96%, and specificity 96%. Another paper by Quang H. Nguyen [14] analysed prediction models using Feature Selection and Ensemble Voting which returned with the accuracy of at least 98%.

Till now, people compared the three-four algorithms [2,5,6] of their choice mainly based on the accuracy. Although accuracy is the main factor, they could get the highest accuracy, not more than 97% even after applying feature selection. So in this paper, We compared seven commonly used algorithms such as Support Vector Machine (SVM), K-Nearest Neighbors (KNN), Decision Tree, Logistic Regression, Gaussian Naive Bayes, Random Forest and XG-boost.

3 Proposed Methodology

We have collected the Breast Cancer malignant growth instances from the benchmark database Wisconsin Breast Cancer diagnosis data set. We compared various ML algorithms like Decision Tree, K Nearest Neighbor, Gaussian Naive Bayes, Random Forests, Logistic Regression, Support Vector Machine. In this paper, we use named XG Boost classifier, which is an ensemble learning algorithm (aggregate of predictive powers of multiple algorithms) for acquiring the best results. Below is the flowchart representing the proposed model Fig. 2.

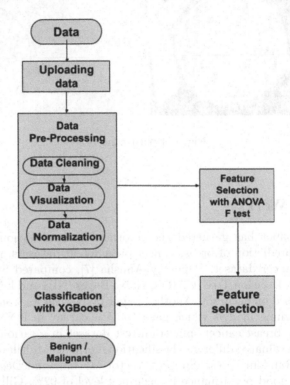

Fig. 2. Proposed model of XG Boost classifier

3.1 Data Collection

Data is collected from the Wisconsin Breast Cancer data set publicly available in UCI Machine Learning Repository [17]. Data set contains 569 occurrences with 30 attributes. It consists of 32 segments, with 'ID number', 'diagnosis' result ("Benign" or "Malignant"), and the 'mean', 'standard deviation' and the 'mean of the worst estimations' of 10 features. The class distribution is shown in the Fig. 3.

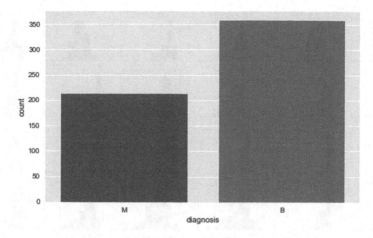

Fig. 3. Class distribution

3.2   Data Processing

Every row in the dataset which is incomplete or has some missing attribute values is removed, and attributes such as 'id' are also deleted as it is of no use.

3.3   Data Manipulation

As the target attribute 'diagnosis' is a categorical data which machine can't read, so is converted into numerical data.

3.4   Data Visualization

After data collection and manipulation, we performed data visualization of all the remaining 31 attributes to identify areas that needed attention or improvement. We can easily interpret data using Fig. 4.

3.5   Feature Selection

Feature selection, it is the most important as the final result values are dependent on the pattern of feature selection. So we choose the feature in such a manner so that we get the best accuracy and other parameters. In this paper, we used ANOVA f-test Feature Selection method.

ANOVA F-Test Feature Selection. ANOVA stands for Analysis of Variance. It is a popular numerical feature selection method. It compares mean between more than two groups. An F-test is a class of statistical tests that compute the relationship between the values of the variances, e.g., the variance of two different samples or the variance that is explained and unexplained by a statistical test here referring to as the ANOVA f-test. We select the features having the best

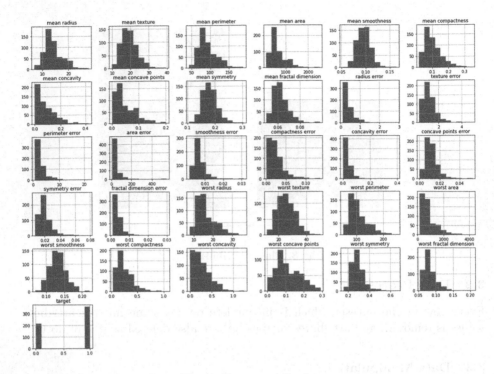

Fig. 4. Features pair-plot

variance using an object and applying the fit. Transform over the features and target variable. The static F is given as Eq. 1

$$F = \frac{Uncertainty\ between\ the\ groups}{Uncertainty\ within\ the\ groups} \tag{1}$$

Z-Score Normalization. The purpose of normalisation is to equalise the scale of all data points such that each attribute is equally important. The Min-Max normalization fail in handling outliers. This outlier issue can be solved by using Z-score normalization. The formula for this technique is given below Eq. 2:

$$Z = \frac{value - \mu}{\sigma} \tag{2}$$

Where μ and σ are mean value and standard deviation value of the feature respectively. The value will be normalised to 0 if it is exactly equal to the mean of all the values of the feature. It will be a negative number if it is below the mean, and a positive number if it is above the mean.

3.6 Data Splitting and Feature Scaling

In this paper, we use 75% training data and 25% of data for testing. Since attributes vary in magnitudes, units and range, we have scaled features using z-score normalization to bring all characteristics to a similar degree level. The feature distribution after feature scaling is shown in the Fig. 4.

4   Background

This paper aims to select the machine learning algorithm that best suits for developing our model to the fullest. Machine learning algorithms classified into two types: Supervised and Unsupervised learning. We need Supervised learning for our breast cancer prediction model [8] (Fig. 5).

Fig. 5. Feature distribution after feature scaling

4.1 Supervised Machine Learning

[7] In this learning, we train the machine using data which is "labelled". This learning algorithm predicts outcomes after learning from the labelled training data. This learning uses regression and other classification techniques to develop predictive models.

- **Logistic Regression:** This algorithm exhibits a direct connection between a dependant (y) and at least one independent (y)'s factor. Since linear regression uncovers a linear relationship, it decides how the dependent variable's value changes with the independent factor's value.
- **KNN:** KNN algorithm which is the short form for K-nearest neighbours, utilizes the given information to predict and allots the new information point dependent on how intently it coordinates with the focuses in the preparation set, i.e., depending on the similarity.
- **SVM:** An SVM model is a data characterization algorithm for predictive analysis, that allocates new data components to one of the known gatherings; it works by defining a straight boundary between two classes. The data points that fall on the right side are considered one class, and the opposite side is regarded as the other.
- **Gaussian Naive Bayes:** Gaussian Naive Bayes is widely used as a classifier and also with few alterations it can be used for regression too. In this algorithm, values are distributed based on Gaussian distribution. And this distribution is also called as a normal distribution. The classification is done based on Bayes Theorem.
- **Decision Tree:** This algorithm identifies different ways to split data. It is used for both classification and regression. Using tree representation, it tries to resolve the error.
- **Random Forest:** Random Forest classifier assembles different decision trees which represent different factual probabilities. Then it combines these decision trees to acquire a steady and precise prediction as shown pictorially in Fig. 10. Trees mapped to a solitary tree known as Classification and Regression Trees (CART) model. This calculation utilized for both regression and classification issues.
- **XGBoost:** XGboost or eXtreme gradient Boosting algorithm is the application of gradient boosted decision trees developed for high speed and better performance. It is an ensemble learning method. Implies that each new model is prepared to rectify the error of the previous model, and the arrangement gets halted when there is no further improvement. In boosting the base learners are weak learners and do not have high predictive power, whereas the final one is a strong learner with high predictive power. The strong learner is a combination of the weak learners that provide some information for prediction (Fig. 6).

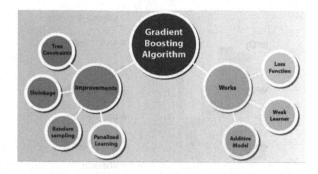

Fig. 6. XGBoost algorithm

5 Experimental Results

In our results, we are considering the confusion matrix which gives a conclusion of the results of the classification problem prediction. It shows how much your model or algorithm classifier is in dilemma when we make predictions. We have also found out Accuracy, Precision, Recall, F1-Score, Sensitivity, Specificity, False Negative Rate and False positive Rate of all the algorithms [6].

- **Accuracy:** XGBoost gives the Highest accuracy of 98.25% which is best for our model, whereas Decision tree gives the lowest accuracy of 88.81% as shown in Fig. 7 (Table 1).

$$Accuracy = \frac{TP + TN}{TP + FP + TN + FN} \tag{3}$$

Here, TP = True Positive
TN = True Negative
FP = False Positive
FN = False Negative

Table 1. Accuracy comparison

Technique	Accuracy
Logistic Regression	94.41%
K-Nearest Neighbors	95.80%
Support Vector Machine	96.50%
Gaussian Naive Bayes	92.31%
Decision Tree	88.81%
Random Forest	95.80%
XGBoost	98.25%

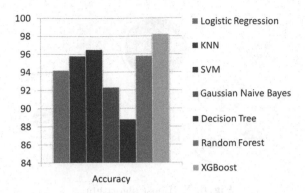

Fig. 7. Accuracy comparison of models

– **Precision:** SVM and Random forest shows the highest precision value of 96.23% and XGBoost has a precision of 95.83%, whereas the decision tree has the lowest precision of 86.23% as shown in Fig. 8 (Table 2).

$$Precision = \frac{TP}{TP + FP} \tag{4}$$

Table 2. Precision comparison

Technique	Precision
Logistic Regression	92.45%
K-Nearest Neighbors	90.57%
Support Vector Machine	96.23%
Gaussian Naive Bayes	86.68%
Decision Tree	86.23%
Random Forest	96.23%
XGBoost	95.83%

– **Recall:** XGBoost has highest recall value of 100%, other algorithms also gave good results but decision tree shows the lowest value of 78.46% as shown in Fig. 9 (Table 3).

$$Recall = \frac{TP}{TP + FN} \tag{5}$$

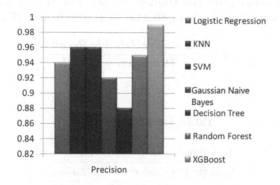

Fig. 8. Precision comparison of models

Table 3. Recall comparison

Technique	Recall
Logistic Regression	92.45%
K-Nearest Neighbors	97.96%
Support Vector Machine	94.44%
Gaussian Naive Bayes	90.38%
Decision Tree	78.46%
Random Forest	92.73%
XGBoost	100%

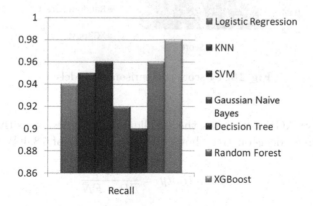

Fig. 9. Recall comparison of models

- **F1-Score:** XGBoost gives the Highest F1 score of 97.87% whereas Decision tree gives the lowest accuracy of 86.44% as shown in Fig. 10 (Table 4).

$$F1 = \frac{2 \times (Precision \times Recall)}{Precision + Recall} \tag{6}$$

Table 4. F1 comparison

Technique	F1
Logistic Regression	92.45%
K-Nearest Neighbors	94.12%
Support Vector Machine	95.33%
Gaussian Naive Bayes	89.52%
Decision Tree	86.44%
Random Forest	96.44%
XGBoost	97.87%

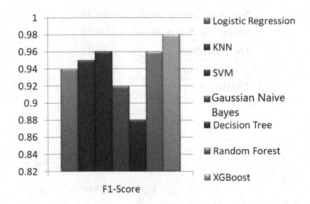

Fig. 10. F1-score comparison of models

- **Sensitivity:** XGBoost shows the excellent and highest sensitivity value of 100%, whereas decision tree shows the lowest value of 78.46% as shown in Fig. 11.

$$Sensitivity = \frac{TP}{TP + FN} \tag{7}$$

- **Specificity:** Support Vector Machine gave the highest specificity of 97.75%, others also gave good results but Gaussian Naive Bayes shows the lowest

Table 5. Sensitivity comparison

Technique	F1
Logistic Regression	92.45%
K-Nearest Neighbors	97.96%
Support Vector Machine	94.44%
Gaussian Naive Bayes	90.38%
Decision Tree	78.46%
Random Forest	92.73%
XGBoost	100%

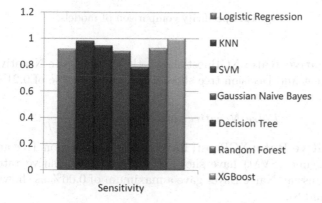

Fig. 11. Sensitivity comparison of models

value of 93.41%. XGBoost has shown 97.06% specificity as shown in Fig. 12 (Tables 5 and 6).

$$Specificity = \frac{TN}{FP+TN} \tag{8}$$

Table 6. Specificity comparison

Technique	Specificity
Logistic Regression	95.56%
K-Nearest Neighbors	94.68%
Support Vector Machine	97.75%
Gaussian Naive Bayes	93.41%
Decision Tree	97.44%
Random Forest	97.73%
XGBoost	97.06%

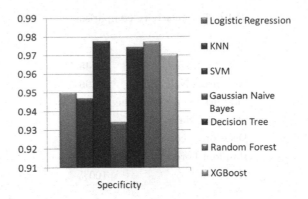

Fig. 12. Specificity comparison of models

- **False Negative Rate:** XGBoost shown the lowest False Negative of 0.00% which is great and Decision tree showed the highest value of 0.21% as shown in Fig. 13.

$$False\ Negative\ Rate = 100 * \frac{FN}{FN + TP} \tag{9}$$

- **False Positive Rate:** XGBoost, Random Forest, Decision Tree and Support Vector Machine (SVM) have shown the least false positive rate of 0.02% whereas Gaussian Naive Bayes gave a maximum of 0.06% as shown in Fig. 14 (Tables 7 and 8).

$$False\ Positive\ Rate = 100 * \frac{FP}{FP + TN} \tag{10}$$

Table 7. False negative rate comparison

Technique	False negative rate
Logistic Regression	0.07%
K-Nearest Neighbors	0.02%
Support Vector Machine	0.05%
Gaussian Naive Bayes	0.09%
Decision Tree	0.21%
Random Forest	0.07%
XGBoost	0.00%

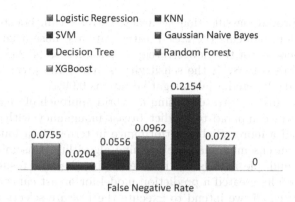

Fig. 13. False negative rate comparison of models

Table 8. False positive rate comparison

Technique	False positive rate
Logistic Regression	0.04%
K-Nearest Neighbors	0.05%
Support Vector Machine	0.02%
Gaussian Naive Bayes	0.06%
Decision Tree	0.02%
Random Forest	0.02%
XGBoost	0.02%

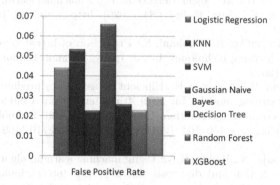

Fig. 14. False positive rate comparison of models

6 Results and Conclusion

This research analysis offered a new plan of action applying Feature Selection based on ANOVA F-test, z-score normalization, and XGBoost classifier algorithm for Prediction of breast cancer. This proposed action offers the following advantages: Improved classification accuracy, better recall, boosted sensitivity, increased precision, reducing the false-positive rate and false-negative rate. The classification accuracy of the new strategy is obtained as 98.25%, the recall is 100%, the f1-score is 97.87%, the sensitivity is 100%, the precision is 95.83%, false positive is 0.02%, and a false-negative rate as 0.00%.

The result of this new strategy using a hybrid approach of 'feature selection and XGBoost' was compared to predict breast malignancy with distinct algorithms. It yielded a more reliable performance in terms of various parameters. As XGBoost is an ensemble machine learning algorithm (Ensemble model is a combination of multiple models). It is able to give the best results. So in this paper we successfully created a prediction model for breast cancer.

For future research, we intend to execute the Feature selection based upon differential evolution algorithm to provide reasonably practical and more precise results. Furthermore, we also plan to deploy the same using different datasets and compare the performance of the hybrid approach using optimal features.

References

1. Breast cancer: prevention and control (2020). https://www.who.int/cancer/detection/breastcancer/en/index1.html
2. Sinha, N.K., Khulal, M., Gurung, M., Lal, A.: Developing a web based system for breast cancer prediction using XGboost classifier. Int. J. Eng. Res. Technol. (IJERT) 9, June 2020. http://dx.doi.org/10.17577/IJERTV9IS060612
3. Jadhav, M., Thakkar, Z., Chawan, P.: Breast cancer prediction using supervised machine learning algorithms. Int. Res. J. Eng. Technol. (IRJET) **07**(08), October 2019. e-ISSN: 2395–0056
4. Karthikeyan, B.: Breast cancer detection using machine learning. Int. J. Adv. Trends Comput. Sci. Eng. **9**, 981–984 (2020). https://doi.org/10.30534/ijatcse/2020/12922020
5. Shravya, C., Pravalika, K., Subhani, S.: Prediction of breast cancer using supervised machine learning techniques. Int. J. Innov. Technol. Explor. Eng. (IJITEE) **8**, 1106–1110 (2019)
6. Naveen, Sharma, R.K., Nair, A.R.: Efficient breast cancer prediction using ensemble machine learning models. In: 2019 4th International Conference on Recent Trends on Electronics, Information, Communication and Technology (RTEICT), Bangalore, India, pp. 100–104 (2019). https://doi.org/10.1109/RTEICT46194.2019.9016968
7. Bharat, A., Pooja, N., Reddy, R.A.: Using machine learning algorithms for breast cancer risk prediction and diagnosis. In: 2018 3rd International Conference on Circuits, Control, Communication and Computing (I4C), Bangalore, India, pp. 1–4 (2018). https://doi.org/10.1109/CIMCA.2018.8739696
8. Teixeira, F., Montenegro, J.L.Z., da Costa, C.A., da Rosa Righi, R.: An analysis of machine learning classifiers in breast cancer diagnosis. In: XLV Latin American Computing Conference (CLEI), Panama, pp. 1–10 (2019). https://doi.org/10.1109/CLEI47609.2019.235094
9. Das, S., Biswas, D.: Prediction of breast cancer using ensemble learning. In: 2019 5th International Conference on Advances in Electrical Engineering (ICAEE), Dhaka, Bangladesh, pp. 804–808 (2019). https://doi.org/10.1109/ICAEE48663.2019.8975544

10. Chandrasegar, T., Nikhilesh Vutukuri, S.B.: Optimized machine learning model using Decision Tree for cancer prediction. In: Innovations in Power and Advanced Computing Technologies (i-PACT), Vellore, India, pp. 1–4 (2019). https://doi.org/10.1109/i-PACT44901.2019.8960129

11. Dhanya, R., Paul, I.R., Sindhu Akula, S., Sivakumar, M., Nair, J.J.: A comparative study for breast cancer prediction using machine learning and feature selection. In: 2019 International Conference on Intelligent Computing and Control Systems (ICCS), Madurai, India, pp. 1049–1055 (2019). https://doi.org/10.1109/ICCS45141.2019.9065563

12. Hungilo, G.G., Emmanuel, G., Emanuel, A.W.R.: Performance evaluation of ensembles algorithms in prediction of breast cancer. In: International Biomedical Instrumentation and Technology Conference (IBITeC). Special Region of Yogyakarta, Indonesia, pp. 74–79 (2019). https://doi.org/10.1109/IBITeC46597.2019.9091718

13. Laghmati, S., Tmiri, A., Cherradi, B.: Machine Learning based system for prediction of breast cancer severity. In: 2019 International Conference on Wireless Networks and Mobile Communications (WINCOM), Fez, Morocco, pp. 1–5 (2019). https://doi.org/10.1109/WINCOM47513.2019.8942575

14. Nguyen, Q.H., et al.: Breast cancer prediction using feature selection and ensemble voting. In: 2019 International Conference on System Science and Engineering (ICSSE), Dong Hoi, Vietnam, pp. 250–254 (2019). https://doi.org/10.1109/ICSSE.2019.8823106

15. Suryachandra, P., Reddy, P.V.S.: Comparison of machine learning algorithms for breast cancer. In: 2016 International Conference on Inventive Computation Technologies (ICICT), Coimbatore, pp. 1–6 (2016). https://doi.org/10.1109/INVENTIVE.2016.7830090

16. Liu, B., et al.: Comparison of machine learning classifiers for breast cancer diagnosis based on feature selection. In: 2018 IEEE International Conference on Systems, Man, and Cybernetics (SMC), Miyazaki, Japan, pp. 4399–4404 (2018). https://doi.org/10.1109/SMC.2018.00743

17. https://archive.ics.uci.edu/ml/datasets/Breast+Cancer+Wisconsin+(Diagnostic)

Natural Language Inference: Detecting Contradiction and Entailment in Multilingual Text

Sai Sree Harsha$^{(\boxtimes)}$ ⓘ, K. Krishna Swaroop ⓘ, and B. R. Chandavarkar ⓘ

Department of Computer Science, National Institute of Technology, Surathkal, Karnataka, India

`{saisreeharsha.181co146,kkrishnaswaroop.181co125,brc}@nitk.edu.in`

Abstract. Natural Language Inference (NLI) is the task of characterising the inferential relationship between a natural language premise and a natural language hypothesis. The premise and the hypothesis could be related in three distinct ways. The hypothesis could be a logical conclusion that follows from the given premise (entailment), the hypothesis could be false (contradiction), or the hypothesis and the premise could be unrelated (neutral). A robust and reliable system for NLI serves as a suitable evaluation measure for true natural language understanding and enables the use of such systems in several modern day application scenarios. We propose a novel technique for the NLI task by leveraging the recently proposed Bidirectional Encoder Representations from Transformers (BERT). We utilize a robustly optimized variant of BERT, integrate a contextualized definition embedding mechanism, and incorporate the use of global average pooling into our proposed NLI system. We use several different benchmark datasets, including a dataset containing premise-hypothesis pairs from 15 different languages to systematically evaluate the performance of our model and show that it yields superior results.

Keywords: Natural Language Processing · Transformers · BERT

The development of artificial intelligence systems possessing capabilities of reasoning and inference has been a long standing goal of research in the domain of Natural Language Processing (NLP). An important step towards achieving such abilities is the NLI task which aims to represent the relationship between a pair of sentences consisting of the premise and hypothesis. Concretely, given a premise P and a hypothesis H the goal of NLI is to perform classification of the sentence pair with the three labels, namely 1) entailment, where the hypothesis H logically follows the given premise P, 2) contradiction, where the hypothesis H is false and 3) neutral, where the premise P and hypothesis H are unrelated. We say that the premise is an entailment of the hypothesis, if a human being who is reading the premise would make an inference that the hypothesis is most probably true [1]. This idea of "human reading" assumes human common sense and

---

The first two authors have equal contribution.

K. R. Venugopal et al. (Eds.): ICInPro 2021, CCIS 1483, pp. 314–327, 2021.
https://doi.org/10.1007/978-3-030-91244-4_25

common background knowledge. Hence, the task of NLI involves modelling such informal reasoning and inference capabilities making it a necessary step towards realising true natural language understanding. In fact, several researchers [2, 3] claim that solving NLI perfectly implies achieving human level understanding of natural language.

The task of NLI facilitates a wide range of NLP applications including question-answering, semantic search, automatic text summarization and evaluation of machine translation systems. As part of a question-answering system, an effective NLI model can be used to evaluate if the target question can be inferred from candidate answers extracted from the source document. A NLI system can also be used to identify semantic equivalence between search queries and sentences in source documents, enabling a superior semantic retrieval of information as opposed to just a keyword based search. In the case of automatic summarization, NLI can be utilised in two crucial ways. NLI can be used to aid the elimination of redundancy by ensuring that the summary does not contain any sentences that can be inferred from the rest of the summary. NLI can also be utilised in ensuring correctness, by checking that the summary is implied by the source document. An efficacious NLI system can also be used to evaluate machine translation systems by assessing if the translation provided by the machine entails, and is entailed by the reference ground truth translation. If such a semantic equivalence exists, then it is most likely a good translation.

Deep learning techniques have resulted in several advancements pertaining to the NLI task. Also, over the recent years several large benchmark datasets such as SNLI [4], MultiNLI [5] have emerged, which have made the training of neural-network based models feasible. Early methods used sequence models such as recurrent neural networks, gated recurrent units and long short term memory models to model NLI. Sequence models suffer from a range of problems such as vanishing gradients and increased computational complexity. Also, they are not able to capture long range dependencies and are not interpretable, making it harder to discern their working.

We propose a new method for the NLI task levering a Transformer [6] based approach. In contrast to sequence models, the Transformer which leverages the self-attention and multi-headed attention mechanism, can capture long-range dependencies. The Transformer also lends itself to parallelization as it does not process input data sequentially leading to significantly faster training and inference times. We utilise Bidirectional Encoder Representations from Transformers (BERT) [7], which is a self-supervised, pre-training technique, that learns highly useful generalizable representations of input text from large amounts of unlabelled data. Specifically we use a robustly optimised variant of BERT whose representations generalise even better to downstream tasks when compared to BERT. We integrate a contextualised definition embedding mechanism, and incorporate the use of global average pooling into our proposed NLI system. We experiment on several benchmark NLI datasets such as MultiNLI and SNLI, in addition to the XNLI dataset which includes premise-hypothesis pairs from

as many fifteen different languages to systematically evaluate the performance
of our model and show that it yields superior results.

The remaining sections are organised in the following manner. Section 1
provides a brief overview of the related NLP concepts. Section 2 describes
benchmark NLI datasets. Section 3 details our proposed methodology. Section 4
describes the experimental setup and Sect. 5 presents the results and analysis.

1   Related Work

1.1   Attention

The attention mechanism was first introduced in [8] where it was utilized to
improve the performance of neural machine translation by determining the relative importance of different words in a sentence.

The attention mechanism receives three vectors as its input, namely the
query, key and value. It performs a dot product of the query and keys and applies
a softmax function over the result of the dot product to get attention scores. It
then performs a sum of the values, weighing each of them using the attention
scores to obtain the final output vector. The query and the keys determine which
values to focus on and can hence be described as 'attending' to the values.

Let $attn(q, K, V)$ represent the attention layer, where q, K and V indicate
query, key and value respectively. Here q is a d_q dimensional vector ($d_q \in Z$).
K and V are two sets with $|K| = |V|$. Each $k_i \in K$ and $v_i \in V$ are d_k and d_v
dimensional vectors, and $i \in [|K|]$. The attention mechanism works as follows:

$$atten(q, K, V) = \sum_{i=1}^{|V|} \alpha_i W_v v_i, \tag{1}$$

$$\alpha_i = \frac{exp((W_q q)^T (W_k k_i))}{Z}, \tag{2}$$

$$Z = \sum_{i=1}^{|K|} exp((W_q q)^T (W_k k_i)), \tag{3}$$

where W_q, W_k and W_v are the parameters to be learned. As proposed by [6] we
implement the attention mechanism in the Transformer as a multi-head attention
model.

1.2   Transformer

Several research works have shown that using the Transformer [6] architecture
leads to superior performance on many NLP tasks. In contrast to the existing
sequence models based on recurrent neural networks, the Transformer which
leverages the self-attention and multi-headed attention mechanism, can capture
long-range dependencies. The Transformer also lends itself to parallelization as

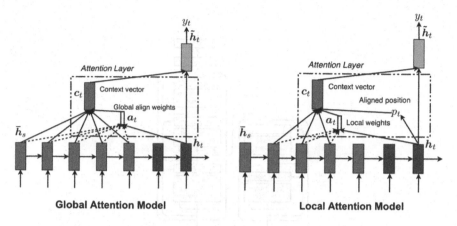

Fig. 1. Attention models [6]

it does not process input data sequentially leading to significantly faster training and inference (Fig. 1).

Figure 2 shows the Transformer model [6] architecture. The left part of the network is the encoder and the right part is the decoder. The encoder consists of N identical layers. Each layer in turn has two blocks within it, the attention block and the feed forward neural network block. The attention block computes the attention function as described in the previous section using h attention heads. The output features of the attention block are passed through a feed-forward layer which learns to represent complex relationships among the vectors. The residual connections or skip connections are also shown in the figure, along with the layer normalization blocks. As the residual connections perform a summation of vectors from different layers, all blocks within the layer always output vectors of the same size, namely d_{model}, to ensure compatibility.

The decoder also consists of N identical layers. Each layer in turn has three blocks. Two of these are same as the ones present in the encoder and the third block computes the attention function using the output of the encoder as input using h attention heads. The decoder also has residual connections that are shown in the figure along with the layer normalization blocks. The self-attention module in the decoder is also modified so the during the processing of sentences it does not take word tokens that occur in the future into consideration.

Unlike RNN based models, the Transformer does not process input data sequentially and hence supports parallel processing resulting in faster training and inference. Therefore, it requires positional encodings to be added to the input which carry information about the relative position of the input code tokens.

1.3   BERT

One of the important challenges in natural language processing is inadequate training data. In order to overcome this problem, BERT [7] is a recently proposed technique for NLP pre-training. It uses a large amount of easily available

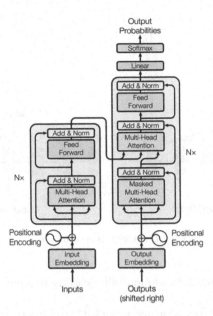

Fig. 2. Transformer model architecture [6]

unlabeled text, and trains a Transformer model to obtain an embedding for input text tokens. In the scenarios where sufficient training data is not available, we can use a Transformer model which is firstly pre-trained using the BERT technique to get deep bidirectional representations of the input tokens. BERT is able to provide such an embedding as it pre-trains the Transformer model to pay attention to both the right and left contexts in the sentence. Using a BERT model (i.e., a Transformer model that has been pre-trained using the BERT technique) and training only a few additional layers on top of the BERT model leads to superior results on a range of NLP tasks.

As shown in Fig. 3, the BERT technique operates in two stages, namely the pre-training stage and the fine-tuning stage. During the pre-training stage, a Transformer model is trained using two objectives. One is the masked language model objective and the other is the next-sentence prediction objective. During masked language modelling, some words from the input sentence are randomly selected and are masked or hidden from the model. The task is then to predict the hidden word token based only on its context. Training the model with such an objective, enables the effective fusion of left and right contexts. In the next-sentence prediction task, the input to the model is a pair of sentences, sampled from a text corpus. The model is trained to predicted whether one of them follows the other. The most important aspect here is that, the pre-training stage uses only unlabeled data, which is easy to obtain in large quantities. In the second stage where the Transformer is fine-tuned, the weights from the previous stage are further adjusted using labeled data that is specific to a particular downstream task such as question-answering or text-summarization.

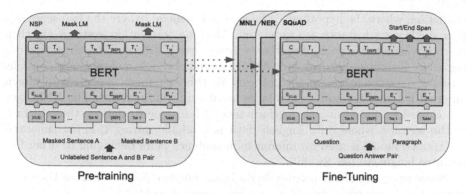

Fig. 3. Pre-training and fine-tuning procedures of BERT [7]

2   NLI Datasets

Several benchmark NLI datasets have emerged over the years, the most prominent of these being the SNLI, MultiNLI and the XNLI datasets. The SNLI dataset [4] comprises of approximately five lakh data instances. Each data instance is made up of a pair of sentences from the English language constituting the premise and the hypothesis statements. Each such pair is also labeled by humans with one of the three labels, namely 1) entailment, where the hypothesis logically follows the given premise, 2) contradiction, where the hypothesis is false and 3) neutral, where the two sentences in a single data instance are unrelated. All of the sentences and labels in the dataset are human written and the subject of these sentences refers to very realistic scenarios. Hence, it is particularly appropriate for measuring the performance of language models on the NLI task.

The MultiNLI dataset [5] contains about four lakh data instances. Each data instance consists of a premise statement and a hypothesis statement. A single data instance represented by one pair of sentences, is labelled as belonging to one of the three classes namely 1) entailment, where the hypothesis logically follows the given premise, 2) contradiction, where the hypothesis is false and 3) neutral, where the two sentences in a single data instance are unrelated. It is very similar to the SNLI dataset, but it contains sentence pairs from a variety of domains. These domains include fiction, government, telephone, travel, face-to-face and letters, etc. and contain data that has been extracted from not only written text but also spoken English. This dataset, hence enables us to measure the performance of language models on the NLI task generalizing to the complete breadth of the language.

The XNLI dataset [9] is made up of data instances that have been randomly sampled from the MultiNLI dataset. Each data instance contains a pair of sentences making up the premise and hypothesis. Eight thousand data instances are sampled and have been labeled as belonging to one of the three classes, namely 1) entailment, where the hypothesis logically follows the given premise, 2) contradiction, where the hypothesis is false and 3) neutral, where the two sentences in a single data instance are unrelated. Machine translation systems are then used to translate the pairs of sentences into fourteen different languages, including many European, African and Asian languages. Each sentence pair contains a premise statement and the hypothesis for this statement is present in fifteen different languages, resulting in a large number of combinations. The XNLI corpus is used for measuring the performance of language models on the NLI task in the scenario where only English data is available during training time and the trained model is used for inference on sentence pairs belonging to other low resource languages like Swahili.

Some examples of the premise-hypothesis sentence pairs as well as their corresponding labels from these datasets are shown in Table 1.

Table 1. Examples of data instances from the SNLI [4], MultiNLI [5] and XNLI [9] datasets

Premise	Hypothesis	Label
A soccer game with multiple males playing.	Some men are playing a sport.	entailment
At the other end of Pennsylvania Avenue, people began to line up for a White House tour.	People formed a line at the end of Pennsylvania Avenue.	entailment
A black race car starts up in front of a crowd of people.	A man is driving down a lonely road.	contradiction
A man inspects the uniform of a figure in some East Asian country.	The man is sleeping.	contradiction
There's so much you could talk about on that I'll just skip that.	I want to tell you everything I know about that!	contradiction
A smiling costumed woman is holding an umbrella.	A happy woman in a fairy costume holds an umbrella.	neutral
An older and younger man smiling.	Two men are smiling and laughing at the cats playing on the floor.	neutral
The Old One always comforted Ca'daan, except today.	Ca'daan knew the Old One very well.	neutral
Το χορίται που μπορεί να με βοηθήσει είναι στον δρόμο προς την πόλη.	Η κοπέλα που θα με βοηθήσει είναι 5 μίλια μακριά.	neutral (Greek)
При измерване на ефективността, съвършенството е недостижимо.	Можете да бъдете перфектни, ако се опитате достатъчно.	entailment (Bulgarian)
In August when everybody's on vacation or something we can dress a little more casual.	August is a black out month for vacations in the company.	contradiction
Your gift is appreciated by each and every student who will benefit from your generosity.	Hundreds of students will benefit from your generosity.	neutral

3 Methodology

We propose a novel approach for the task of natural language inference, using a Transformer model. The input to the Transformer model is a comprises of a sentence pair constituting the premise and the hypothesis. Each sentence in the input sentence pair is comprised of a sequential arrangement of word tokens. The output of the Transformer model is the label of the class, (i.e., either entailment, contradiction or neutral) to which the premise-hypothesis pair belongs to.

The Transformer model is first pre-trained using a robustly optimized variant of the BERT technique [10]. The representations provided by this variant of

BERT are much more superior in terms of their applicability to downstream evaluation through tasks such as question-answering or text summarization. Inspired by [11] and [12] we integrate a contextualized definition embedding mechanism to this Transformer model. Additionally, we also use the global average pooling operation in the last layer of the Transformer model and empirically show that these lead to superior results. In the following sections we describe each component of our proposed approach in detail.

3.1 Robustly Optimized BERT

As discussed in Sect. 1.3, BERT is a technique that is used to pre-train a Transformer model, in a self-supervised manner, using a large amount of unlabeled training data. This pre-trained model can then be fine-tuned, using task-specific datasets, with a small amount of training data and the addition of a few extra layers to obtain better performance. We use a robustly optimised variant of the BERT technique [10] called RoBERTa. The modifications are simple, and include training the model for a larger number of iteration, with a bigger batch size, and using more unlabeled data for training. It also involves removing the objective related to the prediction of suffix sentences from the traditional BERT technique and training the underlying Transformer model on longer sentences. Another key change is altering the pattern of the mask which is applied to the training data, at each iteration as a part of the masked language modelling objective. In the masked language modelling objective some tokens in the input sequence are selected and replaced with the special token [MASK]. The model is then trained using a cross-entropy loss to predict the masked tokens. In the original BERT technique, a single static masking pattern is fixed during the data pre-processing stage. However, in RoBERTa, a new masking pattern is dynamically generated each time a sequence is input to the model. This change is particularly important when training the Transformer model for a larger number of iteration with a bigger batch size.

3.2 Contextualized Definition Embedding

Inspired by [11] and [12], we integrate a contextualized definition embedding mechanism into our Transformer model. In our definition embedding mechanism, we first create a dictionary which consists of word tokens as keys and their associated definitions as values. Then, for all the tokens in the input text sequence which are present in the dictionary, we extract the corresponding definitions. These definitions are also sequences of tokens. Each definition is then passed through a separate pre-trained Transformer model to obtain a fixed size embedding. The pre-trained Transformer model leverages the self-attention and multi-headed attention mechanisms and can capture long-range dependencies. It models complex context-sensitive relationships among the tokens and thus, provides a 'contextualized' embedding for each definition. The embedding of each definition is then added to the embedding of the corresponding input word token. This modified embedding of the input text tokens is used as input to the

RoBERTa pre-trained Transformer model, for the natural language inference task. The use of this contextualized definition embedding mechanism helps us obtain highly meaningful and superior-quality representations of rare or novel words which do not have pre-trained word embeddings.

3.3 Global Average Pooling

Global Average Pooling was first introduced in [13], to replace fully-connected layers in classical convolutional neural networks. Here, one feature map is generated in the last layer corresponding to each category in the classification task. The average of all the values in each feature map is obtained. The softmax function is applied over this vector to obtain the class probabilities. We then rank these class probabilities and consider the highest ranked class as the one predicted by the model. Global average pooling offers several advantages. It enforces strong correspondences between feature maps and categories and thus the feature maps can be easily interpreted as category confidence maps. Also, as global average pooling replaces a fully-connected neural network layer, it decreases the total number of trainable parameters in the network, helps avoid over-fitting and results in lower training and inference time.

We use the global average pooling operation to obtain a tuple of size 3 from the last layer of the Transformer model decoder. The softmax function is applied on the tuple to obtain 3 probability scores corresponding to the classes of entailment, contradiction and neutral. We then rank these class probabilities and consider the highest ranked class as the one predicted by the model.

3.4 Algorithm

Step 1: Use the robustly optimized variant of the BERT method (as described in Sect. 3.1) to pre-train a Transformer model in self-supervised manner.

Step 2: Obtain the sentence pair consisting constituting the premise and the hypothesis.

Step 3: Tokenize the sentence pair into words and obtain word embeddings using the contextualized definition embedding mechanism (as described in Sect. 3.2).

Step 4: Pass the list of word embeddings through the Transformer model obtained in Step 1.

Step 5: Apply global average pooling on the output of the Transformer model (as described in Sect. 3.3), to obtain the predicted label for the input sentence pair as entailment, contradiction or neutral.

4   Experiment

4.1   Datasets

We evaluate our proposed approach using the SNLI, MultiNLI and XNLI datasets. As described in Sect. 2, these benchmark datasets enable us to effectively assess the classification performance of our method. The SNLI dataset [4] comprises of approximately five lakh data instances. Each data instance is made up of a pair of sentences from the English language constituting the premise and the hypothesis statements. Each such pair is also labeled by humans with one of the three labels, namely 1) entailment, where the hypothesis logically follows the given premise, 2) contradiction, where the hypothesis is false and 3) neutral, where the two sentences in a single data instance are unrelated. All of the sentences and labels in the dataset have been written by humans and the subject of these sentences refers to very realistic scenarios.

The MultiNLI dataset [5] contains about four lakh data instances. Each data instance consists of a premise statement and a hypothesis statement. A single data instance represented by one pair of sentences, is labelled as belonging to one of the three classes namely 1) entailment, where the hypothesis logically follows the given premise, 2) contradiction, where the hypothesis is false and 3) neutral, where the two sentences in a single data instance are unrelated. It is very similar to the SNLI dataset, but it contains sentence pairs from ten different domains.

The XNLI dataset [9] is made up of data instances that have been randomly sampled from the MultiNLI dataset. Each data instance contains a pair of sentences making up the premise and hypothesis. Eight thousand data instances are sampled and have been labeled as belonging to one of the three classes, namely 1) entailment, where the hypothesis logically follows the given premise, 2) contradiction, where the hypothesis is false and 3) neutral, where the two sentences in a single data instance are unrelated. Machine translation systems are then used to translate the pairs of sentences into fourteen different languages, including many European, African and Asian languages. Each sentence pair contains a premise statement and the hypothesis for this statement is present in fifteen different languages, resulting in a large number of combinations.

We split each of these datasets into training, validation and test sets. The size of these splits is shown in Table 2 We use the training set for fine-tuning our Transformer model which has been pre-trained using the RoBERTa technique. During this stage we also use the data from the validation set to measure the classification accuracy of the model periodically, in order to tune hyper-parameters and adjust training settings. Once the Transformer model has been trained, we run model inference on the data instances from test split of the data and report the obtained model accuracy and other performance metrics.

Table 2. Statistics of the SNLI [4], MultiNLI [5] and XNLI [9] datasets

	SNLI	MultiNLI	XNLI
Size of train set	5,50,152	3,92,702	1,12,500
Size of validation set	10,000	9,815	2,490
Size of test set	10,000	9,832	5,010

4.2 Training and Evaluation Settings

In our proposed approach, the transformer encoder and decoder, each consist of 12 layers, with a 768 dimensional hidden representation. The attention blocks within the encoder and decoder networks use 12 attention heads, the query vectors have a dimensionality of 128, the key vectors have a dimensionality of 128, and the value vectors have a dimensionality of 128. For the contextualized definition embedding mechanism we use a smaller Transformer model, where the encoder and decoder consist of 6 layers each and use a 512 dimensional hidden representation. The attention blocks within the encoder and decoder layers in this smaller Transformer model use only 8 attention heads, the query vectors have a dimensionality of 64, the key vectors have a dimensionality of 64, and the value vectors have a dimensionality of 64.

We optimise our model using the Adam optimizer [14] with the betas $=$ (0.9, 0.99), epsilon $=$ 1e$-$6 and an L_2 weight decay of 0.01. The learning rate is warmed up over the first 10,000 steps to a peak value of 1e$-$4, and then linearly decayed. We train with a dropout of 0.1 on all layers and attention weights, and the GELU activation function [15].

We implement our proposed approach using the PyTorch framework [16] and utilize the Google Cloud Tensor Processing Unit (TPU) v3 for training our model. We train the model for 200 epochs. We use a batch size of 16. We validate the model performance after each epoch and stop the training process if the classification accuracy of the model on the validation set does not increase for more that 20 successive epochs, to avoid over-fitting.

4.3 Evaluation Metrics

We use several performance metrics to evaluate the classification performance of our model on the test set of each dataset in a holistic manner. These metrics including classification accuracy, precision, recall, F1-score and specificity. Accuracy is the fraction of instances for which our model predicted the correct label. Precision is the percentage of positive predictions that were actually correct and recall is the proportion of actual positives that were identified correctly. F1 score conveys the balance between precision and recall and combines precision and recall scores relative to a specific positive class. Specificity measures the proportion of negatives that were correctly identified.

5 Results and Analysis

Table 3 shows the results obtained from our experiments. It shows the various classification performance metrics including accuracy, precision, recall, F1-score and specificity. From these results we can conclude that our proposed novel approach not only gives a high accuracy for the natural language inference task, but also minimises the number of false positives as well as false negatives, leading to superior performance with respect to other classification metrics as well.

Most importantly, we are able to effectively detect entailment, contradiction and neutrality in multi-lingual text. This is evident from the performance of our model on the XNLI dataset [9] which contains premise-hypothesis pairs from 15 different languages. This shows that our model is able to successfully model the informal reasoning and inference capabilities to some extent, and is therefore a step towards true natural language understanding.

Table 3. Performance of our method on the SNLI [4], MultiNLI [5] and XNLI [9] datasets

	SNLI	MultiNLI	XNLI
Accuracy	92.2	91.8	91.1
Precision	91.2	90.1	89.9
Recall	92.1	91.5	90.4
F1-score	91.6	90.8	90.1
Specificity	91.4	90.6	90.8

We attribute these superior results to the three key design choices that we made in our proposed approach, namely the use of the robustly optimized variant of the BERT technique to pre-train our Transformer model, the integration of the contextualized definition embedding mechanism and the use of global average pooling in the last layer of the Transformer model replacing a fully-connected neural network layer.

In Tables 4 we compare the performance of two models, using classification accuracy on the SNLI [4], MultiNLI [5] and XNLI datasets [9]. In one case the Transformer model is pre-trained using the robustly optimized variant of the BERT technique and in the other the traditional BERT technique is used for pre-training. The results indicate that the modifications in the robustly optimised variant such as training the model for a larger number of iteration, with a bigger batch size, using more unlabeled data for training and dynamically changing masking pattern, helps the model generalize better and leads to better performance.

Table 4. Classification accuracy of our method when different pre-training techniques [10] are used for the Transformer model

	SNLI	MultiNLI	XNLI
Pre-trained with traditional BERT technique	88.4	89.1	88.7
Pre-trained with robustly optimized BERT technique (ours)	92.2	91.8	91.1

In Tables 5 we compare the performance of two models, using classification accuracy on the SNLI [4], MultiNLI [5] and XNLI datasets [9]. In one case we integrate our contextualized definition embedding mechanism and other only input word token embeddings are used. The results indicate that the use of this contextualized definition embedding mechanism helps us obtain highly meaningful and superior-quality representations of rare or novel words which do not have pre-trained word embeddings, which in turn improves performance across all the datasets.

Table 5. Classification accuracy of our method in the presence and absence of the contextualized definition embedding mechanism [11]

	SNLI	MultiNLI	XNLI
Contextualized definition embedding mechanism absent	87.3	88.6	88.2
Contextualized definition embedding mechanism present (ours)	92.2	91.8	91.1

Table 6. Classification accuracy of our method in the presence and absence of global average pooling [13] at the last Transformer layer

	SNLI	MultiNLI	XNLI
Use of fully-connected neural network	90.5	90.1	89.3
Use of the global average pooling operation (ours)	92.2	91.8	91.1

In Tables 6 we compare the performance of two models, using classification accuracy on the SNLI [4], MultiNLI [5] and XNLI datasets [9]. We first obtain the output vectors from the final layer of the Transformer. In one case we apply the global average pooling operation on these features and in the other case we pass these outputs through a fully-connected neural network layer. Both these techniques ultimately obtain a tuple of size 3 from the last layer of the Transformer model decoder. The softmax function is applied on the tuple to obtain 3 probability scores corresponding to the classes of entailment, contradiction and neutral. The model then predicts the label for the given premise-hypothesis pair as the class with the highest probability. The results indicate that the use of global average pooling, decreases the total number of trainable parameters in the network, helps avoid over-fitting, results in lower training and inference time and thus, also leads to better classification performance.

6   Conclusion and Future Work

We propose a novel technique for the task of natural language inference leveraging the Transformer model. We utilize a robustly optimized variant of the

BERT technique for pre-training our model, integrate a contextualized definition embedding mechanism, and incorporate the use of global average pooling into our proposed NLI system. We use several different benchmark datasets, including a dataset containing premise-hypothesis pairs from 15 different languages to systematically evaluate the performance of our model and show that it yields superior results. We also carefully analyse our design choices and empirically show that they are indeed responsible for the superior results obtained. As part of future work, we can explore the adoption of our approach to question-answering, text summarization and other NLP tasks as well as the development of other, perhaps lighter mechanisms for definition embedding.

References

1. Dagan, I., Glickman, O., Magnini, B.: The PASCAL recognising textual entailment challenge. In: Quiñonero-Candela, J., Dagan, I., Magnini, B., d'Alché-Buc, F. (eds.) MLCW 2005. LNCS (LNAI), vol. 3944, pp. 177–190. Springer, Heidelberg (2006). https://doi.org/10.1007/11736790_9
2. Goldberg, Y.: Neural network methods for natural language processing. Synth. Lect. Hum. Lang. Technol. **10**, 1–309 (2017)
3. Nangia, N., Williams, A., Lazaridou, A., Bowman, S.: The RepEval 2017 shared task: multi-genre natural language inference with sentence representations. In: ACL (2017)
4. Bowman, S.R., Angeli, G., Potts, C., Manning, C.D.: A large annotated corpus for learning natural language inference. In: ACL, pp. 632–642 (2015)
5. Williams, A., Nangia, N., Bowman, S.: A broad-coverage challenge corpus for sentence understanding through inference. In: ACL, pp. 1112–1122 (2018)
6. Vaswani, A., et al.: Attention is all you need. In: NeurIPS, pp. 5998–6008 (2017)
7. Devlin, J., Chang, M.-W., Lee, K., Toutanova, K.: BERT: pre-training of deep bidirectional transformers for language understanding. In: ACL, pp. 4171–4186 (2019)
8. Bahdanau, D., Cho, K., Bengio, Y.: Neural machine translation by jointly learning to align and translate. In: ICLR (2015)
9. Conneau, A., et al.: XNLI: evaluating cross-lingual sentence representations. In: ACL (2018)
10. Liu, Y., et al.: Roberta: a robustly optimized BERT pretraining approach. CoRR (2019)
11. Nishida, K., Nishida, K., Asano, H., Tomita, J.: Natural language inference with definition embedding considering context on the fly. In: ACL (2018)
12. Bahdanau, D., Bosc, T., Jastrzebski, S., Grefenstette, E., Vincent, P., Bengio, Y.: Learning to compute word embeddings on the fly. CoRR (2017)
13. Lin, M., Chen, Q., Yan, S.: Network in network. arXiv:1312.4400 (2014)
14. Kingma, D.P., Ba, J.: Adam: a method for stochastic optimization. In: ICLR (2015)
15. Hendrycks, D., Gimpel, K.: Gaussian error linear units (GELUs). arXiv:1606.08415 (2016)
16. Paszke, A., et al.: Automatic differentiation in pytorch. In: NeurIPS Workshop on Autodiff (2017)

Fake News Detection Using Artificial Neural Network Algorithm

U. J. Isha Priyavamtha, G. Vishnu Vardhan Reddy$^{(\boxtimes)}$, P. Devisri,
and Asha S. Manek$^{(\boxtimes)}$

Department of CSE, RV Institute of Technology and Management, Bangalore, India
gadusunuryvishnur_cs19.rvitm@rvei.edu.in

Abstract. As technology is increasing rapidly, nowadays most people prefer social media over newspapers for the news. There are chances of spreading fake news widely on social media because of circulation of news can be done by anyone. The fake news can show adverse effects on economic as well as political issues. It can spread hatred among the community which can cause serious problems to peace in society. Hence, it is important to detect fake news. The developed model using Artificial Neural Network is a solution for the problem of fake news detection and filtered to avoid wide spread. The proposed model detects fake news that includes three main phases 'preprocessing, feature extraction and classification'. Initially input is preprocessed to extract features using clustering algorithms. Subsequently, the model is developed to detect fake news. The Neural Networks and Linear Support Vector Clustering algorithms resulted in 99.90% and 97.5% accuracy respectively. Finally, analysis is carried out for validating the betterment of the presented model in terms of different measures.

Keywords: Artificial Neural Network · Fake News Detection · Preprocessing · Support Vector Clustering

1 Introduction

When newspapers, television and radios were the only source for the news, only authorized people were involved in preparing news [1]. Now-a-days, most of the news is being escalated on internet social community rather than by the news channels or newspapers as it is easy to access and cost effective. This made the spread of fake news much easier. The fake news is spreading at a very large rate, for example, during "U.S presidential election", 2016, more than 25% of Americans has visited a fake news website in the span of six weeks. Fake news is usually created and spread for economic and political benefits [2]. It will show adverse effects when communal hatred is spread by means of fake news. One of the best examples for the destruction caused by the fake news is the "Delhi riots" [1]. Fake news is misleading information that circulates across the internet as news. The rapid escalation of fake news in internet social community is resulting in loss of public trust in social media. Hence, there is an immense need for news detection

© Springer Nature Switzerland AG 2021
K. R. Venugopal et al. (Eds.): ICInPro 2021, CCIS 1483, pp. 328–340, 2021.
https://doi.org/10.1007/978-3-030-91244-4_26

and classification of the news as fake news and genuine news in today's scenario to prevent the spread of fake news. It is also necessary to understand that what news contents come under fake news and what does not come under fake news for detection.

There are three types of fake news:

1) CLICKBAIT: News which displays a sensational headline which is created to obtain more "clicks" which in turn adds to ad revenue.
2) SPONSORED CONTENT: News which appears as genuine news but in fact it is news for advertising.
3) FABRICATED JOURNALISM: News which is completely created for economic or political benefits.

The news which won't fall under fake news category is: satire, reporting mistakes and the news which people won't like (claim actual truth also as "fake" just because people don't agree with it). Over half of the population claims that they regularly come across fake news on social media like Facebook, Twitter. In 2018, at least 20 people were killed as a result of fake news which was spread through social media platforms.

In this work, Machine Learning algorithms i.e. Neural Networks algorithm is used to detect fake news. This algorithm outperformed with 99.90% accuracy whereas Support Vector Clustering algorithm results 97.5% accuracy and Passive Aggressive Classifier gives 96.67% accuracy.

1.1 Research Goal

The objective of the work is to examine the fake news and filter out from the social media containing false and misleading information. Primarily, training data from Kaggle website news.csv [3] is preprocessed and used for training the model. The data from news.csv is then categorized into three characteristics (i) authenticity (ii) intention and (iii) benign news. Further news has been categorized into clickbait, sponsored content and fabricated journalism. Analysis is made based on these characteristics of the news to classify into fake and benign news for purpose of helping users to avoid being lured by clickbait and fabrication.

1.2 Our Contribution

The contribution of the proposed solution is as follows:

- The proposed research work discusses the potential directions to improve fake news detection and reduction capabilities.
- Owing to this, the ANN model is developed that can identify and remove fake news from the results provided to a user by internet social community or a search engine.
- The proposed model provides future directions to detect fake news in social internet media.
- The proposed model performs a binary classification on a news.csv dataset and classifies it into "real" or "fake".

The structure of the research paper is as follows: Sect. 2 shows a literature review of the paperwork. Section 3 discusses the datasets and explains the implementation of the methods for fake news detection. Section 4 depicts the obtained results and compares the results with other existing methods. Section 5 concludes the paper.

2 Literature Survey

2.1 Related Work

Aman Srivatsava [1] has built a website for fake news detection composite of Machine Learning and NLP methods. A combination of five Machine Learning techniques used to built the website are 'Support Vector Machine (SVM)', 'Logistic Regression', Naive Bayes, Random Forest Classifier, Stochastic Gradient Descent (SGD) in combination with Bag of Words, TF-IDF, POSTAG and N-GRAM feature selection. After comparing the accuracy and F1-Score of the model with different combinations of techniques, author concluded the paper with 95% accuracy and 0.95 F1-Score using SVM classifier in combination with TF-IDF and POSTAG.

Waikhom et al., used the publicly available LIAR dataset for fake news detection collected from POLITIFACT.COM. The LIAR dataset provides links to source documents for each case. The work provides ensemble techniques for fake news detection [2].

Xinyi Zhou et al., has used four perspectives for fake news detection surveys based on knowledge, style, propagation and source methods. The survey also highlights potential research tasks based on the news reviews and identify as fundamental theory across various disciplines to encourage interdisciplinary research on fake news [4].

Nicole O'Brien has used Machine Learning technology for the detection of fake news. The accuracy of 95.8% obtained by the model for the dataset of 5,504 fake and real news. The classification of news was based on language patterns using Neural Network classifier [5].

Sharma, K et al., described current problems facing due to fake news in survey paper [6] by focusing the associated technical challenges. The authors focused on the advancements of each method along with advantages and limitations.

2.2 Review

The pros and cons of the existing models regarding the fake news detection system is illustrated in Table 1. Still, at present, the Machine Learning classifier is used in most applications, where it can get impacted by various outer sources. The Artificial Neural Network (ANN) is used in a few and is in the infant stage. Hence more research is deployed in the proposed ANN model. Even the used classifier algorithms have some disadvantages that need a further extension. Some of the pros and cons are explained as follows:

Coh-Metrix-feature extraction tools [7] outperforms with 92% classification accuracy by satisfying readability features considering the Brazilian Portuguese language to detect fake news. But this methodology has to deepen the study on syntactic and semantic features.

Data augmentation technique is used to expand the limited annotated resources for Urdu by applying machine translation from English to Urdu. However, needs improvement in expanding annotated resources for other language pairs with larger parallel resources [8].

WeFEND framework [9] works by integrating three components including the annotator, the reinforced selector and the fake news detector on WeChat dataset consisting

Table 1. Features and challenges of existing models regarding the fake news detection.

Author [citation]	Methodology	Features	Challenges
Roney L. S. Santos et al. [7]	Coh-Metrix-feature extraction tools used for extracting readability metrics	- Analysis of readability features influence on the Brazilian Portuguese language fake news - 92% classification accuracy	Need to deepen study on syntactic and semantic features
Maaz Amjad et al. [8]	Data augmentation technique to expand the limited annotated resources for Urdu	- Investigated whether machine translation from English to Urdu can be applied as a text data augmentation technique to expand the limited annotated resources for Urdu - Findings indicate that character bigrams are the most robust and highest performing feature for the datasets augmented with machine-translated texts	Need to investigate whether text data augmentation through Machine Translated (MT) dataset for other language pairs with larger parallel resources will provide more promising results
YaqingWang et al. [9]	WeFEND framework	- The proposed framework works by integrating three components including the annotator, the reinforced selector and the fake news detector on WeChat dataset consisting of news articles and user feedback with 82.4% accuracy	Need to improve model performance by finding the optimum value for each parameter

of news articles and user feedback with 82.4% accuracy. But still needs betterment in performance accuracy and further methods to improve efficiency.

3 Methodology and Machine Learning Techniques

3.1 Data Collection

The printed and digital media information authenticity is a major concern which badly affects businesses and society. The information reaches and affects with fast pace and gets amplified, distorted, inaccurate on social media. Then this false information acquired enormous potential to cause bad impact on large users in real world. The existing datasets [3] of news articles annotated with veracity is used as a basis for this experimental study. The training dataset trainnews.csv contains six columns and 25116 rows. The description of each of the columns is given in Table 2.

3.2 Preprocessing

The preprocessing is used to process the dataset which consist of noise, missing values and redundancy to obtain a clean dataset. The size of the dataset is reduced because of elimination of unnecessary elements such as over representation of words and repeated occurrences in datasets.

Table 2. Description of the column of dataset [3]

Sl. no.	Column name	Description
1	id	Unique Identification Number of each news article
2	headline	The title of the news
3	news	The full text of the news article
4	serial_no	A serial number
5	written_by	The author of the news article
6	label	Labels - news is fake (1) or not fake (0)

Data is preprocessed at different levels as follows:

- The rows with 'NaN' values are removed from the dataset.
- Label Encoding is used to convert labels into integer values.
- The function 'train_test_split' is imported from 'sklearn.model_selection' to split the entire dataset into testing and training dataset.
- The news data is converted into a CSR sparse matrix based on the Term Frequency - Inverse Document Frequency (TFIDF) Vectorizer.
- The data is converted into 'NumPy' float array.
- This array is given to the input layer of the Artificial Neural Network.

The preprocessed dataset obtained by applying separate python code to the dataset [3] and steps performed are as shown in Fig. 1 below.

3.3 Feature Selection

Term Frequency- Inverse Document Frequency (TF-IDF) method is generally used to find frequent occurring words in classification of documents because it is easy, uncomplicated and high level of processing rate of weighing feature method.

TF-IDF: TF-IDF is a calculation of authenticity of a word by differentiating the number of times a word occurs in a document with the number of documents the word occurs.

$$\mathbf{TF - IDF = TF(term, document)\ X\ IDF\ (term)} \tag{1}$$

Where,

TF-Term Frequency: Number of times term t appears in a document d.

IDF - Inverse Document Frequency is given by:

$$\mathbf{IDF(term) = log\left[\frac{(1+n)}{(1+df(d,t))}\right]} \tag{2}$$

Where,

N - Number of documents.

df (document, term) - Document frequency of the term t.

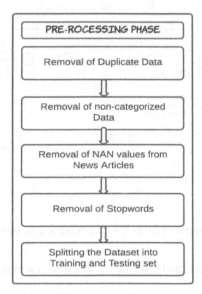

Fig. 1. Steps performed during pre-processing phase.

3.4 Evaluation Metrics

In order to evaluate classification models, accuracy is the one evaluation metric used. The accuracy is defined as the fraction of correct predictions of the proposed model.

$$\mathbf{Accuracy = \frac{Number\ of\ corrected\ predictions}{Total\ number\ of\ predictions}} \tag{3}$$

3.5 Models

Training, preprocessing and post processing are performed in cloud based environment called as 'Collaboratory' i.e. COLAB. The Pandas library, 'Jupyter Notebook' and the Python programming languages were used to develop the proposed model. The flow diagram of the proposed automatic fake news detection model is as shown in Fig. 2.

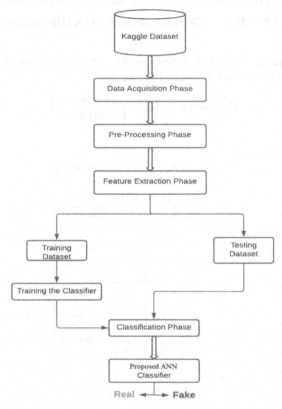

Fig. 2. The flow diagram of the proposed automatic fake news detection system.

Classification problems can be solved by Machine learning in two steps:

1. Learning the model from a training data set.
2. Classifying the hidden data on the basis of the trained model.

The way of organizing Machine learning algorithms is based on the desired result of the algorithm or the type of input available during training of the machine. The following are few of the Machine learning algorithms used for fake news detection and classification of articles for selected dataset [3].

(a) **Logistic Regression:** Logistic Regression algorithm is used to predict the probability of a categorical dependent variable. In Logistic Regression, the dependent variable is a binary variable that contains data coded as 1 (yes, real, etc.) or 0 (no, fake, etc.). In other words, the Logistic Regression model predicts P $(Y = 1)$ as a function of X. The obtained accuracy for the model is 94.4% with Liblinear solver approach as shown in Fig. 3.

(b) **Passive Aggressive Classifier:** Passive Aggressive Classifier (PAC) algorithm learns from the model by staying passive i.e., do nothing if prediction is correct else aggressive i.e. model tries to correct the prediction. The obtained accuracy for the model is 95.15% as shown in Fig. 4.

(c) **Naive Bayes:** Naive Bayes algorithm is a probabilistic classifier based on Bayes Theorem. Naïve Bayes classifiers are highly scalable, requiring a number of parameters linear in the number of variables (features/predictors) in a learning problem. The obtained accuracy for the model is 79.95% as shown in Fig. 5.

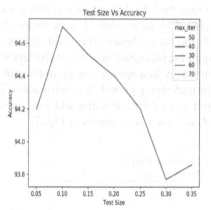
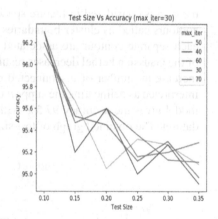

Fig. 3. Accuracy v/s test size in Logistic Regression algorithm.

Fig. 4. Accuracy v/s test size in Passive Aggressive Classifier.

(d) **Stochastic Gradient Descent Classifier:** Stochastic Gradient Descent (SGD) algorithm is a simple yet very efficient approach to fit linear classifiers and regressors under convex loss functions such as (Linear) Support Vector Machines and Logistic Regression. The obtained accuracy for the model is 93.8% as shown in Fig. 6.

(e) **Linear Support Vector Clustering (Linear SVC):** The purpose of a Linear SVC is to fit to the data provided, returning a "best fit" hyper plane that divides, or categorizes the provided data. After getting the "best fit" hyper plane, some features are fed to the classifier to check for predicted class.

In SVC model, data points are outlined from data space to a high dimensional feature space using a Gaussian kernel. The smallest sphere enclosing the image of

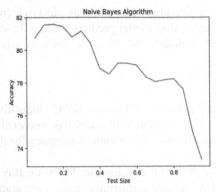

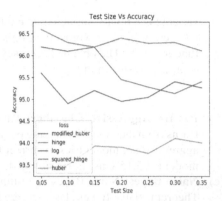

Fig. 5. Accuracy v/s test size in Naive Bayes algorithm.

Fig. 6. Accuracy v/s test size in Stochastic Gradient Descent algorithm.

the data is considered in feature space to map back to data space forming a set of contours called as cluster boundaries enclosed by data points. Points enclosed by each separate contour are associated with the same cluster. The width parameter of the Gaussian kernel decreases resulting in the increased number of clusters with increase in number of disconnected contours in data space. The contours can be interpreted as delineating the support of the underlying probability distribution. The model gives the accuracy 97.5% with split size of 90% training and 10% testing dataset. The obtained graph of test size v/s accuracy is as shown in Fig. 7.

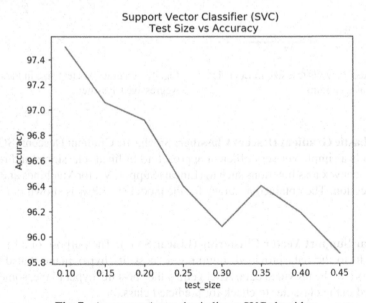

Fig. 7. Accuracy v/s test size in linear SVC algorithm

(f) **Artificial Neural Network (ANN):** Neural networks are the sequence of algorithms which mimic the functioning of the human brain to make predictions. In Neural Networks, the input layer computes the data and transfers it to the output layer to make predictions. The loss of the prediction is calculated by loss function algorithm 'binary cross entropy' which is used for binary classification models. Gradient descent algorithm results in decreasing the loss and directs the loss function towards global minima. The new weights will be updated in the nodes by back propagation as shown in Fig. 8.

The design of Artificial Neural Network architecture and specification of hyper parameters need to be set before training the model. The Table 3 below shows the hyper parameter values which are set before training the model.

In fake news detection, the dataset is label encoded to convert labels into binary data. The whole news text is vectorized using a TFIDF Vectorizer which assigns the values based on statistical weight age of word. The NumPy array input is fed to the Neural Network as input to make predictions.

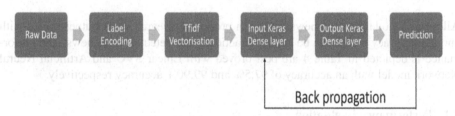

Fig. 8. Back propagation algorithm

The model gives the accuracy 99.90% and loss graphs of Neural Networks in fake news prediction is as shown in Fig. 9.

Table 3. Hyperparameter values of ANN

Name of the hyperparameter	Hyperparameter value
Learning rate	0.001
Batch size	64
Maximum number of epochs	40

While analyzing the relation between accuracy in both training and testing datasets, initially found the model is overfitting over the training dataset. So, added a dropout layer to the Neural Networks to reduce the over fitting.

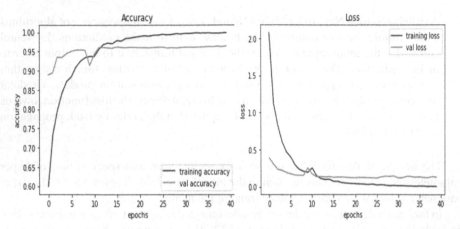

Fig. 9. Accuracy graph and loss graph of Neural Networks

4 Results and Performance Analysis

All trained models for the news dataset in proposed research work outperformed with enough accuracy which can be applied practically for detecting fake news. The performance is depicted in Table 4 are best proved with Linear SVC and Artificial Neural Network model with an accuracy of 97.5% and 99.90% accuracy respectively.

4.1 Performance Evaluation

The dataset is randomly divided into split ratio of 80:20 for training and testing datasets respectively with five-fold cross-validation, among which Linear SVC and ANN performed best compared with the other classifiers (e.g., Logistic Regression; Passive Aggressive Classifier, Stochastic Gradient Descent and Naïve Bayes).

The performance of the classifier is measured in terms of accuracy and result is presented in Table 4.

Table 4. Results of the proposed model for selected Kaggle dataset [3]

Classifier	Accuracy (%)
Logistic Regression (LR)	94.4
Passive Aggressive Classifier (PAC)	96.67
Naïve Bayes (NB)	79.95
Stochastic Gradient Descent (SGD)	96.88
Linear SVC	97.5
Artificial Neural Network (ANN)	99.90

(i) *Comparison of the Proposed Model with Work* [10].

The results show that the proposed model is slightly outperformed in comparison with other similar models for fake news early detection. The proposed Linear SVC and ANN classifiers reports 97.5% and 99.90% accuracy respectively as compared with the XGBoost and Random Forest classifier in work [10] with 89.2% and 84.5% accuracy respectively as depicted in Table 5.

Table 5. Comparison of the proposed model with work [10]

	Work [10]		Proposed model	
Classifier	XGBoost	Random Forest	Linear SVC	ANN
Accuracy	89.2%	84.5%	97.5%	99.90%

(ii) *Comparison of the Proposed Model with Work* [11].

In the proposed research work [11], results obtained by using Neural Network and LSTM classifiers are 93% and 95% accurate with the dataset fetched from Kaggle. As compared to the work proposed in [11], the proposed model outperformed with 97.5% and 99.90% accuracy with Linear SVC and ANN classifier respectively as depicted in Table 6.

Table 6. Comparison of the proposed model with work [11]

	Work [11]		Proposed model	
Classifier	Neural network with Kears	LSTM	Linear SVC	ANN
Accuracy	93%	95%	97.5%	99.90%

5 Conclusion

Fake news data has been categorized into three characteristics as (i) authenticity (ii) intention (iii) benign news. Further it has been categorized into (i) click bait (ii) sponsored content (iii) fabricated journalism. Fake news makes things more complicated and causes a change in people's mind about digital technology. The proposed model concludes by implementing fake news detection model with python coding. The proposed Artificial Neural Network model fits to the data provided. Passive Aggressive Classifier learns staying passive and Naïve Bayes Classifier uses probabilistic classification Bayes theorem. Stochastic Gradient Descent is a very efficient approach in fitting linear classifiers and regressors under Support Vector Machine and Logistic Regression.

Comparison of all the models is tabulated showing the accuracy levels. The proposed model's accuracy score is 99.90% and obtained by applying various NLP and Machine Learning techniques.

The research can be further carried out for various news datasets by using unsupervised machine learning classifiers. The proposed model can be improved by increasing the dataset size.

References

1. Srivastava, A.: Real time fake news detection using machine learning and NLP. Int. Res. J. Eng. Technol. (IRJET) **07**(06) (2020)
2. Waikhom, L., Goswami, R.S.: Fake news detection using machine learning. In: Proceedings of International Conference on Advancements in Computing & Management (ICACM) (2019)
3. https://www.kaggle.com/surekharamireddy/fake-news-detection
4. Zhou, X., Zafarani, R.: A survey of fake news: fundamental theories, detection methods, and opportunities. ACM Comput. Surv. (CSUR) **53**(5), 1–40 (2020)
5. O'Brien, N.: Machine learning for detection of fake news. Ph.D. dissertation, Massachusetts Institute of Technology (2018)
6. Sharma, K., Qian, F., Jiang, H., Ruchansky, N., Zhang, M., Liu, Y.: Combating fake news: a survey on identification and mitigation techniques. ACM Trans. Intell. Syst. Technol. (TIST) **10**(3), 1–42 (2019)
7. Santos, R., et al.: Measuring the impact of readability features in fake news detection. In: Proceedings of the 12th Language Resources and Evaluation Conference, pp. 1404–1413 (2020)
8. Amjad, M., Sidorov, G., Zhila, A.: Data augmentation using machine translation for fake news detection in the Urdu language. In: Proceedings of the 12th Language Resources and Evaluation Conference, pp. 2537–2542 (2020)
9. Wang, Y., et al..: Weak supervision for fake news detection via reinforcement learning. In: Proceedings of the AAAI Conference on Artificial Intelligence, vol. 34, no. 01, pp. 516–523 (2020)
10. Zhou, X., Jain, A., Phoha, V.V., Zafarani, R.: Fake news early detection: a theory-driven model. Digit. Threats Res. Pract. **1**(2), 1–25 (2020)
11. Pavan, M.N., Prasad, P.R., Gowda, T., Vibhakar, T.S., Shidnal, S.: Fake news detection using machine learning. Int. Res. J. Eng. Technol. (IRJET) **07**(12), 2151–2154 (2020)

Path and Information Content-Based Structural Word Sense Disambiguation

Sandip S. Patil$^{(\boxtimes)}$ ⓘ, R. P. Bhavsar ⓘ, and B. V. Pawar

School of Computer Sciences, K.B.C. North Maharashtra University, Jalgaon, MH, India
patil.sandip@sscoetjalgaon.ac.in, {rpbhavsar,bvpawar}@nmu.ac.in

Abstract. Word sense ambiguity is a fundamental characteristic of any natural language. Many words in every natural language have multiple meanings, termed as ambiguous words; ambiguous words impose a big challenge for natural language processing (NLP) applications. Word sense disambiguation (WSD) is the computational approach used to resolve these words sense ambiguities. Path-based and information content-based (IC-based) measures are the popular approaches in the similarity framework. Path-based approaches make use of geometrical distance and information content approaches make use of common information shared and measures similarity between the given concepts. We have studied, implemented, and evaluated various methods under these approaches in WSD over a dataset of 1000 English sentences, randomly picked from Times of India news headlines, this dataset covers 50 ambiguous words of English. The experimental results are presented, discussed and the experimentation has shown that the information content-based Lin similarity WSD approach outperforms all other approaches and yields an average precision of 0.77.

Keywords: Natural language processing · Word sense disambiguation · Word ambiguities · Computational linguistics · Path-based semantic similarity · Information content-based semantic similarity · WordNet

1 Introduction

Word sense ambiguity is a fundamental characteristic of any Natural Language. WSD task refers to identifying the most proper sense of an ambiguous word in given context [1, 14]. If there is an ambiguity in the meanings of a word or sentence, the human mind resolves such ambiguities using their perceived cognition and world knowledge, which is not the case with machines. WSD is an important problem in NLP, due to its impact on the output of NLP applications such as Machine Translation, Information Retrieval, Q&A systems. To illustrate, consider the translation of Marathi sentence: राहूल आणि आशा यांची जोडी अभंग आहे(Rāhūla āṇi āśā yāñcī jōḍī abhaṅga āhē. The pair of Rahul and Asha is unbroken) in English, Here the word in the Marathi sentence 'Abhang', has three senses viz. Type of poetry: man-made, Unbreakable: quality indicator and Intact: statistician each sense can lead to semantically different English sentences. Whereas native speakers can

© Springer Nature Switzerland AG 2021
K. R. Venugopal et al. (Eds.): ICInPro 2021, CCIS 1483, pp. 341–352, 2021.
https://doi.org/10.1007/978-3-030-91244-4_27

disambiguate it to the correct sense of being 'unbreakable', same issue can be found with other Marathi sentences like: पौर्णिमेचा चंद्र सोळा कला युक्त असतो (Paurṇimēcā candra sōḷā kalā yukta asatō) as: The full moon is full of sixteen arts; here the word sola kala assumes the sense of full moon night instead of sixteen types of arts. This multiple meaning phenomenon in the above subjects can also affect the IR and Q&A.

Knowledge-based and data-driven approaches are prescribed in the literature for attempting the computational solution to this ambiguity problem [18]. The knowledge-based WSD approaches include three types; selectional preference, similarity measure, and heuristics WSD. Data driven approaches include three types; supervised, semi-supervised, and unsupervised WSD. Our study focuses on the similarity-based structural WSD to a greater depth. The following section describes the detailed discussion on these followed by the prior art, various path, and IC-based similarity measures experimentation details and results.

2 Prior Art

During the study, we observed that similarity-based WSD is investigated over two decades. Similarity measure has two variations graph-based and structural WSD. Graph-based WSD constructs and exploits graph structures of context and conceptual networks, while structural WSD approach exploits hierarchical properties and relations in the nodes of conceptual networks [14].

Mihalcea et al., 2004 [12] used a graph based ranking algorithm PageRank for Word-Net and assigned a word meaning for open text WSD and achieved average accuracy of 67%. Navigli and Velardi, 2005 [16] explored semantic interconnections in the word sense of the WordNet and performed WSD. Navigli and Lapata, 2007 [15] explored semantic interconnections in the WordNet to perform WSD, they viewed WordNet as a graph in which nodes are synset concepts and edges are the semantic relations and they measured the connectivity for the task of WSD. They have reported average accuracy 68%. Tagarelli et al. 2009 [33] extracted structured features from XML data and generated the semantic relations for WSD. Aggire et al. 2010 [4] constructed a graph from the Unified Medical Language System and used PageRank algorithm for WSD in the metathesaurus. Agirre et al. 2014 [3] used random walks in personalized PageRank algorithm on large lexical knowledge-base for WSD. Pina and Johansson 2016 [29] applied the random walks on sense graph and generated word embedding for context and gloss for the task of WSD. Dongsuk et al. 2018 [17] built a semantic network BableNet and calculated word similarity between selected more important context words and sense definition for the task of WSD. In the literature, it is observed that, graph-based WSD exploits the entire knowledge-base, so it has broader coverage but it increases the computational complexity.

Figure 1 shows the overall taxonomy for structural WSD, it can be observed that structural WSD has three variations, viz. path-based, information content-based, and gloss overlapped-based WSD [14]. Banerjee and Pederson, 2003 [5] extended gloss overlapped in the Lesk algorithm [10] and computed the relatedness between the input synsets from the various relations in WordNet for WSD and achieved maximum accuracy of 34%. Patwardhan et al. 2003 [21], Patwardhan 2003a [19] clustered the context vectors along with semantic similarity and overcome the exact match issue with overlapped approach, achieved maximum accuracy of 39% and 50% respectively.

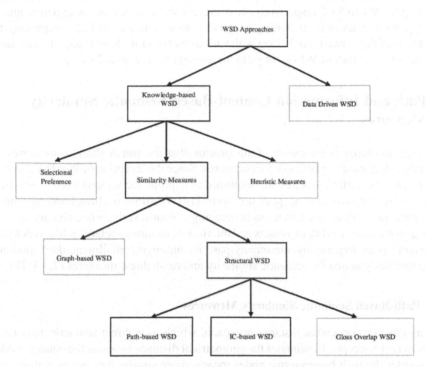

Fig. 1. Taxonomy of similarity-based WSD

Patwardhan et al. 2005 [22] proposed a framework for WSD using the semantic similarity between context bag and sense bag. Patwardhan and Pedersen 2006 [20] describe a WSD using second order gloss vectors along with semantic similarity measures from context words and achieved maximum accuracy of 45%. Pedersen et al. 2007 [27], Patwardhan et al. 2007 [23] used nearest word context selection algorithm for the selection of context bag along with relatedness measures for WSD. Pedersen 2012 [25] used gloss overlapped Lesk similarity measures for WSD [10] and achieved maximum accuracy of 50%. Pedersen 2016 [26] used Wu-palmer and Lesk semantic similarities between WordNet and other dictionaries for WSD and achieved maximum accuracy of 51%. Russer and Pedersen 2016 [32] induced new word sense in the WordNet used Lesk gloss overlap and Wu-Palmer semantic similarity and they achieved maximum accuracy of 59%. Bhingerdive et al. [6] used various features from IndoWordnet and developed API to calculate the path and IC-based semantic similarity and relatedness for Hindi and Marathi words.

In the literature, it is observed that, similarity-based structural WSD explores overlap between context bag and sense bag, which is surface level work and has limited sense coverage. Patwardhan et al. [21], Patwardhan and Pedersen [20], Russer and Pederson [32] etc. used path and IC-based similarity measures between context words and gloss vectors, these have not been semantically explored the context of a target word. As the target words, which has very low frequency of occurrence does not explore the information required to draw the proper sense, it limits the accuracy of path and IC-based WSD up to 39% to 59% respectively. To overcome these issues, in our experimentation strategy, we are not only exploiting the context word relations but also exploiting the synonyms of the context word along with various path and IC-based semantic similarity measures for the task of WSD and yield the average accuracy of 78%.

3 Path and Information Content-Based Semantic Similarity Measures

Semantic similarity is the extent of information that the words share; it measures the hierarchical relationship between the two words. Since the inception of WordNet, number of semantic similarity measures are developed to exploits the various word senses from the lexical relations in the conceptual taxonomy [14]. WordNet is a broad coverage online conceptual network of words in terms of meanings. Nouns, verbs, adjectives and adverbs are organized into network of synsets and interlinked in various relations like synonymy, hyponymy(is-a), hypernymy, meronymy(part of), antonym, entailment, derivation etc., it is particularly suited for semantic similarity and relatedness measures [2, 13, 28].

3.1 Path-Based Semantic Similarity Measures

It is an edge or conceptual distance approach, which uses direct semantic association between two concepts. It estimates the geometrical distance between the concept nodes. The shorter the path between two nodes means, more similar they are as compared to others [7].

Let c_1, c_2 be the two synsets in the WordNet and w_1, w_2 be the two words, Eq. (1) estimates the minimum path between two concepts and using pathlenght between c_1 and c_2 Eq. (2) estimate the path-based similarity between two words w_1 and w_2 [7].

$$sim_{path} = \min\left(\frac{1}{pathlength(c_1.c_2)}\right) \tag{1}$$

$$sim(w_1, w_2) = 2 \times d_{max} - \left[\min_{c1\varepsilon sense(w_1), c2\varepsilon sense(w_2)} pathlength(c_1, c_2)\right] \tag{2}$$

Where d_{max} is the maximum depth of WordNet taxonomy and $sense(w)$ denotes the set of possible senses for word w.

Rada Similarity
Rada et al. [30] proposed a distance metric on the power set of nodes. To calculate similarity measures, it uses a minimum number of edges between the concepts c_1 and c_2 from which Rada similarity score is calculated using Eq. (3)

$$score_{Rada}(c_1, c_2) = d(c_1, c_2) \tag{3}$$

Where, c_1 and c_2 are the concepts in the same context, $score_{Rada}$ is the similarity score between the concepts and $d(c_1, c_2)$ is the shortest distance.

Wu-Palmer Similarity

Wu & Palmer proposed path-based semantic similarity in the conceptual domain, it uses least common subsume node (LCS) for the selection of correct lexical choice. LCS is the common information carrier.

Wu-P similarity is calculated as shown in Eq. (4), which is the ratio of twice the depth of two concepts *lcs* with the depth of individual concepts [34].

$$sim_{Wup} = \frac{2\,X\,depth(lcs)}{sum(Individual\_dept(concepts)} \tag{4}$$

Equation (4) further can be simplified to calculate Wu-P similarity between concepts c_1 and c_2 as given in Eq. (5).

$$sim_{Wup}(c_1, c_2) = \frac{2\,X\,depth(lcs(c_1, c_2))}{depth(c_1) + depth(c_2)} \tag{5}$$

L-Ch Similarity

Leacock & Chodorow proposed hyperlink similarity and incorporated the level of concepts structure; in this, the ratio of shortest path between the two concepts to the twice of complete depth is used to calculate the path similarity [9], which is shown in Eq. (6).

$$sim_{lch} = -log(\frac{minpath(c_1, c_2)}{2\,X\,D}) \tag{6}$$

3.2 Information Content-Based Similarity Measures

It is node-based approach, similarity between the two concepts is the amount to which they share common information, which is called a common information carrier LCS. The similarity value is an information content value and is estimated by the probability of the occurrence of this class in the large text corpus [7].

As per the information theory the information content (IC) of the concept can be quantified as shown in Eq. (7)

$$IC(c) = log^{-1}P(c) \tag{7}$$

Where $P(c)$ is the times instance occurred in the training. It shows as the times of occurrence increase the information content get decreases. Probability of the root node is 1, so its information content will be 0.

The *IC* similarity between the two concepts c_1 and c_2 is calculated as shown in Eq. (8) and (9).

$$sim(c_1, c_2) = \max_{c \varepsilon Sup(c_1, c_2)} [IC(c)] \tag{8}$$

$$sim(c_1, c_2) = \max_{c \varepsilon Sup(c_1, c_2)} [-\log p(c)] \qquad (9)$$

Where, $Sup(c_1, c_2)$ is the common concept between c_1 and c_2.

If the word has more than one sense, word similarity among the multiple senses is calculated using Eq. (10)

$$sim(w_1, w_2) = \max_{c1 \varepsilon sense(w_1) c2 \varepsilon sense(w_2)} [sim(c_1, c_2)] \qquad (10)$$

Where, $senses(w)$ denotes the set of possible senses for *word (w)*.

Resnik Similarity

Resnik P. proposed information content-based measures for semantic similarity in IS-taxonomy, it uses the quantitative characterization of common information between the two words, and a commonality between the words determines the similarity between them [31].

It defines the concept formula as shown in Eq. (11)

$$freq(c) = \sum_{w \varepsilon words(c)} freq(w) \qquad (11)$$

Where $words(c)$ = set of words subsumed by the class c.

$classes(w)$ = classes in which the word is contained, it is the set of possible senses that the word w has in Eq. (12)

$$classes(w) = \{c | w\varepsilon \ words(c)\} \qquad (12)$$

So, $P(c)$ is calculated by Eq. (13)

$$P(c) = \frac{freq(c)}{N} \qquad (13)$$

Resnik similarity between two concepts is calculated from Eq. (12) and (13) as shown in Eq. (14).

$$sim_{res}(c_1, c_2) = IC(LCS(c_1, c_2)) = -\log(P(lcs(c_1, c_2))) \qquad (14)$$

Where lcs is the common information carrier between two concepts c_1 and c_2.

Lin Similarity

Lin proposed an information-theoretic approach for the probability model, this model is used for similarity measures, and it has universality and theoretical justification. The similarity between two words is determined by the ratio of their common information and individual information [11].

Lin similarity between two words w_1 and w_2 is shown in Eq. (15).

$$sim_{(w_1, w_2)} = \frac{2X \ \log P(c_0)}{\log P(c_1) + log P(c_2)} \qquad (15)$$

Where, $w_1 \varepsilon c_1$ and $w_2 \varepsilon c_2$ the commonality between w_1 and w_2 is $w_1 \varepsilon c_0 \hat{} w_2 \varepsilon c_0$, where, c_0 is the subsumes of c_1 and c_2.

So, Lin similarity between two concepts c_1 and c_2 is calculated as shown in Eq. (16).

$$Sim_{lin}(c_1, c_2) = \frac{2 X\ IC(lcs(c_1, c_2))}{IC(c_1) + IC(c_2)} \tag{16}$$

Where *lcs* is the common information carrier between two concepts c_1 and c_2.

J-Cn Similarity

Jiang & Conrath proposed combination of both edge based and node based similarity measures. It uses the link strength and local density of the node and overcomes the limitations in Resnik similarity [7].

The J-Cn similarity between the two concepts c_1 and c_2 is calculated as shown in Eq. (17).

$$sim_{jcn}(c_1, c_2) = IC\ (c_1) + IC(c_2) - 2 X\ IC(lcs(c_1, c_2)) \tag{17}$$

Where *lcs* is the common information carrier between two concepts c_1 and c_2.

4 Experimentation Strategy

We used path-based and information content-based similarity measures on WordNet3.1 along with context synonym generation for the structural WSD. As shown in Fig. 2 and procedure, we firstly extracted topic, ambiguous term from the context of an input sentence; lemmatize the target word followed by stemming using porter stemmer. As shown in Fig. 2 and procedure, in first pass, we extracted the synsets and glosses of the target word, which has the most frequent sense; this pass outputs a sense bag. In the next pass, we extracted the synsets and glosses of context words and their synonyms; this pass outputs the context bag. In last pass, we calculated the fine-grained composite similarity score between the sense bag and context bag, for measuring the similarity, we used path-based measures includes Wu-P, L-Ch similarity and IC-based measures includes Resnik, Lin, and J-Cn similarity and accordingly declared the proper sense and gloss of the given target word. We have conducted these experiments on the data set of 1000 sentences and evaluated the efficiency of each measure.

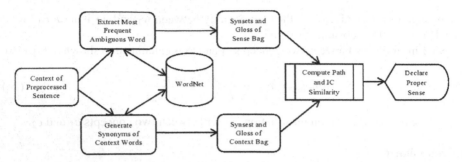

Fig. 2. Process flow of path and IC-based similarity structural WSD

```
Procedure: Structural_WSD (Sentence, Wn, Sim)
Input:   Sentence
Output: Proper_sense_with_similarity(score)
Steps:
    I ← Sentence
    L ← Lemmatize(I)
    Wn ← WordNet_Topology
    Answer    ← ∅
    For Wi in Wn do
        For Ii in L do
            C ← SelectContext(Ii, Context_Words)
            Ci ← Get_Sense_Set(Ii)
            Cj← Get_Sense_Set(C)
    Calculate Sim ← Similarity_Measures(Ci, Cj)
    For Wi in Wn do
        For Ii in L do
            C ← Select_Context(Ii, Context_Words)
            Cj← Get_Synonym(C)
            Wc ← Word_Count(Ci, Wi)
            Answer ← Sim + Wc
```

5 Experimental Setup and Test Beds

For our experimentation, we used Wordnet3.1 as a sense inventory, Punket tokenizer, Porter stemmer, synonyms generator, Wu-Palmer, L-Ch, Resnik, Lin and J-Cn similarity measures, and tested the methodology on the test beds of 1000 sentences randomly picked from India news headlines, which covers 50 ambiguous words and recorded the performances of all path and IC-based similarity measures.

6 Results and Discussion

We have implemented path-based and IC-based WSD for 50 ambiguous words of 1000 English sentences randomly selected from Times of India news headlines [8], which has average context size of 5 words and for evaluating the performance, precision, recall and F1-score for each similarity-based WSD, has been recorded and reported in Table 1 and in Fig. 3, 4 and 5.

Table 1. Performance evaluation of path-based and IC-based similarity measures for WSD

Evaluation parameter (Average value)	Path-based Similarity for WSD		Information content-based similarity for WSD		
	Wu-P	L-Ch	Resnik	Lin	J-Cn
Precision	0.720	0.700	0.710	0.770	0.760
Recall	0.790	0.795	0.765	0.830	0.810
F1-scor	0.741	0.742	0.723	0.789	0.775

Figure 3 shows precision i.e. how many declared senses are relevant (correct), this graph shows that Lin similarity yield a precision of 0.77, which means the highest no of senses declared which are relevant by Lin similarity measures.

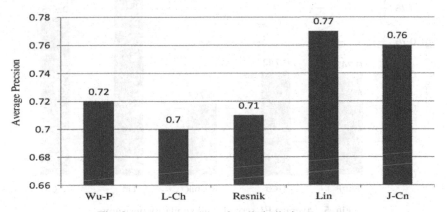

Fig. 3. Average precision for all similarity measures

Figure 4 shows recall i.e. how many relevant (correct) senses are correctly declared, this graph shows that Lin similarity yield recall of 0.83, it means highest number of correct senses are declared by Lin. similarity measures.

Figure 5 shows the F1-score i.e. the weighted average for precision and recall for all similarity measures, it shows Lin similarity has yielded the highest F1-score of 0.789, because it uses information content probability of locally occurring features of the ambiguous terms, which provides a sufficient distinction required to disambiguate.

It is also observed that in the disambiguation for the cases for the terms like; space and development, path similarity provides the highest F1-score of 0.92, it is because, for these specific terms information about the taxonomy is rich in nature as these are not only the leaf terms, but also has the concrete taxonomy in the WordNet, which is not the case

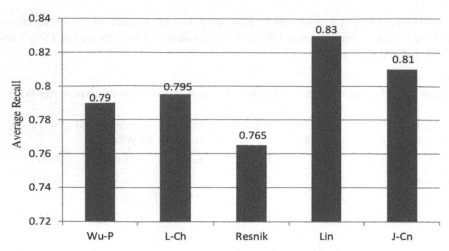

Fig. 4. Average recall for all similarity measures

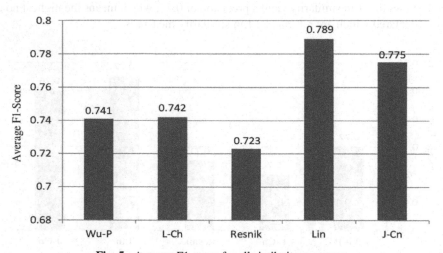

Fig. 5. Average F1-score for all similarity measures

with major terms. Lin similarity targets in quite different angles than other similarities, it uses a probability model, which is theoretically sound and outperforms, while path-based measures are highly dependent on network hierarchy. The information content measures do not require a detailed structure of the taxonomy, so it is not sensitive to the irregular densities of concept links. Pedersen 2010 [24], practically compared path length and information content similarity measures, path length between specific concepts imply smaller distinctions than the general concepts, so path length has limited use for WSD. Once we increase the coverage of concept frequency count, the performance of IC-based measures will get improved for WSD.

7 Conclusion and Future Work

Path and IC-based similarity measures explore lexical structures like WordNet for structural WSD, so it does not require labeled training data. Recorded precision, recall, and F1-score of evaluated experiments indicate that, proposed WSD approach can achieve disambiguation task effectively. As IC-based similarity uses an empirically sound probability model, so it has obtained significantly higher disambiguation accuracy than path-based similarity measures. In the future, IC-based similarity measure can also be used to disambiguate the terms in various Indian languages.

References

1. Agirre, E., Edmonds, P.: Word Sense Disambiguation: Algorithms and Applications, vol. 33. Springer, New York (2007)
2. Agirre, E., Rigau, G.: Word sense disambiguation using conceptual density. In: 16th International Conference on Computational Linguistics, Copenhagen, pp. 16–22 (1996)
3. Agirre, E., de Lacalle, O.L., Soroa, A.: RandomWalks for knowledge-based word sense disambiguation. Comput. Linguist. **40**(1), 58–84 (2014)
4. Agirre, E., Soroa, A., Stevenson, M.: Graph-based Word Sense Disambiguation of biomedical documents. Bioinformetics **26**(22), 2889–2896 (2010)
5. Banerjee, S., Pedersen, T.: Extended gloss overlaps as a measure of semantic relatedness. In: Eighteenth International Joint Conference on Artificial Intelligence, Mexico, pp. 805–810 (2003)
6. Bhingardive, S., Redkar, H., Sappadla, P., Singh, D.: IndoWordNet: similarity – computing semantic similarity and relatedness using IndoWordNet. In: Proceedings of the 8th Global WordNet Conference, Bucharest, Romania, pp. 39–43 (2016)
7. Jiang, J., Conrath, D.: Semantic similarity based on corpus statistics and lexical taxonomy. In: International Conference Research on Computational Linguistics, Taiwan, pp. 1–15 (1997)
8. Kulkarni, R.: Times of India News Headlines. V2 (2020)
9. Leacock, C., Chodorow, M.: Combining local context and WordNet similarity for word sense identification. In: Fellbaum, C. (ed.) WordNet: An Electronic Lexical Database, pp. 265–283. MIT Press, Cambridge, MA (1998)
10. Lesk, M.: Automatic sense disambiguation using machine readable dictionaries: how to tell a pine cone from an ice cream cone. In: Proceedings of the 5th SIGDOC, pp. 24–26. ACM, New York (1986)
11. Lin, D.: An information-theoretic definition of similarity. In: International Conference on Machine Learning, Madison, pp. 296–304 (1998)
12. Mihalcea, R., Tarau, P., Figa, E.: PageRank on semantic networks, with application to word sense disambiguation. In: 20th International Conference on Computational Linguistics, pp. 1126–1132. ACL, Geneva (2004)
13. Miller, G., Beckwith, R., Fellbaum, D., Miller, K.: Introduction to WordNet: an on-line lexical database. Int. J. Lexicogr. **3**, 235–244 (1990)
14. Navigli, R.: Word sense disambiguation: a survey. ACM Comput. Surv. **41**(2), 1–69 (2009)
15. Navigli, R., Lapata, M.: Graph connectivity measures for unsupervised word sense disambiguation. In: 20th International Joint Conference on Artificial Intelligence, IJCAI 2007, Hyderabad, India, pp. 1683–1688 (2007)
16. Navigli, R., Velardi, P.: Structural semantic interconnections: a knowledge-based approach to word sense disambiguation. IEEE Trans. Pattern Anal. Mach. Intell. **27**(2), 1075–1086 (2005)

17. Dongsuk, O., Kwon, S., Kim, K., Ko, Y.: Word sense disambiguation based on word similarity calculation using word vector representation from a knowledge-based graph. In: 27th International Conference on Computational Linguistics, pp. 2704–2714. ACL, Mexico (2018)
18. Patil, S.S., Bhavsar, R.P., Pawar, B.V.: Contrastive study and review of word sense disambiguation techniques. Int. J. Emerg. Technol. **11**(4), 96–103 (2020)
19. Patwardhan, S.: Incorporating Dictionary and Corpus Information into a Context Vector Measure of Semantic Relatedness. University of Minnesota, Duluth (2003)
20. Patwardhan, S., Pedersen, T.: Using WordNet-based context vectors to estimate the semantic relatedness of concepts. In: EACL 2006 Workshop Making Sense of Sense, 4 April 2006, pp. 1–8 (2006)
21. Patwardhan, S., Banerjee, S., Pedersen, T.: Using measures of semantic relatedness for word sense disambiguation. In: Fourth International Conference on Intelligent Text Processing and Computational Linguistics, Mexico City, pp. 241–257 (2003)
22. Patwardhan, S., Banerjee, S., Pedersen, T.: Sense relate::target word – a generalized framework for word sense disambiguation. In: Twentieth National Conference on Artificial Intelligence, Pittsburgh, pp. 73–76 (2005)
23. Patwardhan, S., Banerjee, S., Pedersen, T.: UMND1: unsupervised word sense disambiguation using contextual semantic relatedness. In: SemEval-2007, 23–24 June 2007, pp. 390–393 (2007)
24. Pedersen, T.: Information content measures of semantic similarity perform better without sense-tagged text. In: NAACL HLT 2010, pp. 329–332. ACL, Los Angeles (2010)
25. Pedersen, T.: Duluth: measuring degrees of relational similarity with the gloss vector measure of semantic relatedness. In: First Joint Conference on Lexical and Computational Semantics, Montreal, pp. 497–501 (2012)
26. Pedersen, T.: Duluth at SemEval–2016 task 14 extending gloss overlaps to enrich semantic taxonomies. In: SemEval 2016, pp. 1328–1331, June 2016
27. Pedersen, T., Pakhomov, S., Patwardhan, S., Chute, C.: Measures of semantic similarity and relatedness in the biomedical domain. J. Biomed. Inform. **40**, 288–299 (2007)
28. Pedersen, T., Patwardhan, S., Michelizzi, J.: WordNet::similarity - measuring the relatedness of concepts. In: NAACL 2004, pp. 1024–1025. ACL, Boston (2004)
29. Pina, L.N., Johansson, R.: Embedding senses for efficient graph-based word sense disambiguation. In: 2016 Workshop on Graph-based Methods for Natural Language Processing, pp. 1–5. ACL, San Diego (2016)
30. Rada, R., Milji, H., Bickneli, E., Blettner, M.: Development and application of a metric on semantic nets. IEEE Trans. Syst. Man Cybern. **19**, 17–30 (1989)
31. Resnik, P.: Using information content to evaluate semantic similarity in a taxonomy. In: IJCAI, Montreal, Canada, pp. 448–453 (1995)
32. Russer, J., Pedersen, T.: UMNDuluth at SemEval-2016 task 14: wordnet's missing lemmas. In: SemEval 2016, pp. 1346-1350 June 2016
33. Tagarelli, A., Longo, M., Greco, S.: Word sense disambiguation for XML structure feature generation. In: Aroyo, L., et al. (eds.) ESWC 2009. LNCS, vol. 5554, pp. 143–157. Springer, Heidelberg (2009). https://doi.org/10.1007/978-3-642-02121-3_14
34. Wu, Z., Palmer, M.: Verb sematics and lexical selection. Mach. Transl. 133–138 (1995)

ERDNS: Ensemble of Random Forest, Decision Tree, and Naive Bayes Kernel Through Stacking for Efficient Cross Site Scripting Attack Classification

A. Niranjan[1]($\boxtimes$), K. M. Akshobhya[2], Arun Singh Chouhan[3],
and Praveen Tumuluru[4]

[1] Gokaraju Rangaraju Institute of Engineering and Technology,
Hyderabad 500090, India
[2] SAP Labs Pvt Ltd., Bangalore, India
[3] Malla Reddy University, Hyderabad, India
[4] Koneru Lakshmaiah Education Foundation, Vijayawada, India

Abstract. Cross-Site Scripting (XSS) is a form of client-side code injection attack in which an attacker attempts to execute malicious scripts in the victim's web browser by embedding dangerous code in a legitimate web page or application. During such an attack, the attacker impersonates a victim user and performs any behavior that the user is capable of, as well as accessing any of his data. If the victim user has privileged access to the application, the attacker can take complete control of the app's features and data. XSS attacks are most common on message boards, forums, and websites that accept comments. According to a study from CDN and cloud security provider CDNetworks, attacks on web applications increased by 800% in the first six months of 2020 compared to the same time frame last year. Our approach involves the application of Machine Learning Algorithms for efficient classification of XSS attacks. The MMR and SDMR Algorithms aid in the selection of significant features from a data set without compromising classification results. The proposed ERDNS framework ensures better Detection Accuracy and Precision with least Classification Error values among all the available models.

Keywords: Cross-site scripting attack classification · Ensemble learning · Hybrid feature selection · Managing XSS attacks

1 Introduction

As per the recent Research Reports, XSS constitutes close to 40% of the total cyber attacks. According to a study from CDN and cloud security provider CDNetworks, attacks on web applications increased by 800% in the first six months of 2020 compared to the same time frame last year. Cross-site Scripting

K. R. Venugopal et al. (Eds.): ICInPro 2021, CCIS 1483, pp. 353–365, 2021.
https://doi.org/10.1007/978-3-030-91244-4_28

is a form of client-side code injection attack in which an attacker attempts to execute malicious scripts in the victim's web browser by embedding malicious code in a legitimate web page or application (XSS). During such an attack, the attacker impersonates a victim user and performs any behavior that the user is capable of, as well as accessing any of his data. If the victim user has privileged access to the application, the attacker can take complete control of the app's features and data. XSS attacks are most common on message boards, forums, and websites that accept comments. Based on the origin of malicious script that is responsible for the attack, we have three XSS attack types namely, a) Reflected XSS, b) Stored XSS, and c) DOM-based XSS.

Effective XSS Attack Classification at an earlier stage only can facilitate the XSS Detection process. That is, when a sample could be accurately classified as attack or benign, the XSS Detection Mechanism could stop an XSS attack on its very onset. The application of Machine Learning techniques has been proven to be an effective tool for any classification problem. Use of Hybrid models involving Multi-layered classification schemes is not a new idea in research. The current study looks at and explores a few hybrid models that use multi-layered classification to predict XSS samples effectively. We investigate two Ensemble Schemes, Stacking and Voting, for implementing the Ensemble model, and the most efficient among them is proposed. Until classifying the data set, we investigate various Feature Selection Algorithms to compute the Rank of each feature present in it. Mean and Standard Deviation values for the Ranks generated by various Feature Selection Algorithms are then computed. Finally, a Mean of Mean of Ranks (MMR) and Standard Deviation of Mean of Ranks (SDMR) are calculated for each feature. All features with a Mean and Standard Deviation less than the final MMR and SDMR are removed, leaving behind two Feature subsets with most significant and meaningful features. In addition, these subsets of Features are intersected to determine significant features for classification. As a result, the Time Complexity for classification is greatly reduced. Various built in classifiers are run on the feature subsets obtained after MMR and SDMR and their Intersection. Top three classifiers in terms of chosen performance metrics are identified.

Such classifiers are then subjected to ensemble models such as Stacking and Voting. The performance metrics are again recorded for the two ensemble models. The top performing model between the two, involving the top three Classifiers is finally chosen as our proposed model for efficient classification of XSS attack samples. When a hybrid model combining Random Forest, Decision Tree, and Naive Bayes Kernel is ensembled using Stacking, we expect results that are better in terms of the chosen metrics such as Prediction Accuracy, Classification Error, and Precision. We name the technique as ERDNS (Ensemble of Random Forest, Decision Tree, and Naive Bayes Kernel through Stacking). To demonstrate the efficacy of the proposed ERDNS process, extensive tests were performed on publicly accessible XSS data sets by Fawaz Mokbal et al. [4], namely XSSdataset1engineered.csv and XSSdataset2engineered.csv, which contain 101000 and 140660 instances, respectively. ERDNS outperforms the existing Models, according to the experimental results obtained on the data sets.

A type of Decision Trees that is based on a randomly selected subset of attributes is referred to as a Random Tree. A Decision Tree basically is a set of nodes and branches. A test on an attribute is represented by a node, while the result of a test is represented by a branch. The external nodes (leaves) reflect the final decision. The route from the root to the leaf creates a classification law. A Random Forest is a classifier that is made up of many such individual Random trees. Each Random Tree in a Random Forest is constructed by using Bagging and Feature Randomness to ensure that there is no connection between the trees. Since the Random Forest bases its final prediction on the highest number of votes received for a particular prediction, it is more accurate than any Random Tree.

Naive Bayes is a supervised learning algorithm for solving classification problems that is based on the Bayes theorem. It is a simple and effective classification algorithm that aids in the development of fast machine learning models capable of making quick predictions. It's a probabilistic classifier, which means it makes predictions based on an object's likelihood. The Naive Bayes classifier has the advantage of providing very few training data to estimate the mean and variances of the variables used for classification. Since independent variables are assumed, only the variances of the variables for each label, not the entire covariance matrix, must be calculated. The Naive Bayes Kernel operator, unlike the Naive Bayes operator, can be applied to numerical attributes. In non-parametric estimation techniques, a kernel is a weighting function. The density functions of random variables, as well as the conditional expectation of a random variable, are estimated using kernels. Kernel density estimators belong to the non-parametric class of estimators. Such estimators have no fixed structure and depend on all data points to achieve an approximation, unlike parametric estimators, which have a fixed functional type (structure) and the parameters of this function are the only information we need to store.

Few of the popularly used Ensemble Techniques are Voting, Stacking, StackingC, Bagging, Boosting and Grading. Voting entails creating a variety of submodels and considering the predictions of each of the submodels to decide what the final outcome should be. Stacking involves separate and independent training of heterogeneous learning algorithms on the data, and using their results as additional inputs to the combiner algorithm for the final training. StackingC, a variant of Stacking on the other hand, uses Linear Regression as the Meta Classifier. Linear regression is a method for converting a set of numeric values (x) into an estimated output value (y). Grading is one of the meta classification methods, and it involves finding and correcting any incorrect predictions. Instead of using the predictions of base classifiers as metalevel attributes as in Stacking, Grading uses graded predictions (correct or incorrect).

This article's contributions can be summarised as follows:

- The evaluation of a novel general-purpose classifier architecture based on the Hybrid approach on two data sets is presented.
- MMR and SDMR Algorithms are proposed for efficient Feature Selection.
- Performance Metrics considered and the results obtained for the individual Classifiers and also the Ensemble Models are presented.

The following is how the rest of the article is structured. Section 2 discusses related work in this area, while Sect. 3 introduces the proposed ERDNS structure. The investigative approach is detailed in Sect. 4, and the experimental findings are presented in Sect. 5, along with reviews and discussions. The last section of this paper is the conclusion.

2   Related Work

The focus of the case study presented in [1] is on current methodologies and methods for detecting and mitigating XSS attacks and vulnerabilities. It discusses both client and server-side detection methods, as well as static and dynamic approaches, that have been proposed to detect the attack thus far. It also covers the XSS Defense methods and techniques that are used to secure both the Client and Server. The aim of this paper is to take a close look at how to mitigate XSS attacks using various detection and defence techniques.

The paper proposed by X Zhang et al. [2] discusses about MCTS-T adversarial example generation algorithm that enables the generation model to provide a reward value. The value thus obtained reflects the probability of generative examples bypassing the detector. The authors further optimize their XSS detection model with GAN(Generative Adversarial Network) to enhance its ability to defend against adversarial examples. The disadvantage of MCTS-T algorithm is that it can only generate adversarial examples of XSS traffic at present.

Various Supervised Machine Learning Algorithms such as Support Vector Machine, Decision Tree, Naive Bayes and Un-Supervised Algorithms such as K-Means and Association Rule algorithms along with a Deep Learning Technique called Long Short-Term Memory Algorithms are explored by XiaoLong Chen et al. [3] who have listed out the advantages and disadvantages of these techniques in their work.

The authors in their survey [6] explore the background of XSS attacks, classify the derived types of XSS attacks, the role of cookies and a bunch of tools and methods that can be used for the detection and mitigation of XSS Attacks in detail.

To improve the XSS detection in a low-resource data setting, Mokbal FMM et al. [7] propose a conditional Wasserstein Generative Adversarial Network with a gradient penalty. Their method creates synthetic minority class samples with the same distribution as actual XSS attack scenarios. They use augmented data to train a new boosting model, which is then tested on a real-world data set. Their model produces consistent and accurate samples.

In the work discussed in [8] two approaches are proposed. The first approach employs insecure flow monitoring to filter malicious scripting code inserted into dynamic web pages while the second creates and validates trusted remark for detecting any suspicious activity in static web pages. Finally, the user is presented with the filtered and updated webpage. They evaluate their prototype by testing with a collection of real-world web applications to see whether it could detect and mitigate XSS attacks.

The aim of study carried out in the PG thesis by [9] is to see whether an XSS vulnerability scanner can be rendered more versatile than what is currently available. The aim is to see how reflected parameters can be identified and whether a different approach can be used to enhance XSS vulnerability detection. Mantis methodology basically focuses on the set of characters that could lead to XSS manipulation.

3   ERDNS: Efficient XSS Attack Classification Framework

The Preprocessing and Classification phases will be the focus of the majority of Machine Learning applications. It is a proven fact that a classifier doesn't need all features to make the final prediction. So picking up of only significant features from the data set becomes an important step in the preprocessing phase. We suggest MMR and SDMR algorithms for the selection of Features by leveraging the various advantages of built in Ranking Algorithms that are based on weights. The classification process entails applying the proposed ERDNS framework to the resultant Feature subset through tenfold cross validation using Random Forest, Decision Tree, and Naive Bayes Kernel ensembled via Stacking and various other techniques. The ERDNS framework is depicted in Fig. 1. The data sets XSSdatasetengineered1.csv and XSSdatasetengineered2.csv from Fawaz Mokbal et al. [4] that contain publicly accessible XSS samples collections are used. The experimentation is carried out on these files containing 101000 and 140660 instances, respectively.

The proposed MMR (Mean of Mean of Ranks) Feature Selection algorithm [5] as stated in Algorithm 1 and SDMR (Standard Deviation of Mean of Ranks) as listed in Algorithm 2 employ many existing Feature Selection Algorithms for the determination of Ranks of every Feature. The Mean and Standard Deviations are calculated using the rank values provided by various Algorithms for a Feature. All features with Mean of Ranks less than the Overall Mean value and less than the final Standard Deviation of Mean of Ranks are discarded, leaving only valid and important feature subsets for classification. For the determination of the Ranks R_i, Algorithms such as InfoGain, InfoGainRatio, Rule method, Deviation method, Correlation method, Chi-squared statistics, Gini-Index and Principal Component Analysis are employed. The MMR for XSSdatasetengineered1 is 0.05 and that of XSSdatasetengineered2 is 0.12. The number of Features selected from XSSdatasetengineered1 using MMR is 15 out of 67 while features selected from XSSdatasetengineered2 is 28 out of 77 features. The SDMR for XSSdatasetengineered1 is 0.12 and that of XSSdatasetengineered2 is

0.10. The number of features selected from XSSdatasetengineered1 using SDMR is 14 out of 67 while features selected from XSSdatasetengineered2 is 30 out of 77 features. To further determine the most significant features selected from both the subsets of features obtained from MMR and SDMR, it was decided to perform an intersection operation on these subsets. After this operation, 14 Features were picked from XSSdatasetengineered1 and 26 from XSSdatasetengineered2 as listed in Table 1.

Algorithm 1: MMR (Mean of Mean of Ranks) Feature Selection

1: Input: Data set D having n Features
2: Output: Subset of Most Significant Features (F_1) Present in D
3: **for** every feature $f_i \in D$ **do**
4:     Calculate Ranks R_i using various Feature Weight Calculation Techniques
5: **end for**
6: **for** each feature $f_i \in D$ **do**
7:     Determine Sum ($\sum R_i$) and Mean (m R_i) of Ranks
8:     $\sum R_i = R_1 + R_2 + ... + R_n$ and mR $_i = \sum R_i$ / n
9: **end for**
10: Calculate the Mean of Mean of Ranks M(mR$_i$)
11: Discard all $f_i \in D <$ M(mR$_i$)
12: return F_1

Algorithm 2: SDMR (Standard Deviation of Mean of Ranks) Feature Selection

1: Input: Data set D having n Features
2: Output:The subset of Most Significant Features (F_2) Present in D
3: **for** each feature $f_i \in D$ **do**
4:     Calculate Ranks R_i using various Rank Based Feature Selection Techniques
5: **end for**
6: **for** each feature $f_i \in D$ **do**
7:     Determine Mean (mR$_i$) of the feature
8: **end for**
9: Determine Standard Deviation of Mean of Ranks σ_f (mR$_i$)
10: Discard all $f_i \in D < \sigma_f($ mR$_i$)
11: return F_2

Algorithm 3: Ensemble of Random Forest, Decision Tree, and Naive Bayes through Voting for Efficient Cross Site Scripting Attack Classification (ERDNS)

1: Input: Features subsets F_1 and F_2 obtained from Algorithm 1 and 2 respectively

2: Output: Classification Results

3. Perform the intersection ($\cap$) of the feature subsets F_1 and F_2

3: Subject the result to an Ensemble of Random Forest, Decision Tree,
   and Naive Bayes through Stacking

4: The model is subjected for a ten-fold cross validation
   to estimate the performance metrics of classification

The proposed MMR and SDMR for Feature Selection are presented as Algorithm 1, and Algorithm 2 respectively. Algorithm 3 presents the ERDNS framewoek for efficient XSS Attack Classification. Table 1 lists the Final feature subset obtained after the Intersection of Feature subsets generated by MMR and SDMR.

Figure 1 sketches the System Model of the proposed ERDNS for XSS Attack Classification.

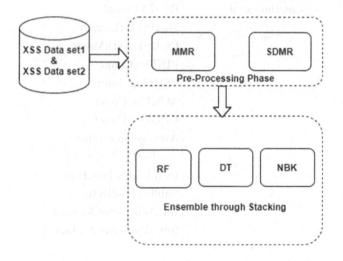

Fig. 1. XSS Attack Classification Model

4   Investigation Methodology

The data sets from publicly available XSS samples collections by Fawaz Mokbal et al. [4] namely XSSdataset1engineered.csv and XSSdataset2engineered.csv consisting of 101000 and 140660 instances respectively are used for the experimentation purpose. The proposed MMR (Mean of Mean of Ranks) as listed in Algorithm 1 and SDMR (Standard Deviation of Mean of Ranks) as listed in

Table 1. Final Feature subset after Intersection operation of MMR and SDMR

Data set-1	Data set-2
url_length	DestinationPort
url_special_characters	FlowDuration
url_tag_script	TotalLengthofFwdPackets
url_tag_iframe	FwdPacketLengthMax
url_attr_src	FwdPacketLengthMean
url_event_onload	BwdPacketLengthStd
url_event_onmouseover	FlowIATMean
url_cookie	FlowIATStd
url_number_domain	FlowIATMax
html_event_onblur	FwdIATTotal
js_file	FwdIATMean
js_pseudo_protocol	FwdIATStd
js_method_alert	FwdIATMax
js_method_eval	BwdIATTotal
	MaxPacketLength
	PacketLengthMean
	FINFlagCount
	PSHFlagCount
	ACKFlagCount
	URGFlagCount
	AveragePacketSize
	AvgFwdSegmentSize
	FwdAvgPackets_Bulk
	SubflowFwdBytes
	Init_Win_bytes_forward
	Init_Win_bytes_backward

Algorithm 2 employ several existing Feature Selection Algorithms for computing the ranks of all the features that are present in the data set. Finally, an overall Mean and Standard Deviation for the Mean of Ranks for all features are computed. All features with Mean of Ranks less than the Overall Mean value and less than the final Standard Deviation of Mean of Ranks are discarded, leaving only valid and important feature subsets for classification. For the determination of the Ranks Ri Algorithms such as InfoGain, InfoGainRatio, Rule method, Deviation method, Correlation method, Chi-squared statistics, GiniIndex and Principal Component Analysis are employed. The MMR for XSSdatasetengineered1 is 0.05 and that of XSSdatasetengineered2 is 0.12. The number of Features selected from XSSdatasetengineered1 using MMR is 15 out of 67 (77.6% Reduction)while features selected from XSSdatasetengineered2 is 28 out of 77 features(63.63% Reduction). The SDMR for XSSdatasetengineered1 is 0.12 and that of XSSdatasetengineered2 is 0.10. The number of features selected from XSSdatasetengineered1 using SDMR is 14 out of 67 (79.1% Reduction)while features selected from XSSdatasetengineered2 is 30 out of 77 features(61% Reduction). To further determine the most significant features selected from both the subsets of features obtained from MMR and SDMR, it was decided to perform an intersection operation on these subsets. After this operation, 14 Features were picked from XSSdatasetengineered1(79.1% Reduction) and 26 from XSS-datasetengineered2 (66.23% Reduction) as listed in Table 1. The performance metrics were recorded after a thorough ten-fold cross validation. A ten-fold cross validation technique typically involves dividing the data set into ten sections, training the model with the nine parts of the data while using the excluded section as the test set, and repeating the process for ten iterations, using each unused test set during each round. Prediction Accuracy, Classification Error, Recall and Precision values are used as the Performance Metrics for determining the efficiency of the classifiers. Two Ensemble approaches namely Voting and Stacking are tried to enhance the Performance of Classification on the top performing classifiers.

If a classifier has a higher true positive rate and a lower false positive rate, it is considered efficient. There are seven efficiency criteria in a traditional classification methodology, which we will discuss below. N_{Ben} is the number of benign samples in the XSS data set, while N_{XSS} is the number of XSS samples. $N_{Ben}rightarrow_{Ben}$ is the number of benign samples correctly identified as benign (TP). True Negative (TN) is the number of XSS samples correctly defined as XSS. $N_{XSS}\rightarrow_{XSS}$ is the symbol for it. False Positive (FP) is a metric for Benign samples that have been mislabeled as XSS. $N_{Ben}\rightarrow_{XSS}$ is the abbreviation, and False Negative (FN) is a metric for XSS incidents mis-classified as Benign. $N_{XSS}\rightarrow_{Ben}$ is the symbol for it. The Detection Rate (DR) refers to the percentage of XSS samples that are detected and correctly identified as XSS.

$$TP = \frac{N_{Ben \rightarrow Ben}}{(N_{Ben \rightarrow Ben} + N_{XSS \rightarrow Ben})} \tag{1}$$

The percentage of benign samples wrongly classified as XSS is known as the False Positive Rate (FPR).

$$FPR = \frac{N_{Ben \rightarrow XSS}}{(N_{XSS \rightarrow XSS} + N_{Ben \rightarrow Ben})} \tag{2}$$

The False Negative Rate (FNR) is the percentage of XSS samples that are wrongly classified as benign when they are not.

$$FNR = \frac{N_{XSS \rightarrow Ben}}{(N_{XSS \rightarrow XSS} + N_{XSS \rightarrow Ben})} \tag{3}$$

The True Negative Rate (TNR) is the percentage of Benign samples that are correctly classified as Benign out of all available Benign samples.

$$TNR = \frac{N_{\text{Ben}\to\text{Ben}}}{(N_{\text{Ben}\to\text{Ben}} + N_{\text{Ben}\to\text{XSS}})} \tag{4}$$

The total number of XSS and Benign samples correctly identified in comparison to the total number of all available instances is referred to as Prediction Accuracy (PA).

$$PA = \frac{(N_{\text{Ben}\to\text{Ben}} + N_{\text{XSS}\to\text{XSS}})}{(N_{\text{Ben}\to\text{Ben}} + N_{\text{XSS}\to\text{XSS}} + N_{\text{Ben}\to\text{XSS}} + N_{\text{XSS}\to\text{Ben}})} \tag{5}$$

The number of true positives divided by the total number of instances listed as positive is Precision.

$$PREC = \frac{N_{\text{XSS}\to\text{XSS}}}{(N_{\text{XSS}\to\text{XSS}} + N_{\text{Ben}\to\text{XSS}})} \tag{6}$$

Recall is defined as the number of true positives divided by the total number of instances that truly belong to the positive class.

$$REC = \frac{N_{\text{XSS}\to\text{XSS}}}{(N_{\text{XSS}\to\text{XSS}} + N_{\text{XSS}\to\text{Ben}})} \tag{7}$$

Figure 5 depicts the performance comparison of various Classifiers in terms of Precision while the other metrics such as Classification Error, Recall and Prediction Accuracy are depicted in Figs. 2, 3, and 4 respectively. From Fig. 2 we can make out that the Classification Error rate is the least in case of Random Forest and Decision Tree while slightly larger in Naive Bayes Kernel. However, these Classifiers record better Recall (as seen in Fig. 3), Prediction Accuracy

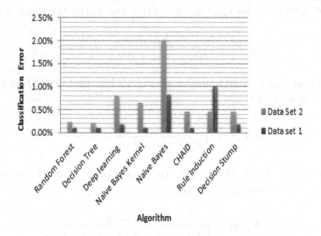

Fig. 2. Classification error comparison of classifiers

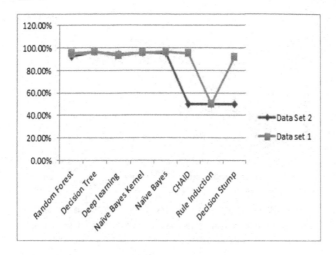

Fig. 3. Recall comparison of classifiers

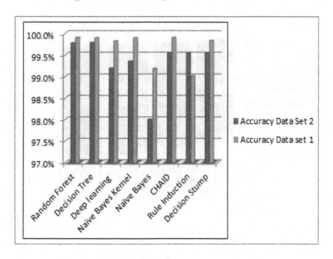

Fig. 4. Prediction accuracy comparison of classifiers

(Fig. 4), and Precision (Fig. 5). Expecting still better results it was decided to combine the three Classifiers using two ensemble models namely Stacking and Voting. Figure 6 indicates the Comparative results of the two Ensemble Models. It can be observed from Fig. 6 that Stacking records lesser Classification Error, with slightly better Accuracy and Precision than the Voting Model. The top performing Ensemble model that is Stacking, involving top three Classifiers Random Forest, Decision Tree, and Naive Bayes Kernel is therefore chosen as our proposed model for efficient classification of XSS attack samples.

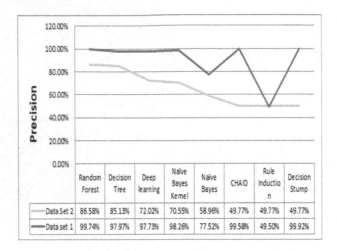

Fig. 5. Precision comparison of classifiers

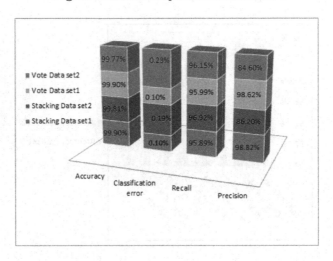

Fig. 6. Comparison of ensemble models

5   Conclusion

The proposed MMR and SDMR Feature Selection Algorithms select 14 Features from XSSdatasetengineered1 out of a total 67 and 26 from XSSdatasetengineered2 out of 77 after the Intersection operation on the Feature subsets is performed during the Pre-Processing phase. This amounts to a dimensionality reduction of 79.1% and 66.23% in the first and second data sets respectively. The resultant Feature subset is subjected to various Classier algorithms and a tenfold cross validation is performed before recording the performance metrics. Two Ensemble approaches namely Voting and Stacking are tried to determine the best Classification Model on the top performing classifiers. The ERDNS model must be checked on real-time data sets and compared with other existing models to see if it can be improved further. The time it takes to complete the whole process must be minimized so that XSS samples can be classified as soon as they are detected.

References

1. Vijayalakshmi, K., Syed Mohamed, E.: Case study: extenuation of XSS attacks through various detecting and defending techniques. Int. J. Innov. Res. Comput. Sci. Technol. (IJIRCST) **9**(1), 1736 (2021). https://doi.org/10.1080/19361610.2020.1735283
2. Zhang, X., Zhou, Y., Pei, S., Zhuge, J., Chen, J.: Adversarial examples detection for XSS attacks based on generative adversarial networks. IEEE Access **8**, 10989–10996 (2020). https://doi.org/10.1109/ACCESS.2020.2965184
3. Chen, X.L., Li, M., Jiang, Yu., Sun, Y.: A comparison of machine learning algorithms for detecting XSS attacks. In: Sun, X., Pan, Z., Bertino, E. (eds.) ICAIS 2019. LNCS, vol. 11635, pp. 214–224. Springer, Cham (2019). https://doi.org/10.1007/978-3-030-24268-8_20
4. Mokbal, F.M.M.: Cross-Site Scripting Attack (XSS) dataset (2020). https://doi.org/10.6084/m9.figshare.13046138.v4
5. Niranjan, A., Nitish, A., Deepa Shenoy, P., Venugopal, K.R.: Security in data mining-a comprehensive survey. Glob. J. Comput. Sci. Technol. **16**(5), 51–72 (2017). https://computerresearch.org/index.php/computer/article/download/1464/1451
6. Rodríguez, G.E., Torres, J.G., Flores, P., Benavides, D.E.: Cross-site scripting (XSS) attacks and mitigation: a survey. Comput. Netw. **166**, 1–21 (2020). https://doi.org/10.1016/j.comnet.2019.106960
7. Mokbal, F.M.M., Wang, D., Wang, X., Fu, L.: Data augmentation-based conditional wasserstein generative adversarial network-gradient penalty for XSS attack detection system. PeerJ Comput. Sci. **6**, e328 (2020). https://doi.org/10.7717/peerj-cs.328
8. Gupta, B.B., Chaudhary, P., Gupta, S.: Designing a XSS defensive framework for web servers deployed in the existing smart city infrastructure. J. Organ. End User Comput. **32**(4) (2020). https://doi.org/10.4018/JOEUC.2020100105
9. Liljebjörn, J., Broman, H.: Mantis The Black-Box Scanner: Finding XSS Vulnerabilities through Parse Errors, Retrieved from Dissertation: http://urn.kb.se/resolve?urn=urn:nbn:se:bth-19566

PRGR-C19: Profiling Rapid Growth Regions of COVID-19 Pandemic, A Data-Driven Knowledge Discovery Approach

G. U. Vasanthakumar[1]($\boxtimes$) (iD), N. Ramu[2] (iD), and M. N. Thippeswamy[1] (iD)

[1] Nitte Meenakshi Institute of Technology, Bangalore, India
vasanth.gu@nmit.ac.in
[2] Huawei Technologies India Pvt. Ltd, Bangalore, India

Abstract. The ongoing Corona Virus Infectious Disease 2019 (COVID-19) Pandemic is spreading across the globe so rapidly that no one could ever imagine its dissemination depth amongst humans. Although the Government is trying its best to curb the spreading of the pandemic in various ways it could, the rate of growth of infected cases is exponentially increasing every hour. This work depicts the rate of growth of Infected COVID-19 cases region-wise across the globe using live data. The collected data is pre-processed and fed to the "Profiling Rapid Growth Regions of COVID-19 (PRGR-C19)" algorithm and processed to determine the region-wise spread of Infected COVID-19 cases regularly and Profile its Rapid Growth Regions. Mathematical Modeling is done to predict its rate of growth during the second wave. The results of this work may predominantly be used by various authorities to optimize the planning and implementation of region-wise intervention strategies like Lockdown/Restriction rules to curtail the further spread of the pandemic during its second wave which has already begun.

Keywords: Corona Virus · COVID-19 · Infected Cases · Pandemic · Profiling · Rapid Growth Regions

1 Introduction

The novel Corona Virus Infectious Disease 2019 (COVID-19) epidemic is affecting enormous humans so quickly that the World Health Organization (WHO) has declared it as a pandemic alerting public health emergency of international concern [1]. Though the disease started to get recognized in Wuhan City of China in December 2019, within a couple of weeks, the disease has spread outside China and now almost to the entire globe. The virus has been discovered upon numerous investigations in local laboratories [2]. The researchers have ascertained that COVID-19 spreads from humans to humans [3], although it was reported to have been found in animals [4]. The severity and the complete extent of this pandemic are yet to be seen as controlling the spread of this infectious disease has become too challenging. In this work, the results of the region-wise assessment of the current Pandemic situation is analyzed and reported.

© Springer Nature Switzerland AG 2021
K. R. Venugopal et al. (Eds.): ICInPro 2021, CCIS 1483, pp. 366–379, 2021.
https://doi.org/10.1007/978-3-030-91244-4_29

The outbreak of the novel Corona Virus spread worldwide is expanding exponentially every day. Although the government has initiated and implemented various regulations, the rate of growth of Infected cases is high in all regions. Various factors like social distancing, lockdown, containment zones, seal down etc., have been enforced for the management of the current Pandemic situation. The effects of the implementation of drastic control measures on the outbreak of this pandemic early in India by the Government indicate a substantially less rate of spread of the pandemic to the society. However, the reduction of the infection rate not only causes a quench of the epidemic peak but also on the mortality rate which could be possible through a disciplined and concerted effort of people as a whole globally.

The worldwide spread of COVID-19 has altered the normal lifestyle of humans and has alerted them to adapt to certain changes for healthier life practices. The pandemic has not only affected the lifestyle but has also pushed humans towards stress with the uncertainty of their lives and has induced tremendous mental trauma motivating to carry out this work. The presented results are useful in understanding and implementing precautionary measures and rules formulation region-wise to reduce further spread of the Pandemic.

The Novel Contributions of this work are to

- Investigate the spread of Infected COVID-19 cases regularly,
- Identify the region-wise spread of Infected COVID-19 cases,
- Mathematically model and predict the rate of growth of Infected COVID-19 cases region-wise during its second wave as well and
- Profile Rapid Growth Regions of Infected COVID-19 cases.

The rest of the paper is so organized with an attempt to keep the readers' attention and interest. Section 2 gives a glimpse of literature and related work carried out by various authors so far. The problem is defined in Sect. 3 whereas the proposed system along with mathematical model and Profiling Rapid Growth Regions of COVID-19 (PRGR-C19) algorithm are presented in Sect. 4. Results and Analysis is discussed in Sect. 5 and the entire work is summarized with Conclusions in Sect. 6.

2 Related Work

As COVID-19 has spread almost to all countries in the world, Lee [5] has discussed the challenges in handling the pandemic situations in both developed and developing countries. He says that developed countries like the United States of America and the United Kingdom can control this epidemic due to the availability of sophisticated health-care systems, resources and facilities. He also put forth the difficulties faced by developing countries in handling this epidemic due to the resource-contained settings, low health care protection infrastructure and lack of investments on health priorities. Yunpeng et al. [6] have discussed the health-care resource availability and preparedness as the better way to handle COVID-19 in all countries by considering two provinces of China as an example. A study by Gallego et al. [7] provides an insight into all international events, games, sports activities etc. canceled due to the pandemic novel coronavirus disease which is spread all over the world. Also, it gives the measures and actions of Japan in handling and protecting the people in its own country and also on protecting the people of Japan staying in other countries.

The measures of Taiwan's government to face and control epidemic coronavirus spread in the country are presented by Wang et al. [8]. This includes sending health tips and other related messages to the public, early recognition of crisis, and so on. The study on SARS-like coronavirus with other beta coronaviruses on amino-acid levels by Wen et al. [9] confirms that there were no beneficial mutations and other mutations identified in the study, thus warranting further investigations. The structural and sequence analysis [10] on angiotensin-converting enzyme 2 (ACE2) raises an alert on interspecies transmission coronavirus and also for cross-species receptor usage. A study of 262 coronavirus positive cases by Sijia et al. [11] on the patients of Beijing found that patients have the highest symptoms like; fever, cough, fatigue, dyspnea and headache respectively.

Comorbidity of novel coronavirus disease is presented by Guan et al. [12] from the study conducted on 686 female patients with an average age of 48.9 years. It provides around 4–5 comorbidity and concludes that Hypertension followed by Diabetics are the most prevalent comorbidities of the novel coronavirus. The widespread SARS-CoV-2 has raised the importance of studying and understanding viral transmission among humans [13]. The first part in humans, which is affected by SARS-CoV-2, is the lower respiratory tract, which may lead to pneumonia or further major injury. Pathogenicity is directly dependent on the interaction between the host antiviral immune system and SARS-CoV-2. However, WHO recommends good ventilation, social distance, avoiding crowds, use of mask, sanitizer are the good practices to stop the spread of novel SARS-CoV-2 virus.

Meghan et al. [14] have proposed a methodology for the development of UN research roadmap that helps the system and countries across the globe to recover from these pandemic losses and to build back stronger. This methodology is based on 5 rapid scoping reviews which also include many social bodies, a highly trained workforce, quick health learners, forecasting research groups, scaling up infrastructure etc,. This proposed roadmap still can be improved by adding various other social and responsible organizations and health researchers who can predict the spread volume and its effectiveness.

The effect masks with valve have been discussed based on various situations by Maryam et al. [15], Incubation period; Asymptomatic disease, Carriers, False-negative cases, and Re-Infection are the 5 situations which depict the ill effects of wearing the mask with valve. The recommendation made by authors to policymakers to avoid people wearing masks with valves.

The study conducted in Territory Care Hospital on Health Care Workers (HCWs) who either direct or indirectly get in contact with confirmed infected or suspected cases. As the number of ICs of HCWs was less as per data due to wearing of PPE kits, IPC measures etc., it is concluded that with adequate precautions and potential training then the risk of being infected can be minimized [16].

To overcome the economic crisis and other business process failures during the pandemic lockdown situation, Pratim et. al. [17], have proposed a continuity planning digital transformation framework with multiple variants which can serve during any pandemic or disaster situations. The five Digital Transformation characteristics presented here are: Modularity, Collaboration with multiple agencies, security and privacy of information, Transparency, digital business integrity.

Walid [18], in his study, addressed the coagulopathy - coagulation disorders, which is clotting disorder or hemorrhage, which is observed in some of the SARS-CoV-2 positive patients. There are several hypotheses to explain this coagulation leading to both increased bleeding and clotting. However, COVID-19 alone cannot be concluded as the sole reason for coagulation since other combination factors also contribute.

3   Problem Definition

Given the data set pertaining to the Infected COVID-19 cases, the objectives are:

i.   To identify and categorize evolving ICs region-wise over time based on the clinically tested and reportedly confirmed cases.
ii. To Mathematically Model and Profile the regions based on the Rate of Growth of ICs.

While profiling the region, a configurable limit of having a minimum number of infected cases count is a prerequisite.

4   Proposed System

The system proposed in this work helps various authorities to optimize the planning and implementation of region-wise intervention strategies like Lockdown/Restriction rules to curtail the further spread of the pandemic.

4.1   Mathematical Model

The exponential growth of the pandemic can be modeled mathematically as follows; Let us consider the following notations in the modeling.

IC   :Infected Case.
IC_d : Count of ICs per day.
ΔIC_d : Count of New ICs for considered day.
ΔIC_{d-1} : Count of ICs for previous day considered.
RG   : Rate of Growth.
RGN   :Rate of Growth of New Cases
RIC_d :Count of Re-ICs per day.

The Rate of Growth of ICs is the ratio of Count of New ICs for the considered day to that of the Count of ICs for the previous day considered and is given by Eq. (1):

$$RG_{IC} = \frac{\Delta IC_d}{\Delta IC_{d-1}} \tag{1}$$

Considering the Re-ICs, the Rate of Growth of New ICs would be computed using Eq. (2):

$$RGN_{IC} = \frac{\Delta\ IC_d - \Delta\ RIC_d}{\Delta\ IC_{d-1}} \tag{2}$$

The ICs will exponentially increase when the Rate of Growth is greater than 1, whereas when it is equal to 1, it is advisable to impose various strict restrictions by the authorities so that exponential increase in ICs can be flattened.

4.2 System Model

The System Model Diagram is as shown in Fig. 1. To find the Rate of Growth of Infected COVID-19 cases, the live data from John Hopkins University [19] is considered for analysis. The data is clogged daily which is then pre-processed to extract meaningful insights like; Patient ID, Sex, Date Reported, Address, Location, COVID-19 symptoms and other ailments reported and so on. Further, based on the region, the ICs are classified to get the region-wise count of ICs. Later, the ICs are categorized to form four groups:

- **Positive** – Patients currently infected,
- **Recovered** – Patients recovered,
- **Death** – Patients died due to COVID-19 and
- **Reinfected** – Patients currently Reinfected with COVID-19.

The rate of growth of ICs is then computed region-wise to profile the rapid growth regions of Infected COVID-19 cases.

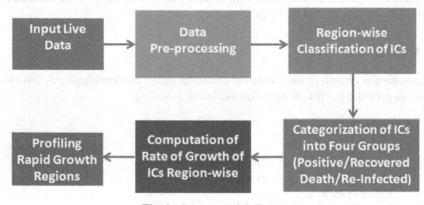

Fig. 1. System model diagram

4.3 Algorithm

The proposed PRGR-C19 algorithm is as shown in Algorithm-1, which is run at frequent time intervals on the data gathered. Every patient record is scanned to identify their geographical location. The patients are then classified location-wise and once the population of the patients grows in the cluster, regions are formed.

Algorithm 1: *Profiling Rapid Growth Regions of COVID-19 (PRGR-C19) Algorithm*

1: **while** (True) **do**
2:    **for** (Every Record) **do**
3:        Identify its Location using Reverse Geo-coding
4:            Classify and Cluster the Patients based on their Location to form the Regions
5:        **end for**
6:    **for** (Every Cluster of Region) **do**
7:        Classify the Patient based on current status as Positive/Recovered/Death/Re-
          -Infected to form respective Clusters.
8:        Compute the Rate of Growth of ICs for the Region
9:    **end for**
10:    Profile the Rapid Growth Regions based on the Rate of Growth of ICs.
11:    **end while**

Now considering every region, the patients are grouped into four categories as either Positive/Recovered/Death/Re-Infected. Then the Rate of Growth of ICs is computed Region-wise. Furthermore, based on the results thus obtained, Rapid Growth Regions are determined and Profiled.

5   Results and Analysis

The data set for analysis is collected from John Hopkins University [19] daily and is stored date-wise. The data set consists of 22 attributes describing Patents' ID, Name, Age, Sex, Address, Location, and so on.

The data set so acquired is considered during the first wave of COVID-19 from 22nd January to 12th July 2020 which is then categorized region-wise into three groups namely; Positive, Recovered and Death. During its second wave considered from 1st February to 10th April 2021, is categorized region-wise into four groups namely; Positive, Recovered, Death and Re-Infected. The proposed PRGR-C19 algorithm is implemented in Python using 'Jupyter', an open-source software development tool upon Intel Pentium i7 system having 4 GB RAM computational capabilities, where Windows 8 supports these configurations.

The live data set is collected from the source and fed to the target system, where the preprocessed data is further processed to get the desired results. In this work, five regions have been considered for analyzing namely; Region-1: USA, Region-2: China, Region-3: Germany, Region-4: Brazil and Region-5: India.

The globally IC status of COVID-19 is as shown in Fig. 2 during its first wave. The graph depicts an exponential increase of reportedly Positive cases from the data over the considered period of analysis during the first wave of COVID-19.

The globally IC status of COVID-19 during its second wave is as shown in Fig. 3. The graph depicts a drastic increase of reportedly Positive cases from the data over the considered period of analysis along with Re-Infected cases.

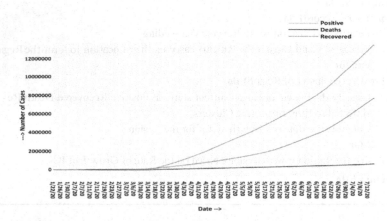

Fig. 2. Globally Infected COVID-19 Case Status during First wave.

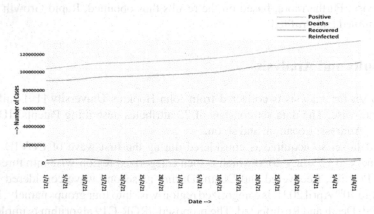

Fig. 3. Globally Infected COVID-19 Case Status during Second wave.

Specific regions are considered and the data is extracted from the categorized data set for further analysis, giving rise to region-wise case status and the percentage of rate of growth of infected cases over the considered period of analysis during the first wave of COVID-19.

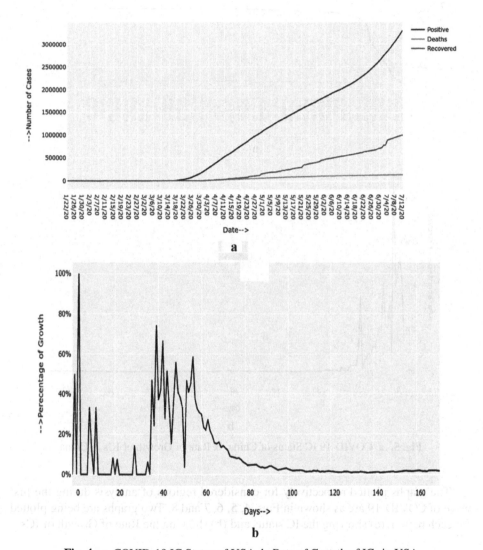

Fig. 4. a. COVID-19 IC Status of USA. b. Rate of Growth of ICs in USA

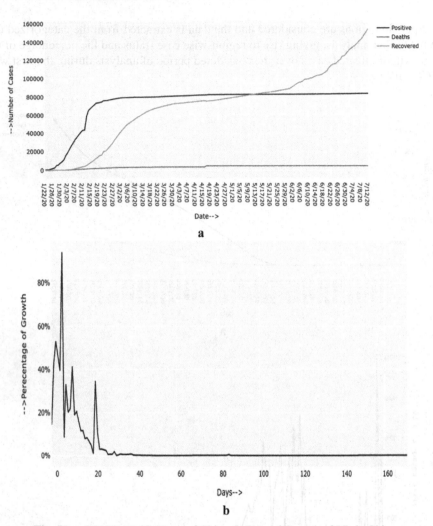

Fig. 5. a. COVID-19 IC Status of China. b. Rate of Growth of ICs in China

The graphs plotted respectively for considered regions of analysis during the first wave of COVID-19 are as shown in Figs. 4, 5, 6, 7 and 8. Two graphs are being plotted for each region, (a) showing the IC status and (b) showing the Rate of Growth of ICs.

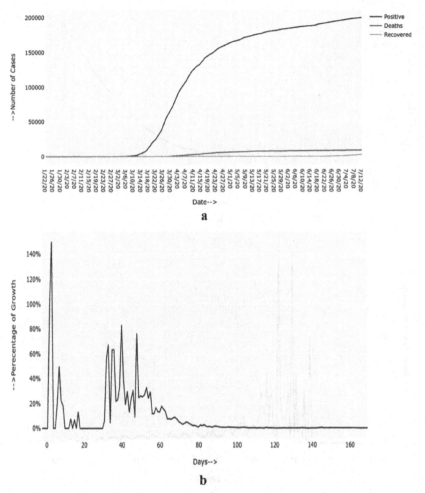

Fig. 6. a. COVID-19 IC Status of Germany. b. Rate of Growth of ICs in Germany

It is observed from the region-wise status depicted from the graphs in Figs. 4, 5, 6, 7 and 8 shown, that in India, however, the ICs are increasing day-by-day, but the Rate of Growth of ICs are substantially low when compared to that of other regions considered for analysis. This substantially low Rate of Growth of ICs in India as depicted from the graph in Fig. 8b, is because of the reason that the Indian Government took early strategic measures like Lockdown/Restriction rules to curtail the spread of pandemic to the society during its first wave. The number of Recovered cases in China is considerably high as depicted from the graph in Fig. 5a.

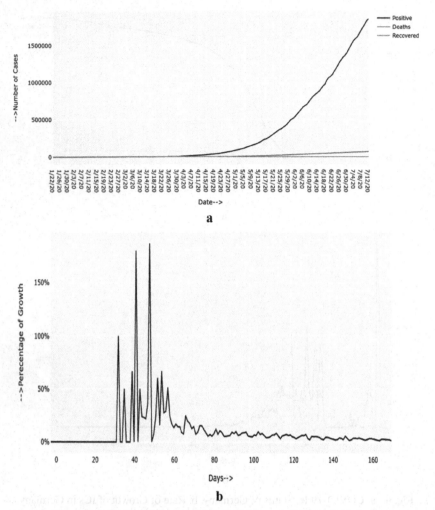

Fig. 7. a. COVID-19 IC Status of Brazil. b. Rate of Growth of ICs in Brazil

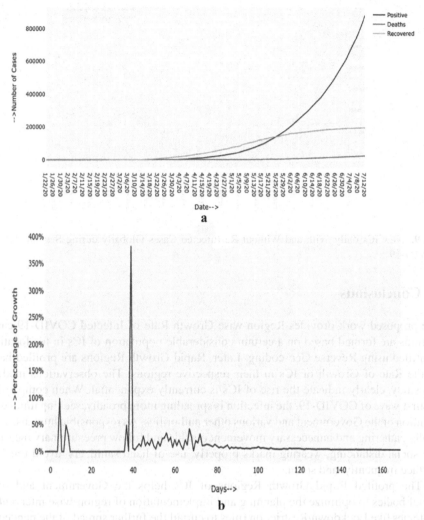

Fig. 8. a. COVID-19 IC Status of India. b. Rate of Growth of ICs in India

The COVID-19 virus during its second wave is spreading faster than in the first wave. As depicted in Fig. 9, the Re-Infected cases are also rising day-by-day, threatening the lives of humans and alerting immediate intervention of authorities. Thus, this work helps authorities to optimize their planning and implementation strategies to control the further spread of the pandemic and prepare themselves with required health care facilities to curb further spread of pandemic.

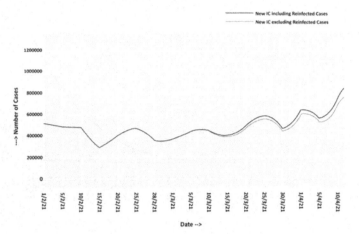

Fig. 9. New ICs daily With and Without Re-Infected Cases Globally during Second Wave of COVID-19

6   Conclusions

The proposed work provides Region-wise Growth Rate of Infected COVID-19 cases. Regions are formed based on a certain considerable population of ICs in the locations identified using Reverse Geo-coding. Later, Rapid Growth Regions are profiled based on the Rate of Growth of ICs in their respective regions. The observations made in this study, clearly indicate the rise of ICs is currently exponential. When compared to the first wave of COVID-19, the infection is spreading more broadly, seeking immediate attention of the Government and various other authorities. As responsible humans, avoid public gathering and unnecessary movement outside and follow precautionary measures like social distancing, wearing marks properly, use of hand sanitizers, avoid touching the face frequently, and so on.

The profiled Rapid Growth Regions of ICs helps the Government and other Social bodies to optimize the planning and implementation of region-wise intervention strategies like Lockdown/Restriction rules to curtail the further spread of the pandemic.

References

1. Biscayart, C., Angeleri, P., Lloveras, S., do Socorro Souza Chaves, T., Schlagenhauf, P., Rodríguez-Morales, A.J.: The next big threat to global health? 2019 novel coronavirus (2019-nCoV): what advice can we give to travellers?–Interim recommendations January 2020, from the Latin-American Society for Travel Medicine (SLAMVI). Travel Med. Infect. Dis. **33**, 1-4 (2020)
2. Zhang, N., et al.: Recent advances in the detection of respiratory virus infection in humans. J. Med. Virol. **92**(4), 408–417 (2020)
3. Ji, W., Wang, W., Zhao, X., Zai, J., Li, X.: Cross-species transmission of the newly identified coronavirus 2019-nCoV. J. Med. Virol. **92**(4), 433–440 (2020)
4. Lu, H., Stratton, C.W., Tang, Y.-W.: Outbreak of pneumonia of unknown etiology in Wuhan, China: the mystery and the miracle. J. Med. Virol. **92**(4), 401–402 (2020)

5. Lee, A.: Wuhan novel coronavirus (COVID-19): why global control is challenging? Public Health, p. 179 (2020)
6. Ji, Y., Ma, Z., Peppelenbosch, M.P., Pan, Q.: Potential association between COVID-19 mortality and health-care resource availability. Lancet Glob. Health **8**(4), 480 (2020)
7. Gallego, V., Nishiura, H., Sah, R., Rodriguez-Morales, A.J.: The COVID-19 outbreak and implications for the Tokyo 2020 summer Olympic games. Travel Med. Infect. Dis. (2020)
8. Wang, C.J., Ng, C.Y., Brook, R.H.: Response to COVID-19 in Taiwan: big data analytics, new technology, and proactive testing. JAMA **323**(14), 1341–1342 (2020)
9. Wen, F., Yu, H., Guo, J., Li, Y., Luo, K., Huang, S.: Identification of the hyper-variable genomic hotspot for the novel coronavirus SARS-CoV-2. J. Infect. **80**, 671–693 (2020)
10. Li, R., Qiao, S., Zhang, G.: Analysis of angiotensin-converting enzyme 2 (ACE2) from different species sheds some light on cross-species receptor usage of a novel coronavirus 2019-nCoV. J. Infect. **4**(80), 469–496 (2020)
11. Tian, S., et al.: Characteristics of COVID-19 infection in Beijing. J. Infect. **80**, 401–406 (2020)
12. Fanelli, D., Piazza, F.: Analysis and forecast of COVID-19 spreading in China Italy and France. Chaos Solitons Fractals **134**, 109761 (2020)
13. Dabbish, A.M., Yonis, N., Salama, M., Essa, M.M., Walid Qoronfleh, M.: Inflammatory pathways and potential therapies for COVID-19: a mini review. Eur. J. Inflamm. **19**, 1–22 (2021)
14. McMahon, M., et al.: The promise of science, knowledge mobilization, and rapid learning systems for COVID-19 recovery. Int. J. Health Serv. **51**, 242–246 (2021)
15. Baradaran-Binazir, M., Heidari, F.: The necessity of prohibiting the masks with exhalation valve during emerging infections Like COVID-19. Asia Pac. J. Public Health (2021)
16. Sharma, S., et al.: Assessment of potential risk factors for 2019-novel coronavirus (2019-nCov) infection among health care workers in a tertiary care hospital, North India. J. Primary Care Commun. Health **12** (2021)
17. Datta, P., Nwankpa, J.K.: Digital transformation and the COVID-19 crisis continuity planning. J. Inf. Technol. Teach. Cases (2021)
18. Alam, W.: Hypercoagulability in COVID-19: a review of the potential mechanisms underlying clotting disorders. SAGE Open Med. **9** (2021)
19. Johns Hopkins University Center for Systems Science and Engineering (2019). https://gisand data.maps.arcgis.com/apps/opsdashboard/index.html

Comparisons of Stock Price Predictions Using Stacked RNN-LSTM

Sheldon Sequeira and P. K. Nizar Banu[✉]

Department of Computer Science, CHRIST (Deemed to be University), Bangalore, India
sequeira.gregory@cs.christuniversity.in,
nizar.banu@christuniversity.in

Abstract. This paper seeks to identify how the RNN-LSTM can be used in predicting the rise and fall in stock markets thereby helping investors to understand stock prices. Therefore, by predicting the nature of the stock market, investors can use different machine learning techniques to understand the process of selecting the appropriate stock and enhance the return investments thereafter. Long Short-Term Memory (LSTM) is a deep learning technique that helps to analyze and predict the data with respect to the challenges, profits, investments and future performance of the stock markets. The research focuses on how neural networks can be employed to understand price changes, interest patterns and trades in the stock market sector. The datasets of companies such as IBM, Cisco, Microsoft, Tesla and GE were used to build the stacked RNN-LSTM model using timesteps of 7 and 14 days. The two layered stacked RNN-LSTM models of the companies such as Microsoft and Tesla achieved their highest model accuracies after being trained over a span of one year whereas the other companies acquired their highest accuracies after being trained over a span of 4 to 5 years which implies that the rate of change of economic factors affecting Microsoft and Tesla over a short span of time is high as compared to the other existing companies.

Keywords: Long Short-Term Memory (LSTM) · Stock market · Machine learning · Investments · Neural networks · Interest patterns · RNN

1 Introduction

Over the past few years, it has been difficult to understand the irregularities in investments and the rise and fall in stock prices not only in the developing but also in the developed stock markets. The country's economy, public news and investments influence the nature of the stock market in that particular country. Therefore, it is exceedingly important that investors and traders look for potential strategies and opportunities to understand the inefficiencies in the stock market and to use technologies in order to gain returns after their investments [1].

It is a tedious task to understand a particular company's Price to Earning's Ratio especially when investors want to invest their money and this can be a huge problem in determining the original price of a stock. However, an algorithm can be created with the help of Machine Learning and Deep Learning that will help to find different patterns

© Springer Nature Switzerland AG 2021
K. R. Venugopal et al. (Eds.): ICInPro 2021, CCIS 1483, pp. 380–390, 2021.
https://doi.org/10.1007/978-3-030-91244-4_30

in the data of a particular stock market company and analyze and compare it with a new set of data of another company in order to make predictions about the stock prices. Machine Learning helps to read and understand data by processing it using algorithms in order to solve a particular type of problem with the help of input data as all the programming is done by the machine itself when it learns new patterns in the existing data. The price of a particular stock can also be predicted if you train the machine to learn the different trends of the stock market by using the Long Short Term Memory Model. You can choose a mathematical algorithm depending on your analytical model to solve a particular problem [2].

The different problems such as irrelevant, noisy, complex or partial information can be solved with the accurate solutions by using the different neural networks such as Long Short-Term Model (LSTM) [3]. There are several uncertainties in the stock markets with respect to their future state in the short and long term. Therefore, the Machine Learning Methods can be used to reduce these uncertainties and predicting stock market returns. The LSTM Model has cell outputs called as the hidden state and the cell state that cause the manipulations to decide the relevant and accurate information from the data that is processed by the model [2, 4].

2 Literature Review

This research seeks to emphasize on how different machine learning techniques help the finance researchers, investors and traders to predict the stock market changes. The objective of this study is to apply the different machine learning models like the LSTM in order to understand which technique can work better to compare different stock market companies and predict the index values and trends in the stocks [5, 6]. This paper is based on the analysis of the following researches conducted to understand the usage and application of machine learning models to predict stock market changes [7].

The research conducted in [8] around 600 datasets are collected from the Dhaka Stock Exchange (DSE) from January 2013-April 2015. The ANFIS and ANN model have been used in order to understand the changes in the closing price of the stock markets based on the company's historical data (Fig. 1).

The ANFIS model helps in getting varied results because of the variation in the membership function and therefore it's important to identify the correct membership function and number. The ANFIS Artificial Neural Network has lesser error because of its adaptive neuro fuzzy inference system to predict DSE's closing price by using six macroeconomic variables and three indices as input variables.

The ANN network is used to predict the indexes in the stock market with regards to trend movements in the forecast in [10] The RNN network is the best when it comes to receiving information in a sequential manner. However, it's drawback is seen when there are too many tasks to be done. The RNN network directs the same kind of tasks for all elements of a sequence and therefore we get the information we need in a sequential manner as the output always depends on the earlier computations. With the help of the memory in the RNN network, the information that is calculated is also captured [10].

The different layer sizes can be tested in order to predict the optimal size with regards to data modelling with the help of Multilayer Perceptron Architecture (MLP). However,

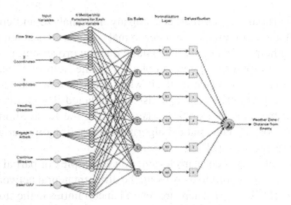

Fig. 1. Structure of ANFIS [9]

in order to obtain a sequential combination of ID convolutional and max pooling layers, the Convolutional Neural Network Architecture (CNN) can be used. Hyperparameters enable to choose the best parameters like the number of neurons, hidden layers, dropout ratio, optimization algorithm as well as number of filters in CNN's for the different architectures. The different neural networks like the MLP, CNN and the RNN can be used to predict and forecast stock price movements [10, 11].

The Technical Analysis (TA) can be applied in the neural networks models like the LSTM in order to help the investors predict the above average profits in the stock market sector as determined in the [12]. The data was selected from various banking, automobile and metal sectors to use the LSTM model to predict the price trends in the stock markets. The model performance can be improved and more profit can be gained when the Stock Indices are used in the application along with the O-LSTM model which is a market independent approach. The application of predicting the price, price trends and the signal for daily trading in the stock market sector [12] (Fig. 2).

In the paper [13], behaviors of the stock markets can be predicted in order to understand which shares can be purchased by the investors and traders in order to receive maximum profits by also analyzing the news and social media as external factors influencing the stock market sector. The accuracy in prediction can be accelerated and improved in the neural network models with the application of deep learning in stock prediction. The data from various social media platforms like Google, Facebook, Twitter and other platforms can be used in order to understand the effects of these platforms on the prices and trends in the stock market sector. The different tweets and news on these platforms serve as the data for Sentimental Analysis especially while using the Stanford NLP model [13].

Data analytics is nothing but analyzing the data in order to find solutions for various problems like understanding why a particular company is having consistent losses, why the sales of a particular commodity have dropped or risen, why the interest rate in banks has reduced, etc. We can understand and interpret various kinds of information with the help of data analytics and therefore future threats and problems can be discussed and solved by understanding and interpreting data. With respect to the rise and fall in stock markets, it is understood that there is no definite pattern. It depends on a variety of factors

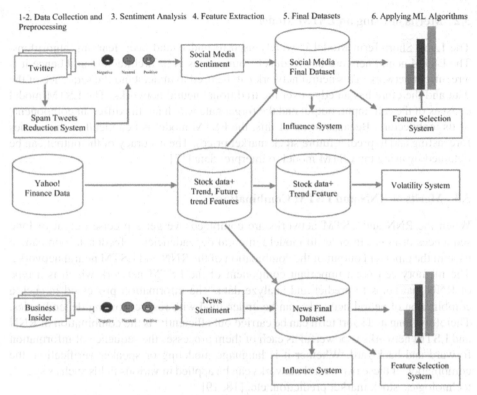

1-2. Data Collection and   3. Sentiment Analysis   4. Feature Extraction   5. Final Datasets   6. Applying ML Algorithms
Preprocessing

Fig. 2. Market Forecasting using financial news and social media [13]

and therefore it is difficult to predict the changes related to the stock market. However, with the help of different machine learning techniques such as Convolutional Neural Network (CNN), Recurrent Neural Network (RNN) and Artificial Neural Networks (ANN), we can predict the rise or fall in stock markets and analyze and investigate the future growth of a company [14, 15].

3 Model Implementation Details

3.1 Merits of Using RNN Model

The internal memory in the RNN Model has an algorithm which analyzes its input and therefore provides sequential data for machine learning problems. Having only one internal memory, RNN can analyze and predict the volatile financial instruments. With the help of fixed size input and output vectors, RNN networks help to analyze the earlier type of networks. One of the benefits of using RNN networks is that the information is obtained in a sequential manner. The output is produced by understanding and activating the hidden and successive layers of input which help to successfully produce the output. RNN networks enable the combination of hidden layers that helps to analyze the relationship between the previous and current input [16].

3.2 Merits of Using an LSTM Model

The Long Short-Term Model is widely used to understand deep learning algorithms. The LSTM neural network has feedback connections in its architecture and hence it is a recurrent network. This neural network can be used to understand the sequence of the data and therefore has an edge over the traditional neural networks. The LSTM model consists of the cell, input, output and the forget gate which are the different components of its architecture. Because of all of this, the LSTM model is beneficial in time series forecasting and to predict future stock market prices. The accuracy of the output can be obtained by using the LSTM model to interpret data [17].

3.3 Merits of RNN and LSTM Combination

When the RNN and LSTM networks are combined, we get a precise output of long sequences of input. In order to model temporal dependencies, a feedback loop can be used in the input and output of the combination of the RNN and LSTM neural networks. The memory cell is an important component of the LSTM network which is a type of RNN that helps to predict and analyze data. The information processed in such a combination of neural networks can be understood with respect to long dependencies. Therefore, long and short term can be carried out efficiently as the combination of RNN and LSTM networks is powerful as each of them processes the sequence of information forward and backward. Whether it is language modeling or speaker verification, the combination of these two neural networks can be applied in various fields such as speech technologies, stock market prediction, etc. [18, 19].

The historical data of companies such as IBM, Cisco, Microsoft, Tesla and GE, were obtained from yahoo finance spanning the years between 2015–2020. The datasets were preprocessed and normalized before training them over 50 RNN-LSTM models. The training sets for each company were divided into 5 categories based on the distinct number of years over timesteps of 7 and 14 days respectively. The first category consisted of the model being trained over the stock price details between the years 2015–19, the second category over 2016–19, the third one over 2017–19, fourth one over 2018–19 and lastly the fifth one over the year 2019. These five types of models were tested over the actual prices of the year 2020 and the prediction accuracy was calculated and compared amongst the different models.

The LSTM architecture used in this experiment consisted of 2 stacked LSTM layers along with 1 dense layer. 20% dropout is configured in both these layers to prevent overfitting. The optimum hidden units in each layer are 50. This was decided after obtaining the least model errors by configuring hidden unit values between 0 to 50. More hidden layers are better but increase the complexity of the model thereby making it more difficult to train [20]. Similarly, two stacked LSTM layers were used to build the final neural network [21] as the model errors increased significantly with the increase in the number of stacked layers after 2. The model was compiled using the Adam optimizer with the loss being mean squared error. Model training was done for 200 epochs with batch size 32 (Fig. 3 and Table 1).

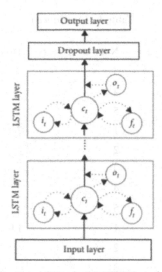

Fig. 3. Architecture of Stacked LSTM Neural Network [21]

Table 1. Final result table with mean squared error as the performance metric

Stock name	Training year/years	Number of years	Time steps	Mean squared error
IBM	2019	1	7	17.41
	2018–19	2		10.36
	2017–19	3		9.40
	2016–19	4		8.71
	2015–19	5		11.74
IBM	2019	1	14	32.28
	2018–19	2		9.37
	2017–19	3		10.26
	2016–19	4		8.18
	2015–19	5		8.27
CISCO	2019	1	7	1.55
	2018–19	2		1.49
	2017–19	3		1.02
	2016–19	4		1.05
	2015–19	5		1.15
CISCO	2019	1	14	1.34
	2018–19	2		0.99

(*continued*)

Table 1. (*continued*)

Stock name	Training year/years	Number of years	Time steps	Mean squared error
	2017–19	3		1.13
	2016–19	4		1.03
	2015–19	5		0.95
MICROSOFT	2019	1	7	48.25
	2018–19	2		62.90
	2017–19	3		54.76
	2016–19	4		98.45
	2015–19	5		92.07
MICROSOFT	2019	1	14	40.64
	2018–19	2		31.59
	2017–19	3		56.65
	2016–19	4		34.01
	2015–19	5		65.37
TESLA	2019	1	7	2984.93
	2018–19	2		6005.25
	2017–19	3		5049.34
	2016–19	4		3575.07
	2015–19	5		4944.27
TESLA	2019	1	14	7605.48
	2018–19	2		14343.07
	2017–19	3		15475.17
	2016–19	4		17281.64
	2015–19	5		50273.49
GE	2019	1	7	0.22
	2018–19	2		0.24
	2017–19	3		0.25
	2016–19	4		0.23
	2015–19	5		0.47
GE	2019	1	14	0.27
	2018–19	2		0.14
	2017–19	3		0.26
	2016–19	4		0.21
	2015–19	5		0.25

4   Result Analysis

The results table summarizes the mean squared error as a performance metric where the data is trained with the model being built in 5 ways for each of the five companies with timesteps of 7 and 14 days. The least MSE for IBM is 8.18 and is observed when the model is trained for 4 years with timesteps of 14 days. The least MSE for Cisco is 0.95 and is observed when the model is trained for 5 years with timesteps of 14 days. The least MSE for Microsoft is 40.64 and is observed when the model is trained for 1 year with timesteps of 7 days. The least MSE for Tesla is 2984.93 and is observed when the model is trained for 1 year with timesteps of 7 days. Lastly, the least MSE for GE is 0.14 and is observed when the model is trained for 2 years with timesteps of 14 days.

IBM and Cisco's deep learning models need to be trained over a span of 4 and 5 years to get better accuracy whereas Microsoft, Tesla and GE have a better accuracy by being trained over a span of 1 to 2 years respectively. Also, Microsoft and Tesla are the only

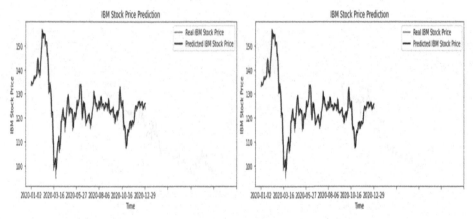

Fig. 4. Lowest MSE of IBM having timesteps of 7 and 14 days

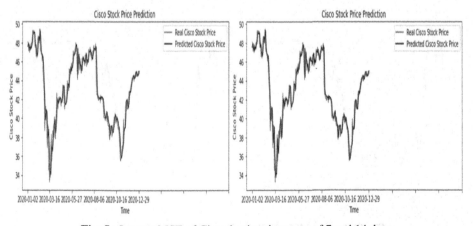

Fig. 5. Lowest MSE of Cisco having timesteps of 7 and 14 days

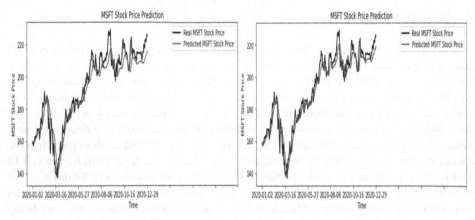

Fig. 6. Lowest MSE of Microsoft having timesteps of 7 and 14 days

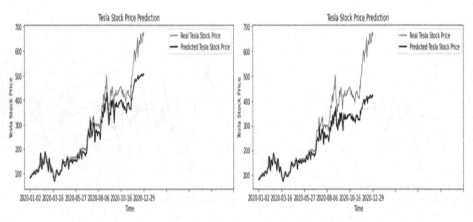

Fig. 7. Lowest MSE of Tesla having timesteps of 7 and 14 days

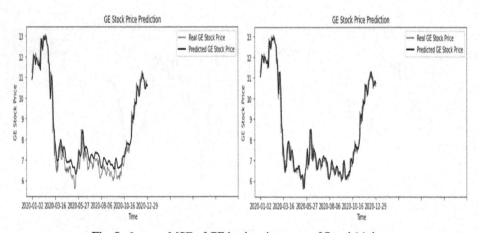

Fig. 8. Lowest MSE of GE having timesteps of 7 and 14 days

two companies which acquired the least error rates when trained over timesteps of 7 days whereas the other companies secured least errors during timesteps of 14 days. The model accuracies obtained by GE seem to have similar values ranging between 0.10 to 0.30 when trained over the past set of years which implies that most of the price fluctuations are captured by all of its models (Figs. 4, 5, 6, 7 and 8).

4.1 Graphs Having Least MSE Values of IBM, Cisco, Microsoft, Tesla and GE

Since the trained RNN-LSTM models were tested and made to predict the future stock prices for the year 2020 which was also the year when the pandemic began, we can observe variabilities in the predictive power of the models. From the above figures, we can see that the models of companies such as IBM, Cisco and GE did a fairly good job of predicting the future stock prices for the year 2020 as the model may have identified and learnt key patterns in its previous set of years whereas the companies such as Microsoft and Tesla showed maximum variation during the test phase as these models may have not been able to capture all the fluctuations in the stock prices while being trained over the previous set of years. Hence, only the past one year seems to contribute in giving the most accurate predictions for both of these models.

5 Conclusion

Forecasting financial stock prices have always been a challenge yet machine learning and deep learning techniques have come a long way in predicting the future prices with greater accuracy. In this experiment, the historical stock price datasets of 5 companies were obtained from yahoo finance. 50 RNN-LSTM models were built where each company's stock data was divided into 5 training sets pertaining to the distinct number of years and tested for the year 2020 having timesteps of 7 and 14 days with the mean squared error being the main performance metric. The stacked RNN-LSTM models of IBM, Cisco and GE performed well having a relatively low error as compared to the models of Microsoft and Tesla. This is mainly due to the trained models of Microsoft and Tesla not being able to learn all the patterns needed so as to effectively predict the future stock prices during the year 2020. Another interesting feature would be that the predicted stock prices of Microsoft and Tesla for the year 2020 were highly affected due to the existence of the Pandemic and hence their models failed to correctly capture all the fluctuations needed to better predict the future prices during the year 2020. Both models achieved their highest accuracies by being trained over the past one year which means that the rate of change of economic factors is high within a short span of time for both Microsoft and Tesla as compared to the other companies.

Limitations of the research include utilizing low computation power as the models were trained on 200 epochs and an increase in the number of epochs may result in a better performance accuracy. Also, a 2 layered stacked LSTM architecture is used for building the models and other existing variations of LSTM architectures can also be implemented. Future scope of this research may include adopting various hyperparameter tuning methods to find the most optimum number of hidden layer units and stacked layers which will significantly contribute in capturing most of the fluctuations in the past data. Using timesteps lower than 7 days may also yield in optimized results.

References

1. Leggio, K.B., Schniederjans, M.J., Cao, Q.: A comparison between Fama and French's model and artificial. Elsevier (2004)
2. Unadkat, V., Sayani, P., Doshi, P., Kanani, P.: Deep Learning for Financial Prediction. IEEE (2018)
3. Grudnitski, G., Osburn, L.: Forecasting S&P and gold futures prices: an application of neural networks. J. Futures Mark. **13**(6), 631–643 (1993)
4. Zhang, G.P.: Neural networks for classification: a survey. IEEE Trans. Syst. Man Cybern.—Part C: Appl. Rev. **30**(4), 451–462 (2000)
5. Strader, T.J., Rozycki, J.J., Root, T.H., Huang, Y.-H.(John): Machine learning stock market prediction studies: review and research directions. J. Int. Technol. Inf. Manag. **28**(4), 1–22 (2020)
6. Lu, W., Li, J., Li, Y., Sun, A., Wang, J.: A CNN-LSTM-based model to forecast stock prices. Artif. Intell. Smart Syst. Simul. **2020**, 1–10 (2020)
7. Gurney, K.: An Introduction to Neural Networks. Taylor & Francis e-Library, UCL Press Limited (2004)
8. Billah, M., Waheed, S., Hanifa, A.: Predicting closing stock price using artificial neural network and adaptive neuro fuzzy inference system (ANFIS): the case of the DHAKA stock exchange. Int. J. Comput. Appl. **129**(11), 1–5 (2015)
9. Keneni, B.M., et al.: Evolving Rule Based Explainable Artificial Intelligence for Unmanned Aerial Vehicles. IEEE (2019)
10. Di Persio, L., Honchar, O.: Artificial Neural Networks architectures for stock price prediction: comparisons and applications. Int. J. Circuits Syst. Sig. Process. **10** (2016)
11. Schmidhuber, J.: Deep Learning in Neural Networks: An Overview. Elsevier Ltd., 13 October 2014
12. Agrawal, M., Khan, A.U., Shukla, P.K.: Stock indices price prediction based on technical indicators using deep learning model. Int. J. Recent Technol. Eng. (IJRTE) **8**(2), 2297–2305 (2019)
13. Khan, W., Ghazanfar, M.A., Azam, M.A., Karami, A., Alyoubi, K.H., Alfakeeh, A.S.: Stock market prediction using machine learning classifiers and social media, news. J. Ambient Intell. Human. Comput. **2020**, 1–24 (2020). https://doi.org/10.1007/s12652-020-01839-w
14. Shen, J., Shafiq, M.O.: Short-term stock market price trend prediction using a comprehensive deep learning system. J Big Data **7**, 1–33 (2020)
15. Guttel, S., Elsworth, S.: Time Series Forecasting Using LSTM Networks
16. Jahan, I.: Stock Price Prediction Using Recurrent Neural Networks (2018)
17. Hands-On Guide To LSTM Recurrent Neural Network For Stock Market Prediction, 27 March 2020. https://analyticsindiamag.com/hands-on-guide-to-lstm-recurrent-neural-network-for-stock-market-prediction/
18. Gimeno, P., Viñals, I., Ortega, A., Miguel, A., Lleida, E.: Multiclass audio segmentation based on recurrent neural networks for broadcast domain data. EURASIP J. Audio Speech Music Process. **2020**(1), 1–19 (2020). https://doi.org/10.1186/s13636-020-00172-6
19. Wang, B., Zhao, C., Qiu, J.: Forecasting stock prices with long-short term memory neural network based on attention mechanism, 3 January 2020
20. Eckhardt, K.: Choosing the right Hyperparameters for a simple LSTM using Keras, 29 November 2018. https://towardsdatascience.com/choosing-the-right-hyperparameters-for-a-simple-lstm-using-keras-f8e9ed76f046
21. Wei, C.-C.: Development of stacked long short-term memory neural networks with numerical solutions for wind velocity predictions. Hindawi – Adv. Meteorol. 18 (2020). Article ID 5462040

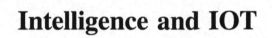

Intelligence and IOT

Sensitivity Analysis of Solar PV System for Different PV Array Configurations

Pallavi Verma$^{(\boxtimes)}$, Priya Mahajan, and Rachana Garg

Delhi Technological University, Bawana Road, Delhi, India

Abstract. Energy demand is increasing day by day and to meet this increasing demand renewable sources-based generating systems are used. Solar energy has many advantages over various renewable resources i.e. it can be harnessed easily, has low initial installation cost, continuous drop in the price of PV panels, no hazardous impact on the surrounding. For a large-scale solar photovoltaic (SPV) based system, an offline characterization study viz. sensitivity analysis needs to be performed at the design stage. Sensitivity study is made to find the behavior of a system due to variation in influential parameter viz insolation, temperature, inductance, duty ratio and capacitance. In the present study, author(s) performed a sensitivity study of SPV system considering various PV array configuration (series (S), parallel (P), series-parallel (SP), bridge linked (BL), honeycomb (HC), total cross-tied (TCT)). Sensitivity functions are developed for SPV cell and boost converter in reference with influential parameters by making use of Taylor's series.

Keywords: Solar photovoltaic system · PV array configuration · PV curve · Sensitivity study

1 Introduction

There are several renewable energy sources available in nature like hydro energy, wind energy, solar energy, geothermal energy, tidal energy, etc. These renewable energy sources are eco-friendly, require less maintenance, are abundant etc. Most of the energy-generating systems developed today are based on solar photovoltaic system (SPV). SPV are semiconductor devices that directly convert solar radiation into electricity. There has been tremendous research and development in the field of PV which has resulted in their reduced cost and increased efficiency [1, 2]. The PV-based projects are cost-effective, space-efficient and have a low operation and maintenance cost, therefore, making them a more reasonable choice when compared to wind or hydro projects. PV array can be interconnected in different configurations compared to conventional (series-parallel) configuration viz. total cross-tied (TCT), bridge linked (BL) and honeycomb (HC) for power enhancement [3]. The output of PV array is highly sensitive to insolation intensity on the PV surface and environmental temperature. Under varying climatological parameters, maximum power point tracking (MPPT) algorithms are used to track global peak from PV array and deliver it to the load [4, 5].

© Springer Nature Switzerland AG 2021
K. R. Venugopal et al. (Eds.): ICInPro 2021, CCIS 1483, pp. 393–403, 2021.
https://doi.org/10.1007/978-3-030-91244-4_31

Sensitivity analysis is performed to get a clear insight into the performance of the system at the design stage itself. Many remarkable contributions by QU Li-nan, Xue-Gui Zhu, Y. Xue, Ruifeng [6, 7, 8] have been made in the field of sensitivity. C. Rodriguez et al. [9] have presented mathematical modeling of grid connected PV system for stability studies. By performing eigen value and vector analysis, dynamic orbits are shown which may help in seeing a problem that may arise due to disturbances. MATLAB/Simulations results show the instances where voltage of PV panel collapses, working near to the maximum power point. Diego Oliva et al. [10] have proposed an artificial bee colony algorithm to estimate the unknown parameters of PV cell for sensitivity analysis. Artificial bee colony algorithm effectively identified the parameters of SPV cell when compared with other algorithms. Prajna et al. [11] performed a sensitivity analysis of current source inverter-based grid-tied SPV system. The uncertainties related to PV array, controller and grid are considered. Sensitivity analysis of system considering different parametric variations was examined by the location of eigenmode. L. Shu et al. [12] showed sensitivity study and voltage stability of grid-connected PV system. The authors studied the impact of disturbing parameters viz. temperature, irradiation, load change on the static and transient behavior of the system. Lei Guo et al. [13] presented cat swarm optimization algorithm for parameter identification and sensitivity analysis of solar cell single and double diode model. The presented results show that with the proposed algorithm higher accuracy of estimated parameters, I–V characteristic of PV has good agreement with experimental I–V data. H. Andrei et al. [14] discussed that the PV cell parameters are dependent on irradiation and temperature. LabVIEW and Matlab software are used to compute the measured and theoretical parameters of the PV cell model. In [15] authors have performed static and dynamic sensitivity analysis on the solar PV plant in Chile to find the effect of various parameters on the financial performance of the plant.

At present, there is limited literature available on the sensitivity concept in engineering applications of solar PV system. The presented manuscript contributes toward the sensitivity analysis of different configurations of solar PV system. The authors have developed sensitivity functions for SPV cell and boost converter in reference of influential parameters viz. insolation, temperature, inductance, duty ratio and capacitance by 1st derivative of Taylor's series. As per our knowledge, till now no literature is available in which sensitivity analysis of different PV array configuration have been performed. This paper shall be helpful to the design engineer for the optimum selection of component and PV array configuration at the design stage of the system. Further, these studies shall be used to develop sensitivity functions for other renewable resources.

2 Proposed System

In the present paper, author(s) have considered solar PV system as shown in Fig. 1, for carrying out the sensitivity analysis. The proposed system works as fluctuating direct current (dc) output is converted into regulated dc of required magnitude by boost converter which is connected to load.

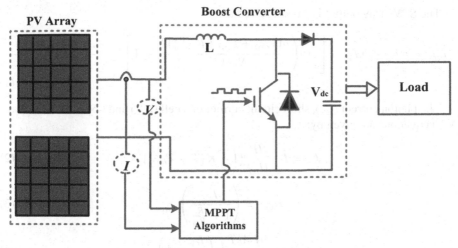

Fig. 1. Block diagram of solar PV system

2.1 Mathematical Modelling of Solar PV System

The solar PV system mainly comprises of PV array, boost converter and load. In the following sections, the design and mathematical modeling are done. Further, the sensitivity of PV cell with various PV array configurations and converter is conducted.

2.1.1 PV Array

In Fig. 2 the ideal and practical equivalent circuit of SPV cell is shown. The SPV cell converts photon light into electrical energy utilizing the principle of photovoltaic effect. The circuit is comprised of a current source with an anti-parallel diode. Series and shunt resistance are incorporated to make ideal PV cell practical.

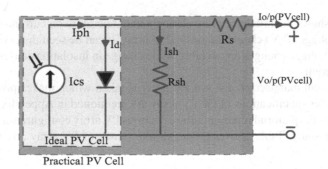

Fig. 2. Equivalent circuit of single diode PV cell

The SPV array output $I_{o/p}$ is expressed as:

$$I_{o/p} = p.I_{cs} - p.I_{sat} * \left[exp\left(\frac{q\left(V_{o/p} + I_{o/p} * R_s\left(\frac{p}{r}\right)\right)}{N_s AkT} \right) - 1 \right] - \frac{V_{o/p} + I_{o/p}R_s\left(\frac{p}{r}\right)}{R_{sh}\left(\frac{p}{r}\right)}$$

(1)

The photon current (I_{cs}), short circuit current of a cell (I_{scc}) and open-circuit voltage (V_{oc}) equations are given by:

$$I_{cs} = I_{scc}\frac{H}{H_{ref}}\left(1 + K\left(T - T_{ref}\right)\right)$$

(2)

$$I_{SCC} = \left(\frac{I_{cs}}{1 + \frac{R_s}{R_{sh}}} \right)$$

(3)

$$V_{OC} = \frac{N_s AkT}{q} ln\left(\frac{I_{cs}}{I_o} + 1 \right)$$

(4)

where, I_{sat} = diode saturation current.
q = elementary charge.
A = idealilty factor,
K = Boltzmann constant.
T_{ac} = actual temperature.
$V_{o/p}$ = output voltage of a PV array,
T = actual operating temperature,
T_{ref} = reference temperature.
p and r = of cells in series and parallel,
R_s and R_{sh} = series and shunt resistance of practical PV cell respectively,
N_s = number of series cells.
H and H_{ref} = solar insolation and reference solar insolation (1kW/m^2).
I_o = saturation current.

Figure 3 shows the influence of insolation and temperature on current-voltage (I-V) and power-voltage (P-V) characteristics of PV array. It can be seen that, with change in temperature, voltage changes appreciably while change in insolation causes appreciable change in current.

In the present manuscript, a 4 × 4 PV array is taken which makes power rating of 1036.8 W. Other specifications of the PV array are mentioned in Appendix A (Table 3). Figure 4 shows the pictorial representation of various PV array configurations considered for sensitivity study. Table 1 gives the maximum power of PV array considered in the study.

2.1.2 Boost Converter

In the proposed system, the boost converter is used as a dc-dc converter as shown in Fig. 5. The different parameters of boost converter [17] are expressed using given below equations.

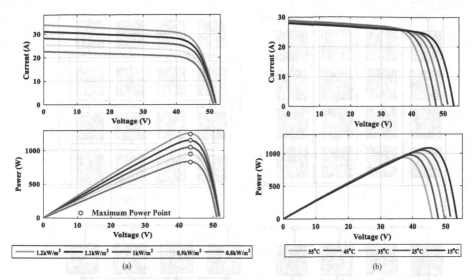

Fig. 3. Current vs Voltage and Power vs Voltage characteristics of system at a) Variable insolation b) Variable temperature.

$$\text{Inductance} \quad L = \frac{V_{i/p} * (V_{o/p} - V_{i/p})}{(\Delta I f_{sw} V_{o/p})} \tag{5}$$

$$\text{Duty ratio } \alpha = 1 - \left(\frac{V_{i/p}}{V_{o/p}}\right) \tag{6}$$

$$\text{Capacitance } C = \frac{I_a(V_{o/p} - V_{i/p})}{(\Delta V V_{o/p} f_{sw})} \tag{7}$$

$$\text{Resistive load } R_l = R_{in}/(1 - \alpha)^2 \tag{8}$$

Where, $V_{i/p}$ = input voltage,
$V_{o/p}$ = output voltage,
α = duty ratio,
ΔI = output ripple current and taken as 10% of the $I_{i/p}$,
f_{sw} = switching frequency,
ΔV = peak ripple voltage and which is 3% of the $V_{o/p}$,
R_{in} = input resistance.
The $V_{o/p}$ of boost converter is 360V which is fed to PV inverter.

3 Sensitivity Analysis

Sensitivity analysis is the performance evaluation technique for evaluating the change in the system's performance with respect to the change in its parameters.

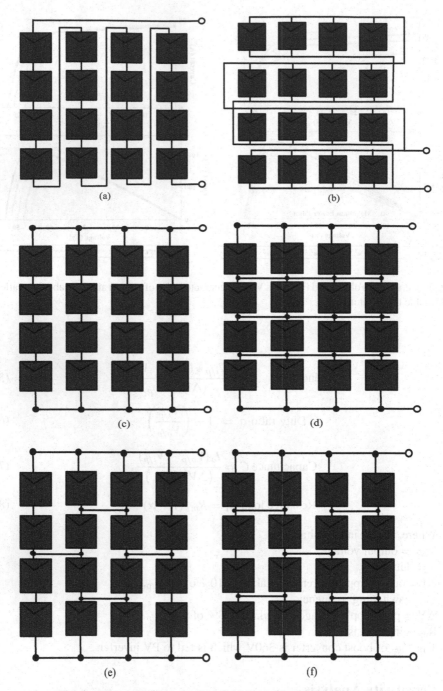

Fig. 4. Schematic diagram of (a) S (b) P (c) SP (d)TCT (e) BL (f) HC

Table 1. Maximum power (P_{nom}), voltage (V_m) and current (I_m) obtained from 4 × 4 PV array under various PV configurations

Configuration	P_{NOM} (W)	V_M (V)	I_M (A)
S	1036.8	288	3.6
P	1036.8	18	57.6
SP	1036.8	72	14.4
TCT	1036.8	72	14.4
BL	1036.8	72	14.4
HC	1036.8	72	14.4

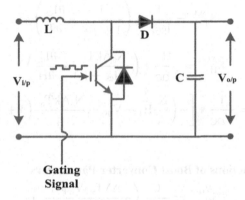

Fig. 5. Equivalent circuit of boost converter

Author(s) have performed a first-order sensitivity analysis of the considered PV system using partial differentiation. The sensitivity functions have been determined and the normalized sensitivity $\left(\widehat{S}_n^B\right)$ values have been computed. Normalised sensitivity is the percentage variation in the value of the system for 1% variation in the value of the influential parameter. The normalized sensitivity, B in reference with 'n' is given by:

$$\widehat{S}_n^B = \frac{\frac{\partial B}{F}}{\frac{\partial n}{n}} = \left(\frac{n}{F}\right)\left(\frac{\partial B}{\partial N}\right) \tag{9}$$

If number of parameters change at the same time, the sensitivity is calculated by Jacobian matrix:

$$J_{O_{ni}}^B = \begin{bmatrix} \frac{\partial B1}{\partial n1} & \frac{\partial B1}{\partial n2} \cdots & \frac{\partial B1}{\partial nN} \\ \vdots & \ddots & \vdots \\ \frac{\partial BA}{\partial n1} & \frac{\partial BA}{\partial n2} & \frac{\partial BA}{\partial nN} \end{bmatrix} \tag{10}$$

In this paper, the sensitivity analysis of solar cell parameters is performed in reference to insolation and temperature. Additionally, boost converter output voltage sensitivity has been determined w.r.t. inductance, capacitance and duty ratio.

The Sensitivity function for solar cell and boost converter parameter are given below:

3.1 Sensitivity Functions of Solar Cell Parameters

$$\hat{S}_H^{I_{CS}} = \frac{H}{I_{CS}} * \left([1 + (T - T_{ref}) * K_i] * \frac{I_{scc}}{H_{ref}} \right) \tag{11}$$

$$\hat{S}_T^{I_{CS}} = \frac{T}{I_{CS}} \left(\frac{I_{scc} * K_i * H}{H_{ref}} \right) \tag{12}$$

$$\hat{S}_H^{I_{SCC}} = \frac{H}{I_{SCC}} \left(\frac{1}{1 + \frac{R_s}{R_{sh}}} * \frac{\partial I_{cs}}{\partial H} \right) \tag{13}$$

$$\hat{S}_T^{I_{SCC}} = \frac{T}{I_{SCC}} * \left(\frac{1}{1 + \frac{R_s}{R_{sh}}} * \frac{\partial I_{cs}}{\partial T} \right) \tag{14}$$

$$\hat{S}_H^{V_{OC}} = \frac{H}{V_{OC}} * \left(\frac{N_s A k T_{ac}}{q(I_{cs} + I_o)} * \frac{\partial I_{cs}}{\partial H} \right) \tag{15}$$

$$\hat{S}_T^{V_{OC}} = \frac{T}{V_{OC}} * \left[-\left(\frac{N_s}{q} E_g - V_{oc} + \frac{N_s A k T_{ac}}{q} \left(3 + \frac{\gamma}{2} \right) \right) \right]$$

3.2 Sensitivity Functions of Boost Converter Parameters

$$\hat{S}_C^{V_{O/P}} = \frac{C}{V_{O/P}} \left(\frac{\Delta V f_{sw} V_{i/P} I_a}{(I_a - C \Delta V f_{sw})^2} \right) \tag{16}$$

$$\hat{S}_\alpha^{V_{O/P}} = \frac{\alpha}{V_{O/P}} \left(\frac{V_{i/p}}{(1 - \alpha)^2} \right) \tag{17}$$

$$\hat{S}_L^{V_{O/P}} = \frac{L}{V_{O/P}} \left(\frac{\Delta I f_{sw} V_{i/p}^2}{(V_{i/P} - L \Delta I f_{sw})^2} \right) \tag{18}$$

Corresponding normalized sensitivity values are calculated using data in appendix (Tables 3 and 4) are shown in Table 2. The sensitivity values have been calculated for different PV array configurations. Further, to validate the results of sensitivity analysis, the sensitivity values obtained from the sensitivity function are compared with the results obtained from difference equations by changing values of parameters of interest by 1%. The two results conform.

It can be observed from Table 2 that all the PV array configurations are equally sensitive to the parameters of interest viz. insolation, temperature, inductance, duty ration and capacitance. The effect of insolation on the V_{oc} is least for all configurations while I_{cs} is most sensitive with respect to insolation. Also, it can be seen from Table 2 that open-circuit voltage is negatively affected by temperature which is also validated from Fig. 3 (b). The V_{oc} decreases with an increase in temperature and increases with an increase in insolation level.

Table 2. Sensitivity function and its values of solar cell and boost converter

Sensitivity function		PV array configuration	Partial differentiation	Difference equation
Solar cell parameters	$\hat{S}_H^{Ics}$	S	1.000	0.999
		P/SP/TCT/BL/HC	1.000	0.999
	$\hat{S}_T^{Ics}$	S	0.097	0.098
		P/SP/TCT/BL/HC	0.097	0.098
	$\hat{S}_H^{Iscc}$	S	0.998	0.998
		P/SP/TCT/BL/HC	0.997	0.997
	$\hat{S}_T^{Iscc}$	S	0.096	0.096
		P/SP/TCT/BL/HC	0.096	0.096
	$\hat{S}_H^{Voc}$	S	0.0040	0.0040
		P/SP/TCT/BL/HC	0.0041	0.0041
	$\hat{S}_T^{Voc}$	S	− 1.05	− 0.99
		P/SP/TCT/BL/HC	− 1.04	− 0.99
Boost converter parameters	$\hat{S}_C^{Vo/P}$	S	0.57	0.56
		P/SP/TCT/BL/HC	0.55	0.55
	$\hat{S}_\alpha^{Vo/P}$	S	0.965	0.965
		P/SP/TCT/BL/HC	0.964	0.964
	$\hat{S}_L^{Vo/P}$	S	0.95	0.96
		P/SP/TCT/BL/HC	0.94	0.95

4 Conclusion

In the present paper, author(s) have carried out a sensitivity study for solar PV system considering different PV array configurations (S, P, SP, TCT, BL, HC). Sensitivity functions and their numerical values for solar cell and boost converter w.r.t. influential parameters viz. insolation, temperature, inductance, duty ratio and capacitance have been calculated. From the present study, it can be concluded that insolation level for solar cell and inductance values for boost converter is the most influential/sensitive parameters.

The sensitivity study shall give a clear insight to the system designers for selecting the optimum components for efficient and reliable operation of the system under steady-state and dynamic conditions.

5 Appendix A

Table 3. Specifications of solar 65W PV module

Components	Values
PV module power	65 W
Rated voltage, current	18 V, 3.6 A
Open-circuit voltage	21.7 V
Series cells, diode ideality factor	72,0.96
Reference insolation, temperature	1000 W/m^2, 298K
Series and parallel resistance	0.56 Ω, 70 Ω
Energy gap	0.7 V

Table 4. Specification of boost converter

Components	Values
Inductor	3.07–9.89 mH
Capacitor	1.25–2.73 μF
Duty ratio	0.5–0.7
Switching frequency	20 kHz

References

1. Gupta, A.K., Chauhan, Y.K., Maity, T.: Experimental investigations and comparison of various MPPT techniques for photovoltaic system. Sādhanā Acad. Proc. Eng. Sci. **43**(8), 132 (2018). https://doi.org/10.1007/s12046-018-0815-0
2. Singh, G.K.: Solar power generation by PV (photovoltaic) technology: a review. Energy **53**, 1–13 (2013)
3. Braun, H., et al.: Topology reconfiguration for optimization of photovoltaic array output. Sustain. Energy, Grids Netw. **6**, 58–69 (2016)
4. Singh, P., Palwalia, D.K., Gupta, A., Kumar, P.: Comparison of photovoltaic array maximum power point tracking techniques. IARJSET. **2**(1), 401–404 (2015)
5. Ramadan, H.S.: Optimal fractional order PI control applicability for enhanced dynamic behavior of on-grid solar PV systems. Int. J. Hydrogen Energy **42**(7), 4017–4031 (2017)
6. Li-Nan, Q., Ling-Zhi, Z., Tao, S., Jing, L.: Identification of photovoltaic power system based on sensitivity analysis. In: Asia-Pacific Power Energy Engineering Conference APPEEC, pp. 1–5 (2013)
7. Zhu, X.-G., Fu, Z.-H., Long, X.-M., Li, X.: Sensitivity analysis and more accurate solution of photovoltaic solar cell parameters. Solar Energy **85**(2), 393–403 (2011). https://doi.org/10.1016/j.solener.2010.10.022

8. Xue, Y., Manjrekar, M., Lin, C., Tamayo, M., Jiang, J.N.: Voltage stability and sensitivity analysis of grid-connected photovoltaic systems. In: IEEE Power Energy Society General Meeting, pp. 1–7 (2011)
9. Rodriguez, C., Amaratunga, G.A.J.: Dynamic stability of grid-connected photovoltaic systems. IEEE Power Soc. Gen. Meet. **2**, 2193–2199 (2004)
10. Oliva, D., Cuevas, E., Pajares, G.: Parameter identification of solar cells using artificial bee colony optimization. Energy **72**, 93–102 (2014)
11. Dash, P.P., Kazerani, M.: Sensitivity analysis of a current-source inverter-based three-phase grid-connected photovoltaic system. In: 2016 IEEE Electric Power Energy Conference EPEC 2016 (2016)
12. Shu, L., Zheng, J., Shen, X., Yin, H., Li, J.: Parameters' sensitivity analysis of grid-connected photovoltaic power generation model under different kinds of disturbances. In: Proceedings of 2012 IEEE Symposium on Electrical and Electronic Engineering EEESYM 2012, pp. 658–661 (2012)
13. Guo, L., Meng, Z., Sun, Y., Wang, I.: Parameter identification and sensitivity analysis of solar cell models with cat swarm optimization algorithm. Energy Convers. Manage. **108**, 520–528 (2016)
14. Andrei, H., Ivanovici, T., Diaconu, E., Ghita, M.R., Marin, O., Andrei, P.C.: Analysis and experimental verification of the sensitivity of PV cell model parameters. In: 2012 International Conference on Synthesis, Modeling, Analysis and Simulation. Methods and Applications to Circuit Design SMACD 2012, pp. 129–132 (2012)
15. Voglitsis, D., Member, S., Baros, D.K., Member, S.: Sensitivity Analysis for the Power Quality Indices of Standalone PV Systems, pp. 25913–25922 (2017)
16. Bellia, H., Youcef, R., Fatima, M.: A detailed modeling of photovoltaic module using MATLAB. NRIAG J. Astron. Geophys. **3**(1), 53–61 (2014)
17. Mudhol, A., Pinto Pius, A.J.: Design and implementation of boost converter for photovoltaic systems. IJIREEICE **4**(2), 110–114 (2016). https://doi.org/10.17148/IJIREEICE/NCAEE.2016.22

Application of Lifting Wavelet Transform in Feature Extraction of Time Series Data

Manju Khanna[1(✉)] and J. K. Mendiratta[2]

[1] MVJ College of Engineering, Bangalore, India
[2] CIIRC, Jyothi Institute of Technology, Bangalore, India

Abstract. A technique of feature extraction used for wind speed, power planning of small grids with hybrid power sources. In the past many techniques, i.e., SVD, FFT and DWT tried for feature extraction and cluster formation. The second-generation wavelet transforms, i.e., Lifting wavelet transform has been in use in diverse application, with much better performance. Motivation to carry out the present work intended for implementation of the algorithm of feature extraction and cluster formation in real time application. The technique applied based upon random number theory with central mean theorem and inequalities to find centroids of clusters of wavelet coefficients for feature extraction. The algorithm implementation carried out on data taken from National Energy for Research Laboratory. The datasets are spread across a period of one year. Results demonstrate the recovery of maximum energy of the signal with minimum error as compared to original data at the third level. It justifies the selection of third level of resolution, aimed at maximum signal energy recovery from the featured data and comparison of statistical results.

Keywords: Feature extraction · Wavelets · Clusters · Inequalities · Binary distribution · Central mean theorem · Standard deviation · Lifting Wavelet Transform (LWT) · Singular Value Decomposition (SVD)

1 Introduction

Wind power generation provides clean energy, which being felt worldwide. However, a commercial utilization of this energy requires prediction of power generation for a long duration. This utilization requires planning and commercial commitments for an integrated power generation. Wind-speed prediction depends upon weather and other environment factors. In the past efforts have been made for a short time wind power generation prediction, i.e. [1] by using data mining algorithms. They predicted the wind power up to 4 h in advance by application of multi-layer perception algorithm. Jinbin Wen et al. [16] used Support Vector Machine (SVM) combined with Lifting Wavelet Transform for a short-term prediction of wind power forecast.

It has been a continuous effort to find attributes/features of the data, while dealing with data mining or knowledge extraction in any application. For time series data, the problem is further aggravated due to its large no of samples loaded with missing values,

© Springer Nature Switzerland AG 2021
K. R. Venugopal et al. (Eds.): ICInPro 2021, CCIS 1483, pp. 404–415, 2021.
https://doi.org/10.1007/978-3-030-91244-4_32

noise, and outliers [18]. Attribute selection defines a task of selecting a subset of them, performing as good as the primitive ones, and named as the features.

For data reduction and feature extraction, different methods, such as, DFT and Discrete Wavelet Transform (DWT) applied [14, 26, 27]. From the above references, it is realized that pre-processing is required before its application to any learning, discovering and visualizing algorithms. In view of this, many algorithms tried, and the best ones selected for the data.

In Time series, data partitioned into sub-groups with similar instances as a group. Clustering algorithms make groups of data close by and separate others with large Euclidean distances. However, high dimensionality of clusters decreases the performance of these algorithms [15].

Other techniques tried for feature extraction, i.e., Li et al. [SIGKDD Explorations, Vol. 4, Issue 2] have surveyed various wavelets and provided methods for selection of various components for feature extraction. Eamonn Keogh et al. [9] have compared indexing measure for the performance analysis for Singular Value Decomposition (SVD), DFT, and Haar Wavelet along with Piecewise Aggregate Approximation (PAA) from their computation times, etc. and concluded the techniques for faster performance compared to other techniques. Adaptive Piecewise Constant Approximation (APCA) performs better than PAA. Christos Faloutsos et al. [10] have shown that generalized L_p performs much better in indexing and signal recovery compared to other techniques, such as SVD, DFT etc. For data, embedded with noise, its dimensionality reduction and feature extraction process improve the mining quality. However, its dimension reduction due to feature extraction reduces further, improves mining quality.

For time series data, with "Curse of Dimensionality" [3], Wu et al. [2000] have found better performance of DWT compared to DFT, in respect of feature extraction and recovery of signal from the original one. They have further presented that DWT performs better in standard deviation, error, and storage space. Fabian Morchen [13] carried out a study on time series data of different applications and compared DFT and DWT in feature extraction and energy recovery and has shown DWT to be superior compared to DFT. For power generation from wind speed, no study has been made in feature extraction and clustering of data, except by Jinbin Wen et al. [16], who have carried out short time power prediction. Recently, Deyun wang et al. [7] have used a Decomposition (VMD) along with wavelet and neural network for optimization by application of Genetic Algorithm (GAWNN) for a multi-step ahead prediction. Their technique employs multi-mode decomposition of signal frequency band, each individually optimized by application of GAWNN. Each of the predicted mode is further combined to predicts overall multi-step wind speed prediction. However, the authors have found features of the signal, which being a sequential data, needs reduction in dimensionality along with feature extraction.

1.1 Motivation

However, above studies and methods are not of much use for planning and sizing of various resources required. To meet the above requirements, a new technique is required for long-term prediction, and classification. With increase of prediction interval, error in prediction increases. This needs reduction of the data recorded, with features of the wind speed recorded for a long duration. Further, generation forecasting and augmentation of power supply, requires control of hybrid power resources, such as solar power, biofuels, and battery banks to feed power to the grid during lean periods. This requires integrated control of power generation resources, when working in tandem. This has led to development of a novel technique for feature extraction, which shall provide long-term prediction without increase of error in prediction. Authors in their paper have carried out long-term prediction of wind speed [29]. Further, authors [17] have shown that lifting wavelet transform performs better in energy recovery, memory requirements as compared to other wavelet Transforms, used for feature extraction and signal recovery. Further, the novel technique used by authors carries out in-place computation of the features, thereby cutting down the computational cost.

1.2 Wavelet Transform

A Fourier Transform being localized in frequency, but not in time. A Short Time Fourier Transform (STFT) or Windowed Fourier Transform (WFT) proposed. It was further generalized to Wavelet Transform, providing a variable sized window. In a Wavelet Transform, higher scales mapped to lower frequencies, thereby providing a better resolution at higher frequencies.

Haar Wavelet Transform: Defining a step function [18]:

$$\chi_{[a,b)}(x) = \{1 \quad if \ a \le x < b$$
$$= \{0 \quad Otherwise \tag{1}$$

Discrete form of Eq. (1), $\varphi(x) = \chi_{[0,1)}(x)$:

$$\varphi_{j,k}(x) = 2^{j/2} \varphi(2^j x - k) \quad j, \ k \in Z \tag{2}$$

Collection $\{\varphi_{j,k}(x)\}_{j,k \in Z}$ named "Haar Scaling function" family in real space $\mathfrak{R}$.

Function $\varphi_{j,k}$ provides translation (shifts) of 'k' integer units and dilating by power of 2, named dyadic dilation.

However, a basic Haar family of wavelet transform is not sufficient to extract features of time series.

1.3 Wavelet Transform-Based Feature Extraction

For efficient computation of features, Mallat designed a family of fast wavelet transform, using Multi-resolution Analysis (MRA) [22], naming them as pyramid algorithm. The process progressively smoothens the data using an iterative procedure. The location of smooth data indicates a low degree of polynomial, with a small value wavelet coefficient and vanishing moment property [4].

However, W. Swelden [24] proposed a Lifting Scheme for construction of a wavelet function, which simplified application of Wavelets. Introduction of this 'Second Generation' concept of wavelet transform [25], provided a generalized method in producing any wavelet. Further, Daubechies I. et al. [6] proved that a function with a Finite Impulse

Response (FIR) filter decomposed by a lifting scheme does not require translations and expansions, thereby avoiding application of Fourier transform. As the Lifting Wavelet carries out in-place calculation it does not require extra memory resource, thus provides a novel and faster calculation by a factor of two.

Application of this so-called 'Lazy wavelet transform has further simplified construction of the lifting scheme [23].

1.4 Construction of Clusters Using Lifting Wavelets

Cluster formation process comprises of:

i. *Data reduction*: A signal "s" when decomposed by lifting wavelet transform (LWT), provides coefficients $(h_0, h_1 \ldots h_N)$, N being total number of terms [21] such that:

$$\sum_{n=0}^{N} (-1)^n n^k h_n = 0 \ldots, \text{ k varying from 0 to } (N-1)/2 \tag{3}$$

$$\sum_{n=0}^{N-2k} h_n h_{n+2k} = 0 \ldots, \text{ k varying from 1 to } (N-1)/2 \tag{4}$$

Taking 'N' terms having two parts of solutions: a (N-1)/2 and d (N-1)/2.

If a signal's 'taken with odd samples L, (n being from 0 to L-1) and taking $a_0 = (a_{0n})$, with $a_1 = (a_{1n})$ and $d_1 = (d_{1n})$, we get:

$$a_{(m+1)n} = \sum_{k=2n}^{2n+N} h_{k-2n} a_{mk} \tag{5}$$

$$d_{(m+1)n} = \sum_{k=(n+1-N)}^{2n+1} (-1)^k h_{2n+1-k} a_{mk} \tag{6}$$

Terms 'a' and 'd' being 'Approximate' and 'Detailed' wavelet coefficients. Taking m = 0, Eqs. (5) and (6) provide total solutions:

Thus, by application of equations, (5), (6), a_0 reconstructed from a_1 and d_1 [S. Pittner, 1994], with energy divided as:

$$\|a_0\|^2 = \|a_1\|^2 + \|d_1\|^2 \tag{7}$$

a_0 can be decomposed as a sequence matrix of $d_1, d_2 \ldots d_M, a_M,$:

Thus, the original signal "s" decomposed in its wavelet coefficients having original wavelet coefficients of signal as a_0 on the top row of the above matrix named "B", Fig. 1 The "B" matrix augmented with zeros, (shaded portion) for rows completion.

The mth row present detailed part "d_m" and coarser portion "a_m" of the signal "s". The sequential decreasing rows consists of from higher to lower frequency regions.

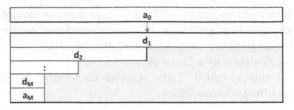

Fig. 1. Lifting wavelet transform coefficients of a sequence a_0

ii) *Feature Extraction and Cluster Formation:*

   Defining matrix B (b_{ij}), with sequence d_1, d_2...d_M and a_m, where 'i being from 1 to.M + 1 and j = 1, 2... (L + N-1)/2) for an even'L' and j = 1, 2... (L + N)/2) for an odd' L'.

   For a random variable Y_k, with variance of VY_k and expectation value EY_k, we get:

$$\rho_n := \sqrt{\sum_{k=1}^{n} VY_k} \quad \text{and} \quad Z_n := \frac{\sum_{k=1}^{n} Y_k - \sum_{k=1}^{n} EY_k}{\rho_n} \tag{8}$$

for $n \in$ N, EY_k , then

$$Z_n \to^D N(0, 1) \text{ as } n \to \infty \text{ and only if } \rho_n \to \infty$$

With $\to^D$ denoting convergence in 'normal' distribution 'N', with values between '0' and '1' [23].

   (Y_k) being an independent sequence with $\gamma \geq e^2$ (*e being Euler number*), for every $\varepsilon > 0$, with N (ε) and expected value EZ_N e, then Z_n:

$$\left| \{ Y_k := Y_k >= \sqrt{2 \ln \frac{N}{\gamma}}, \ k = 1, 2 ..., N \} \right| \tag{9}$$

Matrix B with a binomial distribution, taking '1' for coefficients greater than set value of $Y_{k, \text{ and}}$ '0' for those less than Y_k). For a set of K samples, with $\overline{B} := (|b_{ij}|)$.

1.5 Algorithm for Data Reduction and Cluster Formation

The algorithm for data reduction and feature extraction along with their clusters:

Algorithm 1: Data Reduction

Input: N number of samples, m number of Levels
Output: a_m, d_m coefficients

Step 1: Calculate a_m, d_1, d_2,..d_m by applying LWT
Step 2: If $N > 2^m$ goto Step 1
Step 3: stop

Algorithm 2: Feature Extraction and Cluster Formation

Input: a_m, d_m coefficients
Output Clusters U_i, Extracted Speed

Step 1: Begin
Step 2: Generate a matrix B {B: $=(b_{ij})$}, which contain the approximation a_m and the detail components $d_1...d_m$
Step 3: If there are a smaller number of detail or approximation coefficients pad them with zeros.
Step 4: Calculate T: $=\sqrt{2}(\ln N - \ln \gamma)$ with $\gamma \geq e^2$
Step 5: Do Iteratively
Step 6: IF coefficient value is greater than threshold T replace the coefficient value with 1
Step 7: ELSE Replace it with 0
Step 8: The Elements where there appears a 1 they form a cluster U_i
Step 9: The coefficient values which are replaced by value 1 correspond to peak at specific intervals.
Step 10: Speed is Extracted
Step 11: End

Novelty in Algorithm of Feature Extraction

(1) Application of Lifting Wavelet Transform which reduces number of terms of signal with 2^m (with m indicating level of reduction), further simplifies the computation. For example, if the signal sample collected consists of 'n' terms, in the first step of feature extraction, with m = 1, number of samples shall be reduced to n/2. Similarly, for next step, number of samples reduced to n/4 and so on. This feature is not available in any of the algorithm discussed above, i.e., SVD, FFT etc.
(2) Further, clustering of data helps clubbing of various features available in the data.
(3) Lifting wavelet Transform carries out computation in-place, thereby reducing need for additional computer resources for large data. This allows computation of features of the data in real time.

2 Experiment

The algorithm developed was tested using the wind speed data. The data is taken from NREL site, USA [31] for different periods of one year. The data is comprising of wind speed (m/s) recorded at regular intervals. The data considered individually for 12 months. In each individual month, there are about 4463 samples. These samples are recorded at an interval of every 10 minutes.

2.1 Results

Selection of Level of Resolution
This subsection highlights the reason and the choice of the level of LWT applied for the algorithm of data reduction (Algorithm 1). Feature extraction (Algorithm 2) transforms and application at different levels of resolution. data is collected for different periods of time i.e., one month, 6 months and 1 year. On application of LWT maximum energy of signal achieved, thereby results produced as tabulated in Table 1. A dataset of 1 month contains 4463 samples which are reduced to 558 samples. Similarly, the dataset for 6 months and 1 year are also reduced. Table 1 Shows that at level 3 the energy recovered is more than 99% which implies that after the application of LWT at level 3 on the original data, energy recovery is maximum as compared to level 2 and level 4. The % indicated in the table shows energy recovery with respect to original signal.

Table 1. Energy recovery of signal at different levels of resolution

Time period	Level 2(%)	Level 3(%)	Level 4(%)
1month	96.5	99.8	93.2
6 months	97.2	99.5	96.4
1 year	98.2	99.3	97.4

Data Reduction

The algorithm in Sect. 1.5 for feature extraction with LWT is converted to a graphical user interface tool using MATLAB.

Graphical User Interface

A Graphical User Interface tool developed for ease of application for the above algorithm. The Interface takes as input the wind speed (m/s). The data taken across 12 months with samples spread over periods of 1month,6months and 1 year. For each of the sample period, the corresponding approximation (a) and the detail (d) coefficients are calculated. The coefficients calculated using the Lifting wavelet transform as shown in Figure 2.

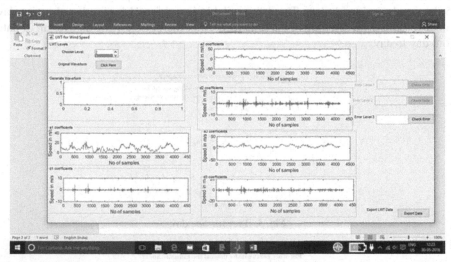

Fig. 2. Shows the 'a' and 'd' coefficients which are generated on the wind speed on application of LWT at Level 3

Table 2 shows the maximum error generated on the recovery of the signal after the inverse transform applied. The results show that the maximum error on recovery is very low. The outcome of the experiment are the a_3, d_1, d_2, d_3 coefficients.

Table 2. Maximum error generated on the recovery of the signal for different time periods.

Time period	Maximum error
1month	$9.91e^{-15}$
6months	$1.09e^{-14}$
1 year	$1.09e^{-14}$

Feature Extraction and Cluster Formation.
The output of Algorithm1 (that is the 'a' and 'd 'coefficients) which is generated acts as input to Algorithm 2. The B matrix produced by Algorithm 1 and 2. From the 'row echelon reduction' only the '1s' are selected in d_3 row while the majority of '1s' were in the a_3 row, indicating majority of clusters created from a_3 coefficients, and so the maximum energy of the signal is contained in this row. Total 11 cluster generated for the signal thereby showing different frequency bands. These clusters are generated for a period

of 1 month. Similar clusters are formed for different periods. Features u_i indicate the wavelet coefficients representing time and frequency information of the signal. Figure 3 shows a comparison of actual and predicted signals of one day. The results show the estimated and the original speed. The number of samples is 144 for a day with an interval of 10min between one sample and the next. Figures 4 and 5 present a comparison for 15 days and one-month data. The data available is not for 24 h of a day, the next day data also starts with resetting of the time recording, thereby showing an overlap of data from day to day.

Fig. 3. Plot for one day

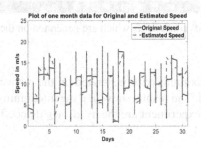

Fig. 4. Plot for one month

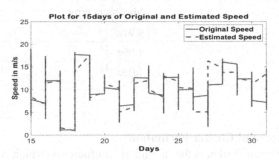

Fig. 5. Plot for 15 days

Mean, Standard Deviation, Root Mean Square (RMS):
The Mean, Standard Deviation and RMS values of Original and reduced data is shown

tabulated for 12 months in Table 3 and 4 respectively. In Table 3 and 4 there are columns (Ori and lwt) which basically indicate the results with respect to the original signal and reduced signal. On careful analysis of the results of each of the individual parameters, it shows that the parameters are either almost equal or close to each other. Hence this indicates that although the number of samples are reduced on application of the Lifting wavelet transform still the parameters (Mean, RMS, SD) remain almost the same. A reduced value in the Standard Deviation and Root Mean Square error indicates that the recovery of the data has been successfully achieved without much error.

Table 3. Comparison of Mean, Standard Deviation, Root Mean Square of Extracted and Original Signal (Jan – June) months.

Measure of the Signal Speed	Jan Ori	Jan lwt	Feb Ori	Feb lwt	Mar Ori	Mar lwt	Apr Ori	Apr lwt	May Ori	May lwt	Jun Ori	Jun lwt
Mean	10.23	10.21	9.563	9.576	9.539	9.543	10.338	10.343	9.494	9.504	7.24	7.276
Standard Deviation (SD)	3.727	3.74	4.539	4.546	4.384	4.365	4.198	4.208	3.754	3.732	3.649	3.623
Root Mean Square (RMS)	10.887	10.873	10.584	10.599	10.498	10.492	11.158	11.165	10.209	10.209	8.108	8.127

Table 4. Comparison of Mean, Standard Deviation, Root Mean Square of Extracted and Original Signal (July-Dec) months

Measure of the signal speed	Jul Ori	Jul lwt	Aug Ori	Aug lwt	Sep Ori	Sep lwt	Oct Ori	Oct lwt	Nov Ori	Nov lwt	Dec Ori	Dec lwt
Mean	6.364	6.367	7.285	7.335	8.455	8.426	7.34	7.326	7.817	7.799	9.197	9.192
Standard deviation (SD)	3.499	3.494	3.933	3.925	4.25	4.249	3.388	3.397	4.638	4.622	4.614	4.619
Root mean square (RMS)	7.262	7.262	8.279	8.317	9.463	9.435	8.084	8.074	9.089	9.064	10.289	10.285

3 Results and Discussion

The results shown in Table 2 above shows energy recovery of the signal, when data is reduced at different levels of resolutions. Above results, demonstrate recovery of signal energy at level '3' of resolution, thereby justifying use of this resolution for all the further experiments. Similarly, Tables 3 and 4 shows comparison of 'mean, 'rms' values and 'standard deviation' of the original and estimated signal, which are very close to each other, thereby justifying validity of the approach followed in feature extraction.

4 Conclusion

As reviewed, different techniques and algorithms are available for feature extraction, for time series data. These algorithms are applied for a short duration. This duration is non-stuffiest to exploit the renewable power resources for commercial utilization. To meet commercial requirements, a long duration prediction is required. To augment this requirement, present work is carried out. Application of Lifting Wavelet Transform (LWT) has justified its application for a long duration prediction. The results produced indicate a level three reduction of data produce very accurate prediction. The conclusion drawn applied for wind power generation, which can be extended for other renewable energy resources. In SVD algorithm, selection of attributes is carried out depending upon their Eigen values in the diagonal Eigen matrix. Similarly, application of DFT algorithm for feature extraction, as shown by research scholars, DWT performs better for different type of data. However, comparison of different techniques is carried out using a set of data applied for short duration prediction. The analysis of the technique presented provides an elegant method of feature extraction. With its simplicity, the algorithm can be implemented through programmable logic controllers for real time applications.

The algorithm developed can be applied for feature extraction and clustering of features for solar radiation and other real-time data in varied applications.

References

1. Kusiak, A., Zheng, H., Song, Z.: Short-term prediction of wind farm power: a data mining approach. IEEE Trans. Energy Conv. **24**(1), 125–136 (2009). https://doi.org/10.1109/TEC. 2008.2006552
2. Yi, B.K., Faloutsos, C.: Fast time sequence indexing for arbitrary norms. In: Proceedings of the 26th International Conference on Very Large Databases, pp. 385–394 (2000)
3. Bellman, R.: Adaptive Control Processes: A Guided Tour. Princeton University Press, Princeton, NJ (1961)
4. Chui, C.K.: An Introduction to Wavelets. Academic Press, Cambridge (1992)
5. Cooley, J.W., Tukey, J.W.: An algorithm for the machine calculation of complex Fourier series. Math. Comput. **19**(90), 297 (1965). https://doi.org/10.1090/S0025-5718-1965-0178586-1
6. Daubechies, I., Sweldens, W.X.: Factoring wavelet transform into lifting steps. J. Fourier. Anal. App. **4**(3), 245–267 (1965)
7. Wang, D., Luo, H., Grunder, O., Lin, Y.: Multi-step ahead wind speed forecasting using an improved wavelet neural network combining variational mode decomposition and phase space reconstruction. Renew. Energy **113**, 1345–1358 (2017)
8. Donoho, D.L.: Unconditional bases are optimal bases for data compression and for statistical estimation. Appl. Comput. Harmonic Anal. **1**(1), 100–115 (1993). https://doi.org/10.1006/ acha.1993.1008
9. Keogh, E., Chakrabarti, K., Pazzani, M., Mehrotra, S.: Dimensionality reduction of fast similarity search in large time series databases. J. Knowl. Inf. Syst. **3**, 263–286 (2000)
10. Faloutsos, C., et al.: Fast time sequence indexing for arbitrary Lp norms. In: Proceedings of 26th International Conference on very large databases, Cairo, Egypt, pp. 297–306, September 2000
11. Keogh, E., Chakrabarti, K., Pazzani, M., Mehrotra, S.: Locally adaptive dimensionality reduction for indexing large time series databases. ACM SIGMOD Record **30**(2), 151–162 (2001). https://doi.org/10.1145/376284.375680

12. Keogh, E., Kasetty, S.: On the need for time series data mining benchmarks: a survey and empirical demonstration. In: Proceedings of the 8th ACM SIGKDD International Conference on Knowledge Discovery and Data Mining, New York, pp. 102–111 (2002)
13. Morchen, F.: Time series feature extraction for data mining using DWT and DFT (2003). www.mybytes.de/papers/moerchen03time.pdf
14. Motoda, H., Liux, H.: Feature Selection, Extraction and Construction: A Data Mining Perspective. Kluwer Academic Publishers Norwell, Kluwer, MA (2003)
15. Zhang, H., Ho, T., Zhang, Y., Lin, M.-S.: Unsupervised feature extraction for time series clustering using orthogonal wavelet transform. Informatics **30**, 305–319 (2006)
16. Wen, J.: Short term wind power forecasting based on lifting wavelet transform and SVM. In: IEEE Conference 2012 paper, 978–1–4577–1600–3/12 (China)
17. Manju Khanna, N.K., Srinath, J., Mendiratta, K.: Feature extraction of time series data for wind speed power generation. In: 2016 IEEE 6th International Conference on Advanced Computing, pp. 169–173 (2016)
18. Schaidnagel, M., Laux, F.: Feature construction for time ordered data sequences. In: The Sixth International Conference on Advances in Databases, Knowledge, and Data Applications (2014)
19. Pittner, S.: Dyadic Orthogonal wavelet bases and related Possibilities for optimal analysis and representation of one-dimensional signals. Ph.D. thesis, Vienna University of Technology (1994)
20. Sen, P.K., Singer, J.M.: Large Sample Methods in Statistics-An Introduction and Applications. Chapman and Hall, New York (1992)
21. Pittner, S., Kamarthi, S.V.: Feature extraction from wavelet coefficients for pattern recognition tasks. IEEE Trans. Pattern Anal. Mach. Intell. **21**(1), 83–88 (1999). https://doi.org/10.1109/34.745739
22. Mallat, S.G.: A Theory of multi-resolution signal decomposition: the wavelet representation. IEEE Trans. Pattern Anal. Mach. Intell. **2**(7), 674–693 (1989)
23. Shodhganga.inflibnet.ac.in/bitstream/10603/4341/7/07
24. Sweldens, W.: The lifting scheme: a custom design construction of biorthogonal wavelet. Appl. Comput. Harmon. Anal. **3**(2), 186–200 (1996)
25. Sweldens, W.: The lifting scheme: a construction of second generation wavelet. SIAM J. Math. Anal. **29**(2), 511–546 (1998)
26. Li, T., Li, Q., Zhu, S., Ogihara, M.: A survey on wavelet applications in data mining. ACM SIGKDD Explor. Newsl. **4**(2), 49–68 (2002). https://doi.org/10.1145/772862.772870
27. Wu, Y.-L., Agrawal, D., El Abbadi, A.: A Comparison of DFT and DWT Based Similarity Search in Time-Series Databases. TRCS00–08 publication, Department of Computer Science, University of California, Santa Barbara (2000)
28. Divyakant Agarwal, Y.-L., El Abbadi, A.: A comparison of DFT and DWT based similarity search in time-series databases. In: Proceedings of the ninth International Conference on Information and Knowledge Management, pp. 488–495 (2000)
29. Khanna, M., Srinath, N.K., Mendiratta, K.: long term wind speed prediction using wavelet coefficients and soft computing. ICTACT J. Soft Comput. **07**(01), 1338–1343 (2016). https://doi.org/10.21917/ijsc.2016.0185
30. LNCS Homepage. http://www.springer.com/lncs. Accessed 21 Nov 2016
31. Logan, D., Neil, C., Taylor, A.: Modeling renewable energy resources in integrated resource planning. In: NREL, June 1994

Design and Implementation of Smart Wheelchair for Disabled/Elderly People

R. Rajesh[✉], P. Manju, R. Ajeethkumar, S. Kanimozhi, R. Karthick,
and T. Mohammed Riyas

Department of Instrumentation and Control Engineering, Sri Krishna College of Technology,
Coimbatore 641042, India
rajesh.r@skct.edu.in

Abstract. People with cognitive/motor/sensory impairment due to disability can use a wheelchair for their locomotive needs. The wheelchair can be manually or mechanically operated. Driving a manual wheelchair is always a challenge since the user always needs the help of another person to move from place to place. This work aims to provide ease in the movement of physically disabled by legs or elderly. For the movement of the wheelchair, 5 different keys are used by which the wheelchair can move forward, reverse, right, left, and stop. The wheelchair is also provided with a physical parameter measuring device with sensors that monitor the primary health conditions like temperature, pulse, and heartbeat as a salient feature for alarming about the emergency conditions of the disabled/elderly. With the help of a Wi-Fi module, the data from the sensors are sent to an IoT device wherein the data is stored and can be monitored periodically. The movement of the wheelchair can also be controlled via the IoT device and is incorporated in such a way that it is user-friendly. This wheelchair is cost-effective and can be widely used for disabled/elderly persons.

Keywords: Atmega 328 · Blynk · ESP 8266 · IoT · Wheelchair · Wi-Fi module

1 Introduction

The number of physically impaired people is increasing rapidly in recent years. Recent statistics show that around 15% of people (700 million) of the total world population are physically and mentally disabled. Among them, 100 million people are physically challenged [1]. Because of physical weakness, many wheelchair users cannot control wheelchairs properly by using their hands. Moreover, automated wheelchairs are not available everywhere in developing countries and the cost is high concerning the economic status of the common people. In India, there are many people with disabilities and the percentage of people with different disabilities varies. Among all the disabilities, the people with physical disability are comparatively high with 41.32% as per the recent statistics. This physical disability can be the impairment of hand or leg and sometimes it can be both hand and leg. The detailed percentage of disabled people in India is given in Fig. 1. To help people with a physical disability, wheelchairs play a vital role.

© Springer Nature Switzerland AG 2021
K. R. Venugopal et al. (Eds.): ICInPro 2021, CCIS 1483, pp. 416–423, 2021.
https://doi.org/10.1007/978-3-030-91244-4_33

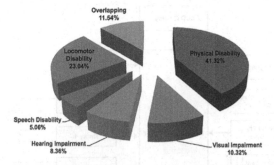

Fig. 1. Percentage of disabled people in India [1].

A wheelchair is a wheeled mobility device that helps the user for shifting from place to place. This device comes in many variations like self-propelled, propelled by the motor, or with the help of an attendee to push [2]. This work focus on the major factors listed below;

- Cost-effective – This wheelchair can benefit the users only if it is affordable and the materials used for the construction of the wheelchair are selected in such a way that it does not need high maintenance in the future.
- User friendly – This facilitates easy learning for the user to operate it and also it achieves maximum efficiency.

In this work, with the help of the keys fitted in the wheelchair, the movement of the wheelchair is controlled. Wheelchair movement is controlled by a direct keypad connected to Arduino or IoT. This wheelchair monitors the health and detects any abnormality of the user using the temperature, heart rate, and humidity rate monitored by the sensors. The data is updated on the server and if any abnormality is found, it will notify the concerned person through cloud data. The user objective is to have the system in a regular wheelchair and alert the caretakers in case of abnormality.

2 Proposed Wheelchair System

This smart wheelchair has a microcontroller that controls the operation of the wheelchair. The inputs for the microcontroller are from the keypad and the sensors. Atmega 28 is used as a microcontroller, 4 relays are used for 4 different movements namely forward, reverse, right, and left. Forward and reverse relays are connected to a gear motor 1. Right, and left relays are connected to a gear motor 2. These 2 motors move the wheelchair according to the input from the keypad or IoT. ESP 8266 Wi-Fi module is connected with a microcontroller and sends the data to the IoT device. Wheelchair movement is controlled by a direct keypad connected toArduino or IoT. The proposed system block diagram given in Fig. 2 and the working flow represnted in Fig. 3.

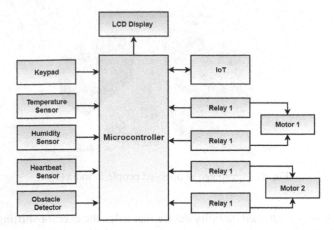

Fig. 2. Block diagram of a smart wheelchair

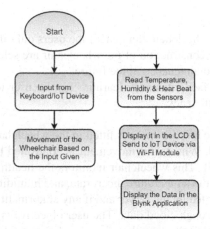

Fig. 3. Flow chart of a smart wheelchair

Obstacle Detector

To detect the obstacle in the path of the wheelchair, an HC-SR04 ultrasonic sensor is used. The sensor transmits ultra-sonic waves continuously and if there are no obstacles the waves are not disturbed [3, 4]. If there is an obstacle in the path, the waves hit the obstacle and get reflected. With the reflection, the obstacle distance can be calculated and it is displayed in the LCD. The wheelchair will automatically stop when the distance between the obstacle and the chair is less than 150 cm. The wheelchair will start to move if the obstacle has moved away from the path of the wheelchair.

Temperature Sensor

LM 35 is used as the temperature sensor. The sensor can measure temperature in the range of 0 to 100°C. This temperature sensor is connected to the Arduino and the Arduino reads the temperature of the person and displays it on the LCD [5]. This reading is also sent to the Blynk application via the Wi-Fi module and it is monitored periodically. If the reading is above the normal human temperature that is when the temperature exceeds 38 °C, the sensor notifies via the application, and hence preventive measures can be taken by the attendee.

Humidity Sensor
Humidity is the amount of water vapor or the amount of moisture in the air sample. The humidity sensor consists of an astable multivibrator in which the capacitance is varied depending on the humidity level [6]. The multivibrator produces the varying pulse signal which is converted into a corresponding voltage signal and the humidity is measured.

Heartbeat Sensor
The pulse rate is a measurement of the heart rate or the number of times the heartbeats per minute. As the heart pushes blood through the arteries, the arteries expand and contract with the flow of the blood. To measure the blood flow rate, an IR transmitter and receiver are used. The IR transmitter continuously transmits the rays and the rays reflect back based on the amount of flow of blood.Depending on the blood flow, the IR rays are interrupted. Due to that IR receiver conduction is interrupted so variable pulse signals are generated. With the pulse signal, the heart rate is measured and displayed in the LCD and the Blynk application. If there is any abnormality, it is notified to the attendee via the Blynk application [7].

Line Following Wheelchair
The line follower has an IR sensor module that has an IR transmitter and a photodiode as a receiver. IR sensor is connected to the Arduino and as a result, the Arduino sends input to the motor, and the wheelchair moves. The path for the destination is taped in black as the color has low reflectance. As the transmitter sends a signal, based on the reflection of the light that is received on the photodiode, the movement of the wheelchair is decided. Four IR sensors are initially placed and the Arduino is coded. To increase the efficiency of the movement of the wheelchair, the number of IR sensors can be increased [8].

The operation of this wheelchair commences when the microcontroller receives input from the keypad or from the IoT device connected to it via a wi-Fi module. The sensors measure the value of temperature, heartbeat, and humidity of the user and send the information to the IoT device via a microcontroller. IoT device is an application named Blynk and this app has control of the movement of the wheelchair and monitors and displays the temperature, heartbeat, humidity of user and alerts the attendee in case of emergency.

3 Result and Discussion

Mild steelhas used for the whellchair design because it can sustain higher stress and its strain & displacement is very low than other material [2]. Figure 4a and b shows the complete setup of the wheelchair with all the circuit connections and the physical parameters measuring sensors. The keypad fitted near the handle of the wheelchair has control of the movement of the wheelchair. The sensors measure the parameters and display them in both the LCD and the mobile application.

Fig. 4. a and b. The hardware setup of the smart wheelchair

The circuit connections are made as per the block diagram wherein the battery supplies power to the setup. The battery is connected to the Arduino via fuse. The Arduino receives input from the sensors and the IoT device. The sensors send the data to both Arduino which is displayed in the LCD and the data is also sent to the blynk application via Wi-Fi module from the Arduino. The readings from the Arduino are sent to the blynk application and input from the blynk application is received. The input from the blynk application is the control of the movement of the wheelchair. The movement of the wheelchair is controlled by both the keypad and also the blynk application. Figure 5 shows the homepage of the Blynk application. This homepage has both the readings of the measured physical parameters and the keys that have control of the movement of the wheelchair. The top portion of the homepage has the reading of temperature, humidity, and pulse rate. The bottom portion of the homepage has the 5 keys which have forward, right, reverse, left, and stop. The keys have control of the movement of the wheelchair.The five random trial has been taken with physical parameters as tabulated in Table 1. Black strip and white strip has been used for analysing the ability of the line or path tracking. There is no reflection from black strip, low or ('0') logic will reach to the microcontroller, and when there is white or reflective surface, there is high reflection and a high or ('1') reaches to the microcontroller. Figure 6 depicts line follower path with respective position of IR sensors. When the Wheelchair is following a black line, the sensor will send the signal to microcontroller and the microcontroller executes these inputs and sends output to PWM pins in which motor driver is connected. In Table 2, the detailed Wheelchair movement according to sensor status has been shown.

Fig. 5. Homepage of blynk application

Table 1. Trail readings of the physical parameters of random persons.

Trial	Person's age	Person's weight in Kg	Temperature in °C	Humidity in $g.m^{-3}$	Heart-rate in bpm
1	40	60	34	30	97
2	30	55	34	30	62
3	35	58	35	32	64
4	20	45	35	35	80
5	15	32	36	33	73

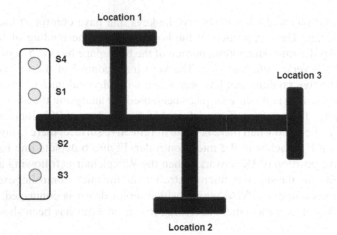

Fig. 6. Line follower path of the wheelchair with its destinations and the sensors.

Table 2. Movement of the wheelchair based on the sensor input.

S4	S2	S1	S3	Movement
0	0	0	0	Stop
1	1	1	1	Forward
1	1	0	0	Left
0	0	1	1	Right

4 Conclusion

This work can also be counted as an initiative for the betterment of physically disabled people who cannot walk so that they can easily handle it with their hands. This wheelchair meets the needs of the elderly and is as economical as one can easily afford it like never before. The benefits of using this wheelchair are that it has increased mobility, increased maneuverability, and increased physical support. As this wheelchair has the physical parameters monitoring sensors along with the Blynk application, it keeps monitoring the health of the person. The person on the wheelchair has control of the movement of the wheelchair and the obstacle detector helps in better operation of the wheelchair. The line follower helps in the easy movement of the wheelchair as it takes the control of the operation of wheelchair movement and thereby reducing the work of the person using it.

References

1. Prakash, I.J.: Aging, disability, and disabled older people in India. J. Aging Soc. Policy **15**, 85–108 (2003)

2. Gaal, R.P., Rebholtz, N., Hotchkiss, R.D., Pfaelzer, P.F.: Wheelchair rider injuries: causes and consequences for wheelchair design and selection. J. Rehabil. Res. Dev. **34**, 58–71 (1997)
3. Zhmud, V.A., Kondratiev, N.O., Kuznetsov, K.A., Trubin, V.G., Dimitrov, L.V.: Application of ultrasonic sensor for measuring distances in robotics. J. Phys. Con. Ser. **1015**, 032189 (2018)
4. Rajesh, R., Baranilingesan, I.: Tilt angle detector using 3-axis accelerometer. Int. J. Sci. Res. Sci. Technol. **4**, 784–791 (2018)
5. Gondchawar, N., Kawitkar, R.S.: IoT based smart agriculture. Int. J. Adv. Res. Comput. Commun. Eng. **5**, 838–842 (2016)
6. Kronenberg, P., Rastogi, P.K., Giaccari, P., Limberger, H.G.: Relative humidity sensor with optical fiber Bragg gratings. Opt. Lett. **27**, 1385–1387 (2002)
7. Kwak, Y.H., Kim, W., Park, K.B., Kim, K., Seo, S.: Flexible heartbeat sensor for wearable device. Biosens. Bioelectron. **94**, 250–255 (2017)
8. Chen, X., Agrawal, S.K.: Assisting versus repelling force-feedback for learning of a line following task in a wheelchair. IEEE Trans. Neural Syst. Rehabil. Eng. **21**, 959–968 (2013)

Solving Maximal Covering Location Problem Using Chemical Reaction Optimization

Md. Shymon Islam[✉], Md. Rafiqul Islam, and Humaira Islam

Computer Science and Engineering Discipline, Khulna University,
Khulna 9208, Bangladesh
shymum1702@cseku.ac.bd

Abstract. The Maximal covering location problem (MCLP) works with a given number of nodes in a network, each node has a demand value and is provided with a fixed number of facilities. Here the target is to maximize the total demands within some constraints. In this article, we have proposed a metaheuristic algorithm based on chemical reaction optimization to solve this problem. We have used two data sets to measure the performance of our proposed method. The proposed method gives better results in almost all the test cases in terms of percentage of coverage as well as computational time.

Keywords: Chemical reaction optimization · Demand maximization · Maximal covering location problem (MCLP) · Meta-heuristic · NP hard

1 Introduction

The covering location problem (CLP) is a well-known type of location problem discussed in location science [1]. The maximal covering location problem (MCLP) is a specific type of CLP that aims at finding an optimal placement of a given number of facilities to each customer or node on a network in such a way that the total demands covered by the served population is maximized. MCLP maintains the constraint that a customer is covered by a facility if it falls within the given constant service area or coverage area [2]. MCLP is a resource constraints problem. The objective of the problem is to serve the demands of customers as much as possible with the limited resources or budget [3]. Maximal covering location problem is also a constraint satisfaction problem that is very useful in locating objective areas to be served in different context. The MCLP has importance not only in private sectors but also in public sectors. Placement of plant, warehouse, telecommunication antennas etc. are the examples of private sectors where MCLP can be applied to find an optimal placement structure. Locating schools, bus-stops, ambulances, parks etc. examples of some public sectors where MCLP is applicable. MCLP is greatly applicable in designing network with constraints. The applications areas of MCLP have motivated us to solve the problem efficiently.

© Springer Nature Switzerland AG 2021
K. R. Venugopal et al. (Eds.): ICInPro 2021, CCIS 1483, pp. 424–437, 2021.
https://doi.org/10.1007/978-3-030-91244-4_34

There are several techniques for solving the maximal covering location problem. Many algorithms were proposed to solve the MCLP by researchers such as simulated annealing ([1] and [3]), lagrangean/surrogate heuristic [4], genetic algorithm [5], greedy heuristic ([6] and [7]), tabu search heuristic [8]. Although many algorithms were proposed and applied to solve MCLP, there is no general algorithm that can give always optimal results. This is the scope to work with MCLP. We have solved the MCLP by using chemical reaction optimization (CRO) algorithm. CRO is a nature based meta-heuristic approach. In recent years, CRO has successfully solved so many optimization problems such as shortest common super-sequence problem [9], RNA structure prediction [10], transportation scheduling in supply chain management with TPL [11], RNA secondary structure prediction with pseudoknots [12], optimization of protein folding [13], flexible job-shop scheduling problems with maintenance activity [14], the distributed permutation flow-shop scheduling problem with makespan criterion [15], the cloud job scheduling [16] etc. with better results than the other existing meta-heuristics algorithms. Due to the performance of CRO, we have chosen this algorithm. We are optimistic that CRO can find the optimal solution in less computational time. The contributions of the proposed work: redesigned the four reaction operators and obtained best results in both the datasets in less computational time compared to the state of the art algorithm (Atta_GA [2]). In this article, we have demonstrated the basic ideas of MCLP in Sect. 1. Section 2 describes the problem statement, Sect. 3 is for related work, Sect. 4 demonstrates our proposed method for solving MCLP using chemical reaction optimization and Sect. 5 concludes the work.

2   Problem Statement

The MCLP does the business with the problem of searching an optimal placement of a given number of facilities on a network in such a way that the total demands of the attended population is maximized [5]. MCLP works with a given number of customers (nodes) in a network. Each customer (node) has three basic properties: demand value, x and y coordinate values. The two coordinate values (x, y) define the position of a customer (node) in the network. Each customer (node) is provided with a fixed number of facilities. Each facility has a service area called the coverage area which is circular in shape and it remains constant throughout the whole process. A facility can be provided to a customer (node) if the Euclidean distance between the concerned node and the facility node is less than the constant service area. The MCLP is NP-hard [17]. The solution is non deterministic in polynomial time as it has to maintain a huge graph or matrix for a small size of customers (nodes) and the total subsets of the customers are also huge in number and the problem is restricted with several conditions. That's why meta-heuristic approaches are applicable to determine efficient solution for this problem.

The mathematical representation of the objective function of MCLP is given by the equation as follows:

$$f(D) = MAX \sum_{x \in D} d_x s_x \tag{1}$$

Expression (1) is taken from [2]. Here x and d_x denotes the index of demand nodes and the demand value at node (customer) x respectively. And s_x is a binary decision variable which becomes 1 when node x is selected and 0 otherwise. The target of the objective function (1) is to maximize the total demands covered by the selected customers. The objective function have two constraints. The first constraint is that each customer is provided with exactly a given fixed number of facilities. And the second one is that a customer (node) is covered or selected when there exists at least one facility within the coverage area of the node.

A sample MCLP example is given in Fig. 1. There are five nodes labeled as C1, C2, C3, C4 and C5. The corresponding demand and coordinate values of each node are presented in the upper and lower portion of each node respectively. The service distance for this network is 100, and the number of facilities to be installed is 2. This is a fully connected graph. For solving this MCLP graph, firstly a distance matrix needs to be created by using the Euclidian distance formula. Then All the possible solutions for five customers having 2 facilities needs to be generated. The total number of sequences is $\frac{5_{p2}}{2!}$ that is 10. Each sequence needs to be traversed in the whole distance matrix and finds the result. So, ten sequences must have ten results, among them the highest value will be the output. In the sample example, the highest value is 83 that is generated for sequence 10100. The final percentage of coverage is calculated by the ratio between sum of all demand values of the customers and the generated result. In Fig. 1, node C1 and C3 are selected because it maximizes the demand values of attended nodes. So, C1 and C3 are colored. Therefore, the estimated result for the graph in Fig. 1 is 83. So, the percentage of coverage is $\frac{83}{103} \times 100$ that is 80.58%.

3   Related Work

Some of the existing algorithms of MCLP are described here.

In 2018, Atta S. et al. proposed an algorithm for solving the maximal covering location problem (MCLP) using genetic algorithm with local refinement [2]. A binary array representation is used as the encoding scheme. The authors designed the selection, crossover and mutation operators respectively along with a local refinement procedure and elitism. Each chromosome goes through to these operators and tries to find out better solutions. They compared their proposed algorithm with other approaches with respect to percentage of coverage and computational time. Most of the cases, their proposed algorithm gives a fair better results in both percentage of coverage and computational time.

Maximo V.R. et al. proposed an intelligent guided adaptive search method for solving the maximum covering location problem (MCLP) in 2016 [8].

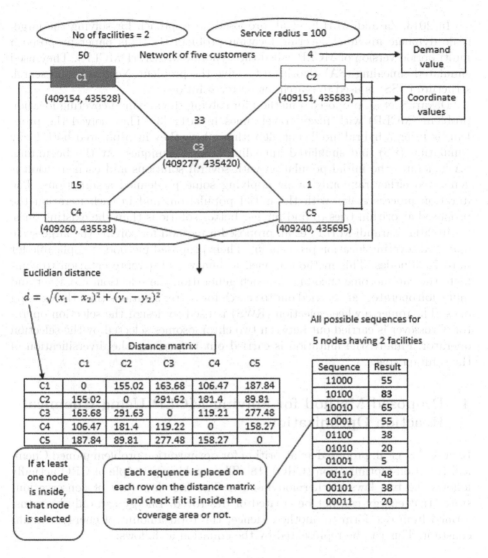

Fig. 1. An example of MCLP

The proposed method by this paper is named as Intelligent-Guided Adaptive Search (IGAS) and it actually follows the Greedy randomized search procedure (GRASP). Artificial neural network is used to build the construction phase of IGAS. This method is specialized for large-sized instances (more than 3000 nodes). An unsupervised machine learning algorithm, GNG (Neural Gas) is used to keep track of the best solutions found in the current iteration. So, the algorithm tries to take decision in such a way that, it can traverse through a promising branch.

In 2013, Zarandi M.H.F. et al. proposed an approach for solving the large-scale dynamic maximal covering location problem [1]. This paper proposed a multi period version of MCLP called the dynamic MCLP (DMCLP). They used simulated annealing (SA) algorithm to solve the problem. Neighborhood search structure (NSS) is used to search for better solutions.

Davari S. et al. proposed a method for solving the maximal covering location problem (MCLP) with fuzzy travel times in 2011 [3]. They solved the problem by using a hybrid intelligent algorithm where they incorporated both fuzzy simulation (FS) and simulated annealing (SA) techniques. At the beginning, SA generates the initial population with several solutions and each solution is generated either randomly or by applying some predefined assumptions. The iteration process starts with the initial population, and in each iteration the proposed algorithm tries to find further better solutions than the existing ones.

In 2011, Zarandi M.H.F. et al. proposed a method for solving the large-scale maximal covering location problem [5]. Their proposed method is applicable for upto 2500 nodes. This method applied a binary vector representation to complete the chromosome encoding. In each generation the selection, crossover and mutation operators are carried out to search for better solutions than the existing ones. The roulette wheel selection (RWS) is used to design the selection operator. Crossover is carried out between two chromosomes selected by the selection operator. Finally, the mutation is carried out to perform the diversification of the solution space.

4   Proposed Method for Solving MCLP Using Chemical Reaction Optimization

Lam A.Y.S. et al. proposed an algorithm for optimization problem named Chemical Reaction Optimization (CRO) [18]. The working principle of CRO actually follows the two laws of thermodynamics. The first law (law of conservation) states that energy can not be created or destroyed. Energy can only be transformed from one form to another. Hence, the total amount of energy remains constant. This can be represented by the equation as follows:

$$\sum_{x=1}^{Popsize(t)} (PE_x(t) + KE_x(t) + buffer(t) = C \tag{2}$$

In Eq. (2), $PE_x(t)$ and $PE_x(t)$ represents the potential energy (PE) and kinetic energy (KE) of molecule x respectively at any time t. $Popsize(t)$ is the total number of molecules, and $buffer(t)$ is the energy in the central buffer at time t. And the total constant energy is denoted by C. The value of C proves that the conservation of energy is maintained in CRO. The second law of thermodynamics ensures transformation of energy among the molecules (PE is converted to KE during iteration stage). CRO follows these two rules to come up with a better solution for optimization problems.

CRO is a population-based metaheuristic. Basically, CRO performs three basic stages. These are the initialization stage, iterations stage and the final stage. In the initialization stage, CRO initializes several attributes (initial parameters) and creates the initial population. Each molecule have several attributes or parameters. Some of these attributes are molecular structure (α), potential energy (PE), kinetic energy (KE), number of hits ($NumHit$) etc. After initializing the initial parameters CRO generates the initial population with popsize (total number of molecules). The second stage is the iterations stage. CRO has four elementary reaction operators. These are the on-wall ineffective collision, intermolecular ineffective collision, synthesis and decomposition. The on-wall ineffective collision and the intermolecular ineffective collision ensures intensification (local search). One the contrary, the synthesis and decomposition operator ensures diversification (global search). These four reaction operators make CRO a better approach for optimization problems compared to other metaheuristics because of its capability of searching. CRO increases or decreases the total number of molecules in each iteration according to the type of operator activated. The third stage is the final stage of CRO. If any of the stopping criteria is met during iterations or the limit of iteration exceeds then CRO will go to the final stage and show necessary outputs.

4.1   Basic Structure of Proposed Algorithm

MCLP_CRO is our proposed method. The initial parameters of CRO is initialized with proper values, then create the initial population randomly. By passing these initial parameters and initial population to the function MCLP_CRO, perform the iterations step. After the termination of MCLP_CRO, measure the outputs. The outputs are saved into two variables named *result* and *time*. The *result* represents the percentage of coverage value and *time* represents the computational time of the proposed method.

4.2   Solution Generation and Initialization

CRO is a population-based metaheuristic. A single unit from the whole population is called a molecule. In the iterations stage, one of the operators is manipulated to come up with new molecules with better objective function value. Here the interesting thing is that, the total number of molecules in each iteration does not remain constant, rather it varries from iteration to iteration. All the molecules with the popsize creates the whole population. Each molecule have several parameters. Some of the initial parameters of CRO are given in Table 1.

Population Generation. The initial population is generated on the basis of random selection. Let there are m customers where f facilities to be located. Each molecule of the initial population is generated by selecting f random indices from the set of customers or nodes $\{1, 2, 3, \ldots, m\}$. Here the value of f is less than m. And each f is generated randomly (random function mod popsize). And by

Table 1. Initial parameters

Symbol	Algorithmic definition
PopSize	Population zize (solution space)
KELossRate	Kinetic energy (KE) loss rate
Molecoll	Decision parameter (unimolecular or bimolecular) of CRO
Buffer	Initial energy in the surroundings
InitialKE	Initial kinetic energy
α and β	Threshold values of CRO
NumHit	Total number of hits a molecule has taken
Minstruct	Structure with minimum potential energy
MinPE	The potential energy when a molecule has minstruct
MinHit	The number of hits when a molecule has minstruct

looping through 0 to popsize, all the molecules are generated accordingly. Thus the initial population is created.

Solution Representation. A solution for MCLP is a set of f potential locations those needs to be chosen from the set of m customers. Let $m = 10$ and $f = 3$. Here, m and f are both are represented by one dimensional array. And initially all the values are 0. Let the randomly selected indices for the potential facility sites are $[3, 5, 9]$. Table 2 represents the indexed f array.

Table 2. Indexed f array

Index	1	2	3
Value	3	5	9

Table 3. Solution representation

Index	1	2	3	4	5	6	7	8	9	10
Value	0	0	1	0	1	0	0	0	1	0

A binary vector representation method is used for solution representation (see Table 3). In the solution, the selected indices of f array are represented by 1 and rest of indices are represented by 0.

4.3 Operator Design

CRO has four reaction operators. These are the on-wall ineffective collision, intermolecular ineffective collision, decomposition and synthesis. These operators are preformed selectively in iterations stage of CRO.

On Wall Ineffective Collision. When one molecule collides with the wall of a container, then the internal structure of the molecule changes. Here molecule m produces a new molecule m'

$$m \to m'$$

Let $m = 10$ and $f = 3$. So, each customer is provided with 3 facilities. The mechanism for on wall ineffective collision is very simple. Randomly select one or more indices from molecule m and change it to create a new molecule m'. In Fig. 2 we can see the 7th and 9th indices of molecule m are changed to form new molecule m'.

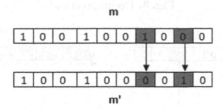

Fig. 2. On wall ineffective collision

Decomposition. In this elementary reaction, two new molecules are generated from a molecule. Two newly generated molecules bring diversity in their structure from the old molecule. Let, molecule m produces two new molecules m_1 and m_2.

$$m \to m_1 + m_2$$

According to our example, we have 10 customers and each customer needs to be provided with 3 facilities. The mechanism is quite simple. Firstly, divide the molecule m into two portions (see Fig. 3) using a random divider function. Then, the first and second portions of molecule m are copied to the beginning of molecule m_1 and m_2 respectively. The rest of the indices of molecule m_1 and m_2 are selected randomly using a random function generator.

Inter-molecular Ineffective Collision. In an inter-molecular ineffective collision, two molecules collide with each other. Let two molecules m_1 and m_2 collide with each other and produce two new molecules m_1' and m_2'.

$$m_1 + m_2 \to m_1' + m_2'$$

This is much similar to On-wall ineffective collision except that the number of molecules is twice here. Molecule m_1' is produced from molecule m_1 and molecule m_2' is produced from m_2. The mechanism is same as on wall ineffective collision and it is shown in Fig. 4. Several indices are changed to form a new molecule.

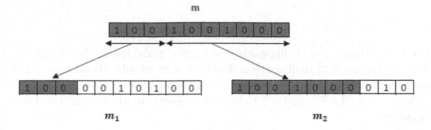

Fig. 3. Decomposition

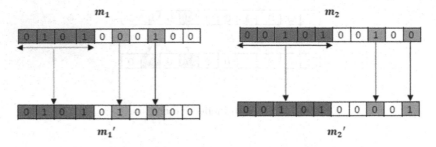

Fig. 4. Inter-molecular ineffective collision

Synthesis. Synthesis operator consolidates two molecules to form a new molecule. It is a reverse procedure of decomposition. Let m_1 and m_2 be two molecules. After the collision, molecule m is created.

$$m_1 + m_2 \rightarrow m$$

Synthesis operator performs diversification to traverse the global solution space and tries to find the optimal solution. One point crossover mechanism (see Fig. 5) is used for synthesis. The first few indices of molecule m are copied from molecule m_1 and rest of the indices are copied from molecule m_2.

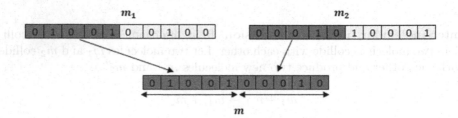

Fig. 5. Synthesis

4.4   Flowchart of Proposed Algorithm

In Fig. 6 we have shown the flowchart of the proposed method. Firstly generate
the initial population with a random function generator. Then check whether the
reaction is unimolecular or intermolecular. If intermolecular reaction occurs, then
either do inter-molecular ineffective collision or synthesis by checking which one
is appropriate. Otherwise perform on-wall ineffective collision or decomposition
in same manner. Then check if a max point is found or not. If any max point
is found then check for the stopping criteria, if it matched then obtain the best
max point and terminates. Otherwise, again check from the beginning, that is
whether the reaction is unimolecular or inter-molecular. Do the same again until
get a max point or reach the iteration limit.

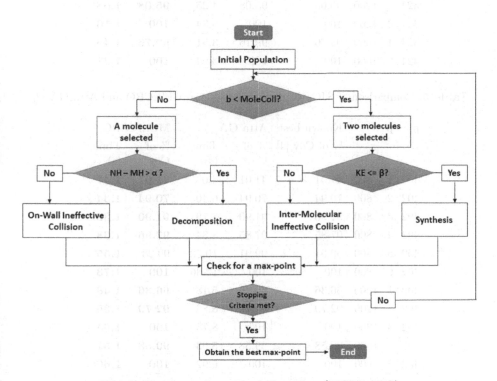

Fig. 6. Flowchart of proposed algorithm (MCLP_CRO)

4.5   Experimental Results and Comparisons

We have taken two real world datasets to measure the performance of our algorithm. These datasets are named as SJC_324 and SJC_402 where 324 and 402
are the total number of nodes in the datasets respectively. The datasets were
collected from *L.A.N.Lorena-instancias - INPE* whose link is available in the

Table 4. Comparison of SJC_324 dataset between MCLP_CRO and Atta_GA [2]

n	p	s	Known best	Atta_GA		MCLP_CRO	
			% of Cov [4]	% of Cov	Time (s)	% of Cov	Time (s)
324	1	800	44.94	44.94	2.42	**44.94**	**1.45**
324	2	800	72.33	72.33	3.79	**72.33**	**1.41**
324	3	800	95.49	95.49	5.49	**95.49**	**1.73**
324	4	800	99.62	99.62	5.07	**99.62**	**1.65**
324	5	800	100	100	6.45	99.80	**1.16**
324	1	1200	81.73	81.73	2.95	**81.73**	**1.47**
324	2	1200	95.08	95.08	4.25	**95.08**	**1.08**
324	3	1200	100	100	5.34	**100**	**1.16**
324	1	1600	99.76	99.76	3.51	**99.76**	**1.46**
324	2	1600	100	100	3.94	**100**	**1.27**

Table 5. Comparison of SJC_402 dataset between MCLP_CRO and Atta_GA [2]

n	p	s	Known best	Atta_GA		MCLP_CRO	
			% of Cov [4]	% of Cov	Time (s)	% of Cov	Time (s)
402	1	800	41.01	41.01	4.03	**41.01**	**1.47**
402	2	800	70.94	70.94	6.40	**70.94**	**1.34**
402	3	800	91.90	91.90	8.79	**91.90**	**1.33**
402	4	800	97.85	97.85	8.34	**97.96**	**1.38**
402	5	800	99.91	99.91	10.52	99.28	**1.55**
402	6	800	100	100	13.36	**100**	**1.73**
402	1	1200	66.36	66.36	5.18	**66.36**	**1.48**
402	2	1200	92.79	92.79	8.86	**92.79**	**1.39**
402	3	1200	100	100	8.75	**100**	**1.64**
402	1	1600	99.58	99.58	5.93	**99.58**	**1.51**
402	2	1600	100	100	6.92	**100**	**1.66**

website (http://www.lac.inpe.br/~lorena/instancias.html). These datasets were used for evaluating recent algorithms for solving maximal covering location problem ([2] and [4]).

We have implemented our proposed algorithm (MCLP_CRO) by using C++ language with device specifications: Processor- Intel(R) Core(TM) i5-7200U CPU @ 2.50 GHz 2.71 GHz, RAM - 8.00 GB (7.90 GB usable), System type - 64-bit operating system, x64-based processor, OS - Windows 10 Pro edition. We have initialized the CRO parameters as iteration = 150, PopSize = 10, KELoss-

Rate = 0.2, MoleColl = 0.4, InitialKE = 1000, $\alpha = 5$ and $\beta = 15,000$. We have recorded the input parameters as n, p and S, where n denotes the total number of customers or nodes in the network, p represents the total number of facilities to be installed and S is the constant service distance or service radius inside where the facilities can be provided to a node or customer. The $\%$ of Cov and Time(s) denotes the percentage of coverage and the computational time respectively. The best known $\%$ of Cov values for both the datasets have been collected from [4]. For a fair comparison, we have implemented Atta_GA [2] also using C++ in the same machine with the same device specification as for MCLP_CRO. The comparison for SJC_324 dataset between MCLP_CRO and Atta_GA [2] is shown is Table 4. Both datasets are tested with 3 different values of S (service distance). These values are 800, 1200 and 1600. There are 10 instances in SJC_324 dataset. By increasing the value of facility by one unit every time, we have measured the percentage of coverage and the computational time. The good results of our proposed method is highlighted by bold sign. For 9 instances out of 10, the proposed method gives better result than Atta_GA [2]. For test input ($n = 324$, $p = 5$ and $S = 800$), our method gave worse result than Atta_GA [2] with respect to percentage of coverage. The comparison for SJC_402 dataset between MCLP_CRO and Atta_GA [2] is shown is Table 5. There are 11 instances in SJC_324 dataset. For 10 instances out of 11, we have got good result compared to Atta_GA [2]. For test input ($n = 402$, $p = 5$ and $S = 800$), our method gave worse result than Atta_GA [2] with respect to percentage of coverage. To observe the efficiency of our proposed method, we have shown two graphs (Fig. 7 and Fig. 8) with service distance 800. Both graphs show that that Atta_GA [2] method takes more time than the proposed method in both datasets.

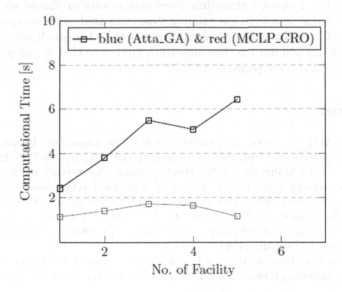

Fig. 7. Time comparison of SJC_324 dataset with $S = 800$

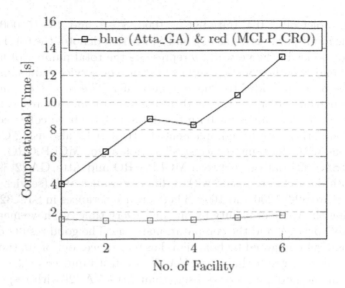

Fig. 8. Time Comparison Of Sjc_402 dataset with $S = 800$

5 Conclusion

In this article, we have proposed a CRO based method for solving the maximal covering location problem. Design the four reaction operators of CRO was a tough task. The proposed algorithm gives best results in almost all cases. We have obtained best results in less running time compared to the state-of-art algorithm (Atta_GA). The proposed algorithm was tested in small scale of MCLP. In the future, we will improve our algorithm further so that it can perform well on both small and large scales of MCLP.

References

1. Zarandi, M.H.F., Davari, S., Sisakht, S.A.H.: The large-scale dynamic maximal covering location problem. Math. Comput. Model. **57**(13), 710–719 (2013)
2. Atta, S., Sinha Mahapatra, P.R., Mukhopadhyay, A.: Solving maximal covering location problem using genetic algorithm with local refinement. Soft. Comput. **22**(12), 3891–3906 (2017). https://doi.org/10.1007/s00500-017-2598-3
3. Davari, S., Zarandi, M.H.F., Hemmati, A.: Maximal covering location problem (MCLP) with fuzzy travel times. Expert Syst. Appl. **38**(12), 14535–14541 (2011). https://doi.org/10.1016/j.eswa.2011.05.031
4. Lorena, L.A.N., Pereira, M.A.: A Lagrangean/surrogate heuristic for the maximal covering location problem using Hillman's edition. Int. J. Ind. Eng. **9**(1), 57–67 (2002)

5. Zarandi, M.H.F., Davari, S., Sisakht, S.A.H.: The large scale maximal covering location problem. Sci. Iran. **18**(6), 1564–1570 (2011). https://doi.org/10.1016/j.scient.2011.11.008

6. Berman, O., Krass, D.: The generalized maximal covering location problem. Comput. Oper. Res. **29**(6), 563–581 (2002). https://doi.org/10.1016/S0305-0548(01)00079-X

7. Máximo, V.R., Nascimento, M.C.V., Carvalho, A.C.: Intelligent-guided adaptive search for the maximum covering location problem. Comput. Oper. Res. **78**, 129–137 (2017). https://doi.org/10.1016/j.cor.2016.08.018

8. Bagherinejad, J., Bashiri, M., Nikzad, H.: General form of a cooperative gradual maximal covering location problem. J. Ind. Eng. Int. **14**(2), 241–253 (2017). https://doi.org/10.1007/s40092-017-0219-5

9. Saifullah, C.M.K., Islam M.R.: Solving shortest common supersequence problem using chemical reaction optimization. In: 5th International Conference on Informatics, Electronics and Vision (ICIEV), Dhaka, pp. 50–55 (2016). https://doi.org/10.1109/ICIEV.2016.7760187

10. Kabir, R., Islam, R.: Chemical reaction optimization for RNA structure prediction. Appl. Intell. **49**(2), 352–375 (2018). https://doi.org/10.1007/s10489-018-1281-4

11. Mahmud M.R., Pritom, R.M., Islam, M.R.: Optimization of collaborative transportation scheduling in supply chain management with TPL using chemical reaction optimization. In: 20th International Conference of Computer and Information Technology (ICCIT), Dhaka, pp. 1–6 (2017). https://doi.org/10.1109/ICCITECHN.2017.8281767

12. Islam, M.R., Islam, M.S., Sakeef, N.: RNA Secondary Structure Prediction with Pseudoknots using Chemical Reaction Optimization Algorithm. IEEE/ACM Trans. Comput. Biol. Bioinf. **18**(3), 1195–1207 (2021). https://doi.org/10.1109/TCBB.2019.2936570

13. Islam, M.R., Smrity, R.A., Chatterjee, S., Mahmud, M.R.: Optimization of protein folding using chemical reaction optimization in HP cubic lattice model. Neural Comput. Appl. **32**(8), 3117–3134 (2019). https://doi.org/10.1007/s00521-019-04447-8

14. Li, J.Q., Pan, Q.K.: Chemical-reaction optimization for flexible job-shop scheduling problems with maintenance activity. Appl. Soft Comput. **12**(9), 2896–2912 (2012). https://doi.org/10.1016/j.asoc.2012.04.012

15. Bargaoui, H., Driss, O.B., Ghédira, K.: A novel chemical reaction optimization for the distributed permutation flowshop scheduling problem with makespan criterion. Comput. Ind. Eng. **111**, 239–250 (2017). https://doi.org/10.1016/j.cie.2017.07.020

16. Zain, A.M., Yousif, A.: Chemical reaction optimization (CRO) for cloud job scheduling. SN Appl. Sci. **2**(1), 1–12 (2019). https://doi.org/10.1007/s42452-019-1758-8

17. Megiddo, N., Zemel, E., Hakim, S.L.: The maximum coverage location problem. SIAM J. Algebraic Discret. Methods **4**(2), 253–261 (1983). https://doi.org/10.1137/0604028

18. Lam, A.Y.S., Li, V.O.K.: Chemical reaction optimization: a tutorial. Math. Comput. Model. **4**(1), 3–17 (2012). https://doi.org/10.1007/s12293-012-0075-1

Geometric Programming (GP) Based Processor Energy Minimization Model for DVS Enabled Real-Time Task Set System

H. T. Manohara[1]([✉]) and B. P. Harish[2]

[1] Department of Electrical Engineering, University Visvesvaraya College of Engineering, Bangalore University, Bangalore 560001, India
manohara.phd@uvce.ac.in
[2] Department of Electronics and Communication Engineering, University Visvesvaraya College of Engineering, Bangalore University, Bangalore 560001, India
bp_harish@bub.ernet.in

Abstract. An analytical processor-level energy model to overcome power walls in communication and computation-intensive mobile applications is proposed. In real-time systems, when peak processing power is not required, judicious selection of supply voltage for energy-efficient operation is critical to address the power-performance tradeoff. The processor is operated at a lower frequency than maximum by dynamically varying its supply voltage to reduce energy. The proposed energy model, driven by a novel control variable called Task Utilization Factor (TUF), generates optimized supply voltage dynamically, within the pre-specified supply volt-age range, for each task of real-time periodic/aperiodic mixed task sets. The Geometric Programming based energy minimization model is proposed, over a range of maximum frequency, on randomly varying mixed task set to derive an optimized operating voltage. Simulation results on synthetic task data show that energy savings of 30% to 38% for randomly generated aperiodic task sets and 19% to 28% for randomly generated mixed task sets.

Keywords: Dynamic voltage-frequency scaling · Energy minimization model · Geometric programming · Real-time mixed task set · Task utilization factor

1 Introduction

The low power requirement critically constrains data-intensive, and computation-intensive applications in mobile communication devices and hence significant work is in progress in recent times to address this issue in real-time task systems. Advances in commercial CMOS chips and power-supply technologies have facilitated processors to run with varying supply voltages at run time, as required by

H. T. Manohara and B. P. Harish—Senior Member IEEE.

logic workload and application timing constraints of periodic and aperiodic real-time task systems, to achieve high energy efficiency. The two primary approaches to decrease energy consumption in mobile computing and communication devices are the slowdown mode and the shutdown mode of the processor. In processor shutdown mode, while the processor is in idle condition, clock generation and timer circuits continue to operate. Shutdown modes exhibit a tradeoff between power savings and mode transition overhead costs. When the required performance is less than its maximum, the processor enters slowdown mode and the speed of the processor is varied by varying the frequency in tandem with the supply voltage. It is observed that in the supply voltage regime of sub-1 V, the processor shutdown mode is found to be less effective than processor slowdown mode in achieving energy savings [1].

A dynamic supply voltage scaling (DVS) technique is presented to decrease aver-age energy consumption at the system level by up to 10x, without sacrificing perceived throughput, for mobile electronic devices [2]. The energy savings are measured by running benchmark applications at constant maximum throughput and by running with the voltage scheduler. The quadratic power dependence on voltage and the inverse linear delay dependence on voltage facilitates the energy-delay tradeoff by deploying DVS techniques. Traditionally, DVS techniques have relied on reducing energy consumption at the processor level instead of energy consumption at the system level. The convex relationship between the energy consumption of a CMOS circuit and the supply voltage, called Dynamic Voltage/Frequency Scaling (DVS/DFS), is an efficient technique to achieve low power and high throughput on-demand in low to moderate speed real-time systems. Energy savings achieved by supply voltage reduction comes at the cost of frequency for periodic tasks, with an approximately linear relation. The processor is placed in a shutdown mode, using a procrastination scheduling algorithm, to maximize the duration of idle intervals and hence find the critical speed as an operating point. Energy savings of 5% to 18% is reported over leakage oblivious DVS [3]. While this work considered the shutdown mode for energy savings, the accompanying irreducible system overhead costs would limit the quantum of energy savings that can be realized with shutdown mode. To guarantee schedulability of task sets and minimum power consumption, an optimal approach, with Earliest Deadline First (EDF) and Rate Monitoring (RM) scheduler, of calculating the Micro-Controller Unit (MCU) resource shutdown is presented and shown that the periodic shutdown method is simple to implement, but not optimal in terms of power savings [4]. A novel dynamic power-aware scheduling scheme is presented for a mixed-criticality real-time tasks system on a uni-core processor to mitigate the tradeoff be-tween minimizing power in Hi-criticality mode and overall average energy. Simulation results demonstrate that the proposed scheme is energy efficient in high critical mode and low critical mode of operation [5].

Several works have been reported in literature involving scheduling of hybrid task sets as an energy minimization technique in a DVS-enabled processor system. A survey on system-level optimization techniques from hierarchical perspectives of task scheduling for 3D processor platforms is presented for thermal analysis [6]. Voltage–frequency Island (VFI) based design paradigm is explored with techniques like ener-gy-aware task scheduling for energy optimization and task adjusting to balance the chip level power for thermal optimization. This facilitates energy-thermal tradeoff under conditions of amplified solution space resulting due to 3-D integration [7]. Multiprocessor systems with DVS architectures having energy-aware, off-line, probability-based scheduling and mapping algorithms are proposed to minimize the processor count while maximizing their utilization, minimizing energy consumption, and addressing the requirements of high performance in the presence of low power constraints [8]. Here, an optimal supply voltage is assigned to the processor, by the operating system, during task execution, and hence, it fails to exploit the slackness in task arrival times. The Downward-Upward Energy Consumption Minimization (DUECM) algorithm is proposed for a heterogeneous system to integrate the conventional upward approach with the downward approach [9]. The Xie model takes the worst-case execution time (WCET) as an optimization variable while setting the max-imum operating frequency. Thus, it fails to leverage on available slackness in task arrival times during task execution and limits the amount of energy savings that can be achieved. To save power for real-time mixed task sets and to ensure event-based aperiodic tasks execution time, loosely cycle conserving Earliest Deadline First (lccEDF) algorithm is proposed and 5% to 10% better response rate and 20% power savings is reported [10].

An online smart allocation policy for a multi-threaded multiprocessor system to schedule aperiodic real-time tasks is proposed and a power model is designed using DVFS constructs at a higher granularity to demonstrate 60% (maximum of 92%) to 30% (maximum of 45%) of energy reduction for synthetic data set and for real data sets, respectively [11]. This scheduling policy fails to consider tasks execution time or task utilization factor for online reduction of operating voltage. Based on slack steal-ing with an energy preserving mechanism, a novel aperiodic task scheduling algorithm for real-time energy harvesting systems is proposed to capitalize on energy surplus and processing time surplus to quickly service the aperiodic tasks. The implications of the energy profile on the aperiodic responsiveness is evaluated and superior responsiveness for the aperiodic activities is observed under different harvested energy pro-files [12]. A hybrid genetic algorithm-based approach is designed which allocates and schedules real-time tasks on a multiprocessor architecture while balancing the load on the processors in the safety-critical applications. Simulation demonstrates that, for a large number of tasks, the hybrid genetic approach is better than the classical GAs in terms of load balancing, response time, and flexibility [13].

A scheduling scheme is formulated as a Linear Programming (LP) problem, based on the schedulability analysis with DVFS for aperiodic tasks to arrive at an optimal number of checkpoints and minimize the worst-case execution time (WCET). The simulations demonstrate that selecting a large number of variable voltages can decrease energy consumption and fewer voltages for tasks, if judiciously selected, to result in significant energy reduction [14]. This work focuses on static slack times and fails to address dynamic slack times. To compute the static slowdown factor under the EDF scheduling scheme for tasks with varying power characteristics, the bisection algorithm is presented for slack identification that results in energy savings of 20% with static slowdown factor and of 40% with dynamic slowdown factor [15]. The bisection algorithms compute only static slowdown factor for computing energy savings and fail to produce the optimal discrete slowdown factor for enhanced energy savings. To achieve energy minimization, computation of optimum constant slow-down factors is formulated as a Geometric Programming model for real-time periodic tasks having varying power characteristics and periods more than the task deadlines [16]. This model applies to periodic and aperiodic tasks only, while mixed periodic & aperiodic task sets are not considered. To overcome the limitation of speed assignment policies for DVS-enabled proces-sor-based embedded systems, for periodic/aperiodic task sets, the bisection algorithm is derived from a frequency-aware methodology that minimizes the overall system energy consumption [17]. A Geometric Programming model is formulated for the computation of CPU speed factor applicable for real-time aperiodic tasks of varying power characteristics and task deadlines greater than periods [17]. By keeping the operating frequency maximum, the GP model for an aperiodic task sets in [16] and [17], uses worst-case execution time (WCET) and worst-case execution cycle (WCEC) as an optimization variable respectively. Thus it fails to utilize slackness in task arrival times and task execution times, thereby limiting energy savings.

An energy minimization model for real-time periodic tasks for uniprocessor systems using Geometric Programming(GP) is proposed by considering Task Utilization Factor (TUF) as a control variable for every task of task sets, to find an appropriate operating frequency and hence to compute the operating voltage from its pre-defined range, while achieving task feasibility. The slackness in task arrival times is exploited to achieve enhanced energy savings. The simulation results indicate 18% to 34% energy savings for standard task sets and 77% for randomly generated task sets, based on their power delay graphs [18]. However, this model is constrained by its shortcoming of not accounting for aperiodic tasks and mixed task set as in a hybrid task system. The present work based on Geometric Programming proposes a novel analytical en-ergy minimization model for periodic, aperiodic, and mixed real-time task sets for dynamic generation of slowdown factors and, in turn, optimal supply voltage levels for the processor in slowdown mode, exploiting the slackness in task arrival times to address the shortcomings of [3, 8, 15–18].

The rest of this paper is structured as follows: Sect. 2 introduces the model and feasibility test of the periodic and aperiodic task sets with related research efforts. The proposed GP-based methodology to generate optimized supply voltage levels for energy minimization is constructed in Sect. 3. While experimental findings are dis-cussed in Sect. 4, Sect. 5 presents conclusions with pointers to future work.

2   System Task Models

An introduction on the real-time hybrid task model, acceptance test, and feasibility tests are presented in brief here:

2.1 Real-Time Periodic and Aperiodic Task Model

A periodic task set of n periodic real time tasks is represented as $N = (t_1, t_2, \ldots, t_n)$. A 3-tuple for $t_i = (T_i, D_i, C_i)$ is used to represent each task, where T_i is the period, D_i is the relative deadline with $D_i \leq T_i$ and C_i is the Worst-Case Execution Time (WCET) for the i_{th} task [15]. Deadline may be implicit ($D_i = T_i$) or constrained ($D_i \leq T_i$) [17]. All tasks are assumed to be independent and preemptive. Further, each aperiodic task is characterized by 3-tuple $t_i = (rt_i, et_i, dt_i)$ where rt_i is the release time, et_i is the execution time and dt_i is the absolute deadline (sum of release time and relative deadline, and is shown in Fig. 1) of the i_{th} aperiodic task.

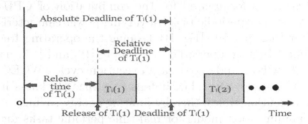

Fig. 1. Characterization of the real-time aperiodic task.

The power consumption value, i.e. the processor power state, is correlated with each task's frequency of execution. Certain tasks are performed before their dead-lines in a system with a fixed processing level, and the processor enters a sleep/slowdown mode for the rest of the time when full processing capacity is not needed. The supply voltage V_{dd}, a constant, is represented as a variable v_{dd} referred to as the operating voltage, and is varied to slowdown the processor. The functional and performance requirements determine the amount of slowdown that the system undergoes. At any time, the slowdown factor 'S' for the i_{th} periodic task is given by:

$$S = \frac{f_i}{f_{max}} \tag{1}$$

where f_i is the desired operating frequency of the i_{th} periodic task and f_{max} the processor's maximum operating frequency. Under slowdown, the execution time of the i_{th} periodic task is given by:

$$e_i = \frac{C_i}{S_i} \tag{2}$$

where C_i is the WCET and S_i is the slowdown factor of the i_{th} task. Thus, by applying slowdown, the execution time is expanded beyond WCET mandated by the task set to achieve energy savings. The speed factor/slowdown factor for the aperiodic task set is taken as 1, the desired operating frequency is the same as the maximum operating frequency for processing aperiodic tasks.

2.2   Acceptance Test for Aperiodic Task Model

The utilization factor is a fraction of time during which the processor is busy in running the tasks, and for a task t_i, without slowdown, is defined as:

$$u_i = \frac{C_i}{T_i} \tag{3}$$

For a task set, the processor utilization (U) is taken as the sum of utilization factors of every task and is given by:

$$U = \sum_{i=1}^{n} u_i \leq 1 \tag{4}$$

When an aperiodic task is released, it must be admitted by an acceptance test to determine whether it can be scheduled to meet its deadline. The bottom line is to ascertain if the aperiodic tasks are feasible under the maximum processor frequency f_{max}. If there are n ready tasks sorted in non-decreasing order of deadlines, the maximum instantaneous utilization of all tasks $U_{aperiodic,max}$ is defined as [17]:

$$U_{aperiodic,max} = max_{(1<i<n)}(u_{i,aperiodic}) \tag{5}$$

where

$$u_{i,aperiodic} = \frac{et_i}{dt_i} \tag{6}$$

is utilization factor for the i_{th} aperiodic task, after slowdown is applied.

2.3   Power Delay Characteristics

The operating voltage, and therefore the operating frequency of the processor vary dynamically at runtime to achieve low energy per task. The power consumption, P, is given by:

$$P = C_{eff} * V_{dd}^2 * f \tag{7}$$

where V_{dd} is the supply voltage, f is the operating frequency, and C_{eff} is the effective switching capacitance. If the processor runs at f_i, the operating frequency and v_{dd} is the operating voltage for a period of t seconds, then the energy consumed during i_{th} task execution is,

$$E(V_{dd}) = C_{eff} * v_{dd}^2 * f_i * e_i \tag{8}$$

With a square-law dependence of energy on operating voltage, and a trade-off be-tween low energy and deadline success having an approximately linear relationship, reasonable selection of operating voltage v_{dd} reduces the necessary maximum operating frequency (resulting in processor slowdown) to achieve the objective of energy minimization. According to an Alpha power law, the task delay is given as:

$$cycletime \propto \frac{1}{f} = t_d = \frac{\kappa * V_{dd}}{(V_{dd} - V_{th})^\alpha} \tag{9}$$

where V_{th} is the threshold voltage of the MOS transistor, α varies in the range 1 to 2, and κ is a process constant. From Eqs. (1) and (9), the supply voltage V_{dd} is related to the slowdown factor S for the i_{th} task as [16]:

$$S_i = \frac{(V_{dd} - V_{th})^\alpha}{\kappa * v_{dd}} \qquad (10)$$

where κ is constant and is given by,

$$\kappa = \frac{(V_{max} - V_{th})^\alpha}{v_{dd}} \qquad (11)$$

where V_{max} is the maximum operating voltage of the processor at the maximum operating frequency.

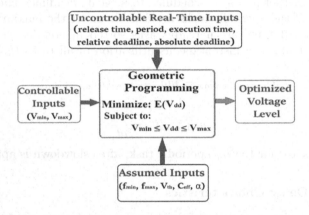

Fig. 2. Geometric Programming based TUF-driven Energy Minimization Model for real-time mixed periodic and aperiodic task system.

3   Geometric Programming Based TUF Driven Energy Minimization Model for Real-Time Task System

The proposed geometric programming [19] based energy minimization model for real-time mixed periodic and aperiodic task system is shown in Fig. 2. The objective is to achieve energy minimization on the execution of periodic and aperiodic task sets having arbitrary deadlines, scheduled in a uniprocessor system. The optimization model for DVS-enabled systems is based on the principle of reducing the operating frequency during periods of low processor utilization, such that the processor executes each task "just-in-time", stretching each task in time to its deadline. The operating frequency is reduced by reducing the operating voltage.

The decrease in frequency, coupled with a quadratic decrease in supply voltage, results in a cubic decrease in power consumption. The time to complete a task in-creases, with decreased frequency, leading to an overall quadratic decrease in the energy consumption for task execution. The uncontrollable inputs for the proposed energy minimization model are the periodic task set characteristics of deadline, period, worst-case execution time, and aperiodic task set characteristics of release time, execution time, and an absolute deadline. For the proposed model, V_{TH}, capacitive load, and the range of operating frequency are assumed. By varying the supply voltage range input (V_{max}, V_{min}), the GP model generates optimized operating voltage levels of v_{dd} for periodic task sets and a single optimized operating voltage for aperiodic task sets to increase energy savings.

3.1   Energy Minimization Model for the Real-Time Periodic Task

Under slowdown, the task utilization factor (TUF) u_i of the i_{th} periodic task is defined as:

$$u_i = \frac{e_i}{T_i} \tag{12}$$

To meet task deadlines, the task utilization factor u_i of the i_{th} periodic task, is defined as the ratio of task execution time to the minimum of period and deadline of the i_{th} periodic task [18], i.e.

$$u_i = \frac{e_i}{min(T_i, D_i)} \tag{13}$$

In the proposed TUF-driven GP-based methodology, the task utilization factor is formulated as a control variable to determine optimized operating voltage level for each task in the task set. In real-time periodic task systems, to achieve energy minimization while meeting the task deadline, energy consumed during the execution of the i_{th} periodic task, in terms of its TUF is given by [18]:

$$E(v_{dd}) = C_{eff} * v_{dd}^2 * f_i * u_i * min(T_i, D_i) \tag{14}$$

3.2   Energy Minimization Model for the Real-Time Aperiodic Task

Similarly, under slowdown, the task utilization factor for the i_{th} aperiodic task is defined as:

$$u_{i,aperiodic} = \frac{et_i}{dt_i} \tag{15}$$

Upon the success of the acceptance test of a new task, the energy minimization model calculates the optimized supply voltage values for all tasks waiting in queue to minimize their total energy consumption. Then, the energy consumed to execute the i_{th} aperiodic task in terms of its utilization factor is given by:

$$E(v_{dd}) = C_{eff} * v_{dd}^2 * f_i * u_{i,aperiodic} * Dt_i \tag{16}$$

where Dt_i is the Relative Deadline of the i_{th} aperiodic task, and it is defined as:

$$Dt_i = dt_i - rt_i \tag{17}$$

To satisfy the task feasibility test, a real-time task system must complete all tasks within their deadlines. If the execution time is considered for the feasibility of the task set, the Devi Test Method (DTM) test equation [20] is given by:

$$\forall i = 1, ..., n : \sum_{k=1}^{i} \frac{C_k}{T_k} + \frac{1}{D_i} \sum_{k=1}^{i} \frac{(T_k - D_k)}{T_k} * C_k \leq 1 \tag{18}$$

To enhance the energy minimization of real-time task system with aperiodic tasks, the energy model in (16) is developed as an objective function of a Geometric Programming formulation with an acceptance test, and the supply voltage and slowdown factor relation in (10) as a constraint and is as follows:

$$minimize : C_{eff} * v_{dd}^2 * f_{max} * u_{i,aperiodic} * Dt_i \tag{19}$$

$$subject\ to\ constraints: \sum_{k=1}^{i} \frac{et_k}{dt_k} + \frac{1}{D_i} \sum_{k=1}^{i} \frac{(dt_k - Dt_k)}{dt_k} * et_k \leq 1 \tag{20}$$

$$\frac{V_{th}}{V_{dd}} + \frac{(\kappa * S * V_{dd})^{\frac{1}{\alpha}}}{V_{dd}} \leq 1 \tag{21}$$

$$e_i = u_{i,aperiodic} * Dt_i \tag{22}$$

$$V_{min} \leq V_{dd} \leq V_{max} \tag{23}$$

4   Discussion: Simulation Results and Energy Savings

The proposed GP-based TUF-driven energy minimization model represented by Eqs. (19) to (23) for aperiodic task sets is solved numerically using a GGP solver referred to as GPPOSY, in the MATLAB platform. The simulation flowchart for the implementation of an energy minimization model based on GP is shown in Fig. 3. While the results for periodic task sets are presented in our earlier work [18], the results for aperiodic task sets are presented here to be followed by those for mixed periodic and aperiodic task sets. The efficacy and adequacy of the proposed energy minimization model are demonstrated for both periodic and aperiodic tasks.

The assessment is performed on a geometric programming platform by selecting the range of operating voltage of uniprocessors to be $V_{min} = 0.6\,\text{V}$, $V_{max} = 1.8\,\text{V}$, $V_{TH} = 0.36\,\text{V}$ and $\alpha = 1.2$. The features of real-time aperiodic task sets are specified: release time range of [0, 1000], relative deadline range [50, 1500], and execution time is 50% of the absolute deadline time where an absolute deadline is the sum of release time and relative deadline. The frequency range is taken to be 2.2 GHz to 3.2 GHz for periodic task sets, and its maximum value, i.e. 3.2 GHz for aperiodic task sets, to utilize the slackness in task arrival to reduce energy consumption/task. While fulfilling the task acceptance test for aperiodic task sets, the proposed TUF-driven GP model minimizes the energy consumption by taking maximum operating frequency and unity slowdown factor to compute the operating voltage. The Xie Model of [9] considers the maximum frequency f_{max} and unity slowdown factor and the execution time of tasks. The proposed work contrasts with the work in [9] incorporating a novel control variable referred to as task utilization factor (TUF). The substantial energy savings achieved in this work are due to the task utilization factor (TUF) driven modeling. To benchmark energy savings delivered by the proposed TUF driven energy minimization model against the Xie Model in [9], the optimized operating voltage and the corresponding energy consumption/task are computed, for identical aperiodic task sets.

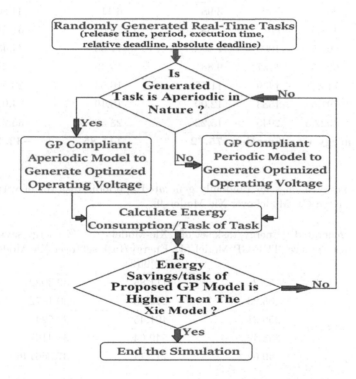

Fig. 3. Simulation Flowchart for GP-based TUF Driven Energy Minimization Model.

The Xie energy model equations of [9] are used for comparison. Simulation experiments are carried out using the TUF driven energy minimization model on randomly generated aperiodic task sets each having 'n' tasks and the results are tabulated in Table 1–Table 2 and the graphical representation of energy savings using dot plots are shown in Fig. 4. It is observed from Table 2 that the TUF driven model achieves 30% to 38% of energy savings on cumulative task sets of different sizes, i.e., n = 10 to 50 tasks, over the Xie energy model in [9], depending on the power delay characteristics of individual task sets.

Table 1. Energy savings for randomly generated aperiodic task set of size n = 10 using TUF-driven GP-Model over Xie Model [9].

Release Time	Execution Time	Absolute Deadline	TUF-GP Model optimized energy/task(J)	Xie model energy/task [9] (J) (V)	% energy saving/task over [9]
0	44.5	89	1.02	1.02	0
92	172.5	345	2.91	3.97	−26.7003
146	241.5	483	3.87	5.55	−30.2703
186	266	532	3.98	6.11	−34.8609
302	478.5	957	7.53	11.00	−31.5455
345	501.5	1003	7.56	11.53	−34.4319
397	628.5	1257	9.88	14.45	−31.6263
417	719.5	1439	11.75	16.54	−28.9601
539	791.5	1583	12.00	18.19	−34.0297
721	1022.5	2045	15.22	23.50	−35.234
Total Energy Consumption			**75.72**	**111.85**	**−32.3082**

Table 2. Energy savings for randomly generated aperiodic task set of varying sizes using TUF-driven GP-Model over Xie Model [9].

Randomly generated aperiodic task set size (n)	Energy/task set of TUF-GP Model (J)	Xie-Model Energy/task set in [9](J)	% energy saving/task over Xie-Model [9]
10	75.72	111.85	−32.3082
20	195.23	279.65	−30.1872
30	250.81	401.45	−32.524
40	338.49	549.64	−38.4167
50	449.62	717.74	−37.356146

Although the proposed work and Xie energy model in [9] generate an identical operating voltage for the maximum operating frequency, the TUF-driven GP algorithms energy savings are significantly better when the task utilization factors and relative deadline of randomly generated aperiodic task sets are introduced in the computation.

This illustrates the significance of 'just-in-time computation' leveraging on the slack-ness in task arrival times. The improved energy savings is not only in terms of better utilization factor in the computation but also in terms of better

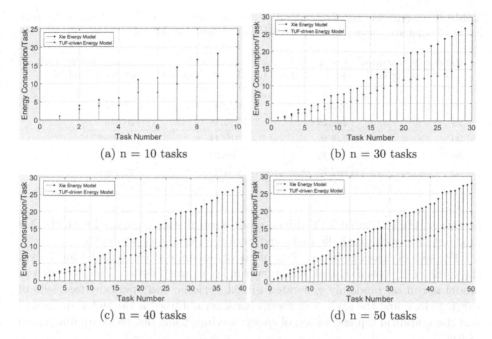

Fig. 4. Dot plots to show energy consumption/task for the proposed TUF driven GP-Model and Xie Model [9] for randomly generated aperiodic task sets of varying sizes.

Table 3. Energy savings for randomly generated mixed periodic and aperiodic task set of size n = 10 using TUF-driven GP-Model over Xie Model [9].

Release Time	Period	Execution Time	Relative Deadline	Absolute Deadline	Frequency using TUF-GP Model (GHz)	Operating Voltage by TUF-GP Model (V)	TUF-GP Model Energy/task (J)	Operating Voltage (V) [9]	Xie Model Energy/task [9] (J)	% Energy savings/task over Xie Model [9]
0	*82*	*44.5*	*89*	*89*	*3.20*	*1.41*	*1.02*	*1.41*	*1.02*	*0*
92	*150*	*172.5*	*253*	*345*	*3.20*	*1.41*	*2.91*	*1.41*	*3.97*	*-26.7003*
146	*275*	*241.5*	*337*	*483*	*3.20*	*1.41*	*3.87*	*1.41*	*5.55*	*-30.2703*
186	487	266	346	532	2.62	1.08	2.08	1.08	3.54	-41.2429
302	1046	478.5	655	957	2.73	1.14	3.84	1.14	7.08	-45.7627
345	1206	501.5	658	1003	2.88	1.23	4.31	1.23	8.63	-50.0579
397	1318	628.5	860	1257	2.89	1.24	6.50	1.24	10.88	-40.2574
417	1321	719.5	1022	1439	2.95	1.27	9.54	1.27	13.20	-27.7273
539	1345	791.5	1044	1583	3.03	1.36	11.80	1.36	15.82	-25.4109
721	1454	1022.5	1324	2045	3.16	1.41	21.12	1.41	23.20	-8.96552
Total Energy Consumption							66.99		92.89	-27.8824

tradeoffs being found. To extend the proposed methodology to real-time mixed periodic and aperiodic task sets, the model generated energy savings/task sets of randomly produced mixed periodic and aperiodic task sets of varying sizes, i.e., n = 10 to 50 tasks is tabulated in Table 3 and Table 4, respectively. While the aperiodic tasks are shown in bold and italic type, periodic tasks are shown in normal type, in Table 3. In the absence of comparable work addressing energy savings for randomly generated mixed periodic and aperiodic task sets in the

Table 4. Energy savings for randomly generated mixed periodic and aperiodic task set of varying sizes using TUF-driven GP-Model over Xie Model [9].

Randomly generated mixed task set size (n)	Energy/task set of TUF-GP Model (J)	Xie-Model Energy/task set in [9] (J)	% energy saving/task over Xie-Model [9]
10	66.99	92.89	−27.8824
20	202.25	275.38	−26.556
30	249.59	319.44	−21.8664
40	330.25	408.25	−19.1059
50	507.77	654.33	−22.3984

literature, the GP-based TUF-driven model's results are compared with the Xie model [9] for both aperiodic tasks and periodic tasks, independently.

The model achieves 19% to 28% of energy savings at the task set level, over the work in [9], based on task sets' power delay characteristics. The TUF-driven model is run on randomly generated mixed periodic and aperiodic task sets of varying sizes, and the savings in energy/task set obtained is tabulated in Table 4 and the graphical representation of energy savings using dot plots are illustrated in Fig. 5.

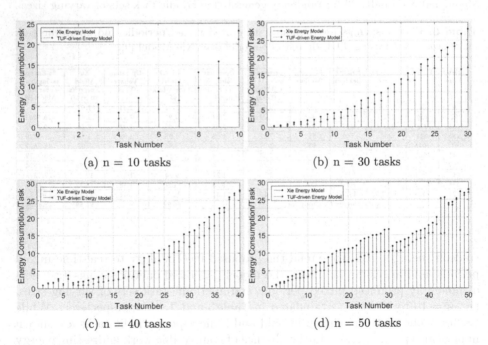

(a) n = 10 tasks

(b) n = 30 tasks

(c) n = 40 tasks

(d) n = 50 tasks

Fig. 5. Dot plots to show energy consumption/task for the proposed TUF driven GP-Model and Xie Model [9] for randomly generated mixed periodic and aperiodic task sets of varying sizes.

5   Conclusions

To maximize the energy efficiency of processors on mobile platforms, the TUF-driven GP-based energy minimization model is proposed and the classical trade-off between performance and power is addressed. In this supervision, the TUF-driven energy minimization model is formulated as a constrained GP optimization problem, and constraints are applied to generate an optimal operating voltage. While meeting the feasibility test for periodic task sets and acceptance test for aperiodic task sets, the model selects the appropriate frequency range from its predefined values for peri-odic task sets and maximum operating frequency for aperiodic task sets to compute the operating voltage level for each task. The impact of the task utilization factor as a major driver for the energy minimization in a uniprocessor environment is evaluated by carrying out real-time simulation experiments on randomly generated synthetic data of aperiodic task sets and mixed sets of periodic and aperiodic tasks of varying task set sizes. Simulation results demonstrate that energy savings vary between 30% to 38% for randomly generated aperiodic task sets and 19% to 28% for randomly generated mixed set of periodic and aperiodic tasks, depending on their power delay characteristics.

References

1. Hong, I., Kirovski, D., Qu, G., Potkonjak, M., Srivastava, M.B.: Power optimization of variable-voltage core-based systems. IEEE Trans. Comput. Aided Des. Integr. Circuits Syst. **18**(12), 1702–1714 (1995)
2. Burd, T.D., Pering, T.A., Stratakos, A.J., Brodersen, R.W.: A dynamic voltage scaled microprocessor system. IEEE J. Solid-State Circuits **35**(11), 1571–1580 (2000)
3. Jejurikar, R., Pereira, C., Gupta, R.: Leakage aware dynamic voltage scaling for real-time embedded systems. In: Proceedings of the Design Automation Conference, pp. 275–280 (June 2004)
4. Lautner, D., Hua, X., DeBates, S., Song, M., Ren, S.: Power efficient scheduling algorithms for real-time tasks on multi-mode microcontrollers. Procedia Comput. Sci. **130**, 557–566 (2018)
5. Ali, I., Jo, Y.I., Lee, S., Lee, W.Y., Kim, K.H.: Reducing dynamic power consumption in mixed-critical real-time systems. An MDPI J. Appl. Sci. **10**(20), 1–19 (2020)
6. Cao, K., Zhou, J., Wei, T., Chen, M., Hu, S., Li, K.: A survey of optimization techniques for thermal-aware 3D processors. J. Syst. Archit. **97**, 397–415 (2019)
7. Jin, S., Wang, Y., Liu, T.: On optimizing system energy of voltage-frequency island based 3-D multi-core SoCs under thermal constraints. J. Integr. VLSI **48**, 36–45 (2014)
8. Anne, N., Muthukumar, V.: Energy-aware scheduling of aperiodic real-time tasks on multiprocessor systems. J. Comput. Sci. Eng. **7**(1), 30–43 (2013)
9. Xie, G., Jiang, J., Liu, Y., Li, R., Li, K.: Minimizing energy consumption of real-time parallel applications using downward and upward approaches on heterogeneous systems. IEEE Trans. Ind. Inform. **13**(3), 1068–1078 (2017)

10. Kim, J.M.: Real-time scheduling for power-saving of mixed tasks with periodic and aperiodic. In: Proceedings IEEE 2nd International Conference on Computer and Communication Systems (ICCCS), pp. 70–73 (July 2017)
11. Ghose, M., Sahu, A., Karmakar, S.: Energy-efficient online scheduling of aperiodic real-time task on large multi-threaded multiprocessor systems. In: Proceedings. IEEE Annual India Conference (INDICON), pp. 1–6 (December 2016)
12. Osta, R.E., Chetto, M., Ghor, H.E.: An optimal approach for minimizing aperiodic response times in real-time energy harvesting systems. In: Proceedings. IEEE/ACS 15th International Conference on Computer Systems and Applications (AICCSA) (January 2019). https://doi.org/10.1109/AICCSA.2018.8612892
13. Gharsellaoui, G.H., Bouamama, S.: A new hybrid genetic algorithm-based approach for critical multiprocessor real-time scheduling with low power optimization. Procedia Comput. Sci. **157**, 1547–1557 (2019)
14. Li, G., Hu, F., Yuan,L.: An energy-efficient fault-tolerant scheduling scheme for aperiodic tasks in embedded real-time systems. In: Proceedings 3rd International Conference on Multimedia and Ubiquitous Engineering, pp. 369–376 (2009)
15. Jejurikar, R., Gupta, R.: Optimized slowdown in real-time task systems. IEEE Trans. Comput. **55**(12), 1588–1598 (2006)
16. Mutapcic, A., Murali, S., Boyd, S., Gupta, R., Atienza, D., Micheli, D.G.: Optimized Slowdown in Real-Time Task Systems via Geometric Programming, Submitted as short note to IEEE Transactions on Computers (July 2007)
17. Zhong, X., Xu, C.Z.: Frequency-aware energy optimization for real-time periodic and aperiodic tasks. In: Proceedings of the ACM SIGPLAN/SIGBED Conference on Languages, Compilers, and Tools for Embedded Systems, pp. 21–30 (June 2007)
18. Manohara, H.T., Harish, B.P.: Dynamic supply voltage level generation for energy minimization real-time tasks using geometric programming. In: Proceedings of the 32nd IEEE International System-On-Chip Conference (SOCC), pp. 376–381 (September 2019)
19. Boyd, S., Kim, S.J., Vandenberghe, L., Hassibi, A.: A tutorial on geometric programming. Educ. Sect. Optim. Eng. **8**(1), 67–127 (2007)
20. Devi, U.: An improved schedulability test for uniprocessor periodic task systems. In: Proceedings of the 15th Euro-Micro Conference on Real-Time Systems, pp. 23–30 (January 2003). https://doi.org/10.1109/emrts.2003.1212723

Process Variability-Aware Analytical Modeling of Delay in Subthreshold Regime with Device Stacking Effects

Anala M.[(✉)] and B. P. Harish

Department of Electronics and Communication Engineering,
University Visvesvaraya College of Engineering, Bangalore University,
Bangalore, India
anala.m@uvce.ac.in, bp_harish@bub.ernet.in

Abstract. In advanced process nodes, circuit design is sensitive to increasing process variability, thereby limiting performance enhancements. As the subthreshold operation is suitable for ultra low power applications, designing process variability-aware circuits is challenging. Accurate, reliable, and fast analytical models to estimate process variability impact on performance are critical in bridging the gap between circuit design and manufacturing. Hence, analytical models are proposed to model delay variability due to variations in device geometry and electrical parameters. The subthreshold current variation at the device level and delay at circuit level are explored due to simultaneous variations in gate length, width, and threshold voltage. The characterization of rising and falling edge delay variability of NAND/NOR structures in 32 nm PTM technology is based on distributions generated using Monte Carlo analysis and validated against SPICE. The model-based mean prediction error is below 0.12%, and the computational burden is reduced by 70X–500X compared to SPICE simulations. Further, it is demonstrated that the models hold good for a wide range of subthreshold operation, with supply voltage ranging from 0.18 V–0.23 V, with 0.2 V being the nominal supply voltage for subthreshold circuit operation.

Keywords: Analytical modeling of delay variation · Device stacking effects · Subthreshold regime · Variability-aware circuit design

1 Introduction

Despite the design maturity of the super-threshold regime, the subthreshold design has gained significant traction due to the rapid growth of battery-powered portable systems. Subthreshold circuits have achieved power reduction of several orders of magnitude in contrast to their super-threshold counterparts. Subthreshold circuits are the technology choice for ultra low power VLSI applications with stringent energy constraints since such circuits offer low-energy solutions with

B. P. Harish—Senior Member, IEEE.

low to moderate performance. Though low supply voltage and low operating current results in significant power savings, the subthreshold circuit design still poses multiple challenges. This can be attributed to its high susceptibility to Process (P), Voltage (V), and Temperature (T) variations that threaten to impact the performance metrics, and in turn, their predictability.

The Maximum Clock Frequency (FMAX) distribution model statistics reveal that significant performance degradation results from systematic within-die parameter fluctuations [1], thereby elucidating device architecture designs and new circuit design methodologies for improving the circuit performance of future Gigascale Integration (GSI). The circuit design is still not very promising in the subthreshold regime, mainly when sub-90 nm nodes are considered since process tolerances during the VLSI fabrication process are not tight enough and can result in significant gate delay variations of up to 300% [2]. The evaluation of process variation impact on propagation delay is modeled analytically and validated against the mixed-mode simulations. Simulation studies have revealed that gate length (L_g), halo dose, and oxide thickness (T_{ox}) are the dominant set of process parameters affecting both device as well as circuit variability [3]. Studies reveal that threshold voltage (V_{th}) and L_g are the significant sources of die-to-die variations and within-die variations, with the variance in L_g and V_{th} being more than 30% and 10%, respectively [4]. Although the corner models have been the mainstay of the IC industry for many years, they assumed a "one-size-fits-all" solution, i.e., within a die, all the devices would exhibit either the worst-case or best-case feature at the same time. While the "safe" method offers guard-bands for process parameter variations, it introduces pessimism in the design, resulting in significant circuit performance loss. Consequently, process parameter variation on delay modeling opens up new design challenges for timing sign-off.

From a design perspective, researchers have attempted to address the variability challenges by analysing the impact of single parameter variations on delay. The subthreshold delay variability due to PVT variations is characterized by first-order analytical models whose model accuracy has been validated through SPICE simulations on a 130 nm CMOS process. The modeling error is within 10%, 12%, and 0.9% for V_{th}, supply voltage (V_{DD}), and temperature variations, respectively [5]. To characterise PVT variations that occur in the subthreshold operation, an analytical model to estimate delay variability is presented with an error of 1% for process, 14.6% for supply voltage, and 6.8% for temperature variations, respectively, with the accuracy improvement achieved through a compensation factor [6]. The absence of modeling both the rising edge and the falling edge delay impacts the completeness of the model. Based on the subthreshold transient current model, output waveform and coupling capacitance equivalence, an analytical delay variability model is derived for the inverter chain under SMIC 40 nm technology node [7]. An analytical model for gate delay variability is proposed, taking into account variations in V_{th} alone and transient variation of device on-current during switching with DIBL effects [8]. A linear compositional model for delay prediction is proposed, i.e., the significant model parameters are first gathered through Monte Carlo simulations, which are then linearly propagated from device to circuit. It is assumed that V_{th} variations follow an Inverse Gaussian Distribution (IGD) [9]. Assuming a log skew-normal distribution for V_{th} variation, delay variability of the stacked gates is evaluated using the bivariate linear model [10]. In [11], the delay variability of a generic logic gate is computed by assuming the correlation between the stacked transistors. Furthermore, the variational model is validated only under specific process corners of V_{th} variation.

The existing models suffer from the following limitations: Though accurate models for subthreshold delay variability have been proposed for static CMOS inverters with V_{th} variations alone in [5–7], the models do not account for the effects introduced by the transistor stack. In [8–11], the stacking effect has been considered, but the delay variability models incorporate V_{th} variations alone. The modeling methodology of [5–11] accounts for the exponential dependence of delay on a single parameter, i.e., V_{th} and hence fail to model concurrent variations in device geometry parameters to depict the linear dependence on delay. The existing works fail to account for concurrent variations in device electrical (V_{th}) and geometry parameters $(L_g$ and $W)$. Though the existing models have effectively considered the single parameter (V_{th}) variation impact on subthreshold delay, they cannot be further extended to model simultaneous variations since the existing models focus only on the exponential relationship of V_{th} with drain current and not linear relationship of geometry parameters.

Process variations significantly impact all the aspects of circuit performance and pose a major challenge to robust IC design in subthreshold regime. Consequently, improving robustness and resolving the performance variability concerns during the early design phase is crucial. This requires accurate and reliable subthreshold analytical models for process variability characterization in modern processes.

To address these shortcomings, the contributions of the proposed work are as follows: The threshold voltage (V_{th}) being a significant parameter contributing to delay variability in the subthreshold regime, a simple and accurate variation-aware delay model is proposed, ensuring a realistic estimation of delay in the presence of its variation. The proposed delay variability model accounts for simultaneous variations in multiple process parameters, namely, V_{th}, L_g, and W ensuring a bridge between design CAD and technology CAD. To the best of the author's knowledge, there have been no variability models to account for the delay dependency on simultaneous variations in V_{th}, L_g, and W. To verify the model scalability with respect to circuit size, the efficacy and adequacy of the models are validated, over a wide range of subthreshold operation, using NMOS and PMOS stacks as in NAND and NOR gate, respectively. This demonstrates that the proposed models are efficacious in estimating the circuit delay variability of any size, complexity, and topology. The results demonstrate that the proposed delay variability models are about 70X–500X computationally more efficient than SPICE simulations in generating the delay distributions, with the mean error being less than 0.12%.

The rest of the paper is structured as follows: Sect. 2 outlines the motivation for the proposed work, principles of modeling methodology, and the subthreshold analytical delay variability models of NMOS stack and PMOS stack in terms of V_{th}, L_g, and W. Section 3 presents the characterization of delay distributions of NMOS stack (NAND gate) and PMOS stack (NOR gate) with SPICE simulation-based validation. Section 4 summarizes the results.

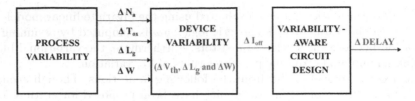

Fig. 1. Block diagram of subthreshold delay modeling in the presence of process variations.

2   Motivation and Principles of Modeling Methodology

The process flow of chip fabrication involves variability issues at every process step. Process variability has a profound impact on delay in the nanometer regime, translating into either parametric or catastrophic chip failure. Hence, there is a need to capture the impact of variability during the early phase of design flow to minimize the "almost-working" chips from being discarded, thereby increasing the chip yield. Further, the need for robust design has led to the development of analytical delay variability models, accounting for process variability. In the subthreshold regime, the propagation delay is ruled by an exponential (V_{th}) and linear dependence on parameters (L_g and W). Thus, the shift in their magnitudes can be very critical, impacting the delay predictability.

In this direction, first an analytical model for delay variability is derived accounting for variations in the dominant factor, V_{th}, in the subthreshold region. Further, the circuit-delay variability is evaluated in the presence of concurrent fluctuations in the electrical and geometry parameters of the transistor.

Figure 1 illustrates the underlying relationship between the device/circuit parameters and process parameters. The device geometry variations and fluctuations in the doping levels incurred by process variations impact the electrical characteristics of the device in unique and often subtle ways, which in turn varies the circuit performance parameters. Thus, it is desirable to map the variations in underlying process parameters such as V_{th}, L_g, and W to the delay variations at the circuit level. To model the statistical variations of gate delay, the variations in electrical and geometry parameters of the transistor are assumed to follow a Gaussian (normal) distribution. According to the experimental results [12], the spread in the parameters are projected to remain at $\pm 3\sigma = \pm 10\%$, with $\pm 3\sigma$ corresponding to variations around the nominal value. The delay variability analysis is executed in three phases:

1. variations in V_{th}: ΔV_{th}
2. variations in V_{th} and L_g: ΔV_{th} and ΔL_g
3. variations in V_{th}, L_g, and W: ΔV_{th}, ΔL_g, and ΔW

The execution of the primary phase involves ΔV_{th} alone while maintaining other parameters constant. The secondary phase entails simultaneous variations in V_{th} and L_g, with width, W, as a constant parameter. The last phase involves simultaneous variations of all the parameters. While the variability analysis in the first phase represents delay fluctuations with respect to a single parameter, the second and last phase characterizes delay variations under multiple parameters. To characterize the variations in rising edge and falling edge gate delay distributions, the proposed model-driven rigorous Monte Carlo analysis is performed. The delay distributions obtained by model calculations are validated against SPICE Monte Carlo simulations.

The proposed analytical framework emphasizes on the absolute amount of variability regardless of the taxonomy of process variability. In prior work [13], a static CMOS inverter was used as a reference circuit to validate the delay variability model. However, the adequacy and efficacy of the models were not evaluated in the case of simultaneous variations in device width and the transistor stacking effect with multiple inputs. To verify the adequacy and scalability of the models with respect to circuit size, there is a need to address the scalability requirements.

2.1 Analytical Delay Distribution Model for NMOS Stack

V_{th} is the most dominant device electrical parameter in the subthreshold region due to the inverse exponential dependence of drive current on V_{th}. A simple and accurate analytical model for computing the delay variability is presented in terms of three dominant parameter variations, V_{th}, L_g, and W. The gate delay model is derived based on the CV/I metric.

To elucidate the DIBL effect, being prominent in short-channel MOSFETs, V_{th} can be modeled as [14]:

$$V_{th} = V_{th0} - \eta V_{ds} \tag{1}$$

where V_{th0} is the zero-bias threshold voltage and η represents the DIBL coefficient.

According to the BSIM4 MOS model, the drain-source current of NMOS in the subthreshold regime is given by [15]:

$$I_{sub} = I_0 e^{\frac{(V_{gs}-V_{th0})}{mV_t}} e^{\frac{\eta V_{ds}}{mV_t}} (1 - e^{\frac{-V_{ds}}{V_t}}) \tag{2}$$

$$\text{As, } e^{\frac{-V_{ds}}{V_t}} \cong 0$$

$$I_{sub} = I_0 e^{\frac{(V_{gs}-V_{th0})}{mV_t}} e^{\frac{\eta V_{ds}}{mV_t}} \tag{3}$$

$$\text{where } I_0 = \mu_n C_{ox} \frac{W}{L_g}(m-1)V_t^2 \tag{4}$$

$V_t(= \frac{kT}{q})$ is the thermal voltage, m is subthreshold slope coefficient, μ_n is the electron mobility, $\frac{W}{g}$ is the transistor aspect ratio, and C_{ox} is the oxide capacitance per unit area. The existing equation from BSIM4 model is considered as the starting point for delay variability modeling.

The propagation delay (τ_d) is given by:

$$\tau_d = \frac{C_L V_{DD}}{I_{sub}} \tag{5}$$

where V_{DD} is the supply voltage and C_L is the load capacitance.

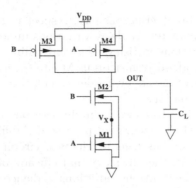

Fig. 2. Stacked NMOS transistors in the NAND gate.

From Eqs. (3)–(5), it follows that I_{sub} has an inverse exponential relationship with V_{th} and linear dependence on L_g and W. This dependence makes it necessary to consider these three parameter variations in order to account for delay variability.

To demonstrate the model scalability with respect to circuit size, the efficacy and adequacy of the models have to be validated using NMOS and PMOS stacks as in NAND and NOR gate, respectively. In [16], the NAND gate delay is computed by replacing the gate with an equivalent inverter. In the subthreshold regime, this method of merging the stacked transistors into a single equivalent transistor is no longer a suitable choice since the intermediate node voltage is non-negligible [8].

The stacking of NMOS transistors in a NAND gate is as shown in Fig. 2. The subthreshold current through the NMOS devices (M1 and M2) in the pull-down network of the NAND gate is expressed as [17]:

$$I_{sub,M1} = I_{01}e^{\frac{(V_{gs,M1}-V_{th0,M1}+\eta V_{ds,M1})}{mV_t}}(1 - e^{\frac{-V_{ds,M1}}{V_t}}) \tag{6}$$

$$I_{sub,M2} = I_{02}e^{\frac{(V_{gs,M2}-V_{th0,M2}-\gamma V_{sb,M2}+\eta V_{ds,M2})}{mV_t}}(1 - e^{\frac{-V_{ds,M2}}{V_t}}) \tag{7}$$

where γ = body bias coefficient, $V_{gs,M1} = V_{DD}, V_{ds,M1} = V_X, V_{gs,M2} = V_{DD} - V_X, V_{sb,M2} = V_X, V_{ds,M2} = V_{DD} - V_X$ and V_X is the intermediate node voltage in the stack.

Equations (6) and (7) can be rewritten as:

$$I_{sub,M1} = I_{01}e^{\frac{(V_{DD}-V_{th0,M1}+\eta V_X)}{mV_t}}(1 - e^{\frac{-V_X}{V_t}}) \tag{8}$$

$$I_{sub,M2} = I_{02}e^{\frac{(V_{DD}-V_X-V_{th0,M2}-\gamma V_X+\eta(V_{DD}-V_X))}{mV_t}}(1 - e^{\frac{-(V_{DD}-V_X)}{V_t}}) \tag{9}$$

The following model assumptions are reasonably valid with acceptable accuracy:

1. Variations in geometrical parameters of M1 and M2 are negligible, i.e., $I_{01} = I_{02} = I_0 = \mu_n C_{ox}\frac{W}{L_g}(m-1)V_t^2$
2. $V_{th0,M1} = V_{th0,M2}$ [8]

3. $e^{\frac{-(V_{DD}-V_X)}{V_t}} \approx 0$ [15]

4. Since $V_X \approx 0.1 V_{DD}$, $\eta V_X \approx 0$ [17]

The resulting expression for V_X is obtained by equating Eqs. (8) and (9) as follows:

$$I_{sub,M1} = I_{sub,M2} \tag{10}$$

$$V_X = V_t ln(e^{\frac{\eta V_{DD}}{mV_t}} + 1) = \frac{kT}{q} ln(e^{\frac{\eta V_{DD}}{mV_t}} + 1) \tag{11}$$

Substituting for V_X in Eq. (8) results in:

$$I_{sub} = I_{sub,M1} = I_0 e^{\frac{(V_{DD}-V_{th0}+\eta V_t(ln(e^{\frac{\eta V_{DD}}{mV_t}}+1)))}{mV_t}} (1 - e^{\frac{-V_t(ln(e^{\frac{\eta V_{DD}}{mV_t}}+1))}{V_t}}) \tag{12}$$

where $I_{sub,M1}$ denotes the worst-case subthreshold current.

Subthreshold Analytical Delay Modeling Under ΔV_{th}

The delay model of τ_d is dependent on ΔV_{th} as depicted by the analytical model of Eq. (5).

Substituting for I_{sub} in Eq. (5) from Eqs. (4) and (12), and by differentiating the equation for τ_d with respect to ΔV_{th} results in:

$$\Delta \tau_d \Delta V_{th,NAND} \equiv \frac{\delta \tau_d}{\delta V_{th}} = \frac{K_1 C_L V_{DD}}{I_0 m V_t e^{\frac{(V_{DD}-\Delta V_{th}+\eta V_t(ln(e^{\frac{\eta V_{DD}}{mV_t}}+1)))}{mV_t}} (1 - e^{\frac{-V_t(ln(e^{\frac{\eta V_{DD}}{mV_t}}+1))}{V_t}})} \tag{13}$$

where $\Delta \tau_d \Delta V_{th,NAND}$ is the NMOS stack delay variability under V_{th} variations alone and K_1 represents the model fitting parameter. Further, V_{th} also depends on T_{ox}, doping concentration, Random Dopant Fluctuation (RDF), and line edge roughness. The analytical model is derived by considering that V_{th} is a composite function of all these parameters.

Subthreshold Analytical Delay Modeling Under ΔV_{th} and ΔL_g

Due to the inverse proportional dependence of I_0 on L_g, gate length variation impacts the delay variation proportionally. Considering simultaneous variations in V_{th} and L_g, the gate delay variability model is expressed as:

$$\Delta \tau_d \Delta V_{th,\Delta L_g,NAND} =$$

$$\frac{K_2 C_L V_{DD}}{\mu_n C_{ox}(m-1)V_t^2 m V_t \frac{W}{\Delta L_g} e^{\frac{(V_{DD}-\Delta V_{th}+\eta V_t(ln(e^{\frac{\eta V_{DD}}{mV_t}}+1)))}{mV_t}} (1 - e^{\frac{-V_t(ln(e^{\frac{\eta V_{DD}}{mV_t}}+1))}{V_t}})} \tag{14}$$

where $\Delta \tau_d \Delta V_{th}$, $\Delta L_{g,NAND}$ is the NMOS stack delay variability under V_{th} and L_g variations, and K_2 represents the model fitting parameter. Continuing to use Eq. 13, ΔL_g is passed to evaluate the impact of gate length variation on delay variability.

This analytical approach of estimating delay variability with concurrent process parameter variations is rigorous compared to the mathematical techniques of linear interpolation and linear superposition as reported in [3].

Subthreshold Analytical Delay Modeling Under ΔV_{th}, ΔL_g, and ΔW

Applying the same above methodology, modeling of gate delay variation under simultaneous fluctuations in V_{th}, L_g, and W takes the form:

$$\Delta \tau_{d\Delta V_{th}, \Delta L_g, \Delta W, NAND} =$$

$$\frac{K_3 C_L V_{DD}}{\mu_n C_{ox}(m-1)V_t^2 m V_t \frac{\Delta W}{\Delta L_g} e^{\frac{(V_{DD} - \Delta V_{th} + \eta V_t (ln(e^{\frac{\eta V_{DD}}{mV_t}} + 1)))}{mV_t}} (1 - e^{\frac{-V_t (ln(e^{\frac{\eta V_{DD}}{mV_t}} + 1))}{V_t}})}$$

(15)

where $\Delta \tau_{d\Delta V_{th}, \Delta L_g, \Delta W, NAND}$ is the NMOS stack delay variability under V_{th}, L_g, and W variations and K_3 represents the model fitting parameter. Continuing to use Eq. (14), ΔW is passed to evaluate the impact of gate width variation on delay variability.

For simultaneous variations in multiple parameters, partial differentiation is not considered for the lack of resulting model accuracy due to complex relationships. Owing to the absence of similar delay variability models in the current literature, the proposed analytical models have not been compared against other works.

2.2 Subthreshold Delay Variability Model for PMOS Stack

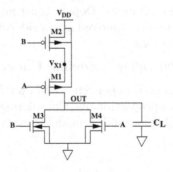

Fig. 3. Stacked PMOS transistors in the NOR gate.

The stacking of PMOS devices in a 2-input NOR gate is depicted in Fig. 3. The intermediate node between M1 and M2, annotated using the voltage V_{X1} is obtained by solving $I_{sub,M1} = I_{sub,M2}$. Thus, the expression for V_{X1} is expressed as:

$$V_{X1} = \frac{(1 + \eta + \gamma)V_{DD}}{(1 + \gamma + 2\eta)}$$

(16)

Table 1. Statistics of τ_{pHL} for $\pm10\%$ variations in V_{th}, L_g, and W for NMOS stack with $V_{DD} = 0.2V$

Delay metrics	$\mu_{\tau,\text{NAND}}$			$\sigma_{\tau,\text{NAND}}$			CT (s)	
	SPICE simulated (ps)	Proposed model (ps)	Proposed model error (%)	SPICE simulated (ps)	Proposed model (ps)	Proposed model error (%)	SPICE	Proposed model
τ_{pHLA}	332.55	332.47	0.02	111.13	95.82	+13.77	536.9	7.9
τ_{pHLB}	263.74	263.40	0.12	95.28	75.91	+20.32	511.9	7.6
τ_{pHLAB}	295.77	295.58	0.06	101.13	85.19	+15.76	560.1	7.9

Applying the same methodology as in NMOS stack (NAND gate), the subthreshold delay variability of a NOR gate with respect to variations in V_{th} alone, V_{th} and L_g, and V_{th}, L_g, and W are derived.

The ability to compute the delay variability of subthreshold digital circuits of any complexity, size, and topology demonstrates the versatility of the proposed analytical models. The model fitting parameters can be easily computed for different NMOS/PMOS stack depths using a curve-fitting tool.

3   Results and Discussion

3.1   NMOS Stack Delay Distributions for Subthreshold Operation

Predictive Technology Model (PTM) [18] file of a 32 nm process node is used to design a two-input NAND gate with V_{th0} (NMOS) $= 0.3558$ V and V_{th0} (PMOS) $= -0.24123$ V, and an optimal supply voltage of $V_{DD} = 0.2$ V for subthreshold circuit operation. The NAND gate is simulated by setting the pulse inputs with rise and fall times of 50 ps, and $C_L = 1$ aF [6], accounting for the area and sidewall capacitances of NMOS and PMOS devices. The falling edge gate delay, due to A input, is denoted as τ_{pHLA}. Similarly, τ_{pHLB} and τ_{pHLAB} denote falling edge delay due to B input and when both A and B inputs are tied together, respectively. The nominal delay values of τ_{pHLA}, τ_{pHLB}, and τ_{pHLAB} obtained by SPICE simulations are 314.33 ps, 246.71 ps, and 277.95 ps, respectively.

Assuming that the parameters V_{th}, L_g, and W follow Gaussian (normal) distribution with $\pm3\sigma$ corresponding to $\pm10\%$ of their nominal, the delay distributions and their corresponding statistics are evaluated. Extensive Monte Carlo simulations of 5000 runs are performed to generate delay distributions, and mean ($\mu_{\tau,NAND}$), standard deviation ($\sigma_{\tau,NAND}$) and Computational Time (CT) are computed. The NMOS stack delay distribution plots generated by the proposed models are superimposed on the SPICE distributions.

Statistics of Delay Distributions with Variability in V_{th}, L_g, and W

The comparison of model predicted distribution statistics to SPICE simulations for simultaneous variations in V_{th}, L_g, and W of $\pm10\%$ each, are summarized in Table 1. The model predicted delay mean error is less than 0.02% for τ_{pHLA}, 0.12% for τ_{pHLB}, and 0.06% for τ_{pHLAB}, respectively, against the SPICE simulated data. As indicated by the standard deviation statistics, the model generated

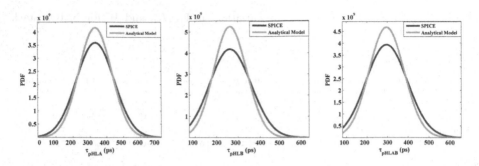

Fig. 4. Comparison of SPICE simulated and analytical modeled delay distributions of (A) τ_{pHLA}, (B) τ_{pHLB}, and (C) τ_{pHLAB} with $\pm 10\%$ ΔV_{th}, ΔL_g, and ΔW, and $V_{DD} = 0.2\,V$

Table 2. Delay variability (σ/μ) statistics for NMOS stack with $V_{DD} = 0.2V$

Delay metrics	V_{th}		V_{th} and L_g		V_{th}, L_g, and W	
	SPICE (σ/μ) %	Proposed model (σ/μ) (%)	SPICE (σ/μ) %	Proposed model (σ/μ) (%)	SPICE (σ/μ) %	Proposed model (σ/μ) (%)
τ_{pHLA}	32.04	29.71	33.22	32.38	33.41	28.82
τ_{pHLB}	33.49	28.81	35.11	32.38	36.12	28.81
τ_{pHLAB}	31.62	28.81	33.10	32.38	34.19	28.82

distributions are at least $+13.77\%$ tighter than SPICE distributions, demonstrating better delay predictability. Figure 4 compares the analytical model delay distributions of τ_{pHLA}, τ_{pHLB}, and τ_{pHLAB} with those of SPICE simulations, for simultaneous variations in V_{th}, L_g, and W of $\pm 10\%$ each.

The error in the analytical model predicted standard deviation, when in error, can be considered to be within an acceptable limit for the 32 nm process, given the high accuracy for the predicted subthreshold mean delay. In contrast to the SPICE simulations, the model generated standard deviation error, in reality, depicts tighter distribution, thereby enhancing predictability and design robustness. Further, a good fit is shown between the model and SPICE generated distributions, thus validating the analytical model approach against SPICE simulations. The results demonstrate that the proposed analytical model is about 70X computationally more efficient than SPICE simulations in generating the delay distributions for all cases. The analytical model approach predicts the delay statistics accurately with significantly less computational complexity.

Delay spread or variability is expressed as the ratio of the standard deviation to the mean, i.e., (σ/μ). Table 2 shows the variability to be 33.41%, 36.12%, and 34.19% with SPICE simulations and 28.82%, 28.81%, and 28.82% with the model

Table 3. Statistics of τ_{pHL} for $\pm 10\%$ variations in V_{th}, L_g, and W for NMOS stack with $V_{DD} = 0.18\,\text{V}$

Delay metrics	$\mu_{\tau,\text{NAND}}$			$\sigma_{\tau,\text{NAND}}$			CT (s)	
	SPICE simulated (ps)	Proposed model (ps)	Proposed model error (%)	SPICE simulated (ps)	Proposed model (ps)	Proposed model error (%)	SPICE	Proposed model
τ_{pHLA}	478.34	478.40	0.00	166.51	137.88	+17.19	1646.8	8.1
τ_{pHLB}	401.45	401.35	0.00	151.07	115.67	+23.43	1637.0	7.6
τ_{pHLAB}	513.49	513.19	0.00	195.00	147.91	+24.14	1632.0	8.1

Table 4. Statistics of τ_{pHL} for $\pm 10\%$ variations in V_{th}, L_g, and W for NMOS stack with $V_{DD} = 0.23\,\text{V}$

Delay metrics	$\mu_{\tau,\text{NAND}}$			$\sigma_{\tau,\text{NAND}}$			CT (s)	
	SPICE simulated (ps)	Proposed model (ps)	Proposed model error (%)	SPICE simulated (ps)	Proposed model (ps)	Proposed model error (%)	SPICE	Proposed model
τ_{pHLA}	190.09	189.96	0.00	58.52	54.75	+6.44	1847.0	8.3
τ_{pHLB}	145.72	145.49	0.00	47.87	41.93	+12.40	1619.2	7.9
τ_{pHLAB}	165.11	165.07	0.00	50.11	47.57	+5.06	1730.5	8.2

for τ_{pHLA}, τ_{pHLB}, and τ_{pHLAB}, respectively, under $\pm 10\%$ variations in V_{th}, L_g, and W each. The analytical model based design generates delay distributions tighter than those of SPICE simulations, resulting in better predictability and increased performance gains.

In order to ensure that the proposed model operates over a wide range of supply voltage in the subthreshold regime, the accuracy and efficacy of the models are validated at $V_{DD} = 0.18\,\text{V}$ and $V_{DD} = 0.23\,\text{V}$. The delay variability statistics of τ_{pHLA}, τ_{pHLB}, and τ_{pHLAB} for an NMOS stack are reported in Tables 3 and 4 for simultaneous variations in V_{th}, L_g, and W of $\pm 10\%$ each, at $V_{DD,min}$ $= 0.18\,\text{V}$ and $V_{DD,max} = 0.23\,\text{V}$, respectively. The worst-case model computed standard deviation statistics of $+24.14\%$ depict better predictability. From the statistics, it is evident that the model predicted mean perfectly matches with the SPICE evaluated mean, with tighter model distributions enhancing predictability and computational efficiency.

3.2 PMOS Stack Delay Distributions for Subthreshold Operation

The nominal delay values of τ_{pLHA}, τ_{pLHB}, and τ_{pLHAB} evaluated using SPICE simulations are 62.85 ps, 90.93 ps, and 85.31 ps, respectively. The estimated mean $(\mu_{\tau,NOR})$, standard deviation $(\sigma_{\tau,NOR})$, and CT are based on the assumption that the device parameter variations follow Gaussian (normal) distribution with $\pm 3\sigma$ variation of $\pm 10\%$ of their nominal value.

The statistics for rising edge delay distributions are presented in Table 5 for $\pm 10\%$ variations in V_{th}, L_g, and W, each. The proposed model generated rising edge delay distributions superimposed on those of SPICE distributions

Table 5. Statistics of τ_{pLH} for ±10% variations in V_{th}, L_g, and W for PMOS stack with $V_{DD} = 0.2\,\mathrm{V}$

Delay metrics	$\mu_{\tau,NOR}$			$\sigma_{\tau,NOR}$			CT (s)	
	SPICE simulated (ps)	Proposed model (ps)	Proposed model error (%)	SPICE simulated (ps)	Proposed model (ps)	Proposed model error (%)	SPICE	Proposed model
τ_{pLHA}	63.82	63.80	0.03	13.40	11.23	+16.19	1568.8	5.2
τ_{pLHB}	92.31	92.31	0.00	19.25	16.25	+15.58	1567.0	5.0
τ_{pLHAB}	86.52	86.50	0.02	16.82	15.23	+9.45	1556.3	5.2

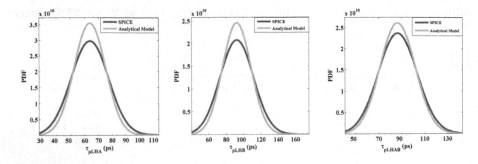

Fig. 5. Comparison of SPICE simulated and analytical modeled delay distributions of (A) τ_{pLHA}, (B) τ_{pLHB}, and (C) τ_{pLHAB} with ±10% ΔV_{th}, ΔL_g, and ΔW, and $V_{DD} = 0.2\,\mathrm{V}$

of τ_{pLHA}, τ_{pLHB}, and τ_{pLHAB} for ±10% variations in V_{th}, L_g, and W each is shown in Fig. 5. From the standard deviation statistics, it is evident that, the distributions get tighter by at least +9.45%. Further, it is demonstrated that the computational overhead of the proposed models is about 300X lower than that of SPICE counterparts.

Table 6 shows delay variability for PMOS stack of 20.99%, 20.85%, and 19.44% for τ_{pLHA}, τ_{pLHB}, and τ_{pLHAB}, respectively, with SPICE simulations and 17.6% with analytical model based design. The distribution tightness offered by the model results in better predictability and sustained performance gains.

To ensure that the proposed model is comprehensive, the PMOS stack model is validated with V_{DD} ranging from 0.18 V–0.23 V ($|V_{th}| = 0.24123$ V). The model predicted delay variability statistics match well with the SPICE statistics as reported in Table 7 for $V_{DD,min} = 0.18$ V and Table 8 for $V_{DD,max} = 0.23$ V for ±10% variations in V_{th}, L_g, and W each, with significantly higher computational efficiency of around 500X. While the model exhibits tighter distributions by a minimum of +17.60% at $V_{DD} = 0.18$ V, the model distributions are slightly wider, indicating the worst case standard deviation error of −8.29% at $V_{DD} = 0.23$ V.

Table 6. Delay variability (σ/μ) statistics for PMOS stack with $V_{DD} = 0.2\,\mathrm{V}$

Delay Metrics	V_{th}		V_{th} and L_g		V_{th}, L_g, and W	
	SPICE (σ/μ) %	Proposed model (σ/μ) (%)	SPICE (σ/μ) %	Proposed model (σ/μ) (%)	SPICE (σ/μ) %	Proposed model (σ/μ) (%)
τ_{pLHA}	17.46	17.60	20.60	21.00	20.99	17.60
τ_{pLHB}	17.27	17.60	20.45	21.01	20.85	17.60
τ_{pLHAB}	16.39	17.60	19.08	21.01	19.44	17.60

Table 7. Statistics of τ_{pLH} for $\pm 10\%$ variations in V_{th}, L_g, and W for PMOS stack with $V_{DD} = 0.18\,\mathrm{V}$

Delay metrics	$\mu_{\tau,\mathrm{NOR}}$			$\sigma_{\tau,\mathrm{NOR}}$			CT (s)	
	SPICE simulated (ps)	Proposed model (ps)	Proposed model error (%)	SPICE simulated (ps)	Proposed model (ps)	Proposed model error (%)	SPICE	Proposed model
τ_{pLHA}	85.37	85.36	0.00	20.03	15.03	+24.96	2729.9	5.3
τ_{pLHB}	122.91	122.87	0.00	28.30	21.63	+23.56	2856.7	5.4
τ_{pLHAB}	116.97	116.97	0.00	24.99	20.59	+17.60	2763.2	5.5

Table 8. Statistics of τ_{pLH} for $\pm 10\%$ variations in V_{th}, L_g, and W for PMOS stack with $V_{DD} = 0.23\,\mathrm{V}$

Delay metrics	$\mu_{\tau,\mathrm{NOR}}$			$\sigma_{\tau,\mathrm{NOR}}$			CT (s)	
	SPICE simulated (ps)	Proposed model (ps)	Proposed model error (%)	SPICE simulated (ps)	Proposed model (ps)	Proposed model error (%)	SPICE	Proposed model
τ_{pLHA}	43.61	43.60	0.00	7.57	7.67	-1.32	2935.8	6.0
τ_{pLHB}	62.85	62.84	0.00	11.04	11.06	-0.18	2840.3	6.1
τ_{pLHAB}	58.58	58.57	0.00	9.52	10.31	-8.29	2744.8	5.9

4 Conclusions

In the nanoscale regime, the impact of process variability on circuit design is becoming extremely critical to performance estimation and get further accentuated in the subthreshold regime. To mitigate the undesirable effects induced by process variability, analytical delay variability models based on the CV/I metric are proposed in the design phase, to map the variations in device geometry and electrical parameters to the delay variations at the circuit level. To validate the adequacy and scalability of the models with circuit size, analytical models are derived to compute the delay variability of NAND and NOR structures, incorporating the stacking effect. The delay distributions generated by the proposed models in 32 nm PTM technology are superimposed on the distributions generated using the SPICE simulator, and an excellent fit is achieved, thus validating the model's delay prediction against the SPICE results over a wide range of subthreshold operation. The accuracy, adequacy, and simplicity of the delay variability models offer enhanced predictability and design robustness at the system design level. The proposed delay variability models are about 70X - 500X computationally more faster than SPICE simulations.

Process variation-aware analytical models can be extended to incorporate variations in supply voltage and temperature, demanding equal attention in the design of digital subthreshold circuits, resulting in comprehensive PVT-aware circuit design.

References

1. Bowman, K.A., Duvall, S.G., Meindl, J.D.: Impact of die-to-die and within-die parameter fluctuations on the maximum clock frequency distribution for gigascale integration. IEEE J. Solid-State Circuits **37**(2), 183–190 (2002)
2. Zhai, B., Hanson, S., Blaauw, D., Sylvester, D.: Analysis and mitigation of variability in subthreshold design. In: Proceedings of the 2005 International Symposium on Low Power Electronics and Design, CA, USA, pp. 20–25. IEEE (2005)
3. Harish, B.P., Bhat, N., Patil, M.B.: Analytical modeling of CMOS circuit delay distribution due to concurrent variations in multiple processes. Solid State Electron. **50**(7–8), 1252–1260 (2006)
4. Boning, D.S., Nassif, S.: Models of process variations in device and interconnect. In: Design of High Performance Microprocessor Circuits, pp. 98–115. IEEE Press (2000)
5. Lin, T., Chong, K., Gwee, B., Chang, J.S., Qiu, Z.: Analytical delay variation modeling for evaluating sub-threshold synchronous/asynchronous designs. In: Proceedings of the 8th IEEE International NEWCAS Conference, Montreal, Canada, pp. 69–72. IEEE (2010)
6. Kim, S., Agarwal, V.: Analytical delay and variations modeling in the subthreshold region. In: Proceedings of the National Conference on Undergraduate Research (NCUR), Asheville, North Carolina (2016)
7. Guo, J., et al.: Analytical inverter chain's delay and its variation model for subthreshold circuits. IEICE Electron. Expr. **14**(11), 1–12 (2017)
8. Frustaci, F., Corsonello, P., Perri, S.: Analytical delay model considering variability effects in subthreshold domain. IEEE Trans. Circuits Syst. II: Expr. Briefs **59**(3), 168–172 (2012)
9. Chen, J., et al.: Linear compositional delay model for the timing analysis of sub-powered combinational circuits. In: Proceedings of the 2014 IEEE Computer Society Annual Symposium on VLSI, FL, USA, pp. 380–385. IEEE (2014)
10. Cao, P., et al.: An analytical gate delay model in near/subthreshold domain considering process variation. IEEE Access **7**, 171515–171524 (2019)
11. Zhang, Y., Calhoun, B.H.: Fast, accurate variation-aware path timing computation for sub-threshold circuits. In: Proceedings of the 15th International Symposium on Quality Electronic Design, CA, USA, pp. 243–248. IEEE (2014)
12. Packan, P., et al.: High performance 32nm logic technology featuring 2nd generation high-k + metal gate transistors. In: Proceedings of the 2009 IEEE International Electron Devices Meeting (IEDM), MD, USA, pp. 1–4. IEEE (2010)

13. Anala, M., Harish, B.P.: Analytical modeling of process variability in subthreshold regime for ultra low power applications. In: Proceedings of the 2018 IEEE Asia Pacific Conference on Circuits and Systems, Chengdu, China, pp. 257–260. IEEE (2018)
14. Consoli, E., Giustolisi, G., Palumbo, G.: An accurate ultra-compact I-V model for nanometer MOS transistors with applications on digital circuits. IEEE Trans. Circuits Syst. I: Regular Pap. **59**(1), 159–169 (2012)
15. Taur, Y., Ning, T.H.: Fundamentals of Modern VLSI Devices, 2nd edn. Cambridge University Press, New York (1998)
16. Chandra, N., Yati, A.K., Bhattacharyya, A.B.: Extended-Sakurai-Newton MOS-FET model for ultra-deep-submicrometer CMOS digital design. In: Proceedings of the 2009 22nd International Conference on VLSI Design, New Delhi, India, pp. 247–252. IEEE (2009)
17. Keane, J., Eom, H., Kim, T., Sapatnekar, S., Kim, C.: Stack sizing for optimal current drivability in subthreshold circuits. IEEE Trans. VLSI Syst. **16**(5), 598–602 (2008)
18. Predictive Technology Model (PTM) [Internet], 29 October 2007 [updated 6 Jan 2012; cited Jan 2021]. http://ptm.asu.edu/

Video Stabilization for Drone Surveillance System

N. Aswini[1]([⊠]) and S. V. Uma[2]

[1] Department of Electronics & Communication, CMR Institute
of Technology (Affiliated To VTU), Bengaluru, India
[2] Department of Electronics & Communication, RNS Institute
of Technology (Affiliated To VTU), Bengaluru, India

Abstract. Now-a-days, drones are very commonly used in various civil applications like disaster management, traffic monitoring, crowd monitoring, infrastructure inspection and search and rescue. Drones have different types of sensors, among which cameras are normally used in these type of real time scenarios. The low flying small drones commonly have fixed cameras. The video/images from these cameras are not smooth due to the propeller vibrations and environmental effects. These unstable videos need to be smoothened without adding extra payloads. This will improve the quality of the videos and will help in further data/image processing. In this paper, we propose an optical flow-based motion estimation and further smoothening of drone videos to improve the Inter Frame Transformation Fidelity (ITF) when compared to the existing feature extraction methods.

Keywords: Drones · Image processing · Computer vision · Surveillance

1 Introduction

Fully automated Drones are in high demand for security and surveillance due to their greatest advantage – no threat to human life!. They fly precisely with the help of various types of sensors like gyroscopes, accelerometers, barometers, global positioning systems, inertial measurement units and vision sensors. The sensors are like eyes and ears to the drones with the help of which they can navigate safely. The sensors other than vision sensors help them to determine its location and orientation. The vision sensors can be utilized to detect and avoid obstacles in the path of the drone. There are significant works for stabilizing drone videos for moving object tracking and surveillance for aerial videos. Motion estimation is performed by extracting key feature points and then applying various filtering techniques. All these methods revolve around the three steps of Motion estimation, Motion smoothing and Compensation for generating stabilized videos. To achieve instantaneous stabilization of video frames for real time applications, optical flow-based motion estimation can perform much better in terms of the Inter Frame Transformation Fidelity (ITF) when compared to feature extraction methods. This video pre-processing step is very much essential prior to obstacle detection.

© Springer Nature Switzerland AG 2021
K. R. Venugopal et al. (Eds.): ICInPro 2021, CCIS 1483, pp. 468–480, 2021.
https://doi.org/10.1007/978-3-030-91244-4_37

In this proposed research work, Motion estimation is performed by finding the Euclidean transformation which incorporates the motion effects due to translation, rotation, and scaling. To find the motion between frames, optical flow is measured before performing transformation. Once the Motion is estimated, Motion smoothing is done by cumulatively adding the differential motion calculated. To get a smooth trajectory out of the differential motion obtained, a moving average filter is applied to smoothen the curve. The final step is to reconstruct the stabilized video. This is performed in two steps: -

a) Finding the difference between the smoothened trajectory and the original one.
b) The difference is added back to the original one.

The block diagram in Fig. 1 shows the basic steps in video stabilisation.

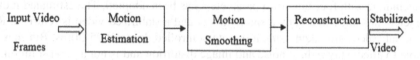

Fig. 1. Video stabilization process

2   Literature Review

Drone video cameras can be fixed or moving. In large drones, there is a provision for video camera and gimbals, but for small size drones, there is not much space to install both. They have fixed cameras and the vibration of the engine and the environmental conditions can produce unstable videos. So, videos need to be smoothened to reduce the effect of camera motion in the final video which is to be processed. As an initial step, the image transformation parameters between consecutive video frames need to be derived. Second, unwanted motion needs to be filtered out and third, the stabilized video need to be reconstructed.

Video frames are stabilized by Speeded Up Robust Feature extraction, matching and motion compensation by Mingkhwan et al. in [1]. Feature matching is performed for consecutive video frames. A 2D image model can be represented using various transformation models such as Projective, Affine and Similarity based model. An affine transformation model which uses lesser parameters compared to projective transformation has been used for motion compensation in this paper. They have evaluated the peak signal to noise ratio and mean square error for four test videos taken using a fixed camera mounted on a small multirotor flying at a height of 20 m above sea level with 1080-pixel resolution at 30 frames per second.

Malagi et al. [2] has removed the undesired effects of a moving camera, by performing image registration as an initial step. They have used Rotation Invariant Fast Feature detection technique along with RANSAC matching for tracking aerial images. Once matched, the mapping of the images is performed using affine transformation.The unwanted motion of camera causes difficulty in moving object detection and tracking.

This problem has been addressed with the help of Scale Invariant Feature Transform (SIFT) and affine transformation model by Walha et al. in [3]. Kalman filter and Median filter is used to remove noise from the video frames. Real time aerial videos covering buildings, cars and people moving around is taken with a resolution of 960×540 pixels. Even though very accurate, SIFT is computationally extensive and time consuming and therefore a balance between the time complexity and processing time is required here. Another SIFT based motion estimation is performed by Rehman et al. in paper [4]. Once SIFT features are extracted, the outliers are removed using RANSAC algorithm, followed by projective transformation technique. A moving averaging filter is used for smoothing the video frames. Finally, the stabilized video is reconstructed.

A real time video stabilisation and motion estimation for unmanned aerial video is attempted by FAST feature extraction and affine transformation technique by Yue Wang et al. in [5]. The matching between consecutive video frames has been performed using Sum of Squared Differences (SSD). A 2D rotation, translation based affine transformation technique with four unknown parameters has been adopted. The estimated motion parameters are smoothened by a spline model and a stabilized video is returned. Real time aerial videos are taken using a remote-controlled drone at 25-30fps. But they are not able to effectively reduce noise and image distortion and is not a good solution. By extracting ORB features and performing histogram clustering, a relatively faster video stabilisation is proposed by Baotong Li et al. in [6]. Histogram clustering is used to calculate the parameters of an affine transformation matrix. Gaussian filtering is done to remove undesired motion effects. Four videos are experimented, two videos taken using a mobile camera and the other two are aerial videos with a resolution of 640×480 pixels. The efficiency of the method is affected when the feature point matching fails due to lack of enough features in some situations. A real time video stabilisation for unmanned aerial vehicles using optical flow measurement is given in Lim et al. [7]. They use simpler transformation models such as rigid and similarity transformation (hybrid model) rather than perspective transformation. The method works by initially finding the corners in image frames using optical flow measurement followed by a 2D motion estimation model. An averaging window is used to smoothening the video. The resolution of the video frames is 640×480 pixels.

A combination of Oriented Rotation BRIEF (ORB) feature extraction with affine transformation is proposed by Xu et al. [8]. Global motion estimation of the input video frames is initially performed by ORB which is a combination of FAST corner detector and BRIEF descriptor. Once motion is estimated, smoothing is performed using Gaussian filter followed by image warping. The validation is performed by recording videos using a handheld camera. Kulkarni et al. [9], have used also used FAST corner detection to extract key features points. After acquiring the video frames, the Red – Cyan color composite is produced to illustrate the pixel wise difference between consecutive frames. For each extracted key point, a FREAK descriptor [10] centered around the point is calculated. Feature matching is done by calculating the Hamming distance between the descriptor points. Using RANSAC algorithm, the outliers are removed to improve the efficiency. Finally, affine transformation techniques are used followed by stabilized video output. Shaky videos using handheld cameras with 320×240 pixel resolution at 30 fps has been used by them. An aerial video stabilization for micro air vehicles is proposed by

Wu et al. in [11]. They use SURF feature extraction for motion estimation. The feature points in adjacent frames are matched using nearest neighbour search method. RANSAC and Least squares method is used to remove the outliers followed by Kalman filtering to smoothen the motion parameters. They have used a set of air-to-ground videos of 640 × 480 pixel resolution.

3 Motion Estimation

Video stabilization can be 3Dimensional which will reconstruct the camera trajectory or can be 2Dimensional which will estimate the motion between successive video frames. In this research work, a 2Dimensional motion estimation and compensation is proposed which combines Lucas Kanade optical flow [12] measurement and Euclidean Transformation. Optical flow is a pattern of apparent motion between the drone and the obstacle. Before estimating the motion between successive video frames, we must efficiently detect the corner points. The Shi Tomasi corner detector [13] shows better results when compared to its predecessor, the Harris corner detector [14]. In Shi Tomasi, a small window is made to scan the whole image, checking for corners in the image. If the window is located on a corner, shifting will cause a great change in appearance when compared to flat regions or edges. In Python OpenCV, goodFeaturesToTrack function [15] will find the Shi Tomasi corners. Once the good features in the previous frame is calculated, next Lucas Kanade Optical flow is measured between successive frames. Once we get the location of features in the previous and current frame, then a mapping is done using Image transformation. For this a Euclidean transformation which will map the successive frames is used (Fig. 2). Euclidean model is more restrictive than Homography and Affine transforms, but it is adequate for the motion estimation in cases where the camera movement is usually small between successive video frames. The estimateRigidTransform function [16] in OpenCV gives a limited combination of translation, rotation, and scaling. Once the motion has been estimated, the translation and rotation values are extracted and stored in an array. These stored values are used for smoothing.

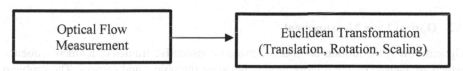

Fig. 2. Motion Estimation

This method of motion estimation uses textured regions with lots of corner. This is an ideal method of object tracking. If the feature point in the current frame is occluded by the features in the previous frame, then it is taken care of while implementing Lucas Kanade optical by keeping the status flag = 1.

3.1 Corner Point Detection

Corner locations are identified for each frame by applying Shi Tomasi corner detector. If W is a window located at location (x, y) having pixel intensity I(x, y), Shi Tomasi corner detection steps are:-

1. Formation of auto correlation matrix M

$$M(x, y) = \sum_{u,v} w(u, v) * \begin{bmatrix} I_x^2(x, y) & I_x I_y(x, y) \\ I_x I_y(x, y) & I_y^2(x, y) \end{bmatrix} \tag{1}$$

2. Window is either a rectangular window or Gaussian. I_x and I_y are image derivatives in (x, y) directions. $W(u, v)$ denotes weighting window over (u, v) area.
3. Find λ_1, λ_2, the Eigen values of the auto-correlation matrix M.
4. Find the corner determination factor, $C(x, y) = \min(\lambda_1, \lambda_2)$. If it is greater than a threshold value, then it is considered a corner.

All the corners below the quality level will be rejected. The function will take the first strongest corner, removes all adjacent corners which are at a distance lesser than the range of Euclidean distance specified, and returns M corners. The Fig. 3 shows the corners detected for previous frame (Left) and the current frame (Right).

Fig. 3. Corners for frame #1280 (left), #1281 (right)

Once the corner points are detected, the next step performed is to find the motion flow between consecutive video frames using optical flow measurement.

3.2 Optical Flow Measurement

Optical flow is the motion of objects between consecutive frames of a video sequence which are caused by the relative motion between the object and camera. The problem of optical flow can be expressed as in Fig. 4. Consider a pixel $I(x, y, t)$ in the first frame where (x, y) are the spatial co-ordinates and time t. In the next frame, it moves by (dx, dy) over time dt. If the pixel intensities are constant between consecutive video frames then,

$$I(x, y) = I(x + dx, y + dy, t + dt) \tag{2}$$

Assuming the movement to be small, take Taylor series expansion on the right-hand side and divide by dt to get the optical flow equation as:

$$\frac{\partial I}{\partial x} u + \frac{\partial I}{\partial y} v + \frac{\partial I}{\partial t} = 0 \tag{3}$$

Where $u = dx/dt$ and $v = dy/dt$.

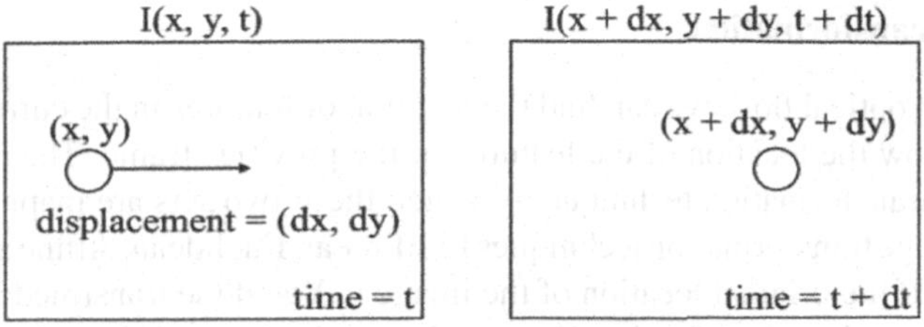

I(x, y, t)

(x, y)
○———————▶

displacement = (dx, dy)

time = t

I(x + dx, y + dy, t + dt)

(x + dx, y + dy)
○

time = t + dt

Fig. 4. Optical flow measurement principle

In Eq. 3, $\partial I/\partial x$ and $\partial I/\partial y$ are the image gradients and $\partial I/\partial t$ is the gradient along time. The term (u, v) is unknown to us. To find this unknown term, Lucas Kanade method is applied. The assumption is that all the neighbouring pixels are having similar motion. Take a 5×5 patch around the pixel under consideration. This gives 25 points having the same motion. Next find $\partial I/\partial x$, $\partial I/\partial y$ and $\partial I/\partial t$ for all these 25 points. Now it becomes 25 equations with two unknowns. Representing these equations as a matrix is of the form $Ad = B$ as given in Eq. 4. A least squares principle is used to solve these equations and find the value of the matrix $\begin{pmatrix} u \\ v \end{pmatrix}$.

$$\begin{bmatrix} I_x(\mathbf{p_1}) & I_y(\mathbf{p_1}) \\ I_x(\mathbf{p_2}) & I_y(\mathbf{p_2}) \\ \vdots & \vdots \\ I_x(\mathbf{p_{25}}) & I_y(\mathbf{p_{25}}) \end{bmatrix} \begin{bmatrix} u \\ v \end{bmatrix} = - \begin{bmatrix} I_t(\mathbf{p_1}) \\ I_t(\mathbf{p_2}) \\ \vdots \\ I_t(\mathbf{p_{25}}) \end{bmatrix} \qquad \begin{matrix} A & d = b \\ {}_{25\times2} & {}_{2\times1} \ {}_{25\times1} \end{matrix}$$

Least squares solution for d given by $\qquad (A^T A)\, d = A^T b$ $\hfill$ (4)

$$\begin{bmatrix} \sum I_x I_x & \sum I_x I_y \\ \sum I_x I_y & \sum I_y I_y \end{bmatrix} \begin{bmatrix} u \\ v \end{bmatrix} = - \begin{bmatrix} \sum I_x I_t \\ \sum I_y I_t \end{bmatrix}$$

$\qquad\qquad A^T A \qquad\qquad\qquad\qquad A^T b$

The matrix $\begin{pmatrix} u \\ v \end{pmatrix}$. can be calculated by taking $(A^T A)^{-1} A^T b$. The implementation of Lucas Kanade optical flow measurement is done for all the video frames to estimate the motion. The optical flow we measured for a moving car is given in Fig. 5 (shown for 2 video frames).

Fig. 5. Optical flow measurement for two video frames

3.3 Image Transformation

By performing optical flow, we can find the location of features in the current frame and we already know the location of the features of the previous frame. The next step is to use an image transformation technique by which these two sets are mapped. There are various 2D image transformation techniques like Linear, Euclidean, Affine and Projective Homography. If the original location of the image is X and the transrmed location is X' with a translation distance of t, then the translation, rotation and scaling operation can be represented as:

1. Translation motion - $X' = X + t$
2. Rotation and Translation – $X' = RX + t$ where R is the rotation matrix with an angle θ given by,

$$R = \begin{pmatrix} cos\theta & -sin\theta \\ sin\theta & cos\theta \end{pmatrix} \tag{5}$$

3. Scaled rotation - $X' = SRX + t$, where S represents the scaling.

A combination of these three operations are achieved by using a Rigid or Euclidean transformation. After motion estimation, it is decomposed into x and y translation along with rotation angle. These values are then stored in an array for further operations.

4 Motion Smoothing

A smooth trajectory of motion is created by cumulatively adding the transformation parameters $(dx, dy, d\theta)$. In this step, add up the motion between the frames to calculate the trajectory. Three curves which show how the motion (x, y and angle) are changing over time is obtained by performing cumulative addition. This is performed using cumsum (cumulative sum) in numpy. These three curves are then smoothened using a moving average filter. If the curve is stored in an array 'a', so that the points of the curve are $a[0]....a[n-1]$, then the kth element of the smoothing curve can be obtained by averaging filter of width 5 as: -

$$f[k] = \frac{a[n-2] + a[n-1] + a[n] + a[n+1] + a[n+2]}{5} \tag{6}$$

This will smooth a noisy curve. The smoothened trajectory for four test videos is included in the results section.

5 Reconstruction

The stabilised video reconstruction is the final step. This is done by finding the difference between smoothened trajectory and original trajectory and the difference is added back to the original transform. For the motion (x, y, θ), the transformation gives a 2×3 matrix of the form:

$$T = \begin{bmatrix} \cos(\theta) & -\sin(\theta) & x \\ \sin(\theta) & \cos(\theta) & y \end{bmatrix} \qquad (7)$$

The video is reconstructed back according to the new transformation matrix values. For a 2×3 matrix input, OpenCV provides cv2.warpAffine [17] function. The reconstructed transformation matrix, which is stored in an array, is used for generating stabilized image frames. While reconstructing back the video, it is seen that there are black boundary artifacts. So, the border needs to be fixed to mitigate this problem. This is performed by scaling and rotating the image without moving the center of the image. With an upscaling of 4% and zero rotation, the function getRotationMatrix2D [18] in OpenCV is used for this. Once again image warping is done to reconstruct a smooth, stabilized video.

The step-by-step operation performed for stabilizing the video frames is given in Fig. 6. Frame by frame the video is given to an optical measurement block followed by transformation. Once transformed, the videos are smoothened using a moving averaging filter and the stabilized video is reconstructed. This method of video pre-processing has been inspired from the huge set of library resources available in Python OpenCV [19] which can be used to develop real time computer vision applications.

Proposed Algorithm Steps:-

1. Input Video frames converted to grey scale.
2. For each video frame from the input sequence, find corners by Shi Tomasi corner detector.
3. For each pair of consecutive frames, Calculate Lucas Kanade Optical flow.
4. Find transformation matrix (translation, scaling, and rotation).
5. Store the transformation matrix and repeat the steps for all video frame pairs.
6. Cumulative adding of transformation parameters dx, dy and $d\theta$.
7. Apply moving average filter on the added trajectory.
8. Reconstruction of video sequence by applying affine warping.

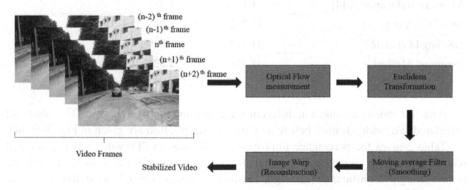

Fig. 6. Proposed method of motion estimation and smoothing for video stabilization

6 Simulation Results

To evaluate the performance of the proposed technique, the Peak Signal to Noise Ratio (PSNR), the Mean Squared Error (MSE) and Inter Frame Transformation Fidelity (ITF) is calculated. The MSE between frames is calculated as: -

$$MSE = \frac{1}{mn} \sum_{i=0}^{m-1} \sum_{j=0}^{n-1} [I(i,j) - K(i,j)]^2 \tag{8}$$

Where m and n are frame dimensions. The MSE should be lesser for stabilized videos. The PSNR calculated between two consecutive video frames is defined as: -

$$PSNR(in\,dB) = 10 \log_{10} \frac{IMAX}{MSE} \tag{9}$$

Where IMAX is the maximum value of an image. Higher PSNR value represents good quality of stabilized video. The average value of the PSNR between consecutive frames is defined as ITF given by,

$$ITF = \frac{1}{Nframe - 1} \sum_{k=1}^{Nframe-1} PSNR(k) \tag{10}$$

The ITF value is used to obtain a rough estimate of the quality of stabilized video. Higher the value, the better the quality. The ITF values of the proposed technique is compared with the existing techniques for a shaky video dataset (Video Sequence 1) taken at 25 fps. The ITF values obtained and the percentage improvement with respect to the original video dataset is given in Table 1.

Table 1. Comparison with existing techniques

Technique (Video Sequence 1)	ITF	Percentage improvement
Original Video	30.608	–
M. Awais Rehman et al. [4]	31.123	1.68%
Mingkhwan et al. [1]	31.282	2.22%
Baotong Li et al. [6]	31.313	2.30%
Proposed Method	31.378	**2.52%**

Four test videos are taken at different scenarios and their PSNR and ITF values are calculated. The video frames before and after reconstruction are given in Fig. 7(a)–(d).

Table 2 gives the percentage improvement obtained in ITF value when compared with original video sequence. The scenarios considered are of the test cases where drone is navigating in a traffic area (low light condition), moving towards a car, towards a tree and a navigating in a corridor.

Fig. 7. (a): Video frames of Sequence 1 (before and after stabilization). (b): Video frames of Sequence 2 before and after stabilization. (c): Video frames of Sequence 3 before and after stabilization. (d): Video frames of Sequence 4 before and after stabilization

Table 2. Improvement in ITF values

	Original video	Stabilized video	Percentage improvement in ITF
Video Sequence 1	30.608	31.378	2.52%
Video Sequence 2	32.337	33.362	3.17%
Video Sequence 3	29.534	30.652	3.79%
Video Sequence 4	34.877	36.217	3.84%

The PSNR plot, smoothened trajectory, and the frame-to-frame transformation plots for all the four video sequences are given in Fig. 8(a) and Fig. 9(b). The stabilized video (shown in green) shows much improvement in the PSNR value when compared to the unstabilized one.

Sequence 1 & 2- Traffic Video and Moving Car Video

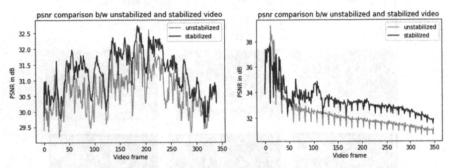

Fig. 8. (a) PSNR graph plot before (red) and after (green) stabilization (Color figure online)

Sequence 3 & 4: Tree Video and Corridor Video

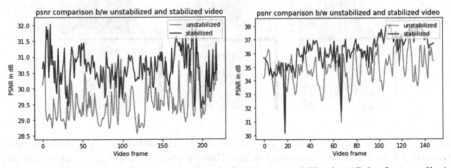

Fig. 9. (b) PSNR graph plot before (red) and after (green) stabilization (Color figure online)

7 Conclusions and Future Enhancements

Due to the environmental and structural limitations, drones are highly susceptible to atmospheric effects which introduces jitter and the video becomes highly unstable. It will be difficult to detect obstacles from these unstable videos. So, there is a necessity to introduce a real time video stabilization before detecting obstacles. To achieve this goal, an optical flow-based motion estimation and image transformation method is performed in this paper. The evaluation metrics used are the Peak Signal to Noise Ratio and Inter Frame Transformation Fidelity (ITF). The higher the values of PSNR and ITF, the better the quality of the videos. The performance is much improved when compared with the existing methods. In future, we would like to take more different test case scenarios like in windy and foggy environment and check the improvement in ITF values. The entire stabilization should also be performed onboard which will help in developing much improved obstacle detection algorithms for autonomous navigation of the drones.

References

1. Mingkhwan, E., Khawsuk, W.: Digital image stabilization technique for fixed camera on small size drone. In: 2017 Third Asian Conference on Defence Technology (ACDT) (2017). https://doi.org/10.1109/acdt.2017.7886149
2. Malagi V.P., Ramesh Babu, D.R.: Rotation-invariant fast feature based image registration for motion compensation in aerial image sequences. In: Guru, D.S., et al. (eds.) Springer Nature Singapore Pte Ltd. 2018, Proceedings of International Conference on Cognition, and Recognition, Lecture Notes in Networks and Systems, vol. 14. Springer, Singapore (2018). https://doi.org/10.1007/978-981-10-5146-3_20
3. Walha, A., Wali, A., Alimi, A.M.: Video stabilization with moving object detecting and tracking for aerial video surveillance. Multimed. Tools Appl. **74**(17), 6745–6767 (2014). https://doi.org/10.1007/s11042-014-1928-z
4. Rehman, M.A., Raza, M.A., Ahmed, M., Ijaz, M.A., Ahmad, M., Mahmood, M.H.: Comparative analysis of video stabilization using SIFT flow and optical flow. In: 2019 International Conference on Electrical, Communication, and Computer Engineering (ICECCE), Swat, Pakistan, 2019, pp. 1–6 (2019). https://doi.org/10.1109/ICECCE47252.2019.8940784
5. Wang, Y., Hou, Z., Leman, K., Chang, R.: Real-time video stabilization for unmanned aerial vehicles. In: MVA2011 IAPR Conference on Machine Vision Applications, Nara, Japan, 13–15 June 2011
6. Li, B., Chen, Y., Ren, J., Cheng, L.: A fast video stabilization method based on feature matching and histogram clustering. In: Balas, V.E., et al. (eds.) Information Technology and Intelligent Transportation Systems, Advances in Intelligent Systems and Computing, vol. 455. Springer International Publishing, Switzerland (2017). https://doi.org/10.1007/978-3-319-38771-0_31
7. Lim, A., Ramesh, B., Yang, Y., Xiang, C., Gao, Z., Lin, F.: Real-time optical flow-based video stabilization for unmanned aerial vehicles. J. Real-Time Image Process. **16**(6), 1975–1985 (2017). https://doi.org/10.1007/s11554-017-0699-y
8. Xu, J., Chang, H., Yang, S., Wang, M.: Fast feature-based video stabilization without accumulative global motion estimation. IEEE Trans. Consum. Electron. **58**(3), 993–999 (2012). https://doi.org/10.1109/TCE.2012.6311347
9. Kulkarni, S., Bormane, D.S., Nalbalwar, S.L.: Video stabilization using feature point matching. IOP Conf. Ser.J. Phys. Conf. Ser. **787**(2017), 012017 (2017)

10. Alahi, A., Ortiz, R., Vandergheynst, P.: FREAK: fast retina keypoint. In: 2012 IEEE Conference on Computer Vision and Pattern Recognition, Providence, RI, 2012, pp. 510–517 (2012). https://doi.org/10.1109/CVPR.2012.6247715

11. Hao, W., He, S.-Y.: An aerial video stabilization method based on SURF feature. ITM Web Conf. **477**, 05004 (2016)

12. Lucas, B.D., Kanade, T.: An iterative image registration technique with an application to stereo vision. In: Proceedings of the Imaging Understanding Workshop, pp. 121–130 (1981)

13. Shi, J., Tomasi, C.: Good features to track. In: 9th IEEE Conference on Computer Vision and Pattern Recognition, pp. 593–600 (June 1994)

14. Harris, C., Stephens, M.: A combined corner and edge detector. In: Proceedings of the Alvey Vision Conference (1988)

15. https://docs.opencv.org/2.4/modules/imgproc/doc/feature_detection.html#goodfeaturestotrack

16. https://docs.opencv.org/2.4/modules/video/doc/motion_analysis_and_object_tracking.html#estimaterigidtransform

17. https://docs.opencv.org/2.4/modules/imgproc/doc/geometric_transformations.html?highlight=warpaffine#warpaffine

18. https://docs.opencv.org/2.4/modules/imgproc/doc/geometric_transformations.html#getrotationmatrix2d

19. https://learnopencv.com/

MQTT-SN Based Architecture for Estimating Delay and Throughput in IoT

V. Tejashree, N. Vidhyashree, S. Anusha, K. Anu, P. S. Akshatha$^{(\boxtimes)}$, and S. M. Dilip Kumar

Department of CSE, University Visvesvaraya College of Engineering, Bangalore, India

Abstract. MQTT-SN provides an energy efficient data transfer under undependable wireless sensor networks (WSNs), one of the major empowering technologies for Internet of Things (IoT). WSNs are facing challenges like packet loss during delivery, small lifetime of participating nodes and end-to-end delay. Hence, to overcome such problems a new data-centric communication method called Message Queuing Telemetry Transport for Sensor Nodes (MQTT-SN) protocol which is a publish/subscribe messaging system have been proposed in this paper. A MQTT-SN based architecture for dynamic gateway selection is proposed to minimize delay and increase throughput. Mobility aware Hybrid Clustering Algorithm has been discussed for transferring data between clusters of mobile sensor nodes in WSNs.

Keywords: MQTT-SN · Publish/Subscribe · Delay · Throughput · IoT

1 Introduction

Wireless sensor networks (WSNs) has gained huge attention from past few years in both commercial point of view such as crop field monitoring, industrial automation and transportation business to technical point of view such as smart home monitoring, etc. [15]. A data centric communication approach which is a pub/sub messaging system to the above mentioned WSNs are proposed. MQTT-SN protocol has been used to achieve the data transmission and communication between mobile sensor nodes, gateway and MQTT broker. Message queuing telemetry Transport (MQTT) protocol has been implemented for machine-to-machine communication which is data centric protocol of IoT [16]. MQTT protocol has been designed to optimize the communication over WSNs where the network connection is very dynamic and wireless links are quite likely to fail and bandwidth is at a prime choice. MQTT allows for a more reliable link in a network that is too difficult for wireless devices with low battery life and limited storage. In this paper, a lightweight pub/sub MQTT-SN protocol

© Springer Nature Switzerland AG 2021
K. R. Venugopal et al. (Eds.): ICInPro 2021, CCIS 1483, pp. 481–490, 2021.
https://doi.org/10.1007/978-3-030-91244-4_38

based architecture for estimating delay and throughput in IoT has been proposed. MQTT-SN is an another form of MQTT, which is designed specifically for low-power, low-cost devices that use bandwidth-constrained wireless sensor networks like Zigbee [14]. In a WSN environment a network frequently changes due to failure of wireless links between the sensor devices. Once the sensor node fails it is changed by a new node to keep the service alive instead of getting it repaired.

In this paper a dynamic gateway selection method with which sensor nodes communicate through an application in wireless media is presented. A mobility aware hybrid clustering algorithm has been discussed, which is three tier architecture. A sensor node publishes its data to cluster head and that data is transferred to gateway. The gateway publishes that data to MQTT broker which it will deliver to subscribers. A subscriber can subscribe to any related data through topic ids of particular topic names and get information through the gateway once it has been published by a publisher. In this environment, most of the events occur without knowing the source address. Instead of information about the publisher, applications are concerned with the data that must be published. Therefore the link between mobile sensor nodes and the gateway is highly concerned. The proposed smart dynamic gateway selection method reduces the delay and increases the throughput.

1.1 Paper Organization

The rest of the paper is structured as follows: Motivation and contributions of the paper is discussed in Sect. 2. The relative survey papers on MQTT-SN protocol is discussed in Sect. 3. The Sect. 4 discusses on MQTT-SN architecture. In Sect. 5, the proposed architecture is discussed. Finally, in Sect. 6, conclusion of the paper is presented.

2 Motivation and Contributions

2.1 Motivation

MQTT-SN based IoT system is made up of several mobile sensor nodes. Data transfer between publishers and subscribers takes place via gateway and broker. Gateway plays an important role as an intermediate node between mobile sensor node and broker. Thus if it is unable to select an appropriate gateway among several gateways leading to increase in delay and high power consumption. The motivation is to deal with the above mentioned problems by proposing a mobility aware hybrid clustering algorithm and dynamic gateway selection algorithm in this paper.

2.2 Contributions

The major contributions of this paper include the following:

1. Provides strengths and weaknesses of recent publications on MQTT-SN.
2. Provides MQTT-SN architecture.
3. Provides proposed MQTT-SN architecture for WSNs.
4. Provides details of clustering and data transmission.
5. The proposed architecture concentrates mainly on how to reduce end-to-end delay and increase the throughput.

3   Literature Survey

In this section the existing methods that uses mqtt-sn protocol for various applications are discussed.

Table 1. Summary of literature survey

Reference	Objectives	Advantages	Disadvantages
K Govindan et al. [1]	End to end service assurance	End-end to delay estimation derivation	Does not consider real time implementation
E F Silva et al. [2]	Efficient network services.	An improved Version of MQTT	Does not verify data loss in real scenario
Hunkeler et al. [12]	Performance comparison of MQTT and CoAP	Adaption of MQTT to the constraints of WSNs	Does not support sleeping clients
S Lee et al. [3]	Loss and delay estimation	Realistic network environment results	Does not consider various levels of QoS
D G Roy et al. [4]	End to end delay and message loss estimation	Improved performance of WSNs	Does not consider delay of inside WSNs
Kai-Hung Liao et al. [6]	MQTT & MQTT-SN based applications	Comparison of both the versions of protocol	Usage of high energy devices
S Raza et al. [5]	Secure IoT architecture for end to end communication	Implementation of IPsec	Energy Overhead
J E Luzuriaga et al. [7]	Improving data delivery in mobile scenario	Enhanced reliability	Does not measured performance for other applications
Bilal Jan et al. [8]	Survey of an Energy efficient hierarchical clustering	Battery powered sensor nodes	Recharging is not possible
S Zafar et al. [9]	Mobility aware hybrid clustering algorithm	Reduces energy utilization	Short battery life
Jani Lalkkakorpi et al. [10]	Dynamic gateway placement Algorithm	Reduced delay	Packet loss
Edoardo Longo et al. [11]	Spanning tree protocol for MQTT brokers	Message replication among brokers	Lack of routing strategies
Joohwan Kim et al. [13]	Minimizing Delay and Maximizing Lifetime for WSNs	Reduced delay	Non-Poisson wake-up process is not extended

An end to end service assurance in IoT has been proposed by kannan govindan et al. [1] using MQTT-SN protocol. Edelberto franco silva et al. [2] have explained how MQTT-SN can be used to provide an efficient network services. A comparative study is made between MQTT-SN and others protocols which proved that MQTT-SN is more efficient for battery operated devices by Hunkeler et al. [12]. Shinho Lee et al. [3], estimated MQTT loss and delay according to QoS levels through correlation analysis. An application aware end to end delay and message loss estimation in IoT driven by MQTT-SN is made by Deepsubhra Guha Roy et al. [4]. An MQTT and MQTT-SN based applications of IoT have been implemented by Kai-Hung Liao et al. [6]. S Raza et al. [5], has proposed an secure architecture in IoT for end-to-end communication. J E Luzuriaga et al. [7], has proved using MQTT in mobile scenarios improved data delivery considering the results from realistic test bed. An energy efficient hierarchical clustering approaches in WSNs has been proposed by Bilal Jan et al. [8]. An efficient mobility aware hybrid clustering algorithm for mobile wireless sensor networks to improve energy efficiency and scalability in WSNs has been proposed by S Zafar et al. [9]. Jani Lalkkakorpi et al. [10], has proposed an algorithm for dynamic gateway placement to minimize end-to-end delay in mobile networks. A spanning tree protocol has been proposed by Edoardo Longo et al. [11], for MQTT brokers to reduce latency. Joohwan Kim et al. [13], has proposed efficient method to reduce delay and maximize lifetime for WSNs and to optimize the forwarding schemes for minimizing the packet-delivery delays between sensor nodes and the sink. Table 1 describes the related works with objectives, advantages and disadvantages.

4   MQTT-SN Architecture

4.1   For WSNs

MQTT-SN is an extended version of MQTT and is extensively used in wireless sensor networks (WSNs).

It is mainly designed to operate on low cost and low power sensor devices. It is used to connect the IoT devices on the top of Zigbee which is an open industrial association with the aim to provide single communication method for WSNs. MQTT-SN has some additional features compared to MQTT such as, using simple UDP connection for faster and light weighted message transferring over WSNs. The architecture of MQTT-SN is given in Fig. 1. The architecture mainly has three components such as MQTT-SN clients, MQTT-SN gateways and forwarders.

The steps for the connectivity procedure of MQTT-SN are as follows:

1. *Discovery of active gateways*: In WSN scenario, one or more active gateways can announce their presence by using ADVERTISE message. The mobile sensor nodes will get to know about the active gateways with the help of GWINFO message and can publish the data. The only limitation here is that each mobile sensor node can connect to only one gateway at a time.

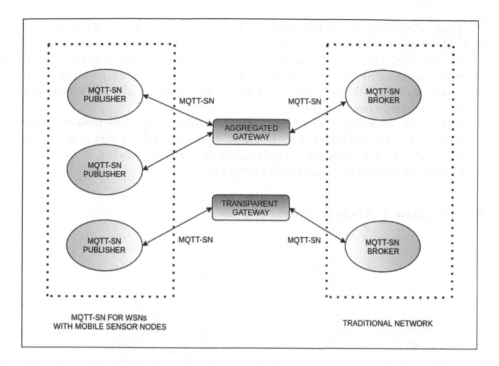

Fig. 1. MQTT-SN for WSNs.

2. *Topic-Id registration*: One or more additional features of MQTT-SN is it uses short *Topic-Ids* in place of long topic names. Registration occurs prior to publishing the data through PUBLISH message, where all the subscribers, gateways and clients come to know about the list of topic-ids parallel to topic names.

4.2 For Gateway Selection Strategy

MQTT-SN is known for its efficiency compared to MQTT protocol. The main aim here is to optimise the protocol for low cost, battery driven IoT devices with limited processing capacity. Figure 2 shows the architecture for gateway selection strategy. It mainly comprises of three components they are: MQTT-SN clients, MQTT-SN forwarder and broker. There are three types of gateways as follows:

1. *Transparent Gateway:* In this method each mobile MQTT-SN node connects with only one gateway. The number of connections between MQTT-SN clients and gateway is same as the connections between gateway and broker. It is exclusively used for end to end and transparent message exchanges between clients and broker. This will perform a syntax translation between MQTT and MQTT-SN protocols. Its implementation is simpler but it requires MQTT broker to support a separate connection for each active client.

2. *Aggregating Gateway:* In this method all the mobile sensor nodes can connect to the same gateway. All messages are exchanged between mobile sensor nodes and gateway. The gateway decides which data to be transferred to the broker. Its implementation is more complex compared to transparent gateway but it is more useful for WSNs because it reduces the number of MQTT connections that the broker has to support.

3. *MQTT-SN Forwarder:* The MQTT-SN forwarder is an intermediate node between broker and mobile sensor nodes. The main role of this is to encapsulate the data from mobile sensor nodes and the forwarding it to broker without changing the content and vice versa.

5   Proposed Architecture

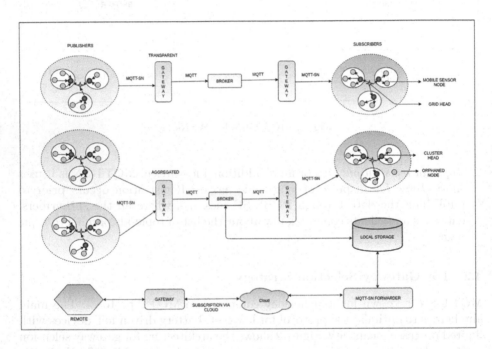

Fig. 2. WSN with mobile sensor nodes.

5.1   Problem Statement

Designing an architecture for Estimating end-to-end delay and throughput in IoT using MQTT-SN protocol, dynamic gateway selection algorithm and mobility aware hybrid clustering algorithm (MHCA) for mobile sensor nodes.

5.2   System Model

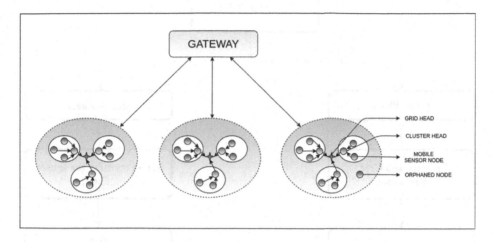

Fig. 3. MQTT-SN enabled WSN with mobile sensor nodes.

Figure 3 depicts architecture of MQTT-SN enabled WSN. MQTT-SN is a lightweight messaging protocol and it uses topic IDs instead of topic names. These features of MQTT-SN reduces the overhead, delay and helps in increasing throughput in communication. The mobile WSNs(MSWN) made of mobile sensor nodes(MSNs) which are initially deployed at random positions in an area for sensing and transferring the data to a gateway deployed at some distance from the MSNs deployed area. The assumptions made for proposed MHCA are as follows:

1. The sensor nodes are mobile with variable speeds and pause times.
2. The gateway is stationary.
3. The MSNs can move at anytime.
4. All MSNs are aware of their location, residual energy and velocity.
5. Initially all the nodes have same energy.
6. The transmitting power of the nodes are adjustable accordingly depending on the distance of the cluster head and gateway.

   In Fig. 4, the device classification is as follows:

- GW: Stationary gateway at the top of the hierarchy which is placed at some distance from clusters.
- GH2: Layer-2 grid heads.
- CH1: Layer1 cluster heads and at layer-0 MSNs are present.

The flow of data is from bottom to top that is from MSNs to gateway. The mobility of the MSNs is marked in the Fig. 1 as a node going out of the cluster. The procedure for the proposed method is given. The description of the procedure is as follows. There are two phases namely:

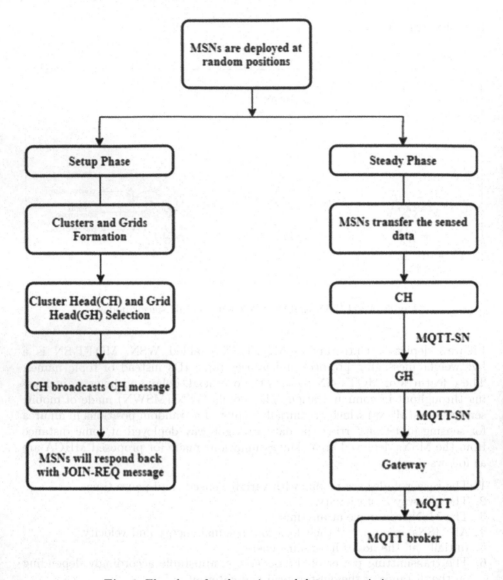

Fig. 4. Flowchart for clustering and data transmission

1. Setup-state phase
2. Steady-state phase

In the setup-state phase clusters and grids are formed and respective cluster heads and grid heads are selected and all MSNs are associated with their respective clusters and cluster heads.

The layer-2 is designed using Fuzzy c-means method based centralized clustering and layer-1 is designed using LEACH mobile inspired distributed clustering which considers MSNs energy and velocity. The cluster heads broadcast the CH message and MSNs that are closer to CH will respond back with JOIN-REQ message. Each cluster head creates its own TDMA schedule and that will be shared with MSNs of respective clusters.

In steady-state phase layer-0 MSNs transfer the sensed data to its respective CHs following the advertised TDMA using MQTT-SN protocol. The CH collects and aggregates the data from all MSNs and sends that data to layer-2 grid heads via MQTT-SN protocol, this data is further transferred to static gateway through MQTT-SN protocol. The mobility of MSNs is managed as follows: Every layer-2 GH maintains a list of layer-1 CHs and every layer-1 CHs maintains a list of MSNs associated with it. Upon not hearing from the MSNs in two consecutive frames, the MSN is considered to be moved out of its range. Hence it is removed from the list and TDMA schedule is updated accordingly. For managing the mobility we have considered cumulative acknowledgment which is broadcast by CH after every frame. GH acknowledges on receiving the data from CH and CH acknowledges on receiving the data from MSNs. If a CH does not receive acknowledgement from its GH then it considers itself to be orphaned due to its mobility. This orphaned node moves nearer to another cluster and sends JOIN-REQ message. Based on the degree of association and Cluster range the CH accepts the request and updates its TDMA schedule and node list.

6   Conclusions

In this paper, The Hierarchical clustering algorithm Mobility aware Hybrid Clustering Algorithm(MHCA) is discussed for cluster head selection in WSNs with mobile sensor nodes. An Architecture for dynamic gateway selection has been proposed. The proposed MQTT-SN architecture can be implemented as proposed to reduce end-to-end delay and increase throughput. The future plan is to implement the proposed method and evaluate the results.

References

1. Govindan, K., Azad, A.P.: End-to-end service assurance in IoT MQTT-SN. In: 2015 12th Annual IEEE Consumer Communications and Networking Conference (CCNC), pp. 290–296 (2015)
2. Silva, E.F., Dembogurski, B.J., Vieira, A.B., Ferreira, F.H.C.: IEEE P21451-1-7: providing more efficient network services over MQTT-SN. In: IEEE Sensors Applications Symposium (SAS) 2019, pp. 1–5 (2019)
3. Lee, S., Kim, H., Hong, D.K., Ju, H.: Correlation analysis of MQTT loss and delay according to QoS level. In: The International Conference on Information Networking 2013 (ICOIN), pp. 714–717 (2013)
4. Roy, D.G., Mahato, B., De, D., Buyya, R.: Application-aware end-to-end delay and message loss estimation in Internet of Things (IoT) -MQTT-SN protocols. Future Gener. Comput. Syst. 89, 300–316 (2018). ISSN 0167–739X

5. Raza, S., Duquennoy, S., Höglund, J., Roedig, U., Voigt, T.: Secure communication for the internet of things - a comparison of link-layer security and IPsec for 6LoW-PAN. In: Security and Communication Networks, vol. 7, no. 12, pp. 2654–2668 (2014)
6. Liao, K.-H., Lin, C.-Y.: International Journal of Design, Analysis and Tools for Integrated Circuits and Systems, 6(1), 48–49 (2017)
7. Luzuriaga, J.E., Perez, M., Boronat, P., Cano, J.C., Calafate, C., Manzoni, P.: Improving MQTT data delivery in mobile scenarios: results from a realistic testbed. Mobile Inf. Syst. 3(1), 11–17 (2016)
8. Jan, B., Farman, H., Javed, H., Montrucchio, B., Khan, M., Ali, S.: Energy efficient hierarchical clustering approaches in wireless sensor networks: a survey. Wirel. Commun. Mobile Comput. 2017, Article ID 6457942, 14 p. (2017)
9. Zafar, S., Bashir, A., Chaudhry, S.A.: Mobility-aware hierarchical clustering in mobile wireless sensor networks. IEEE Access 7, 20394–20403 (2019)
10. Lakkakorpi, J., Flinck, H., Heinonen, J., Korja, P., Partti, T., Soranko, K.: Minimizing delays in mobile networks: with dynamic gateway placement and active queue management, pp. 1–3 (2016)
11. Longo, E., Redondi, A.E., Cesana, M., Arcia-Moret, A., Manzoni, P.: MQTT-ST: a spanning tree protocol for distributed MQTT brokers. In: ICC 2020–2020 IEEE International Conference on Communications (ICC), pp. 1–6 (2020)
12. Hunkeler, U., Truong, H.L., Stanford-Clark, A.: MQTT-S - a publish/subscribe protocol for wireless sensor networks. In: 2008 3rd International Conference on Communication Systems Software and Middleware and Workshops (COMSWARE '08), pp. 791–798 (2008)
13. Kim, J., Lin, X., Shroff, N.B., Sinha, P.: Minimizing delay and maximizing lifetime for wireless sensor networks with anycast. IEEE/ACM Trans. Netw. 18(2), 515–528 (2010)
14. Stanford-Clark, A., Truong, H.L.: International business machines (IBM) Corporation version, 1(2) (2013)
15. da Rocha, H., Monteiro, T.L., Pellenz, M.E., Penna, M.C., Alves Junior, J.: An MQTT-SN-based QoS dynamic adaptation method for wireless sensor networks. In: Barolli, L., Takizawa, M., Xhafa, F., Enokido, T. (eds.) AINA 2019. AISC, vol. 926, pp. 690–701. Springer, Cham (2020). https://doi.org/10.1007/978-3-030-15032-7_58
16. Light, R.A.: Mosquitto: server and client implementation of the MQTT protocol. J. Open Source Softw. 2(13), 265 (2017)

Smart Retail Goods Delivery: A Novel Approach for Delivery of Online Store Products Using IoT

N. Ravinder[✉], S. Hrushikesava Raju, G. Kantha Rao, N. Rajesh, V. Murali Mohan, and Venkata Naresh Mandhala

Department of Computer Science and Engineering, Koneru Lakshmaiah Education Foundation, Green Fields, Vaddeswaram, Guntur 522502, India
ravindernellutla@kluniversity.in

Abstract. Nowadays, the delivery of online products became a more hectic task for any shopping store. To reduce the cost spent on delivery persons as well as time incurred to reach the customer to be considered for originating the proposed approach. To reduce floated expenditures that vary on the location of delivery as well to save the time of delivery of the product to the right customer, the proposed approach identified is the usage of drones for delivering the products. The drones are attached with IoT that accepts the location of the customer as input and ships the product to the customer very quickly as output. In between the input and output, the processing could be instructed thoroughly to reach the customer. The drones are considered as the easiest mode of transportation as well as the attached IoT over the drones is useful to accept payments digitally. The drones are directed by the authorized employees and will accept input from those authorized personalities. Once the shipment is made by the drone, the next nearest customer location to be feed to deliver the next product, and like this continued till the number of customers is serviced. The benefit of this is 24/7 service to be done which means not only on the day but also on the night based on customer availability and interest.

Keywords: Drones · IoT · Shortest routes · Audio · Report submission

1 Introduction

In our daily lives, we are purchasing products in the online mode. Whenever purchased in online, the time taken to receive the product is varying depending on the delivery persons and their mode of transport. After the product reached to the native location of the customer, it takes more time also because in-proper management of the deals. The number of drones to use depends on the demand of that online market in the society. Once drones are purchased, that is one time investment is done, the maintenance cost for the operation of drones requires software as well as authentic employees and is less compared to manual delivery. The drones to be purchased based on amount of weight they will bear as well as service to the number of customers. Accordingly, varying type drones to be purchased. In the proposed approach, how the drones are operated and how much they weigh, how long they operate once charged plays success of the proposed.

© Springer Nature Switzerland AG 2021
K. R. Venugopal et al. (Eds.): ICInPro 2021, CCIS 1483, pp. 491–500, 2021.
https://doi.org/10.1007/978-3-030-91244-4_39

The kind of service that drones will do is based on the kind of shorted path routing is used. The route of the customers are given as input, the picking of the location is done internally by shortest path routing algorithm, and transports the item to the customer till the last location in the route is read. The objective of this is automating the process and reducing the cost for transportation as well as cost on man power. In this, drones are not only for capturing videos or images of the meetings or celebrations at conventional centers, but also demonstrated as they are used to social service activities and business oriented. In this, the time taken to execute service to the customers is more because many factors involved for the delay and consumes the time are traffic in the way on the road, waiting for the customer sometimes, doesn't meet expectations of the deadlines due to many external factors, and etc. In order to pay attention in delivering the goods, a dynamic mechanism is required that will deliver the goods with the specified time, and will update the status instantly. This may be applicable to the post offices also, where posting of the letters or packets could be done and statuses are updated automatically after each activity. The motto to initiate the proposed ideology is to reduce time and enhance the revenue for the service provider merchants. In future, many applications work to be simplified because of revolution in technology. How to deliver the goods to the clients efficiently is a challenging task. In this, what studies are there so far to delivery the goods, and how the proposed methodology is going to be simply the task and will perform the task efficiently. The objective is to deliver the goods with dynamic nature using efficient dynamic routing approaches and automatic updating of the status. As technology is upgrading, bringing that to the user level and getting customer satisfaction is the motto of this work.

2 Literature Review

In this, there were many methods used in delivering the goods. All those existing methods are discussed here along with their limitations. Such pitfalls are overcome using proposed method. The proposed method will use the drones using IoT. The benefits provided are to be listed in the end paragraph of this chapter. In listed references [1, 2] describe that the characteristics of the drones to be needed in various applications and their service in that aspect, and future requirements in those drones, and how much pay load the drone to support while delivering to the user. Also, the style of delivering will be made simpler to the all parts such as urban, and rural. In references such as [3, 4] states that goods are transported from one place to another place far away using horse drawn drones, and from there, railways are used to transport to other places, then trucks, vans, cars are used to ship the goods to the required store place. Also, many benefits such as increased efficiency and fastest delivery are guaranteed. In addition, there flaws and few other challenges need to be faced. In [5], and [6] works, first study is focused on challenges, strategies, possible solutions are brought from a many studies and concluded that crones are best possible option for speedy delivery, and second work is focused on evolution of drones and their need in the present market, and pinned the benefits such as green emissions compared to vehicles, and speedy transport using this aerial kind. In [7, 8] studies, first study is on analysis the people over cities when delivery to be happened on the same day using drones, and second study is focused on cargo drones which are capable of shipping the goods over a large distance within a country and these drones are made in variety of options that meet user-friendliness and possess dynamic environment. In analysis of [9, 10], first study is on route to be predicted in order to process few components related to chemistry and physics and second study is on shorted route in the process of few chemical components respectively. Such routes are to be predicted during the analysis of components over the boards. In [11, 12] studies, first study is focused on integrated delivery for real time application using evolutionary and simulation approaches over semiconductor manufacturing designs. and second is focused on novel approach for delivering the vaccine, its features, and motion capturing in the video clip is demonstrated. In [13], the study is focused on constraints based routing by unmanned aerial vehicles in a muti-route planning and multi-obstacle avoidance model's dynamic environment. In [14] study, the effective power delivery is done by analysis of genetic algorithm and its output is best upsurge is demonstrated. In [15] study, there is a novel framework and proposal for premature delivery of conceived ladies using WHSN is demonstrates and is concluded. In [16] study, there is analysis on delivering the packets in maximizing the throughout in the wireless sensor networks using cluster approach is demonstrated and is concluded. In [17] study, the watchword is guiding the travel route through a novel approach is demonstrated and is explored in a reliable manner. In study [18], the aim is focused on travelling sales problem where shortest route is found from source to destination using java macaque algorithm is demonstrated and is explored in a reliable manner.

In the sources mentioned in [19], the demonstration is on usage of Internet of Technology and sensors, their interaction in order to communicate and generate automatic reports as per the application.

In the regard of information mentioned in [20], the present pandemic is to be restricted using the digital mask that reports the virus in the environment that the user is currently staying. The mask designed will provide statistics about the objects in the present environment. With respect to source specified in [21], the IoT is used in detecting the location and automatically takes its currency and converts that into the user's currency. This user flexibility is provided in this context.

As per the study mentioned in [22], the IOT is used in the power banks and portable devices in order to exchange charging power in the user-friendly atmosphere. The customized way of charging is done through the designed app and IOT technology. With respect to the description given in [23], the IOT is used in communicating the weighted objects falling to the other devices in order to catch it and send it gently to the ground using automated net.

As the information of [24], the IOT is used in the industries where level of gas is monitored and detects the leakage if any such is identified during the passage of gas over the pipes that are placed from the source to destination. This detection avoids harmful incidents over the people. With respect to the source specified in [25], the IOT is useful over the users in such a way that users health bulletin to be monitored and provides a guide to maintain the fitness based on food diet. With the view of source mentioned in [26], the IOT and GSM are used in determining the popular places when a user wants to make a trip in the world. The guidance is to be provided about the top places and ranked places in those cities along with route map. As per the source demonstrated in [27], the GSM and IOT are used to monitor the garbage bins and alert the nearest the municipal office in order to clear it which avoids wastage of visiting many times of that bin. In the regard of [28], any intrusion is detected in the IOT based internet environment in the homes, which should be alerted and avoids future inconveniences. In the aspect description given in [29], the detection of premature bosom irregularity in the images related to especially personal healthcare systems is discussed and the role modality is explored in processing the system. With regard of demonstration of info in [19], the human effort is maximum minimized and increases automation through the app based on IOT devices and checks the eye sight remotely. The significance of IOT is clearly depicted and would be useful in making the proposed system. In the perspective of demonstration from [30], the location in the ocean is detected that is capable of creating cyclones as like normal cyclones using IoT and would generate such detail using controller station. The data from [31], the detection of entities are detected suing ML approach and IoT and would generate report on objects that were identified. From the collected information from [20], the nearest locations and objects were identified as infected were detected and prescriptions are suggested using IoT and capable methods.

In these studies, these analyzations are taken and are considered for the building the ideology of proposed work. These works help to use their terminologies and features are directly or indirectly provided as inputs for the proposed ideology. In proposed ideology, the route is predicted using dynamic shorted route algorithm and is automated to reach the target to serve the user to whom delivery to be done. This approach will help to deliver the item in less time by keeping track of user location and update in the app or software after the delivery is done.

3 Proposed Approach

In this proposed approach, the authentic employees will give a route to the drone, the plan to visit the customers is chosen by the shorted route algorithm. Sometimes, the selection of customer may vary depends on the customer time available. The functionality of this proposed is as follows:

The characteristics of this proposed approach falls into input, processing, and output.

a) **Input:** In this, the number of customers to be serviced by the drone and the locations are given. The drone reads the locations, call to the customer. If the customer is willing to take the product, will include in the route plan. If not willing on that day, will reschedule that customer in next plan in near time.

b) **Processing:** The customer will receive products by the drone. During call to the customer, if customer will need the product on different location than static location, that destination location to be feed and will ship the product dynamically. Like this, remaining customers are to be serviced. The shorted route algorithm is used here to pick the next customer and if the customer is ready to accept, the drone will fix that customer location and ships the product. This process is repeated till last location in the route is processed and stops the day activity once all customers of that day plan are serviced.

c) **Output:** The products to be delivered to the customers according to the scheduled day, collects payment and service review from that customer there and then.

The architecture of the shipping of items to the users is done using drones is as follows (Fig. 1):

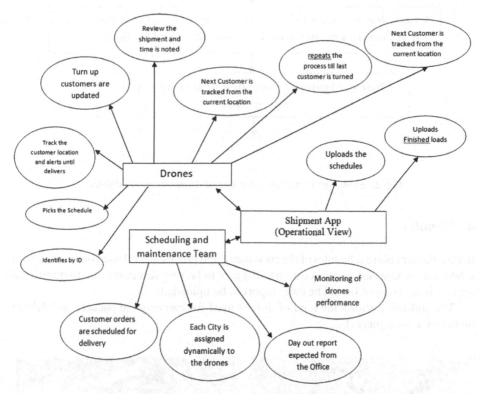

Fig. 1. Architecture of usage of drones in delivering the goods

There were two modules identified such as Drones where these are main vehicles that will carry goods and ship the items to the customers, and Scheduling and Maintenance Team where shortest routing is loaded based on customer places marked in a schedule and estimated time is noted to complete the schedule of that day, the drones performance and its activities are also monitored by this team.

The pseudo procedure for above method is defined as follows:

Step4: Like step3, other drones schedules are also monitored and their reports are updated. Once schedules of a day are finished, schedules for the next day are planned and are assigned to the drones.

Step5: If any service or maintenance is required for certain drones, the technician and admin will be reported about the issue and are resolved.

The flowchart for this proposed approach is defined as follows. In which, scheduled and maintenance team is the first module and drones is another module (Fig. 2).

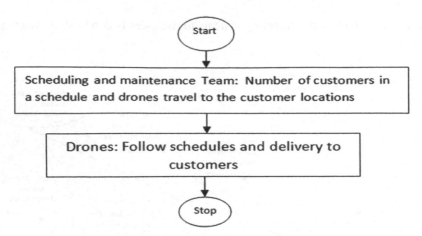

Fig. 2. Flowchart of modules of usage of drones in goods delivery

4   Results

In this, the inputs are schedules of the customers, and processing is based on shortest path scheduling where customer to customer distance to be very less compared to traditional approach, and output should be daily report to be uploaded.

The following shows the kind of drones used for carrying the packets and deliver them to the customers (Fig. 3).

Fig. 3. Kinds of drones for carrying the goods from one place to another location

These drones vary in number of wheels or fans they consist of, the speed they support, and payload that they carry, and technology they support depending on the hardware.

In logistics and goods delivery, the sales will be increased whenever the item to be delivered on the same day. Here, every customer is treated as valuable and service to be provided as they expected. The following is the survey made about certain cities and customers in those expecting delivery of products on the same day. The revenue will be decreased because of lack of that facility. The statistics were recorded and depicted in Fig. 4.

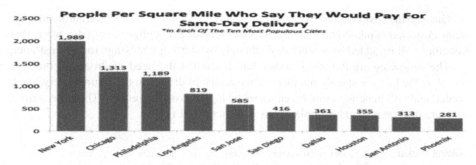

Fig. 4. Survey over same day delivery and pay on the spot

Assume location Tirupati location, where customer locations are marked and their product names and price to pay is also marked. This schedule is given as input to the drone. The auto-generated schedule for delivering the goods is mentioned in Fig. 5.

The sample Schedule is as follows:

```
Schedule1 - D001(Drone ID)
1. Markandeyulu, Tiruchanoor - Rager 2.0 - 1100
2. Mastan, ThummalaGunta - Spectacle - 499
3. Radha, S V ZOO Park - Portable Mini Fan - 599
4. Veera Sastry, Sathyanarayna Colony - Shoes - 499
5. Shiva, Leelamahal - Toys - 1099
6. Ramaiah, Auto Nagar - Window Curtains - 999
7. Latha, Mahila University - Scrubs, Hair Pins - 99
```

Fig. 5. Schedule of customers

From the above schedule, the customer places are loaded on a map, small distance locations are tracked, and customers of that location are alerted and get serviced. The picking up of users in the shorter distances were automatically selected by the defined algorithm from the proposed system (Fig. 6).

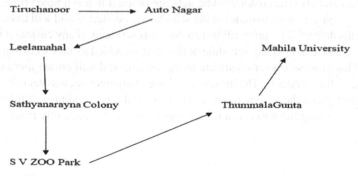

Fig. 6. Scheduling of customer places according to the distance

Once all the customers are serviced, the report of that day to be updated automatically. If any customer is missed on that day because of customer negligence or busy work, that customer will be added to another day schedule by alerting a message to that customer.

The following are the applications that demanded the need of drones in order to complete the tasks in speedy manner. Hence, usage of drones in the future is very much needed and will generate more revenues to the logistics organizations. The expectations of usage of drones in many applications is depicted in Fig. 7.

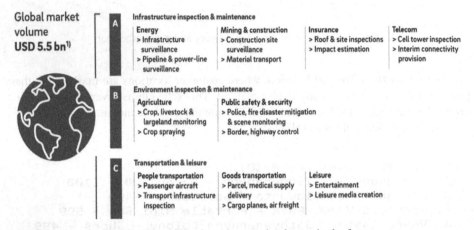

Fig. 7. The applications that require drones in the future

The revenue to be generated using drones is unpredictable compared to traditional method. The aim of this is to satisfy the customers and automate the process of delivering the goods.

5 Conclusion

In this, people are customers who ordered the products. They are serviced in less time using drones where drones are air vehicles that are tracked by the schedule and maintenance team and also the customer who gets the product in a estimated time. Once the customer is serviced, next customer in the schedule to be alerted and will track the drone location. This delivery continues till last customer is serviced. If any customer is missed, that could be added to next day schedule or that may be added to another drone to take up that task. This process leads to generate more revenue and will attract more customers towards the online products. The drones also have challenges such as security, and their powered throughout the assigned day work. Hence, the novel approach titles usage of drones for delivering the goods is a future demanding and is presently in updating.

References

1. Hader, M., Baur, S.: Cargo Drones: the future of parcel delivery (2020). https://www.roland berger.com/en/Point-of-View/Cargo-drones-The-future-of-parcel-delivery.html#:~:text=in% 20North%20Carolina.-,Transportation%20of%20air%20freight,more%20often%20with% 20more%20flexibility
2. High payload drone for delivery and transport, Airborne Drones. https://www.airbornedrones. co/delivery-and-transport/
3. Willems, D.: Cargo transportation takes drone technologies to the next level (2020). https:// umsskeldar.aero/cargo-transportation-takes-drone-technologies-to-the-next-level/
4. Cargo drones, Drones for tomorrows air cargo. https://www.iata.org/en/programs/cargo/ cargo-drones/
5. Kellermann, R., Biehle, T., Fischer, L.: Drones for parcel and passenger transportation: a literature review (2020). https://doi.org/10.1016/j.trip.2019.100088
6. Gain, B.C.: Drone deliveries begin a slow take off (2020). https://www.ttnews.com/articles/ drone-deliveries-begin-slow-takeoff
7. Robinson, A.: Transport drones & autonomous vehicles: the transportation mode dance card is getting Full. https://cerasis.com/transport-drones/
8. Kumar, F.S.: Giant cargo drones will deliver packages farther and faster (2019). https://www. theverge.com/2019/6/10/18657150/autonomous-cargo-drones-delivery-boeing-aircraft-faa-regulation
9. Chandrika, M., Ravindra, A.V., Rajesh, C., Ramarao, S.D., Ju, S.H.: Studies on structural and optical properties of nano ZnFe2O4 and ZnFe2O4 TiO2 composite synthesized by co-precipitation route. Mater. Chem. Phys. https://doi.org/10.1016/j.matchemphys.2019.03.059
10. Regalia, V.R., Addada, R.R., Puli, V.S., Saxena, A.S., Chatterjee, A.: A short and concise route to total synthesis of Dendrodolide. Tetrahedron Lett. (2017). https://doi.org/10.1016/j. tetlet.2017.04.097
11. Manupati, V.K., Revanth, A.S., Srikanth, K.S.S.L., Maheedhar, A., Sreekara Reddy, M.B.S.: Real-time rule-based scheduling system for integrated delivery in a semiconductor manu-facturing using evolutionary algorithm-based simulation approach. In: Dash, S.S., Arun, M., Panigrahi, B.K., Das, S. (eds.) Artificial Intelligence and Evolutionary Computations in Engi-neering Systems. AISC, vol. 394, pp. 981–989. Springer, New Delhi (2016). https://doi.org/ 10.1007/978-81-322-2656-7_90
12. Rajeswari, P.R., Raju, S.V., Ashour, A.S., Dey, N.: Insilico approach for epitope prediction toward novel vaccine delivery system design. Feature Detect. Motion Detect. Video Process. 256–266 (2016). https://doi.org/10.4018/978-1-5225-1025-3.ch012
13. Prathyusha, K., Sastry, A.S.C.S., Chaitanya, S.M.: Dynamic constraint based multi-route planning and multi-obstacle avoidance model for unmanned aerial vehicles. J. Adv. Res. Dyn. Control Syst. 9(1), 348–361 (2017)
14. Abdul, A.M., Cherukuvada, S., Soujanya, A., Srikanth, R., Umar, S.: Analysis of genetic algorithm for effective power delivery and with best upsurge. In: Satapathy, S.C., Bhateja, V., Raju, K.S., Janakiramaiah, B. (eds.) Data Engineering and Intelligent Computing. AISC, vol. 542, pp. 1–9. Springer, Singapore (2018). https://doi.org/10.1007/978-981-10-3223-3_1
15. Chintala, R.R., Narasinga Rao, M.R., Venkateswarlu, S.: A proposal for observing conceived ladies having high risk of premature delivery using WHSN. Int. J. Eng. Technol. (UAE) 7(2), 53–56 (2018). https://doi.org/10.14419/ijet.v7i2.32.13524
16. Karthikeyan, T., Brindha, V., Manimegalai, P.: Investigation on maximizing packet delivery rate in WSN using cluster approach. Wirel. Pers. Commun. 103(4), 3025–3039 (2018). https:// doi.org/10.1007/s11277-018-5991-z

17. Mandhala, V.N., Ganesh, N., Sai Shirini, K., Sri, C.A.: A novel approach on suggesting travel route by efficient watchword. Int. J. Eng. Technol. (UAE) **7**(2.32 Special Issue 32), 228–232 (2018)
18. Karunanidy, D., Amudhavel, J., Datchinamurthy, T.S., Ramalingam, S.: A novel JAVA macaque algorihtm for travelling salesman prolem. IIOAB J. (2017)
19. Hrushikesava Raju, S., Burra, L.R., Waris, S.F., Kavitha, S., Dorababu, S.: Smart eye testing. In: Bhattacharyya, S., Nayak, J., Prakash, K.B., Naik, B., Abraham, A. (eds.) International Conference on Intelligent and Smart Computing in Data Analytics. AISC, vol. 1312, pp. 173–181. Springer, Singapore (2021). https://doi.org/10.1007/978-981-33-6176-8_19
20. Tumuluru, P., Hrushikesava Raju, S., Sai Baba, C.H.M.H., Dorababu, S., Venkateswarlu, B.: ECO friendly mask guide for corona prevention. In: IOP Conference Series Materials Science and Engineering, vol. 981, no. 2. https://doi.org/10.1088/1757-899X/981/2/022047
21. Sai Baba, C.H.M.H., Hrushikesava Raju, S., Santhi, M.V.B.T., Dorababu, S., Saiyed Faiayaz Waris, E.: International currency translator using IoT for shopping. In: IOP Conference Series Materials Science and Engineering, vol. 981, no. 4. https://doi.org/10.1088/1757-899X/981/4/042014
22. Sunanda, N., Hrushikesava Raju, S., Waris, S.F., Koulagaji, A.: Smart instant charging of power banks. In: IOP Conference Series Materials Science and Engineering, vol. 981, no. 2. https://doi.org/10.1088/1757-899X/981/2/022066
23. Mothukuri, R., Hrushikesava Raju, S., Dorababu, S., Waris, S.F.: Smart catcher of weighted objects. In: IOP Conference Series Materials Science and Engineering, vol. 981, no. 2. https://doi.org/10.1088/1757-899X/981/2/022002
24. Kavitha, M., Hrushikesava Raju, S., Waris, S.F., Koulagaji, A.: Smart gas monitoring system for home and industries. In: IOP Conference Series Materials Science and Engineering, vol. 981, no. 2. https://doi.org/10.1088/1757-899X/981/2/022003
25. Hrushikesava Raju, S., Burra, L.R., Waris, S.F., Kavitha, S.: IoT as a health guide tool. In: IOP Conference Series, Materials Science and Engineering, vol. 981, no. 4. https://doi.org/10.1088/1757-899X/981/4/042015
26. Hrushikesava Raju, S., Burra, L.R., Koujalagi, A., Waris, S.F.: Tourism enhancer app: user-friendliness of a map with relevant features. In: IOP Conference Series, Materials Science and Engineering, vol. 981, no. 2. https://doi.org/10.1088/1757-899X/981/2/022067
27. Kavitha, M., et al.: Garbage bin monitoring and management system using GSM. Int. J. Innov. Technol. Explor. Eng. **8**(7), 2632–2636 (2019)
28. Kavitha, M., Anvesh, K., Arun Kumar, P., Sravani, P.: IoT based home intrusion detection system. Int. J. Recent Technol. Eng. **7**(6), 694–698 (2019)
29. Kavitha, M., Venkata Krishna, P., Saritha, V.: Role of imaging modality in premature detection of bosom irregularity. In: Venkata Krishna, P., Gurumoorthy, S., Obaidat, M.S. (eds.) Internet of Things and Personalized Healthcare Systems. SAST, pp. 81–92. Springer, Singapore (2019). https://doi.org/10.1007/978-981-13-0866-6_8
30. Subbarao, G., Hrushikesava, S., Burra, L.R., Mandhala, V.N., Seetha Rama Krishna, P.: Smart cyclones: creating artificial cyclones with specific intensity in the dearth situations using IoT. In: Saha, S.K., Pang, P.S., Bhattacharyya, D. (eds.) Smart Technologies in Data Science and Communication. LNNS, vol. 210, pp. 349–358. Springer, Singapore (2021). https://doi.org/10.1007/978-981-16-1773-7_28
31. Lalitha, V.L., Hrushikesava Raju, S., Sonti, V.K., Mohan, V.M.: Customized smart object detection: statistics of detected objects using IoT. In: 2021 International Conference on Artificial Intelligence and Smart Systems (ICAIS), pp. 1397–1405 (2021). https://doi.org/10.1109/ICAIS50930.2021.9395913

Innovative Farming: An Automated IoT Based Multipurpose Vehicle for Farmers

S. Vasanti Venkateshwar[1], Mohammad Mohiddin[1(✉)], Vineeta[2], and Asha S. Manek[3]

[1] Department of ECE, GNIT, Hyderabad, India
mohiddinmd.ecegnit@gniindia.org
[2] Department of CSE, AMC Engineering College, Bangalore, India
[3] Department of CSE, RVITM, Bangalore, India

Abstract. Agriculture is a major occupation in India and the farmers are mostly from villages those are not aware of recent technologies and their usage. Rising cost of production, accessibility of proficient laborers, deficiency of water resources and monitoring of crop etc., are the major problems faced by the farmers in India. The proposed cost effective automated sensor based multipurpose vehicle solves the above problems to improve productivity of the crops by maintaining its nutrition value. The proposed system combines robotics with agriculture; is integrated with Arduino Mega, Sensors, Motors, Relays and GPS & GSM Modules etc. and capable of moving around the field like a farmer.

Keywords: Agriculture · Arduino Mega · Multipurpose vehicle · Robot · Sensors

1 Introduction

Agriculture is a vital source for humans to survive. The farmers who do agriculture are 70% of the population of India and they spend most of the time in cultivating the farm land. They are mostly in villages and are not well-versed with the availability of recent technologies and their usage. Agriculture is becoming a high-tech industry giving rise to new skilled agriculturalists. The rapid developing technology in agriculture not only helps to improve the production competencies of farmers but also expands automation technology such as robotics development in agriculture. As per UN Department of Economic and Social Affairs estimation, the world population will rise from 7.3 billion today to 9.7 billion in 2050 [1]. The requirement of food and farmers will be a great challenge in near future. The demand of agricultural robot will increase by 20% until 2022; is the speculation by many researchers which will in turn outgrow a market capitalization by $13 billion. Today, there is a massive growth of farm robotics particularly in indoor machines and for milking cows. Some initiatives have been taken like automated tractors or drones but these are still initiatives not fully automated and have a more complicated mechanism. Hence it is need of today to develop fully automated robot which can roam through the farms like a farmer [2].

© Springer Nature Switzerland AG 2021
K. R. Venugopal et al. (Eds.): ICInPro 2021, CCIS 1483, pp. 501–512, 2021.
https://doi.org/10.1007/978-3-030-91244-4_40

Major problems faced by the farmers are raised input costs, non-accessibility of trained laborers, deficiency of water sources and crop monitoring, increase in production yield, etc. The proposed system can be boon to agriculturist which is fully automated vehicle and is capable of moving around the field like a farmer. The proposed robot is integrated with Arduino Mega, Soil Moisture Sensor, Rain Sensor, DC Motor, Stepper Motor, GPS Module, GSM Module, Relays etc. It can be used for ploughing the farm land, seeding, and leveler to close with the soil, water sprayer to spray water with least changes in accessories with minimum cost and weed controlling. The farmers will be provided with a wireless remote control using which they may turn on or off the vehicle robot and control its movement. Farmers can be trained for operating the system without having any knowledge about sensors.

1.1 Societal Relevance

The Societal Relevance of the proposed multipurpose vehicle is as follows:

- Village Farmers will be benefited with the availability of recent technologies and their usage.
- Agriculture is one such area where identification and classification of diseased crops can increase the crop production significantly with the help of user-friendly Multipurpose Vehicle.
- Diseased crop identification and alerting the farmers can help agriculture as well as the food industry.
- Complete automated kit for the farmers by applying IoT techniques to yield more crops by generating agriculture data.

The organization of the paper is as follows: Sect. 2 deals with related work in the area. Section 3 describes the background and justification. Sections 4 and 5 describes objectives and status of multipurpose vehicles. Section 6 illustrates the approach to design multipurpose agriculture vehicle and use of hardware and software components to build the prototype. Section 7 deals with the results and discussion. Section 8 describes the performance analysis and Sect. 9 concludes the paper.

2   Related Work

In research work [3], M. A. Abdurrahman, et al., proposed a product for watering the farmland in the area where scarcity of water problem is more. The simple circuitry system automatically controls the flow of water and has been built with low cost sensors for watering plants as per the requirement sensed by soil moisture level. The system consists of LCD to display humidity and temperature levels.

Rahul Dagar et al., implemented sensor based agriculture system. Air temperature sensor, soil pH sensor, soil moisture sensor, humidity sensor, water volume sensor etc., are used in the model. The model is a simple architecture of IoT sensors that collect information and send it over the Wi-Fi network to the server, and then server can take actions depending on the information [4].

Saqib et al., used a tree -based foundation to expand the network area by adding central divisions to it [5]. Each sensor node included cell, a microcontroller, a humidity sensor and a network unit. Small data points from the central node sent to the cloud every day for future analysis. After a thorough analysis, the contact distance was measured to be between 250 m between the two points and increased to 750 m by the addition of two beams. The results showed that the method can be used as a model to meet the requirements for soil surveying, transportation and storage in a large area of farm.

The system proposed in [6] automatically irrigates when the soil is dry and sends alert to mobile via SMS. The farmer's phone gets alert if the field needs irrigation, on the other hand the farmer sends an SMS to the manager to start watering and then sends another SMS to stop after receiving an SMS from the microcontroller for a well full of water.

The paper [7] explains an IoT-based intelligent irrigation system to detect underground and control water plants. The microcontroller is used to receive information from the underwater system sensor and to transfer information about the GPRS network.

3 Background and Justification

The global population will reach 9.15 million by 2050 [1]. Therefore, the challenge for the next few decades is to meet the needs of a growing global population by developing productive management systems while protecting the environment. Many developing countries, including India, are facing an acute shortage of agriculture. Large numbers of young people are moving from rural to urban areas in search of a better life. As a result, farmers are working longer hours than the average hours due to a shortage of workers. Human resources, animals and agricultural products are used in seed preparation, planting, planting or sowing, fertilizers and the use of chemicals, culture and traditions. Despite of using available tractors, many farmers use traditional methods of using animals for plow, producing small crops and require a lot of cultivation. Farmers need to use modern tools and techniques that will help in reaching food to everyone in this world. Currently, traditional and hand-held tools such as wheels and sawmill are well-known for planting in paddy fields. All devices are designed for humans. On the other hand, management is the key to make agriculture more efficient and maintain its sustainability.

4 Objectives

The main objectives of the proposed works are as follows:

- To develop modern agricultural model consisting of related sensors to aid the farmers with the requirements like when to plant the seeds, adequate water & manure supply and protecting the crop from diseases & weeds.
- To design robot using sensors which collect crop related data, manage the crop and monitor it.
- To program proposed multipurpose farming vehicle which performs elementary farming jobs such as crop harvesting much quicker than the manual labors.

- To introduce digital agriculture system that provides maximum farm yield by minimum effects on environment.
- To generate crop data using multiple sensors used for analysis. This analyzed data further help in decision making.

5 National and International Status

"Farmers are the eyes of our country". India being first and the United States is second most has ploughing land in the world. Due to urbanization, the average size of farm land increasing but the number of farm lands in the United States are decreasing as many farmers sold their land. As a result, involvement of population in agriculture is reducing. Only 1% of the population is involved in the farming. In India, comparatively the size of farms is smaller ranging between 10 and 400 acres. Another problem that India is facing is lack of sufficient laborer, cost and time requirement for several field jobs. The farmers are experiencing substantial losses. The diminishing farm workforce needs an automated solution and sustainable situation. If the necessary steps have not been taken, food taste might change or loose nutritional value as in current scenario there is a use of more chemicals for cultivating more crops. Many youngsters hesitate to choose farming as a career option because of unpredictability of crop production in India. Modernization and upgrading Indian agriculture system and food industry is the need of an hour. In future, there will be huge requirement of many such farm robots interacting with human farmers in big farm land. Presently, motivating smart youngsters to choose farming as career option so that they can improve farming in India with latest technology is very much needed.

The proposed automated multipurpose vehicle design suits for the current Indian farming scenario. The goal is to design agriculture robot that uses various sensors along with visual cameras, GPS, Light Detection and Ranging (LiDAR) and other on-board sensors which can help to collect data autonomously. The internet connectivity has improved in rural areas due to digitization of India. Hence, this multipurpose vehicle will be connected to cloud which will store collected data for further use and can be easily accessible.

6 Approach to Design Automated IoT Based Multipurpose Vehicle for Agriculture

The aim of the work is to develop a multipurpose vehicle robot model to solve problems faced by the farmers such as changes in weather, environment, when to plant the seeds, adequate water and manure supply in the farmland, protecting the crop from weeds to improve productivity of the crops. The goal is to develop model as an entire solution kit for the agriculturalists to assist them with the agricultural activities from initial stage to ending step. Multifunctional system elements mean faster return on investment. The smart mechanization system of agriculture robots helps the people to establish an efficient agriculture system.

The IoT-Agriculture Robot consists of following Modules:

- Manage and monitor crop and soil, weather conditions, adequate water and manure supply.
- Identification of position of seeds dropping on land and protecting the crops from diseases & weeds.
- Scanning of soil, light, water, humidity, temperature management using sensing technologies and storing data on cloud.
- Including GPS using positioning technologies and communication through mobile using communication technologies.
- Software development which aids automation and IoT based solution required for farm land.
- Providing necessary information to the farmers using SMS.

Figure 1 shows architecture of proposed model developed for multiple farming operations. The proposed system is used for digging the soil by controlling the movement of Servo Motor and DC motor is used for sowing the seeds which opens and closes the plate to release the seeds at the desired distance. As the multipurpose agricultural robot moves forward after dropping the seeds, the digged place needs to be covered with soil again. This is done by using the attached plate. The moisture in the soil is checked by the humidity and soil moisture sensors. Whenever moisture level is decreased beyond the reference level, the water pump is turned on automatically. LCD display connected to the Arduino displays the ON/OFF status of the water pump. GPS module is connected to the Arduino to update the status of the water pump to the farmer through the Mobile App messages. Data sensed by different sensors can be uploaded to the cloud using Raspberry Pi and through the USB camera attached to the Raspberry Pi, images of weeds developed in the farm can be collected for weed detection and further analyzed

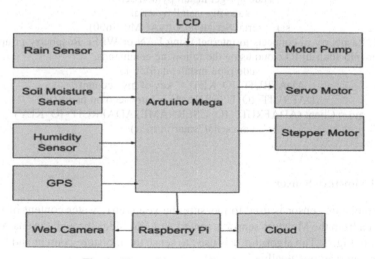

Fig. 1. The architecture of proposed model.

for the necessary actions. All these sensors gather the respective information for which they have been built. All these sensors are connected through the WiFi module.

This data captured by sensors is recorded and stored in the WiFi module and sent to the Base Station (Website). The same data is sent to the farmer's smart phone through the GSM module which is connected to the Raspberry pi3 board.

The system proposed is cost effective and does not require the costly equipment for its navigation; it is designed to be automatic and light weight. These advantages make it a real aid to the farmers.

Proposed multipurpose vehicle is integrated with "Arduino Mega", "Soil Moisture Sensor", "Rain Sensor", "DC Motor", "Stepper Motor", "Servo Motor", "GPS Module", "GSM Module", "Relays" etc. are explained below.

6.1 Data Collection from Sensors

The proposed model stores retrieved sensor data on cloud to be further used in analyzing the crop, soil, condition of weather, when to plant the seeds, adequate water and manure supply in the farmland, protecting the crop from diseases & weeds. The retrieved sensor values will be aggregated into one-units of blocks to transmit to the server with the help of Raspberry pi3 module. Below are mentioned steps for data collection from sensors which are connected to Arduino uploaded to cloud through raspberry pi.

Step 1: Arduino read the sensor data and store in a variable.
int y = analogRead(A0);
Step 2: This data sent to Raspberry pi using serial communication.
Serial.println(y);
Step3: Raspberry pi receives the data serially using the following libraries and commands.
sudo apt-get install python-serial
sudo pip install pyserial
ser = serial.Serial('/dev/ttyACM0',9600)
Step 4: Raspberry pi connects to internet using LAN or Wi-Fi. Raspberry pi uploads the data into adafruit IO cloud using the following commands.
sudo pip3 install adafruit_io
ADAFRUIT_IO_KEY = ' key of my account '
ADAFRUIT_IO_USERNAME = ' my account name '
aio = Client (ADAFRUIT_IO_USERNAME, ADAFRUIT_IO_KEY)
aio.send("smartfarm",r)

6.2 Soil Moisture Sensor

The soil moisture sensor is used to measure the volumetric water content of soil. The analog data from the Moisture sensor is sent to Arduino Mega through the analog pin A0 as shown in Fig. 2. The algorithm 1 is used to sense the moisture content and to control the water pump automatically.

6.3 Water Pump

Small motor water pump 220 V as shown in the Fig. 3 is connected to the Arduino with a relay module that mechanically moves water through a pipe which supplies water to the soil. The water pump gets switched on as soon as it receives command from Arduino pin 2 to irrigate the green area and switches off to deactivate the irrigation process.

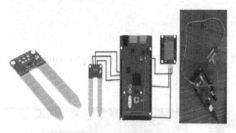 

Fig. 2. Soil moisture sensor.   **Fig. 3.** Water pump.

6.4 Rain Sensor and LCD Interfacing

The proper use of water is an important application of the proposed system. Therefore, the rain sensor as shown in Fig. 4 used in the proposed model alerts the farmer to take some measures such as stopping the irrigation process in the farmland.

Algorithm 1:   **Algorithm 2:**

if (moisture_content>900)   if (rain_content<750)
    {   {
            Turn ON Water pump   Turn ON stepper
    }   }
    else   else
    {   {
            Turn Off Water pump   Turn Off stepper motor
    }   }

Rain water detector [8] is an analog sensor which is suitable for irrigation fields, home automation, communication; automobiles etc. detects the rain to alert users. The analog data from the rain sensor is sent to Arduino Mega through the analog pin A1. The algorithm 2 is used to sense the rain and to control the stepper motor as shown in Fig. 5 automatically. A liquid-Crystal Display (LCD) as shown in Fig. 6 is a flat-panel display or other electronically modulated optical device that uses the light-modulating properties of liquid crystals combined with polarizer. Liquid crystals do not emit light directly, instead using a backlight or reflector to produce images in color or monochrome.

Fig. 4. Rain sensor interfacing to Arduino.   **Fig. 5.** Stepper motor interfacing Arduino.

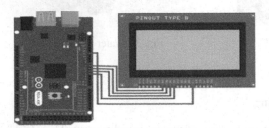

Fig. 6. LCD interfacing to Arduino Mega.

6.5 GPS

GPS stands for Global Positioning System as shown in Fig. 7 allows determining the location and GPS receiver time anywhere on earth.

GPS systems are used all over the world and it can be available on a wide range of devices, from cell phones to cars to smart watches. Due to the low prices, low power consumption and small size, GPS Neo 6M is used in available smart phones as devices [9].

6.6 DC Motor Interfacing to Arduino Mega

Figure 8 shows the interfacing of DC motor to Arduino Mega through relay. A DC motor is used to move the plate in which seeds are present. As the plate moves, seeds fall on to the sowing area.

6.7 Interfacing Arduino Mega and USB Camera to Raspberry Pi 3B

Data sensed by different sensors connected to Arduino Mega can be transmitted serially to Raspberry Pi for uploading data to the cloud which is shown in Fig. 9.

USB cable can be used between the Arduino and Raspberry Pi boards. USB camera as shown in Fig. 10 is connected to one of the USB ports of Raspberry Pi. A USB camera is used to capture images which will help in weed detection.

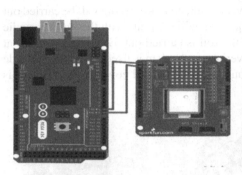

Fig. 7. GPS interfacing to Arduino Mega.   **Fig. 8.** DC motor interfacing to Arduino Mega.

Fig. 9. Arduino Mega interfacing to Raspberry Pi 3B.   **Fig. 10.** USB camera interfacing to Raspberry Pi 3B.

7 Results and Discussion

The implementation of IoT based multipurpose agricultural vehicle enable agriculturist to perform their farming process in a timely manner and increase productivity. The following are the impacts of IoT based multipurpose vehicle proposed for agricultural purpose:

- The net impact on jobs and the quality of work is positive to complement and augment labor activities.
- Automation provides the opportunity for humans to focus on higher-skilled, higher-quality and higher-paid tasks.
- The automated multipurpose vehicle can improve overall production of farmland maintaining its nutrition value by mechanizing whole repetitive and slow processes of farmers by performing all the tasks efficiently.
- Planting, spraying, harvesting and inspecting processes will be economically cheaper.

The results of three mechanisms ploughing, seeding and covering mechanisms can be achieved on irrigated land by using the proposed model of multipurpose agriculture vehicle. Maximum achieved depth to dig the soil for soaking seed is 7 cm, this is controlled by the Arduino Mega and motor driver circuit, and the next dig will be carried out after a distance of 10 cm. This ploughing process will continue with the same periodic interval. After ploughing, the levelling of the soil is carried out. Each seed will fall at the interval of 9.81 cm. The level of the covering mechanism is fixed 0.5 cm deep inside the soil. Height leveler mechanism of ploughed, seeding and covering mechanism is achieved.

Fig. 11. Final working model of a multipurpose vehicle for farmers.

The proposed multipurpose vehicle performs all the functions such as removing weeds, sowing seeds, sprinkling water & fertilizers, cultivating and leveling operation into a one single vehicle. Manual testing on the field also gives enough idea about the vehicle. The operations like ploughing, seed sowing and harvesting are done in the same vehicle, so the cost is reduced. The existing seed sowing machine weighs more and has a complex working metering mechanism. But in this sowing machine, the weight is reduced and the working method is simple by connecting a separate motor. The final working model of a multipurpose vehicle for farmers is as shown in Fig. 11.

8 Performance Analysis

Unmanned aerial robots have been developed to control pest and diseases spread in farmland. But these aerial robots are not able to reach roots of a plant. Crops are required to monitor for sprinkling water and fertilizers as well as for weeding from leaf to stalk levels. Survival and maintenance of such roots is another important challenge and is required to be addressed. Battery life of multirotor ranges from a few minutes to half an hour which is not sufficient for monitoring large farmland without recharging breaks.

The proposed model provides economically effective way to perform all farming functions automatically as compared to the other work presented in [10–13] of Table 1. Table 1 illustrates analysis of other related research works with the proposed system. The system is incorporated with Lithium Ion batteries, every one second the system operates in idle mode, so the system consumes less power. The performance of the model is tested with a prototype developed with the required depth to sow the seeds.

Table 1. Comparison of proposed model with other works.

Comparison with the other work	Proposed work	Mondal and Rehena [10]	Vijay et al. [11]	Hsu et al. [12]	Puranik et al. [13]
Parameters	Temperature, humidity, rain sensor, soil moisture, seed dispenser	Temperature, soil moisture	Moisture, temperature	Soil moisture	Temperature, humidity, soil moisture, pH sensor
Microcontroller	Arduino Mega, Raspberry Pi-3	Arduino UNO	NodeMCU	Arduino	Arduino
LCD display	Yes	No	No	No	Yes
GPS	Yes	No	No	No	No
Smart system	Yes	Yes	Yes	Yes	No
Cloud platform	Yes	Yes	Yes	Yes	No

9 Conclusion and Future Scope

The proposed multipurpose vehicle robot for farming can be used practically for several purposes such as sowing, fertilizing, cultivating the farmland and removing the weed. The proposed system can be a benefit to the farmers to integrate farming with robotics. This vehicle moves around the field like a farmer with a complete kit from beginning to end process required for farming. The whole idea of multipurpose vehicle is a new concept which can be successfully implemented in real life situations. The Multipurpose Agricultural Vehicle is cheaper compared to other farming vehicles commonly used and is suitable for small land owners. The use of electricity to run a multipurpose agricultural vehicle is affordable only for smaller vehicles which can be used for small land areas. The use of battery run vehicles will help in reducing pollution up to a large extent.

The vehicle can be modified by adding more operations such as the seed sower can be modified according to the type of crops grown in the region; the tyres of the vehicle can be changed according to type of land; solar panels can be used to power the vehicle. Mobile Application can be developed for the farmers' business network which will make use of the Machine Learning algorithm and the analytic tools to drive the results of data on pricing.

The sensors and robot working status and battery life can be updated to the user through GSM. This concept can be designed with drones. High resolution cameras and thermal cameras can be incorporated into systems for image processing and detection of pests.

References

1. https://www.un.org/development/desa/en/news/population/world-population-prospects-2019.html

512 S. Vasanti Venkateshwar et al.

2. https://www.nist.gov/blogs/manufacturing-innovation-blog/4-types-robots-every-manufactu
   rershould-know
3. Abdurrahman, M.A., Gebru, G.M., Bezabih, T.T.: Sensor based automatic irrigation management system. Int. J. Comput. Inf. Technol. **04**(03) (2015). (ISSN: 2279–0764)
4. Dagar, R., Som, S., Khatri, S.K.: Smart farming – IoT in agriculture. In: International Conference on Inventive Research in Computing Applications (ICIRCA 2018), pp 1052–1056
   (2018)
5. Saqib, M., Almohamad, T.A., Mehmood, R.M.: A low-cost information monitoring system
   for smart farming applications. Sensors **20**(8), 2367 (2020)
6. Yasin, H.M., Zeebaree, S.R.M., Zebari, I.M.I.: Arduino based automatic irrigation system:
   monitoring and SMS controlling. In: 2019 4th Scientific International Conference Najaf
   (SICN), pp. 109–114. IEEE (2019)
7. Anitha, A., Sampath, N., Jerlin, M.A.: Smart irrigation system using Internet of Things.
   In: 2020 International Conference on Emerging Trends in Information Technology and
   Engineering (ic-ETITE), pp. 1–7. IEEE (2020)
8. https://www.elprocus.com/rain-sensor-working-and-its-applications/
9. https://www.electroschematics.com/neo-6m-gps-module/
10. Ashifuddin Mondal, Md., Rehena, Z.: IoT based intelligent agriculture field monitoring sys-
    tem. In: 8th International Conference on Cloud Computing, Data Science & Engineering
    (Confluence), pp. 625–629 (2018)
11. Puranik, V., et al.: Automation in agriculture and IoT. In: 2019 4th International Conference
    on Internet of Things: Smart Innovation and Usages (IoT-SIU) (2019)
12. Vijay, et al.: An IoT instrumented smart agricultural monitoring and irrigation system. In: 2020
    International Conference on Artificial Intelligence and Signal Processing (AISP) (2020)
13. Hsu, W.-L., et al.: Application of Internet of Things in smart farm watering system. Sens.
    Mater. **33**(1), 269–283 (2021)

Author Index

Printed in the United States
by Baker & Taylor Publisher Services